DERIVATIVES

1. $\dfrac{dau}{dx} = a\dfrac{du}{dx}$

2. $\dfrac{d(u+v)}{dx} = \dfrac{du}{dx} + \dfrac{dv}{dx}$

3. $\dfrac{d(uv)}{dx} = u\dfrac{dv}{dx} + v\dfrac{du}{dx}$

4. $\dfrac{d(u/v)}{dx} = \dfrac{v(du/dx) - u(dv/dx)}{v^2}$

5. $\dfrac{d(u^n)}{dx} = nu^{n-1}\dfrac{du}{dx}$

6. $\dfrac{d(u^v)}{dx} = vu^{v-1}\dfrac{du}{dx} + u^v(\log u)\dfrac{dv}{dx}$

7. $\dfrac{d(e^u)}{dx} = e^u\dfrac{du}{dx}$

8. $\dfrac{d(e^{au})}{dx} = ae^{au}\dfrac{du}{dx}$

9. $\dfrac{da^u}{dx} = a^u(\log a)\dfrac{du}{dx}$

10. $\dfrac{d(\log u)}{dx} = \dfrac{1}{u}\dfrac{du}{dx}$

11. $\dfrac{d(\log_a u)}{dx} = \dfrac{1}{u(\log a)}\dfrac{du}{dx}$

12. $\dfrac{d \sin u}{dx} = \cos u\dfrac{du}{dx}$

13. $\dfrac{d \cos u}{dx} = -\sin u\dfrac{du}{dx}$

14. $\dfrac{d \tan u}{dx} = \sec^2 u\dfrac{du}{dx}$

15. $\dfrac{d \cot u}{dx} = -\csc^2 u\dfrac{du}{dx}$

16. $\dfrac{d \sec u}{dx} = \tan u \sec u\dfrac{du}{dx}$

17. $\dfrac{d \csc u}{dx} = -(\cot u)(\csc u)\dfrac{du}{dx}$

18. $\dfrac{d \arcsin u}{dx} = \dfrac{1}{\sqrt{1 - u^2}}\dfrac{du}{dx}$

19. $\dfrac{d \arccos u}{dx} = \dfrac{-1}{\sqrt{1 - u^2}}\dfrac{du}{dx}$

20. $\dfrac{d \arctan u}{dx} = \dfrac{1}{1 + u^2}\dfrac{du}{dx}$

21. $\dfrac{d \operatorname{arccot} u}{dx} = \dfrac{-1}{1 + u^2}\dfrac{du}{dx}$

22. $\dfrac{d \operatorname{arcsec} u}{dx} = \dfrac{1}{u\sqrt{u^2 - 1}}\dfrac{du}{dx}$

23. $\dfrac{d \operatorname{arccsc} u}{dx} = \dfrac{-1}{\sqrt{u^2 - 1}}\dfrac{du}{dx}$

24. $\dfrac{d \sinh u}{dx} = \cosh u\dfrac{du}{dx}$

25. $\dfrac{d \cosh u}{dx} = \sinh u\dfrac{du}{dx}$

26. $\dfrac{d \tanh u}{dx} = \operatorname{sech}^2 u\dfrac{du}{dx}$

27. $\dfrac{d \coth u}{dx} = -(\operatorname{csch}^2 u)\dfrac{du}{dx}$

28. $\dfrac{d \operatorname{sech} u}{dx} = -(\operatorname{sech} u)(\tanh u)\dfrac{du}{dx}$

29. $\dfrac{d \operatorname{csch} u}{dx} = -(\operatorname{csch} u)(\coth u)\dfrac{du}{dx}$

30. $\dfrac{d \sinh^{-1} u}{dx} = \dfrac{1}{\sqrt{1 + u^2}}\dfrac{du}{dx}$

31. $\dfrac{d \cosh^{-1} u}{dx} = \dfrac{1}{\sqrt{u^2 - 1}}\dfrac{du}{dx}$

32. $\dfrac{d \tanh^{-1} u}{dx} = \dfrac{1}{1 - u^2}\dfrac{du}{dx}$

33. $\dfrac{d \coth^{-1} u}{dx} = \dfrac{1}{u^2 - 1}\dfrac{du}{dx}$

34. $\dfrac{d \operatorname{sech}^{-1} u}{dx} = \dfrac{-1}{u\sqrt{1 - u^2}}\dfrac{du}{dx}$

35. $\dfrac{d \operatorname{csch}^{-1} u}{dx} = \dfrac{-1}{|u|\sqrt{1 + u^2}}\dfrac{du}{dx}$

INTEGRALS (AN ARBITRARY CONSTANT MAY BE ADDED TO EACH INTEGRAL.)

1. $\displaystyle\int x^n\,dx = \frac{1}{n+1}\,x^{n+1}\quad (n \neq -1)$

2. $\displaystyle\int \frac{1}{x}\,dx = \log|x|$

3. $\displaystyle\int e^x\,dx = e^x$

4. $\displaystyle\int a^x\,dx = \frac{a^x}{\log a}$

5. $\displaystyle\int \sin x\,dx = -\cos x$

6. $\displaystyle\int \cos x\,dx = \sin x$

7. $\displaystyle\int \tan x\,dx = -\log|\cos x|$

8. $\displaystyle\int \cot x\,dx = \log|\sin x|$

9. $\displaystyle\int \sec x\,dx = \log|\sec x + \tan x| = \log\left|\tan\left(\tfrac{1}{2}x + \tfrac{1}{4}\pi\right)\right|$

10. $\displaystyle\int \csc x\,dx = \log|\csc x - \cot x| = \log\left|\tan\tfrac{1}{2}x\right|$

11. $\displaystyle\int \arcsin\frac{x}{a}\,dx = x\arcsin\frac{x}{a} + \sqrt{a^2 - x^2}\quad (a > 0)$

12. $\displaystyle\int \arccos\frac{x}{a}\,dx = x\arccos\frac{x}{a} - \sqrt{a^2 - x^2}\quad (a > 0)$

13. $\displaystyle\int \arctan\frac{x}{a}\,dx = x\arctan\frac{x}{a} - \frac{a}{2}\log(a^2 + x^2)\quad (a > 0)$

14. $\displaystyle\int \sin^2 mx\,dx = \frac{1}{2m}(mx - \sin mx \cos mx)$

15. $\displaystyle\int \cos^2 mx\,dx = \frac{1}{2m}(mx + \sin mx \cos mx)$

16. $\displaystyle\int \sec^2 x\,dx = \tan x$

17. $\displaystyle\int \csc^2 x\,dx = -\cot x$

18. $\displaystyle\int \sin^n x\,dx = -\frac{\sin^{n-1} x \cos x}{n} + \frac{n-1}{n}\int \sin^{n-2} x\,dx$

19. $\displaystyle\int \cos^n x\,dx = \frac{\cos^{n-1} x \sin x}{n} + \frac{n-1}{n}\int \cos^{n-2} x\,dx$

20. $\displaystyle\int \tan^n x\,dx = \frac{\tan^{n-1} x}{n-1} - \int \tan^{n-2} x\,dx\quad (n \neq 1)$

21. $\displaystyle\int \cot^n x\,dx = -\frac{\cot^{n-1} x}{n-1} - \int \cot^{n-2} x\,dx\quad (n \neq 1)$

22. $\displaystyle\int \sec^n x\,dx = \frac{\tan x \sec^{n-2} x}{n-1} + \frac{n-2}{n-1}\int \sec^{n-2} x\,dx\quad (n \neq 1)$

23. $\displaystyle\int \csc^n x\,dx = -\frac{\cot x \csc^{n-2} x}{n-1} + \frac{n-2}{n-1}\int \csc^{n-2} x\,dx\quad (n \neq 1)$

(Continued at the back of the book)

BASIC
MULTIVARIABLE
CALCULUS

BASIC
MULTIVARIABLE
CALCULUS

Jerrold E. Marsden
California Institute of Technology

Anthony J. Tromba
University of California—Santa Cruz

Alan Weinstein
University of California—Berkeley

 Springer

 W. H. Freeman and Company

Cover photograph courtesy of H. Armstrong Roberts, Inc., New York, NY.
Text illustrations from *Vector Calculus* used with permission of W.H. Freeman and Company.
Figure 6.4.4 used with permission of Cordon Art B.V., Baarn, The Netherlands.

Library of Congress Cataloging-in-Publication Data
Marsden, Jerrold E.
  Basic multivariable calculus/Jerrold E. Marsden, Anthony J.
 Tromba, Alan Weinstein.
    p. cm.
  Includes index.
  ISBN 3-540-97976-X (Springer) ISBN 0-7167-2443-X (W.H. Freeman)
  1. Calculus. I. Tromba, Anthony. II. Weinstein, Alan, 1943-
 III. Title.
 QA303.M336 1993
 515--dc20   92-38049

Printed on acid-free paper.

Under the co-publishing agreement between Springer-Verlag and W.H. Freeman and Company, the text is
available in North America exclusively from W.H. Freeman and Company and outside North America
exclusively from Springer-Verlag.

Photocomposed copy from the authors' files using LaTeX.
Printed and bound by Hamilton Printing Co., Rensselaer, NY.
Printed in the United States of America.

9 8 7 6 5 4 3 (Corrected third printing, 2000)

ISBN 3-540-97976-X Springer-Verlag Berlin Heidelberg New York
ISBN 0-7167-2443-X W.H. Freeman and Company New York   SPIN 10783503

To Barbara, Inga, Margo, and the memory of Murray Weinstein.

Preface

This text is intended for a one-semester sophomore-level course in the calculus of functions of several variables, including vector analysis. Such a course is sometimes preceded by a beginning course in linear algebra, but this is not an essential prerequisite. We use only the rudiments of matrix algebra, and the necessary concepts are developed in the text. We do assume a knowledge of the fundamentals of one-variable calculus—differentiation and integration of the standard functions.

Computational skills and intuitive understanding are, for many students, more important than theory at this level, a need we have tried to meet by making the book as concrete and student-oriented as possible. We do this in two ways. First, we include a large number of physical illustrations from such areas as fluid mechanics, gravitation, and electromagnetic theory, although prior knowledge of these subjects is not presumed. Second, we endeavor to present the material as straightforwardly and simply as possible. For example, although we formulate the definition of the multidimensional derivative properly, we do so in terms of matrices of partial derivatives rather than abstract linear transformations. This device alone can save one or two weeks of teaching time and avoids the necessity of a linear algebra prerequisite.

We have isolated many important items in boxes, which should help the student identify key concepts. A few of the *really important* results are in shaded boxes. We have also included some historical and other notes for the student.

This text is a synthesis of our books **Vector Calculus** (W.H. Freeman) and **Calculus III** (Springer-Verlag). The former gives a more thorough treatment at a higher level, whereas the latter is multivariable and vector calculus in the context of a standard calculus course.

The Student Guide and the Instructor's Guide

The **student guide** by K. Pao and F. Soon that accompanies this text is commercially available for student use. It contains solutions to every other odd-numbered problem, sample exams, and helpful hints for the student. The **instructor's guide** is available to teachers from W.H. Freeman. It contains extra examples worked out for classroom use as well as additional sample exams.

The Role of the Computer

The computer is becoming essential as a tool for doing mathematics, including calculus and its applications. Computers are becoming cheaper, and software for doing symbolic calculations (*i.e.,* with formulas rather than numbers) and drawing graphs is becoming easier to use. The availability of this software may make you wonder why it is necessary to learn the theory of calculus and to do calculations and graphing "by hand." There are several reasons for this.

- In doing a problem by hand, you get more than the answer. You may learn something from the intermediate steps which gives you insight into the problem.

- You need to know how to set up problems in a form the computer can accept.

- Computer programs make mistakes. By knowing the theory, and having done numerous examples by hand, you are more likely to be able to recognize an unreasonable answer from the computer.

- Problem-solving ability improves with practice. Furthermore, many problems require simple computations which are still easier to do by hand.

- What you learn in doing problems by hand helps you solve other problems.

Many universities have computer labs that can provide very valuable supplements for vector calculus. We especially recommend the supplement "Laboratory Manual for Multivariable and Vector Calculus" by B. Felsager and B.H. West (Cornell University), which was written in consultation with the authors of this book. Additional computer-related material to accompany this text is in preparation.

Acknowledgments

We thank all the users of the previous books **Vector Calculus** and **Calculus III**, on which this book is based, who took the trouble to tell us of corrections and gave us suggestions for improvements. We especially thank Fred Soon, Karen Pao, Mike Hoffman, Frederick Hoffman, Jerry Kazdan, Martin Shim, Takashi Toriguchi, Kin Lee, Alex Baptista. Joanne Seitz, Asha Weinstein, Beverly West, Zhang-ju Liu, Stephen Miller, Kok-Wui Cheong, and Tsit-Yuen Lam, who assisted us in various ways. We also thank the students in our classes who generously informed us of errors. We are very grateful to Barbara Marsden for her superb preparation of the original LaTeX file. Finally, we thank all the people at Springer-Verlag and W.H. Freeman who worked in such a dedicated way on all aspects of this project.

Corrections

Despite our efforts and extensive class testing to eliminate errors from this book, some will invariably remain. Please send your corrections and other remarks to: Jerrold Marsden (marsden@cds.caltech.edu) or Control and Dynamical Systems, Caltech, Pasadena, CA 91125 or Alan Weinstein (alanw@ math.berkeley.edu) at the Department of Mathematics, University of California, Berkeley, CA 94720 or Anthony Tromba at the Department of Mathematics, University of California, Santa Cruz, CA 95064.

Jerrold E. Marsden
Anthony J. Tromba
Alan Weinstein

Prerequisites and Notation

This section summarizes some concepts and notation used throughout the book. Students can read through it quickly now, then refer back later if the need arises.

The collection of all real numbers is denoted $\mathbb{R}$. Thus, $\mathbb{R}$ includes the integers, that is, the collection $\ldots, -3, -2, -1, 0, 1, 2, 3, \ldots$; the *rational numbers* p/q, where p and q are integers ($q \neq 0$); and the *irrational numbers*, such as $\sqrt{2}, \pi$, and e. When we write $a \in \mathbb{R}$, we mean that a is a *member* (or *element*) of the set $\mathbb{R}$, in other words, that a is a real number. Members of $\mathbb{R}$ may be visualized as points on the real-number line, as shown in Figure 0.0.1.

Given two real numbers a and b with $a < b$ (*i.e.*, with a less than b), we can form the *closed interval* $[a, b]$, consisting of all x such that $a \leq x \leq b$, and the *open interval* (a, b), consisting of all x such that $a < x < b$, as well as the *half-open intervals* $(a, b]$ and $[a, b)$ (Figure 0.0.2).

The *absolute value* $|a|$ of a number a is defined to be a if $a \geq 0$ and $-a$ if $a \leq 0$. For example, $|3| = 3, |-3| = 3, |0| = 0$, and $|6| = 6$. The *triangle*

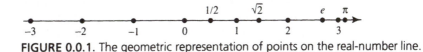

FIGURE 0.0.1. The geometric representation of points on the real-number line.

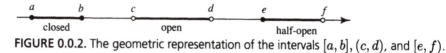

FIGURE 0.0.2. The geometric representation of the intervals $[a, b]$, (c, d), and $[e, f)$.

inequality states that the following always holds: $|a + b| \leq |a| + |b|$. The **distance from** a **to** b is $|a - b|$. Thus, the distance from 6 to 10 is 4 and from -6 to 3 is 9.

For two collections of objects (*i.e.*, sets) A and B, $A \subset B$ means that A is a **subset** of B; that is, every member of A is also a member of B. For example, if $B = \mathbb{R}$, the set of integers $\{\ldots, -3, -2, -1, 0, 1, 2, 3, \ldots\}$ is a subset of B, as is the set $\mathbb{Q}$ of rational numbers or all of $\mathbb{R}$, but the set $\mathbb{C}$ of complex numbers is not.

The symbol $A \cup B$ means the **union** of A and B, the collection whose members are members of either A or B. Thus,

$$\{\ldots, -3, -2, -1, 0\} \cup \{-1, 0, 1, 2, \ldots\} = \{\ldots, -3, -2, -1, 0, 1, 2, \ldots\}.$$

Similarly, $A \cap B$ means the **intersection** of A and B; *i.e.*, this set consists of those members of A and B that are in *both* A and B. The intersection of the two sets above is $\{-1, 0\}$.

We write $A \backslash B$ for the set of those members of A that are not in B. Thus,

$$\{\ldots, -3, -2, -1, 0\} \backslash \{-1, 0, 1, 2, \ldots\} = \{\ldots, -3, -2\}.$$

We can also specify sets as in the following examples:

$$\{a \in \mathbb{R} \mid a \text{ is an integer }\} = \{\ldots, -3, -2, -1, 0, 1, 2, \ldots\}$$
$$\{a \in \mathbb{R} \mid a \text{ is an even integer }\} = \{\ldots, -2, 0, 2, 4, \ldots\}$$
$$\{x \in \mathbb{R} \mid a \leq x \leq b\} = [a, b].$$

A **function** $f : A \to B$ is a rule that assigns to each $a \in A$ one specific member $f(a)$ of B. We call A the **domain** of f and B the **target** of f. The set $\{f(x) \mid x \in A\}$ consisting of all the values of $f(x)$ is called the **range** of f. Denoted by $f(A)$, the range is a subset of the target B. (It may be all of B, in which case f is said to be **onto** B.) The fact that the function f sends a to $f(a)$ is denoted by $a \mapsto f(a)$. For example, the function $f(x) = x^3/(1 - x)$ that assigns the number $x^3/(1 - x)$ to each $x \neq 1$ in $\mathbb{R}$ can also be defined by the rule $x \mapsto x^3/(1 - x)$. Functions are also called **mappings**, **maps**, or **transformations**. The notation $f : A \subset \mathbb{R} \to \mathbb{R}$ means that A is a subset of $\mathbb{R}$ and that f assigns a value $f(x)$ in $\mathbb{R}$ to each $x \in A$. The **graph** of f consists of all the points $(x, f(x))$ in the plane (Figure 0.0.3).

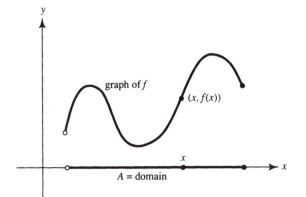

FIGURE 0.0.3. The graph of a function with the half-open interval A as its domain.

The notation $\sum_{i=1}^{n} a_i$ means $a_1 + \cdots + a_n$, where $a_1, \ldots, a_n$ are given numbers. For instance, the sum of the first n integers is

$$1 + 2 + \cdots + n = \sum_{i=1}^{n} i = \frac{n(n+1)}{2}.$$

The **derivative** of a function $f(x)$ is denoted $f'(x)$ or, if $y = f(x)$,

$$\frac{dy}{dx}.$$

Occasionally, we will commit what is called an "abuse of notation" and write the derivative as df/dx or $y'(x)$. The **definite integral** of f from a to b is written $\int_a^b f(x)\, dx$.

We assume that the reader is familiar with the functions used in one-variable calculus, such as $\sin x$, $\cos x$, $\exp x = e^x$, $\arcsin x = \sin^{-1} x$, and $\log x$ (we write $\log x$ for the natural logarithm, which is sometimes denoted $\ln x$ or $\log_e x$). Students are expected to know, or to review as the course proceeds, the basic rules of differentiation and integration for these functions of one variable, as well as the chain rule, the quotient rule, integration by parts, and so forth.

Note

Note especially the use of exp and log. For example:
$\exp(x + y)$ means e^{x+y};
$\log(3x)$ means $\ln(3x)$.

Short tables of derivatives and integrals, adequate for the needs of this text, are printed at the front and back of the book along with other tables useful in multivariable and vector calculus.

The end of a proof is denoted by the symbol ■, whereas the end of an example or remark is denoted by the symbol ♦. A reference such as §**3.2** means Section 2 of Chapter 3.

Contents

Preface   **vii**

1   Algebra and Geometry of Euclidean Space   **1**

   1.1   Vectors in the Plane and Space   2

   1.2   The Inner Product and Distance   22

   1.3 2×2 and 3×3 Matrices and Determinants   39

   1.4   The Cross Product and Planes   46

   1.5 n-Dimensional Euclidean Space   60

   1.6   Curves in the Plane and in Space   73

       Review Exercises   83

2   Differentiation   **91**

   2.1   Graphs and Level Surfaces   92

   2.2   Partial Derivatives and Continuity   109

   2.3   Differentiability, the Derivative Matrix, and Tangent Planes   . . .   124

   2.4   The Chain Rule   133

   2.5   Gradients and Directional Derivatives   146

   2.6   Implicit Differentiation   160

       Review Exercises   166

3 Higher Derivatives and Extrema   **171**

   3.1   Higher Order Partial Derivatives 172

   3.2   Taylor's Theorem 182

   3.3   Maxima and Minima 190

   3.4   Second Derivative Test 201

   3.5   Constrained Extrema and Lagrange Multipliers 211

         Review Exercises 221

4 Vector-Valued Functions   **227**

   4.1   Acceleration 228

   4.2   Arc Length 235

   4.3   Vector Fields 241

   4.4   Divergence and Curl 249

         Review Exercises 263

5 Multiple Integrals   **269**

   5.1   Volume and Cavalieri's Principle 270

   5.2   The Double Integral Over a Rectangle 280

   5.3   The Double Integral Over Regions 291

   5.4   Triple Integrals 306

   5.5   Change of Variables, Cylindrical and Spherical Coordinates . . . 318

   5.6   Applications of Multiple Integrals 339

         Review Exercises 350

6 Integrals Over Curves and Surfaces   **355**

   6.1   Line Integrals 356

   6.2   Parametrized Surfaces 374

   6.3   Area of a Surface 382

   6.4   Surface Integrals 398

         Review Exercises 411

7 The Integral Theorems of Vector Analysis   **415**

   7.1 Green's Theorem 416

   7.2 Stokes' Theorem 429

   7.3 Gauss' Theorem 446

   7.4 Path Independence and the Fundamental Theorems of Calculus 458

       Review Exercises 473

Epilogue   **479**

Practice Examination 1   **481**

Practice Examination 2   **485**

Answers to Odd-Numbered Exercises   **489**

Index   **521**

1

Algebra and Geometry of Euclidean Space

Corresponding to any system of vector analysis is its practical unity. This was Gibbs' point of view in building up his system. He used it in his courses on electricity and magnetism and on the electromagnetic theory of light.

E.B. Wilson *(from **Vector Analysis**, 1901, the first vector calculus text for students)*

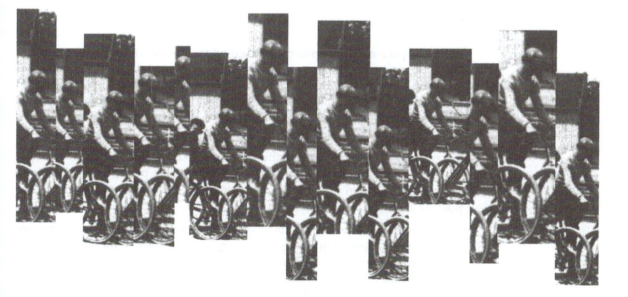

In this chapter we explain the basic operations performed on vectors in three-dimensional space: vector addition, scalar multiplication, and the dot and cross products. In §**1.5** we generalize some of these notions to Euclidean n-space, and in §**1.6** we study curves and their tangent vectors.

1.1
Vectors in the Plane and Space

Points P in the plane are represented by ordered pairs of real numbers (a_1, a_2); the numbers a_1 and a_2 are called the ***Cartesian coordinates of*** P. We draw two perpendicular lines, label them as the x and y axes, and then drop perpendiculars from P to these axes, as in Figure 1.1.1. After designating the intersection of the x and y axes as the origin and choosing units on these axes, we produce two directed distances a_1 and a_2 as shown in the figure; a_1 is called the x ***component*** of P, and a_2 is called the y ***component***.

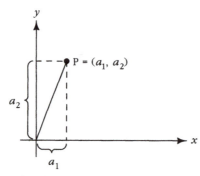

FIGURE 1.1.1. Cartesian coordinates in the plane.

Points in space may be similarly represented as ordered triples of real numbers. To construct such a representation, we choose three mutually perpendicular lines that meet at a point in space. These lines are called the x ***axis***, y ***axis***, and z ***axis***, and the point at which they meet is called the ***origin*** (this is our reference point). We choose a scale on these axes, as shown in Figure 1.1.2.

The triple $(0, 0, 0)$ corresponds to the origin of the coordinate system, and the arrows on the axes indicate the positive directions. For example, the triple $(2, 4, 4)$ represents a point 2 units from the origin in the positive direction along the x axis, 4 units in the positive direction along the y axis, and 4 units in the positive direction along the z axis (Figure 1.1.3).

Because we can associate points in space with ordered triples in this way, we often use the expression "the point (a_1, a_2, a_3)" instead of the longer phrase "the point P that corresponds to the triple (a_1, a_2, a_3)." We say that a_1 is the x ***coordinate*** (or first coordinate), a_2 is the y ***coordinate*** (or second coordinate), and a_3 is the z ***coordinate*** (or third coordinate) of P. It is also common to denote points in space with the letters $x, y,$ and z in place of $a_1, a_2,$ and a_3. Thus the triple (x, y, z) represents a point whose first coordinate is x, second coordinate is y, and third coordinate is z.

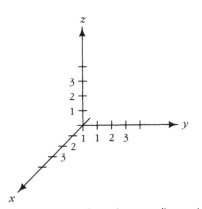

FIGURE 1.1.2. Cartesian coordinates in space.

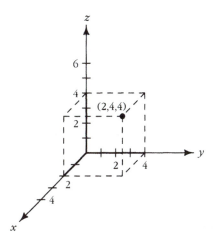

FIGURE 1.1.3. Geometric representation of the point $(2, 4, 4)$ in Cartesian coordinates.

We employ the following notation for the line, the plane, and three-dimensional space.

(i) The real number line is denoted $\mathbb{R}^1$ or simply $\mathbb{R}$.

(ii) The set of all ordered pairs (x, y) of real numbers is denoted $\mathbb{R}^2$.

(iii) The set of all ordered triples (x, y, z) of real numbers is denoted $\mathbb{R}^3$.

When speaking of $\mathbb{R}^1, \mathbb{R}^2$, and $\mathbb{R}^3$ simultaneously, we write $\mathbb{R}^n$, where $n = 1, 2$, or 3; or $\mathbb{R}^m$, where $m = 1, 2, 3$. Starting in §**1.5** we will also study $\mathbb{R}^n$ for $n = 4, 5, 6, \dots$, but the cases $n = 1, 2, 3$ are closest to our geometric intuition and will be stressed throughout the book.

The operation of addition can be extended from $\mathbb{R}$ to $\mathbb{R}^2$ and $\mathbb{R}^3$. For $\mathbb{R}^3$, this is done as follows. Given the two triples (a_1, a_2, a_3) and (b_1, b_2, b_3), we define their **sum** to be

$$(a_1, a_2, a_3) + (b_1, b_2, b_3) = (a_1 + b_1, a_2 + b_2, a_3 + b_3).$$

Example 1

$$
\begin{aligned}
(1,1,1) + (2,-3,4) &= (3,-2,5), \\
(x,y,z) + (0,0,0) &= (x,y,z), \\
(1,7,3) + (a,b,c) &= (1+a, 7+b, 3+c). \quad \blacklozenge
\end{aligned}
$$

The element $(0,0,0)$ is called the **zero element** (or just **zero**) of $\mathbb{R}^3$. The element $(-a_1, -a_2, -a_3)$ is the **additive inverse** (or **negative**) of (a_1, a_2, a_3), and we write $(a_1, a_2, a_3) - (b_1, b_2, b_3)$ for $(a_1, a_2, a_3) + (-b_1, -b_2, -b_3)$.

There are several important product operations that we will define on $\mathbb{R}^3$. One of these, called the *inner product,* assigns a real number to each pair of elements of $\mathbb{R}^3$. We shall discuss it in detail in §**1.2**. Another product operation for $\mathbb{R}^3$ is called *scalar multiplication* (the word "scalar" is a synonym for "real number"). This product combines scalars (real numbers) and elements of $\mathbb{R}^3$ (ordered triples) to yield elements of $\mathbb{R}^3$ as follows: given a scalar α and a triple (a_1, a_2, a_3), we define the **scalar multiple** by

$$\alpha(a_1, a_2, a_3) = (\alpha a_1, \alpha a_2, \alpha a_3).$$

Example 2

$$
\begin{aligned}
2(4, e, 1) &= (2 \cdot 4, 2 \cdot e, 2 \cdot 1) = (8, 2e, 2), \\
6(1,1,1) &= (6,6,6), \\
1(u,v,w) &= (u,v,w), \\
0(p,q,r) &= (0,0,0). \quad \blacklozenge
\end{aligned}
$$

As a consequence of their definitions, addition and scalar multiplication for $\mathbb{R}^3$ satisfy the following identities:

(i)	$(\alpha\beta)(a_1, a_2, a_3) = \alpha[\beta(a_1, a_2, a_3)]$	(associativity)
(ii)	$(\alpha + \beta)(a_1, a_2, a_3)$	(distributivity)
	$= \alpha(a_1, a_2, a_3) + \beta(a_1, a_2, a_3)$	
(iii)	$\alpha[(a_1, a_2, a_3) + (b_1, b_2, b_3)]$	(distributivity)
	$= \alpha(a_1, a_2, a_3) + \alpha(b_1, b_2, b_3)$	
(iv)	$\alpha(0,0,0) = (0,0,0)$	(property of zero)
(v)	$0(a_1, a_2, a_3) = (0,0,0)$	(property of zero)
(vi)	$1(a_1, a_2, a_3) = (a_1, a_2, a_3)$	(property of the unit element)

The identities are proven directly from the definitions of addition and scalar multiplication. For instance,

$$\begin{aligned}
(\alpha + \beta)(a_1, a_2, a_3) &= ((\alpha + \beta)a_1, (\alpha + \beta)a_2, (\alpha + \beta)a_3) \\
&= (\alpha a_1 + \beta a_1, \alpha a_2 + \beta a_2, \alpha a_3 + \beta a_3) \\
&= \alpha(a_1, a_2, a_3) + \beta(a_1, a_2, a_3).
\end{aligned}$$

For $\mathbb{R}^2$, addition and scalar multiplication are defined just as in $\mathbb{R}^3$, with the third component of each vector dropped off. All the properties above still hold.

Let us turn to the geometry of these operations in $\mathbb{R}^2$ and $\mathbb{R}^3$. For the moment, we define a **vector** to be a directed line segment beginning at the origin, that is, a line segment with specified magnitude and direction, and initial point at the origin. Figure 1.1.4 shows several vectors, drawn as arrows beginning at the origin. In print, vectors are usually denoted by boldface letters: **a**. By hand, we usually write them as $\vec{a}$ or simply as a, possibly with a line or wavy line under it.

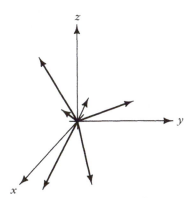

FIGURE 1.1.4. Geometrically, vectors are thought of as arrows emanating from the origin.

Using this definition of a vector, we may associate with each vector **a** the point (a_1, a_2, a_3) where **a** terminates, and conversely, with each point (a_1, a_2, a_3) in space we can associate a vector **a**. Thus, we shall identify **a** with (a_1, a_2, a_3) and write $\mathbf{a} = (a_1, a_2, a_3)$. For this reason, the elements of $\mathbb{R}^3$ not only are ordered triples of real numbers, but are also regarded as vectors. The triple $(0, 0, 0)$ is denoted **0**. We call a_1, a_2, and a_3 the **components** of **a**, or when we think of **a** as a point, its **coordinates**.

Two vectors $\mathbf{a} = (a_1, a_2, a_3)$ and $\mathbf{b} = (b_1, b_2, b_3)$ are equal if and only if $a_1 = b_1, a_2 = b_2$ and $a_3 = b_3$. Geometrically this means that **a** and **b** have the same direction and the same length (or "magnitude").

Geometrically, we define vector addition as follows. In the plane containing the vectors $\mathbf{a} = (a_1, a_2, a_3)$ and $\mathbf{b} = (b_1, b_2, b_3)$ (see Figure 1.1.5), form the parallelogram having **a** as one side and **b** as its adjacent side. The sum $\mathbf{a} +$

b is the directed line segment along the diagonal of the parallelogram. This geometric view of vector addition is useful in many physical situations, as we shall see later.

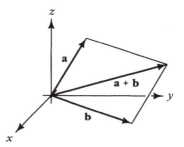

FIGURE 1.1.5. The geometry of vector addition.

To show that our geometric definition of addition is consistent with our algebraic definition, we demonstrate that $\mathbf{a} + \mathbf{b} = (a_1 + b_1, a_2 + b_2, a_3 + b_3)$. We shall prove this result in the plane and leave the proof in three-dimensional space to the reader. Thus, we wish to show that if $\mathbf{a} = (a_1, a_2)$ and $\mathbf{b} = (b_1, b_2)$, then $\mathbf{a} + \mathbf{b} = (a_1 + b_1, a_2 + b_2)$.

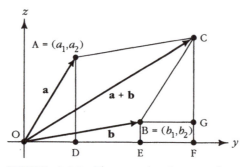

FIGURE 1.1.6. The construction used to prove that $(a_1, b_1) + (a_2, b_2) = (a_1 + b_1, a_2 + b_2)$.

In Figure 1.1.6 let $\mathbf{a} = (a_1, a_2)$ be the vector ending at the point A, and let $\mathbf{b} = (b_1, b_2)$ be the vector ending at point B. By definition, the vector $\mathbf{a} + \mathbf{b}$ ends at the vertex C of parallelogram OBCA. To verify that $\mathbf{a} + \mathbf{b} = (a_1 + b_1, a_2 + b_2)$, it suffices to show that the coordinates of C are $(a_1 + b_1, a_2 + b_2)$. The sides of the triangles OAD and BCG are parallel, and the sides OA and BC have equal lengths, which we write as OA = BC. The triangles are congruent, so BG = OD; since BGFE is a rectangle, EF = BG. Furthermore, OD = a_1 and OE = b_1. Hence, EF = BG = OD = a_1. Since OF = EF + OE, it follows that OF = $a_1 + b_1$. This shows that the x coordinate of $\mathbf{a} + \mathbf{b}$ is $a_1 + b_1$. The proof that the y coordinate is $a_2 + b_2$ is analogous. This argument assumes A and B to be in the first quadrant, but similar arguments hold for the other quadrants.

Figure 1.1.7(a) illustrates another way of looking at vector addition: in terms of triangles rather than parallelograms. That is, we translate (without rotation) the directed line segment representing the vector **b** so that it begins at the end of the vector **a**. The endpoint of the resulting directed segment is the endpoint of the vector **a**+**b**. We note that when **a** and **b** are collinear, the triangle collapses to a line segment, as in Figure 1.1.7(b).

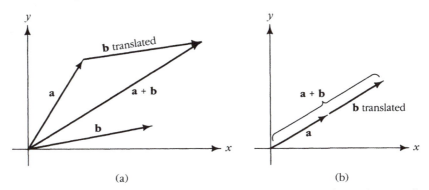

FIGURE 1.1.7. (a) Vector addition may be visualized in terms of triangles as well as parallelograms. (b) The triangle collapses to a line segment when **a** and **b** are collinear.

In Figure 1.1.7 we have placed **a** and **b** *head to tail*. That is, the tail of **b** is placed at the head of **a**, and the vector **a**+**b** goes from the tail of **a** to the head of **b**. If we do it in the other order, **b** + **a**, we get the same vector by going around the parallelogram the other way.

This figure teaches us something useful — it is a good idea to let vectors "glide" or "slide," keeping the same magnitude and direction. We want, in fact, to regard two vectors as the *same* if they have the same magnitude and direction. When we insist on vectors beginning at the origin, we will say that we have ***bound vectors***. If we allow vectors to begin at other points, we will speak of ***free vectors*** or just ***vectors***.

Vectors

Vectors (also called free vectors) are directed line segments in [the plane or] space represented by directed line segments with a beginning (tail) and an end (head). Line segments obtained from each other by translation (but not rotation) represent the same vector.

The components (a_1, a_2, a_3) of **a** are the (signed) lengths of the projections of **a** along the three coordinate axes; equivalently, they are defined by placing the tail of **a** at the origin and letting the head be the point (a_1, a_2, a_3). We write $\mathbf{a} = (a_1, a_2, a_3)$.

Two vectors are added by placing them head to tail and drawing the vector from the tail of the first to the head of the second, as in Figure 1.1.7.

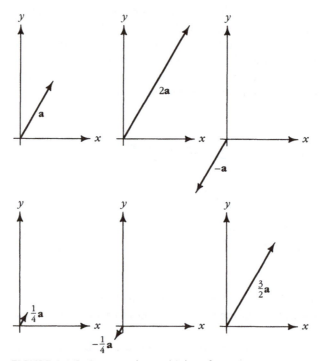

FIGURE 1.1.8. Some scalar multiples of a vector **a**.

Scalar multiplication of vectors also has a geometric interpretation. If α is a scalar and **a** a vector, we define $\alpha\mathbf{a}$ to be the vector that is $|\alpha|$ times as long as **a**, with the same direction as **a** if $\alpha > 0$, but with the opposite direction if $\alpha < 0$. Figure 1.1.8 illustrates several examples.

Using an argument based on similar triangles, one finds that if $\mathbf{a} = (a_1, a_2, a_3)$,

$$\alpha\mathbf{a} = (\alpha a_1, \alpha a_2, \alpha a_3).$$

That is, the geometric definition coincides with the algebraic one.

Given two vectors **a** and **b**, how do we represent the vector $\mathbf{b} - \mathbf{a}$ geometrically, *i.e.*, what is the geometry of vector subtraction? Since $\mathbf{a} + (\mathbf{b} - \mathbf{a}) = \mathbf{b}$, we see that $\mathbf{b} - \mathbf{a}$ is the vector that one adds to **a** to get **b**. In view of this, we may conclude that $\mathbf{b} - \mathbf{a}$ is the vector parallel to, and with the same magnitude as, the directed line segment beginning at the endpoint of **a** and terminating at the endpoint of **b** when **a** and **b** begin at the same point (see Figure 1.1.9).

Example 3   In Figure 1.1.10, which vector is (a) $\mathbf{u} + \mathbf{v}$? (b) $3\mathbf{u}$? (c) $-\mathbf{v}$?

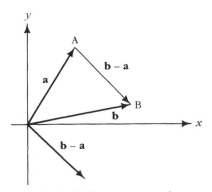

FIGURE 1.1.9. The geometry of vector subtraction.

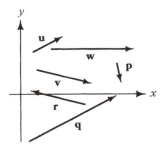

FIGURE 1.1.10. Find $\mathbf{u} + \mathbf{v}, 3\mathbf{u}$ and $-\mathbf{v}$.

Solution

(a) To construct $\mathbf{u} + \mathbf{v}$, we represent $\mathbf{u}$ and $\mathbf{v}$ by directed line segments so that the head of the first coincides with the tail of the second. We fill in the third side of the triangle to obtain $\mathbf{u} + \mathbf{v}$ (see Figure 1.1.11). Comparing with Figure 1.1.10 we find that $\mathbf{u} + \mathbf{v} = \mathbf{w}$.

(b) $3\mathbf{u} = \mathbf{q}$ (see Figure 1.1.12).

(c) $-\mathbf{v} = (-1)\mathbf{v} = \mathbf{r}$ (see Figure 1.1.12). ◆

Example 4   Let $\mathbf{v}$ be the vector with components $(3, 2, -2)$ and let $\mathbf{w}$ be the vector from the point $(2, 1, 3)$ to the point $(-1, 0, -1)$. Find $\mathbf{v} + \mathbf{w}$. Illustrate with a sketch.

Solution   Since $\mathbf{w}$ has components $(-1, 0, -1) - (2, 1, 3) = (-3, -1, -4)$, we find that $\mathbf{v} + \mathbf{w}$ has components $(3, 2, -2) + (-3, -1, -4) = (0, 1, -6)$, as illustrated in Figure 1.1.13. ◆

To describe vectors in space, it is convenient to introduce three special vectors along the $x, y,$ and z axes:

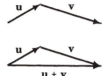

FIGURE 1.1.11. The geometric construction of $\mathbf{u} + \mathbf{v}$.

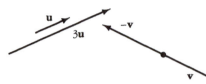

FIGURE 1.1.12. To find $3\mathbf{u}$, draw a vector in the same direction as $\mathbf{u}$, three times as long; $-\mathbf{v}$ is a vector having the same length as $\mathbf{v}$, pointing in the opposite direction.

$\mathbf{i}$:   the vector with components $(1, 0, 0)$
$\mathbf{j}$:   the vector with components $(0, 1, 0)$
$\mathbf{k}$:   the vector with components $(0, 0, 1)$.

These ***standard basis vectors*** are illustrated in Figure 1.1.14. In the plane one has, analogously, $\mathbf{i}$ and $\mathbf{j}$ with components $(1, 0)$ and $(0, 1)$.

Let $\mathbf{a}$ be any vector, and let (a_1, a_2, a_3) be its components. Then

$$\mathbf{a} = a_1\mathbf{i} + a_2\mathbf{j} + a_3\mathbf{k}$$

since the right-hand side is given in components by

$$
\begin{aligned}
a_1(1, 0, 0) + a_2(0, 1, 0) + a_3(0, 0, 1) &= (a_1, 0, 0) + (0, a_2, 0) + (0, 0, a_3) \\
&= (a_1, a_2, a_3).
\end{aligned}
$$

Thus we can express every vector as a sum of scalar multiples of $\mathbf{i}, \mathbf{j}$, and $\mathbf{k}$.

The Standard Basis Vectors

1. The vectors $\mathbf{i}, \mathbf{j}$, and $\mathbf{k}$ are unit vectors along the three coordinate axes, as shown in Figure 1.1.14.
2. If $\mathbf{a}$ has components (a_1, a_2, a_3), then

$$\mathbf{a} = a_1\mathbf{i} + a_2\mathbf{j} + a_3\mathbf{k}.$$

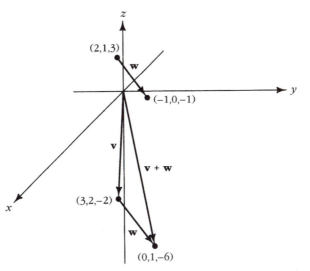

FIGURE 1.1.13. Adding $\mathbf{v} = (3, 2, -2)$ to $\mathbf{w}$, the vector from $(2, 1, 3)$ to $(-1, 0, -1)$.

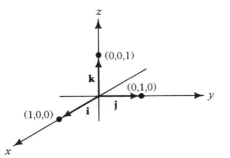

FIGURE 1.1.14. The standard basis vectors.

We have used the term "standard basis" because it is commonly used in linear algebra. For the moment, do not worry about the precise meaning of the term "basis."

Example 5   The vector $(2, 3, 2)$ is $2\mathbf{i} + 3\mathbf{j} + 2\mathbf{k}$, and the vector $(0, -1, 4)$ is $-\mathbf{j} + 4\mathbf{k}$. Figure 1.1.15 shows $2\mathbf{i} + 3\mathbf{j} + 2\mathbf{k}$; the student should draw in the vector $-\mathbf{j} + 4\mathbf{k}$. ◆

Addition and scalar multiplication may be written in terms of the standard basis vectors as follows:

$$(a_1\mathbf{i} + a_2\mathbf{j} + a_3\mathbf{k}) + (b_1\mathbf{i} + b_2\mathbf{j} + b_3\mathbf{k}) = (a_1 + b_1)\mathbf{i} + (a_2 + b_2)\mathbf{j} + (a_3 + b_3)\mathbf{k}$$

and

$$\alpha(a_1\mathbf{i} + a_2\mathbf{j} + a_3\mathbf{k}) = (\alpha a_1)\mathbf{i} + (\alpha a_2)\mathbf{j} + (\alpha a_3)\mathbf{k}.$$

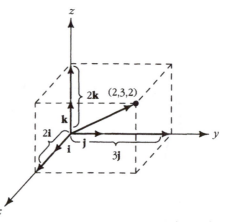

FIGURE 1.1.15. Representation of $(2, 3, 2)$ in terms of the standard basis vectors, $\mathbf{i}, \mathbf{j}$, and $\mathbf{k}$.

To apply vectors to geometric problems, it is useful to assign a vector to a *pair* of points in the plane or in space, as follows. Given two points P and P' we can draw the vector $\mathbf{v}$ with tail P and head P', as in Figure 1.1.16, where we write $= \overrightarrow{PP'}$ for $\mathbf{v}$.

FIGURE 1.1.16. The vector from P to P' is denoted $\overrightarrow{PP'}$.

If $P = (x, y, z)$ and $P' = (x', y', z')$, then the vectors from the origin to P and P' are $\mathbf{a} = x\mathbf{i} + y\mathbf{j} + z\mathbf{k}$ and $\mathbf{a}' = x'\mathbf{i} + y'\mathbf{j} + z'\mathbf{k}$, respectively, so the vector $\overrightarrow{PP'}$ is the difference $\mathbf{a}' - \mathbf{a} = (x' - x)\mathbf{i} + (y' - y)\mathbf{j} + (z' - z)\mathbf{k}$. (See Figure 1.1.17.)

The Vector Joining Two Points

If the point P has coordinates (x, y, z) and P' has coordinates (x', y', z'), then the vector $\overrightarrow{PP'}$ has components $(x' - x, y' - y, z' - z)$.

Example 6   (a) Find the components of the vector from $(3, 5)$ to $(4, 7)$.

(b) Add the vector $\mathbf{v}$ from $(-1, 0)$ to $(2, -3)$ and the vector $\mathbf{w}$ from $(2, 0)$ to $(1, 1)$.

(c) Multiply the vector $\mathbf{v}$ in (b) by 8. If this vector is represented by the directed line segment from $(5, 6)$ to Q, what is Q?

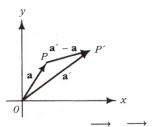

FIGURE 1.1.17. $\overrightarrow{PP'} = \overrightarrow{OP'} - \overrightarrow{OP}$.

Solution

(a) As in the preceding box, we subtract the ordered pairs $(4, 7) - (3, 5) = (1, 2)$. Thus the required components are $(1, 2)$.

(b) The vector $\mathbf{v}$ has components $(2, -3) - (-1, 0) = (3, -3)$ and $\mathbf{w}$ has components $(1, 1) - (2, 0) = (-1, 1)$. Therefore, the vector $\mathbf{v} + \mathbf{w}$ has components $(3, -3) + (-1, 1) = (2, -2)$.

(c) The vector $8\mathbf{v}$ has components $8(3, -3) = (24, -24)$. If this vector is represented by the directed line segment from $(5, 6)$ to Q, and Q has coordinates (x, y), then $(x, y) - (5, 6) = (24, -24)$, so $(x, y) = (5, 6) + (24, -24) = (29, -18)$. ◆

Many of the theorems of plane geometry can be proved by vector methods. Here is one example.

Example 7   Use vectors to prove that the diagonals of a parallelogram bisect each other.

Solution   Let OPRQ be the parallelogram, with two adjacent sides represented by the vectors $\mathbf{a} = \overrightarrow{OP}$ and $\mathbf{b} = \overrightarrow{OQ}$. Let M be the midpoint of the diagonal OR, N the midpoint of the other, PQ. (See Figure 1.1.18.)

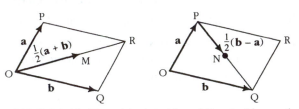

FIGURE 1.1.18. The midpoints M and N coincide, so the diagonals OR and PQ bisect each other.

Then $\overrightarrow{OR} = \overrightarrow{OP} + \overrightarrow{OQ} = \mathbf{a} + \mathbf{b}$, by the parallelogram rule for vector addition, so $\overrightarrow{OM} = \frac{1}{2}\overrightarrow{OR} = \frac{1}{2}(\mathbf{a} + \mathbf{b})$. On the other hand,

$$\overrightarrow{PQ} = \overrightarrow{OQ} - \overrightarrow{OP} = \mathbf{b} - \mathbf{a}, \text{ so } \overrightarrow{PN} = \frac{1}{2}\overrightarrow{PQ} = \frac{1}{2}(\mathbf{b} - \mathbf{a}),$$

and hence

$$\overrightarrow{ON} = \overrightarrow{OP} + \overrightarrow{PN} = \mathbf{a} + \frac{1}{2}(\mathbf{b} - \mathbf{a}) = \frac{1}{2}(\mathbf{a} + \mathbf{b}).$$

Since $\overrightarrow{OM}$ and $\overrightarrow{ON}$ are equal vectors, the points M and N coincide, so the diagonals bisect each other. ◆

Another geometric application of vectors is in finding the equations of lines and other figures in the plane and in space.

Later it will be especially useful for us to know the equation of the line that passes through a given point P in the direction of a given vector $\mathbf{d}$. A point R lies on the line (see Figure 1.1.19) if and only if the vector $\overrightarrow{PR}$ is a multiple of $\mathbf{d}$. Thus, we can describe all points R on the line by $\overrightarrow{PR} = t\mathbf{d}$ for some number t. As t varies, R moves on the line; when $t = 0$, R coincides with P.

If the coordinates of the given point P are (x_1, y_1, z_1) and those of the general point R are (x, y, z), then $\overrightarrow{PR} = (x - x_1, y - y_1, z - z_1)$. If the vector $\mathbf{d}$ has components (a, b, c), then we have the equation

$$(x - x_1, y - y_1, z - z_1) = t(a, b, c).$$

Equating components of these vectors, we get the following result:

Parametric Equation of a Line: Point–Direction Form

The equation of the line l through the point $P = (x_1, y_1, z_1)$ and pointing in the direction of the vector $\mathbf{d} = a\mathbf{i} + b\mathbf{j} + c\mathbf{k}$ is $\overrightarrow{PR} = t\mathbf{d}$, where $R = (x, y, z)$, is the general point on l and the parameter t takes on all real values. In coordinate form, the equations are

$$\begin{aligned} x &= x_1 + at, \\ y &= y_1 + bt, \\ z &= z_1 + ct. \end{aligned}$$

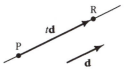

FIGURE 1.1.19. If the line through P and R has the direction of the vector **d**, then the vector from P to R is a multiple of **d**.

For lines in the xy plane, the z component is not present; otherwise, the results are the same.

Example 8   (a) Find the equations of the line in space through the point $(3, -1, 2)$ in the direction $2\mathbf{i} - 3\mathbf{j} + 4\mathbf{k}$.

(b) Find the equation of the line in the plane through the point $(1, -6)$ in the direction of $5\mathbf{i} - \pi\mathbf{j}$.

(c) In what direction does the line $x = -3t + 2, y = -2(t - 1), z = 8t + 2$ point?

Solution

(a) Here $P = (3, -1, 2) = (x_1, y_1, z_1)$ and $\mathbf{d} = 2\mathbf{i} - 3\mathbf{j} + 4\mathbf{k}$, so $a = 2, b = -3$, and $c = 4$. From the box above, the equations are

$$x = 3 + 2t, \quad y = -1 - 3t, \quad z = 2 + 4t.$$

(b) Here $P = (1, -6)$ and $\mathbf{d} = 5\mathbf{i} - \pi\mathbf{j}$, so the line is

$$R = (1, -6) + (5t, -\pi t) = (1 + 5t, -6 - \pi t)$$

or

$$x = 1 + 5t, \quad y = -6 - \pi t.$$

(c) Using the preceding box, we construct the direction $\mathbf{d} = a\mathbf{i} + b\mathbf{j} + c\mathbf{k}$ from the coefficients of t: $a = -3, b = -2, c = 8$. Thus the line points in the direction of $\mathbf{d} = -3\mathbf{i} - 2\mathbf{j} + 8\mathbf{k}$. ◆

Example 9   Do the lines $(x, y, z) = (t, -6t + 1, 2t - 8)$ and $(x, y, z) = (3t + 1, 2t, 0)$ intersect?

Solution   If the lines intersect, there must be numbers t_1 and t_2 such that the corresponding points are equal:

$$(t_1, -6t_1 + 1, 2t_1 - 8) = (3t_2 + 1, 2t_2, 0);$$

that is

$$
\begin{aligned}
t_1 &= 3t_2 + 1, \\
-6t_1 + 1 &= 2t_2, \\
2t_1 - 8 &= 0.
\end{aligned}
$$

From the third equation, $t_1 = 4$. The first equation then becomes $4 = 3t_2 + 1$ or $t_2 = 1$. We must check whether these values satisfy the middle equation:

$$
\begin{aligned}
-6t_1 + 1 &\stackrel{?}{=} 2t_2, \quad i.e., \\
-6 \cdot 4 + 1 &\stackrel{?}{=} 2 \cdot 1, \quad i.e., \\
-24 + 1 &\stackrel{?}{=} 2.
\end{aligned}
$$

The answer is no, so the lines do not intersect.   ♦

A line through the point P can also be specified by giving another point Q. From this, we can determine the direction $\mathbf{d}$ as $\overrightarrow{PQ}$ (see Figure 1.1.20). If $P = (x_1, y_1, z_1)$ and $Q = (x_2, y_2, z_2)$, then $\mathbf{d} = (x_2 - x_1)\mathbf{i} + (y_2 - y_1)\mathbf{j} + (z_2 - z_1)\mathbf{k}$, and so the equations of the line are

$$
\begin{aligned}
x &= x_1 + (x_2 - x_1)t, \\
y &= y_1 + (y_2 - y_1)t, \\
z &= z_1 + (z_2 - z_1)t.
\end{aligned}
$$

Parametric Equation of a Line: Point–Point Form

The parametric equations of the line l through the points $P = (x_1, y_1, z_1)$ and $Q = (x_2, y_2, z_2)$ are

$$
\begin{aligned}
x &= x_1 + (x_2 - x_1)t, \\
y &= y_1 + (y_2 - y_1)t, \\
z &= z_1 + (z_2 - z_1)t,
\end{aligned}
$$

where $R = (x, y, z)$ is the general point of l, and the parameter t takes on all real values.

Example 10   Find the equation of the line through $(2, 1, -3)$ and $(6, -1, -5)$.

Solution   Using the preceding box, we choose $(x_1, y_1, z_1) = (2, 1, -3)$ and $(x_2, y_2, z_2) = (6, -1, -5)$, so the equations are

$$
\begin{aligned}
x &= 2 + (6 - 2)t = 2 + 4t, \\
y &= 1 + (-1 - 1)t = 1 - 2t, \\
z &= -3 + (-5 - (-3))t = -3 - 2t. \quad ♦
\end{aligned}
$$

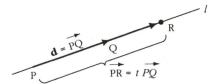

FIGURE 1.1.20. R is on the line through P and Q when $\overrightarrow{PR} = t \, \overrightarrow{PQ}$ for some t.

The description of a line *segment* requires that the domain of the parameter t must be restricted.

Example 11 Find the "equation" of the line segment between $(1,1,1)$ and $(2,1,2)$.

Solution The *line* through $(1,1,1)$ and $(2,1,2)$ is described in parametric form by $(x,y,z) = (1+t,1,1+t)$, as t takes on all real values. When $t = 0$, the point (x,y,z) is $(1,1,1)$, and when $t = 1$, the point (x,y,z) is $(2,1,2)$. Thus, the point (x,y,z) lies between $(1,1,1)$ and $(2,1,2)$ when $0 \le t \le 1$, so the line *segment* is described by the equations

$$
\begin{aligned}
x &= 1+t, \\
y &= 1, \\
z &= 1+t,
\end{aligned}
$$

together with the inequalities $0 \le t \le 1$. ◆

We can also give parametric descriptions of geometric objects other than lines.

Example 12 Describe the points that lie within the parallelogram whose adjacent sides are the vectors **a** and **b** based at the origin ("within" includes points on the edges of the parallelogram).

Solution Consider Figure 1.1.21. If P is any point within the given parallelogram and we construct lines l_1 and l_2 through P parallel to the vectors **a** and **b**, respectively, we see that l_1 intersects the side of the parallelogram determined by the vector **b** at some point $t\mathbf{b}$, where $0 \le t \le 1$. Likewise, l_2 intersects the side determined by the vector **a** at some point $s\mathbf{a}$, where $0 \le s \le 1$.

Note that P is the endpoint of the diagonal of a parallelogram having adjacent sides $s\mathbf{a}$ and $t\mathbf{b}$; hence, if **v** denotes the vector $\overrightarrow{OP}$, we see that $\mathbf{v} = s\mathbf{a} + t\mathbf{b}$. Thus, all the points in the given parallelogram are endpoints of vectors of the form $s\mathbf{a} + t\mathbf{b}$ for $0 \le s \le 1$ and $0 \le t \le 1$. Reversing our steps we see that all vectors of this form end within the parallelogram. ◆

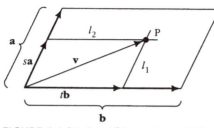

FIGURE 1.1.21. Describing points within the parallelogram formed by vectors **a** and **b**, with vertex **0**.

Since two different lines through the origin determine a plane through the origin, so do two nonparallel vectors. If we apply the same reasoning as in Example 12, we see that the entire plane formed by two nonparallel vectors **v** and **w** consists of all points of the form $s\mathbf{v} + t\mathbf{w}$ where s and t can be any real numbers, as in Figure 1.1.22. We will come back to this type of representation when we study parametric surfaces in Chapter **6**.

Historical Note

Until around 1900, many scientists resisted the use of vectors in favor of more complicated objects called quaternions. The book that popularized vector methods was ***Vector
Analysis*** by E. B. Wilson (reprinted by Dover in 1960), which was based on lectures
delivered by J. W. Gibbs at Yale in 1899 and 1900. Wilson was reluctant to take Gibbs'
course, since he had just completed a full-year course in quaternions at Harvard under
J. M. Peirce, a champion of quaternionic methods, but was forced by a dean to add
the course to his program. (For more details, see M. J. Crowe, ***A History of Vector
Analysis***, Dover, 1967, 1985.)

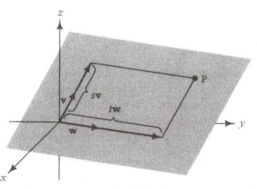

FIGURE 1.1.22. Describing points P in the plane formed from vectors **v** and **w**.

We have thus described the points in the plane by two parameters. For this reason, we say the plane is ***two-dimensional***. Similarly, a line is called ***one-dimensional*** whether it lies in the plane or in space or is the real number line itself.

The plane determined by $\mathbf{v}$ and $\mathbf{w}$ is called the plane *spanned by* $\mathbf{v}$ and $\mathbf{w}$. When $\mathbf{v}$ is a scalar multiple of $\mathbf{w}$ and $\mathbf{w} \neq \mathbf{0}$, then $\mathbf{v}$ and $\mathbf{w}$ are parallel and the plane degenerates to a straight line. When $\mathbf{v} = \mathbf{w} = \mathbf{0}$ (that is, both are zero vectors), we obtain a single point.

There are three particular planes that arise naturally in a coordinate system and which will be useful to us later. We call the plane spanned by vectors $\mathbf{i}$ and $\mathbf{j}$ the xy plane, the plane spanned by $\mathbf{j}$ and $\mathbf{k}$ the yz plane, and the plane spanned by $\mathbf{i}$ and $\mathbf{k}$ the xz plane. These planes are illustrated in Figure 1.1.23.

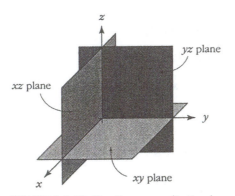

FIGURE 1.1.23. The three coordinate planes.

Exercises for §1.1

Plot the points in Exercises 1−4.

1. $(1, 0, 0)$

2. $(0, 2, 4)$

3. $(3, -1, 5)$

4. $(2, -1, \frac{1}{2})$

Complete the computations in Exercises 5−8.

5. $(1, 2) + (3, 7) =$

6. $(-2, 6) - 6(2, -10) =$

7. $(6, 0, 5) + (5, 0, 6) =$

8. $(1, 3, 5) + 4(-1, -3, -5) =$

Solve for the unknown quantities, if possible, in Exercises 9−12.

9. $(1, 2) + (0, y) = (1, 3)$

10. $a(1, 1) + b(1, -1) = (3, 5)$

11. $a(2, -1) = (b, -c)$

12. $0(3, a) = (3, a)$

In Exercises 13−16, sketch the given vectors $\mathbf{v}$ and $\mathbf{w}$. On your sketch, draw in $-\mathbf{v}, \mathbf{v} + \mathbf{w}$, and $\mathbf{v} - \mathbf{w}$.

13. $\mathbf{v} = (2, 1)$ and $\mathbf{w} = (1, 2)$

14. $\mathbf{v} = (0, 4)$ and $\mathbf{w} = (2, -1)$

15. $\mathbf{v} = (2, 3 - 6)$ and $\mathbf{w} = (-1, 1, 1)$

16. $\mathbf{v} = (2, 1, 3)$ and $\mathbf{w} = (-2, 0, -1)$

Exercises 17−20 refer to Figure 1.1.24.

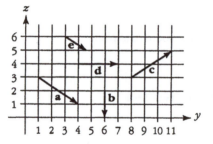

FIGURE 1.1.24. Compute with these vectors in Exercises 17−20.

17. Which vector is (a) $\mathbf{a} - \mathbf{b}$, (b) $\frac{1}{2}\mathbf{a}$?

18. Find the number r such that $\mathbf{c} - \mathbf{a} = r\mathbf{b}$.

19. Draw the vectors (a) $\mathbf{c} + \mathbf{d}$ (b) $-2\mathbf{c} + \mathbf{a}$. What are their components?

20. Draw the vectors (a) $3(\mathbf{c} - \mathbf{d})$, (b) $-\frac{2}{3}\mathbf{c}$. What are their components?

In Exercises 21−24, express the given vectors in terms of the standard basis.

21. The vector with components $(7, 2, 3)$.

22. The vector with components $(-1, 2, \pi)$.

23. The vector from $(0, 1, 2)$ to $(1, 1, 1)$.

24. The vector from $(3, 0, 5)$ to $(2, 7, 6)$.

Find the equations of the lines described in Exercises 25–28.

25. The line passing through $(1, -1, -1)$ in the direction of $\mathbf{j}$.

26. The line passing through $(0, 2, 1)$ in the direction of $2\mathbf{i} - \mathbf{k}$.

27. The line passing through $(-1, -1, -1)$ and $(1, -1, 2)$.

28. The line passing through $(-5, 0, 4)$ and $(6, -3, 2)$.

Describe the points that lie in the given configuration for Exercises 29–32.

29. The parallelogram whose adjacent sides are the vectors $\mathbf{i} + 3\mathbf{k}$ and $-2\mathbf{j}$.

30. The parallelogram whose adjacent sides are the vectors $2\mathbf{j} + 2\mathbf{k}$ and $4\mathbf{i}$.

31. The plane spanned by $\mathbf{v}_1 = (2, 7, 0)$ and $\mathbf{v}_2 = (0, 2, 7)$.

32. The plane spanned by $\mathbf{v}_1 = (3, 7, 1)$ and $\mathbf{v}_2 = (0, 3, 4)$.

Prove the statements in Exercises 33–36.

33. The line segment joining the midpoints of two sides of a triangle is parallel to and has half the length of the third side.

34. For any constant a and vector $\mathbf{v} = (x, y, z)$, $a\mathbf{v} = (ax, ay, az)$. (Use an argument based on similar triangles.)

35. If PQR is a triangle in space and $b > 0$ is a number, then there is a triangle with sides parallel to those of PQR and side lengths b times those of PQR.

36. The medians of a triangle intersect at a point, and this point divides each median in a ratio of $2 : 1$.

Problems 37 and 38 require some knowledge of chemical notation.

37. Write the chemical equation $CO + H_2O = H_2 + CO_2$ as an equation in ordered triples (x_1, x_2, x_3) where x_1, x_2, x_3 are the number of carbon, hydrogen, and oxygen atoms, respectively, in each molecule.

38.(a) Write the chemical equation $pC_3H_4O_3 + qO_2 = rCO_2 + sH_2O$ as an equation in ordered triples with unknown coefficients $p, q, r,$ and s.

  (b) Find the smallest positive integer solution for $p, q, r,$ and s.

1.2
The Inner Product and Distance

In this chapter we discuss two products of vectors: the inner product and the cross product. These products are useful in physical applications and have interesting geometric interpretations.

We begin in this section with the **inner product** (the names **dot product** and **scalar product** are often used instead), giving an algebraic definition of the inner product and then showing how it is related to the geometric concepts of length and angle. Let $\mathbf{a} = a_1\mathbf{i} + a_2\mathbf{j} + a_3\mathbf{k}$ and $\mathbf{b} = b_1\mathbf{i} + b_2\mathbf{j} + b_3\mathbf{k}$. We define the **inner product** of $\mathbf{a}$ and $\mathbf{b}$, written $\mathbf{a} \cdot \mathbf{b}$, to be the real number

$$\mathbf{a} \cdot \mathbf{b} = a_1 b_1 + a_2 b_2 + a_3 b_3.$$

Note that the inner product of two vectors is a scalar.

Example 1   (a) If $\mathbf{a} = 3\mathbf{i} + \mathbf{j} - 2\mathbf{k}$ and $\mathbf{b} = \mathbf{i} - \mathbf{j} + \mathbf{k}$, calculate $\mathbf{a} \cdot \mathbf{b}$.

(b) Calculate $(2\mathbf{i} + \mathbf{j} - \mathbf{k}) \cdot (3\mathbf{k} - 2\mathbf{j})$.

Solution

(a) $\mathbf{a} \cdot \mathbf{b} = 3 \cdot 1 + 1 \cdot (-1) + (-2) \cdot 1 = 3 - 1 - 2 = 0$.

(b) $(2\mathbf{i} + \mathbf{j} - \mathbf{k}) \cdot (3\mathbf{k} - 2\mathbf{j}) = (2\mathbf{i} + \mathbf{j} - \mathbf{k}) \cdot (0\mathbf{i} - 2\mathbf{j} + 3\mathbf{k})$
$= 2 \cdot 0 - 1 \cdot 2 - 1 \cdot 3 = -5.$ ♦

The inner product has several important algebraic properties. If $\mathbf{a}, \mathbf{b}$ and $\mathbf{c}$ are vectors in $\mathbb{R}^3$ and α and β are real numbers, then

(i) $\mathbf{a} \cdot \mathbf{a} \geq 0$; $\mathbf{a} \cdot \mathbf{a} = 0$ if and only if $\mathbf{a} = \mathbf{0}$;

(ii) $(\alpha \mathbf{a}) \cdot \mathbf{b} = \alpha(\mathbf{a} \cdot \mathbf{b})$ and $\mathbf{a} \cdot \beta\mathbf{b} = \beta(\mathbf{a} \cdot \mathbf{b})$;

(iii) $\mathbf{a} \cdot (\mathbf{b} + \mathbf{c}) = \mathbf{a} \cdot \mathbf{b} + \mathbf{a} \cdot \mathbf{c}$ and $(\mathbf{a} + \mathbf{b}) \cdot \mathbf{c} = \mathbf{a} \cdot \mathbf{c} + \mathbf{b} \cdot \mathbf{c}$; and

(iv) $\mathbf{a} \cdot \mathbf{b} = \mathbf{b} \cdot \mathbf{a}$.

To prove the first of these properties, observe that if $\mathbf{a} = a_1\mathbf{i} + a_2\mathbf{j} + a_3\mathbf{k}$, then $\mathbf{a} \cdot \mathbf{a} = a_1^2 + a_2^2 + a_3^2$. Since a_1, a_2, and a_3 are real numbers, we know that a_1^2, a_2^2 and a_3^2 are non-negative; hence, so is their sum. Moreover, if $a_1^2 + a_2^2 + a_3^2 = 0$, then $a_1 = a_2 = a_3 = 0$; therefore $\mathbf{a} = \mathbf{0}$ (zero vector). The proofs of the other properties of the inner product are also easily obtained from the formula that defines it.

It follows from the Pythagorean theorem that the **length** of the vector $\mathbf{a} = a_1\mathbf{i} + a_2\mathbf{j} + a_3\mathbf{k}$ is $\sqrt{a_1^2 + a_2^2 + a_3^2}$ (see Figure 1.2.1). This length is denoted by $\|\mathbf{a}\|$ and is sometimes called the **norm** of $\mathbf{a}$. Since $\mathbf{a} \cdot \mathbf{a} = a_1^2 + a_2^2 + a_3^2$, it follows that

$$\|\mathbf{a}\| = (\mathbf{a} \cdot \mathbf{a})^{\frac{1}{2}}.$$

Vectors with norm 1 are called **unit vectors**. For example, the basis vectors $\mathbf{i}, \mathbf{j}, \mathbf{k}$ discussed in §1.1 are unit vectors. Observe that for any nonzero vector $\mathbf{a}$, the vector $\mathbf{a}/\|\mathbf{a}\|$ is a unit vector. When we divide $\mathbf{a}$ by $\|\mathbf{a}\|$, we say that we have **normalized** $\mathbf{a}$.

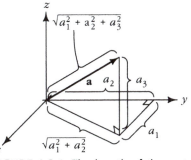

FIGURE 1.2.1. The length of the vector $\mathbf{a} = (a_1, a_2, a_3)$ is given by the Pythagorean formula: $\|\mathbf{a}\| = (a_1^2 + a_2^2 + a_3^2)^{\frac{1}{2}}$.

Example 2

(a) Normalize $\mathbf{v} = 2\mathbf{i} + 3\mathbf{j} - \frac{1}{2}\mathbf{k}$.

(b) Find unit vectors $\mathbf{a}, \mathbf{b}$, and $\mathbf{c}$ in the plane such that $\mathbf{b} + \mathbf{c} = \mathbf{a}$.

Solution

(a) We have $\|\mathbf{v}\| = \sqrt{2^2 + 3^2 + (1/2)^2} = (1/2)\sqrt{53}$, so the normalization of $\mathbf{v}$ is

$$\mathbf{u} = \frac{1}{\|\mathbf{v}\|}\mathbf{v} = \frac{4}{\sqrt{53}}\mathbf{i} + \frac{6}{\sqrt{53}}\mathbf{j} - \frac{1}{\sqrt{53}}\mathbf{k}.$$

(b) Since all three vectors are to have length 1, a triangle with sides $\mathbf{a}, \mathbf{b}$, and $\mathbf{c}$ must be equilateral as in Figure 1.2.2. Orienting the triangle as in the figure, we take $\mathbf{a} = \mathbf{i}$, then $\mathbf{b} = \frac{1}{2}\mathbf{i} + (\sqrt{3}/2)\mathbf{j}$, and $\mathbf{c} = \frac{1}{2}\mathbf{i} - (\sqrt{3}/2)\mathbf{j}$. You should check that $\|\mathbf{a}\| = \|\mathbf{b}\| = \|\mathbf{c}\| = 1$ and that $\mathbf{b} + \mathbf{c} = \mathbf{a}$. ◆

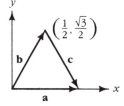

FIGURE 1.2.2. The vectors $\mathbf{a}, \mathbf{b}$, and $\mathbf{c}$ are represented by the sides of an equilateral triangle.

If $\mathbf{a}$ and $\mathbf{b}$ are vectors based at the same point, we have seen that the vector $\mathbf{b} - \mathbf{a}$ is parallel to and has the same magnitude as the directed line segment from the endpoint of **a** to the endpoint of **b**. It follows that the distance from the endpoint P of **a** to the endpoint Q of **b** is $\|\mathbf{b} - \mathbf{a}\|$ (or $\|\overrightarrow{PQ}\|$) (see Figure 1.2.3).

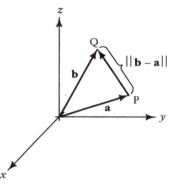

FIGURE 1.2.3. The distance between the tips of **a** and **b** is $\|\mathbf{b} - \mathbf{a}\|$.

Inner Product, Length, and Distance

Letting $\mathbf{a} = a_1\mathbf{i} + a_2\mathbf{j} + a_3\mathbf{k}$ and $\mathbf{b} = b_1\mathbf{i} + b_2\mathbf{j} + b_3\mathbf{k}$,

$$\mathbf{a} \cdot \mathbf{b} = a_1b_1 + a_2b_2 + a_3b_3,$$

$$\|\mathbf{a}\| = \sqrt{a_1^2 + a_2^2 + a_3^2}.$$

To **normalize** a vector **a**, form the vector

$$\frac{\mathbf{a}}{\|\mathbf{a}\|}.$$

The **distance between** the endpoints of **a** and **b** is $\|\mathbf{a} - \mathbf{b}\|$, and the **distance between** P and Q is $\|\overrightarrow{PQ}\|$.

Example 3   Find the distance from the endpoint of the vector **i** to the endpoint of the vector **j**.

Solution

$$\|\mathbf{j} - \mathbf{i}\| = \sqrt{(0 - 1)^2 + (1 - 0)^2 + (0 - 0)^2} = \sqrt{2}. \quad \blacklozenge$$

Example 4   Let $P_t = t(1, 1, 1)$.

    (a) What is the distance from P_t to $(3, 0, 0)$?

    (b) For what value of t is the distance shortest?

(c) What is the shortest distance?

Solution

(a) By the distance formula, the distance is

$$\sqrt{(t-3)^2 + (t-0)^2 + (t-0)^2} = \sqrt{t^2 - 6t + 9 + t^2 + t^2}$$
$$= \sqrt{3t^2 - 6t + 9}.$$

(b) The distance is shortest when its square, namely $3t^2 - 6t + 9$, is least, that is, when $(d/dt)(3t^2 - 6t + 9) = 6t - 6 = 0$, or $t = 1$.

(c) For $t = 1$, the distance in (a) is $\sqrt{6}$. ◆

Trigonometry tells us that we can find the angles of a triangle if we know the lengths of its sides. Since lengths are given by the dot product, we should also be able to use the dot product to compute angles. The relationship is given as follows.

Angles and the Inner Product

Let **a** and **b** be two vectors in $\mathbb{R}^3$ and let θ be the angle between them, where $0 \leq \theta \leq \pi$ (see Figure 1.2.4). Then

$$\mathbf{a} \cdot \mathbf{b} = \|\mathbf{a}\| \, \|\mathbf{b}\| \cos \theta.$$

Thus we may express the angle between nonzero vectors **a** and **b** as

$$\theta = \cos^{-1}\left(\frac{\mathbf{a} \cdot \mathbf{b}}{\|\mathbf{a}\| \, \|\mathbf{b}\|}\right).$$

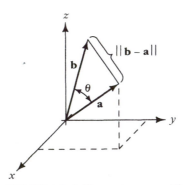

FIGURE 1.2.4. The vectors **a** and **b** and the angle θ between them.

To establish this, we recall the law of cosines from trigonometry. It states that for the triangle with adjacent sides determined by the vectors **a** and **b**, we have

$$\|\mathbf{b} - \mathbf{a}\|^2 = \|\mathbf{a}\|^2 + \|\mathbf{b}\|^2 - 2\|\mathbf{a}\| \cdot \|\mathbf{b}\| \cos\theta.$$

Since $\|\mathbf{b} - \mathbf{a}\|^2 = (\mathbf{b} - \mathbf{a}) \cdot (\mathbf{b} - \mathbf{a})$, $\|\mathbf{a}\|^2 = \mathbf{a} \cdot \mathbf{a}$, and $\|\mathbf{b}\|^2 = \mathbf{b} \cdot \mathbf{b}$, we can rewrite the preceding equation as

$$(\mathbf{b} - \mathbf{a}) \cdot (\mathbf{b} - \mathbf{a}) = \mathbf{a} \cdot \mathbf{a} + \mathbf{b} \cdot \mathbf{b} - 2\|\mathbf{a}\| \|\mathbf{b}\| \cos\theta.$$

By the distributive law for the dot product,

$$\begin{aligned}(\mathbf{b} - \mathbf{a}) \cdot (\mathbf{b} - \mathbf{a}) &= \mathbf{b} \cdot (\mathbf{b} - \mathbf{a}) - \mathbf{a} \cdot (\mathbf{b} - \mathbf{a}) \\ &= \mathbf{b} \cdot \mathbf{b} - \mathbf{b} \cdot \mathbf{a} - \mathbf{a} \cdot \mathbf{b} + \mathbf{a} \cdot \mathbf{a} \\ &= \mathbf{a} \cdot \mathbf{a} + \mathbf{b} \cdot \mathbf{b} - 2\mathbf{a} \cdot \mathbf{b}.\end{aligned}$$

Thus,

$$\mathbf{a} \cdot \mathbf{a} + \mathbf{b} \cdot \mathbf{b} - 2\mathbf{a} \cdot \mathbf{b} = \mathbf{a} \cdot \mathbf{a} + \mathbf{b} \cdot \mathbf{b} - 2\|\mathbf{a}\| \|\mathbf{b}\| \cos\theta.$$

That is,

$$\mathbf{a} \cdot \mathbf{b} = \|\mathbf{a}\| \|\mathbf{b}\| \cos\theta,$$

proving our assertion.

Example 5   Find the angle between the vectors $\mathbf{i} + \mathbf{j} + \mathbf{k}$ and $\mathbf{i} + \mathbf{j} - \mathbf{k}$ (see Figure 1.2.5).

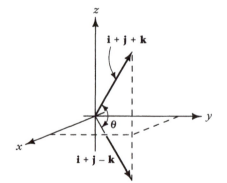

FIGURE 1.2.5. Finding the angle between $\mathbf{a} = \mathbf{i} + \mathbf{j} + \mathbf{k}$ and $\mathbf{b} = \mathbf{i} + \mathbf{j} - \mathbf{k}$.

Solution   Using the preceding box, we have

$$(\mathbf{i} + \mathbf{j} + \mathbf{k}) \cdot (\mathbf{i} + \mathbf{j} - \mathbf{k}) = \|\mathbf{i} + \mathbf{j} + \mathbf{k}\| \|\mathbf{i} + \mathbf{j} - \mathbf{k}\| \cos\theta$$

and so $1+1-1 = (\sqrt{3})(\sqrt{3}) \cos \theta$. Hence, $\cos \theta = 1/3$. That is, $\theta = \cos^{-1}(1/3) \approx 1.23$ radians ($\approx 71°$). ♦

Notice that if **a** and **b** are nonzero, then $\mathbf{a} \cdot \mathbf{b} = 0$ if and only if $\cos \theta = 0$. Hence we get:

Perpendicular Vectors

The inner product of two nonzero vectors is zero if and only if the vectors are perpendicular. Often we say that perpendicular vectors are ***orthogonal***.

The standard basis vectors **i**, **j**, and **k** are mutually orthogonal and of length 1; any such system is called ***orthonormal***.

Example 6   The vectors $\mathbf{i}_\theta = (\cos \theta)\mathbf{i} + (\sin \theta)\mathbf{j}$ and $\mathbf{j}_\theta = -(\sin \theta)\mathbf{i} + (\cos \theta)\mathbf{j}$ are orthonormal, since

$$\begin{aligned}
\mathbf{i}_\theta \cdot \mathbf{j}_\theta &= -\cos \theta \sin \theta + \sin \theta \cos \theta = 0, \\
\|\mathbf{i}_\theta\| &= \sqrt{\cos^2\theta + \sin^2\theta} = 1, \\
\|\mathbf{j}_\theta\| &= \sqrt{(-\sin \theta)^2 + \cos^2\theta} = 1
\end{aligned}$$

(see Figure 1.2.6). ♦

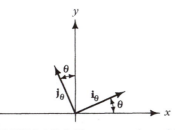

FIGURE 1.2.6. The vectors $\mathbf{i}_\theta$ and $\mathbf{j}_\theta$ are orthogonal.

Example 7   Find a unit vector in the xy plane that is orthogonal to $\mathbf{v} = \mathbf{i} - 3\mathbf{j}$.

Solution   If $\mathbf{w} = a\mathbf{i} + b\mathbf{j}$ is perpendicular to $\mathbf{i} - 3\mathbf{j}$, then $0 = \mathbf{v} \cdot \mathbf{w} = a - 3b$; so $a = 3b$. A solution is $3\mathbf{i} + \mathbf{j}$, but this is not a unit vector. Dividing by the length $\sqrt{3^2 + 1^2} = \sqrt{10}$, we find the solution $\mathbf{w} = (3\mathbf{i} + \mathbf{j})/\sqrt{10}$. Another solution is $-(3\mathbf{i} + \mathbf{j})/\sqrt{10}$ (see Figure 1.2.7). ♦

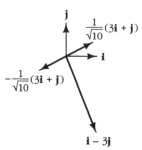

FIGURE 1.2.7. There are two unit vectors orthogonal to $\mathbf{i} - 3\mathbf{j}$.

The inner product of two vectors is the product of their lengths times the cosine of the angle between them. This relationship is often of value in geometric problems. It also leads to the following important inequality.

Cauchy–Schwarz Inequality

For any two vectors $\mathbf{a}$ and $\mathbf{b}$,

$$|\mathbf{a} \cdot \mathbf{b}| \leq \|\mathbf{a}\| \, \|\mathbf{b}\|.$$

Equality holds if and only if either $\mathbf{a}$ is a scalar multiple of $\mathbf{b}$, or one of $\mathbf{a}$ or $\mathbf{b}$ is $\mathbf{0}$.

To prove this, we use the formula $\mathbf{a} \cdot \mathbf{b} = \|\mathbf{a}\| \, \|\mathbf{b}\| \cos\theta$ to give $|\mathbf{a} \cdot \mathbf{b}| = \|\mathbf{a}\| \, \|\mathbf{b}\| \, |\cos\theta| \leq \|\mathbf{a}\| \, \|\mathbf{b}\|$ (since $-1 \leq \cos\theta \leq 1$). If either $\mathbf{a}$ or $\mathbf{b}$ is zero, then both sides are zero. Otherwise, equality holds in $|\mathbf{a} \cdot \mathbf{b}| \leq \|\mathbf{a}\| \, \|\mathbf{b}\|$ exactly when $|\cos\theta| = 1$; *i.e.*, $\theta = 0$ or π, which means that $\mathbf{a}$ is a (positive or negative) scalar multiple of $\mathbf{b}$.

Example 8   Verify the Cauchy–Schwarz inequality for $\mathbf{a} = -\mathbf{i} + \mathbf{j} + \mathbf{k}$ and $\mathbf{b} = 3\mathbf{i} + \mathbf{k}$.

Solution   The dot product is $\mathbf{a} \cdot \mathbf{b} = -3 + 0 + 1 = -2$ so $|\mathbf{a} \cdot \mathbf{b}| = 2$. Also, $\|\mathbf{a}\| = \sqrt{1 + 1 + 1} = \sqrt{3}$ and $\|\mathbf{b}\| = \sqrt{9 + 1} = \sqrt{10}$, and it is true that

$$2 \leq \sqrt{3} \cdot \sqrt{10}$$

because $\sqrt{3} \cdot \sqrt{10} > \sqrt{3} \cdot \sqrt{3} = 3 \geq 2$.   ◆

The inner product helps us compute the ***projection*** of a vector in a given direction. If $\mathbf{v}$ is a vector, and l is the line through the origin in the direction of a vector $\mathbf{a}$, then the ***orthogonal projection*** of $\mathbf{v}$ *on* $\mathbf{a}$ is the vector whose tip is obtained by dropping a perpendicular line to l from the tip of $\mathbf{v}$, as in Figure 1.2.8.

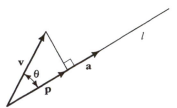

FIGURE 1.2.8. **p** is the orthogonal projection of **v** on **a**.

Referring to the figure, we see that **p** is a multiple of **a** and that **v** is the sum of **p** and a vector that is perpendicular to **a**. Thus we can write the equation

$$\mathbf{v} = c\mathbf{a} + \mathbf{q},$$

where $\mathbf{p} = c\mathbf{a}$ and $\mathbf{a} \cdot \mathbf{q} = 0$. Taking the dot product of **a** with both sides of $\mathbf{v} = c\mathbf{a} + \mathbf{q}$, we find $\mathbf{a} \cdot \mathbf{v} = c\mathbf{a} \cdot \mathbf{a}$, so $c = (\mathbf{a} \cdot \mathbf{v})/(\mathbf{a} \cdot \mathbf{a})$, and hence

$$\mathbf{p} = \frac{\mathbf{a} \cdot \mathbf{v}}{\|\mathbf{a}\|^2}\mathbf{a}.$$

The length of **p** is

$$\|\mathbf{p}\| = \frac{|\mathbf{a} \cdot \mathbf{v}|}{\|\mathbf{a}\|^2}\|\mathbf{a}\| = \frac{|\mathbf{a} \cdot \mathbf{v}|}{\|\mathbf{a}\|} = \|\mathbf{v}\|\cos\theta.$$

Orthogonal Projection

The ***orthogonal projection*** of **v** on **a** is the vector

$$\mathbf{p} = \frac{\mathbf{a} \cdot \mathbf{v}}{\|\mathbf{a}\|^2}\mathbf{a}.$$

Example 9   Find the orthogonal projection of $\mathbf{i} + \mathbf{j}$ on $\mathbf{i} - 2\mathbf{j}$.

Solution   With $\mathbf{a} = \mathbf{i} - 2\mathbf{j}$ and $\mathbf{v} = \mathbf{i} + \mathbf{j}$, the orthogonal projection of **v** on **a** is

$$\frac{\mathbf{a} \cdot \mathbf{v}}{\mathbf{a} \cdot \mathbf{a}}\mathbf{a} = \frac{1-2}{1+4}(\mathbf{i} - 2\mathbf{j}) = -\frac{1}{5}(\mathbf{i} - 2\mathbf{j})$$

(see Figure 1.2.9).   ◆

A useful consequence of the Cauchy–Schwarz inequality, called the ***triangle inequality***, relates the lengths of vectors **a** and **b** and of their sum $\mathbf{a} + \mathbf{b}$.

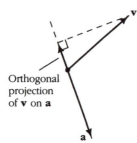

FIGURE 1.2.9. The orthogonal projection of $\mathbf{v}$ on $\mathbf{a}$ equals $-\frac{1}{5}\mathbf{a}$.

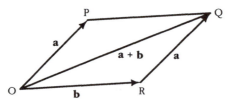

FIGURE 1.2.10. This geometry shows that $\|OQ\| \leq \|OR\| + \|RQ\|$, or, in vector notation, that $\|\mathbf{a} + \mathbf{b}\| \leq \|\mathbf{a}\| + \|\mathbf{b}\|$, which is the triangle inequality.

Geometrically, it says that the length of any side of a triangle is no greater than the sum of the lengths of the other two (see Figure 1.2.10).

Triangle Inequality

For vectors $\mathbf{a}$ and $\mathbf{b}$ in space,

$$\|\mathbf{a} + \mathbf{b}\| \leq \|\mathbf{a}\| + \|\mathbf{b}\|.$$

To demonstrate the triangle inequality, we consider the square of the left-hand side:

$$\|\mathbf{a} + \mathbf{b}\|^2 = (\mathbf{a} + \mathbf{b}) \cdot (\mathbf{a} + \mathbf{b}) = \|\mathbf{a}\|^2 + 2\mathbf{a} \cdot \mathbf{b} + \|\mathbf{b}\|^2.$$

By the Cauchy–Schwarz inequality, we have

$$\begin{aligned}
\|\mathbf{a}\|^2 + 2\mathbf{a} \cdot \mathbf{b} + \|\mathbf{b}\|^2 &\leq \|\mathbf{a}\|^2 + 2\|\mathbf{a}\| \|\mathbf{b}\| + \|\mathbf{b}\|^2 \\
&= (\|\mathbf{a}\| + \|\mathbf{b}\|)^2.
\end{aligned}$$

Thus,

$$\|\mathbf{a} + \mathbf{b}\|^2 \leq (\|\mathbf{a}\| + \|\mathbf{b}\|)^2;$$

taking square roots proves the triangle inequality.

Example 10   (a) Verify the triangle inequality for $\mathbf{a} = \mathbf{i} + \mathbf{j}$ and $\mathbf{b} = 2\mathbf{i} + \mathbf{j} + \mathbf{k}$.

(b) Prove that $\|\mathbf{u} - \mathbf{v}\| \leq \|\mathbf{u} - \mathbf{w}\| + \|\mathbf{w} - \mathbf{v}\|$ for any vectors $\mathbf{u}, \mathbf{v}$, and $\mathbf{w}$. Illustrate with a figure in which $\mathbf{u}, \mathbf{v}$, and $\mathbf{w}$ are based at the same base point.

Solution

(a) We have $\mathbf{a} + \mathbf{b} = 3\mathbf{i} + 2\mathbf{j} + \mathbf{k}$, so $\|\mathbf{a} + \mathbf{b}\| = \sqrt{9 + 4 + 1} = \sqrt{14}$. On the other hand, $\|\mathbf{a}\| = \sqrt{2}$ and $\|\mathbf{b}\| = \sqrt{6}$, so the triangle inequality asserts that $\sqrt{14} \leq \sqrt{2} + \sqrt{6}$. The numbers bear us out: $\sqrt{14} \approx 3.74$, while $\sqrt{2} + \sqrt{6} \approx 1.41 + 2.45 = 3.86$.

(b) We find that $\mathbf{u} - \mathbf{v} = (\mathbf{u} - \mathbf{w}) + (\mathbf{w} - \mathbf{v})$, so the result follows from the triangle inequality with $\mathbf{a}$ replaced by $\mathbf{u} - \mathbf{w}$ and $\mathbf{b}$ replaced by $\mathbf{w} - \mathbf{v}$. Geometrically, we are considering the shaded triangle in Figure 1.2.11.

♦

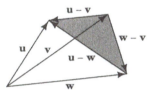

FIGURE 1.2.11. Illustrating the inequality $\|\mathbf{u} - \mathbf{v}\| \leq \|\mathbf{u} - \mathbf{w}\| + \|\mathbf{w} - \mathbf{v}\|$.

We now turn to some physical applications of vectors, beginning with an application to navigation. Suppose that, on a part of the earth's surface small enough to be considered flat, we introduce coordinates so that the x axis points east, the y axis points north, and the unit of length is the kilometer. If we are at a point P and wish to get to a point Q, the ***displacement vector*** $\mathbf{d} = \overrightarrow{PQ}$ joining P to Q tells us the direction and distance we have to travel. If x and y are the components of this vector, the displacement of P to Q is "x kilometers east, y kilometers north".

Example 11   Suppose that two navigators who cannot see each other but can communicate by radio wish to determine the relative position of their ships. Explain how they can do this if they can determine their displacement vectors to the same lighthouse.

Solution   Let P_1 and P_2 be the positions of the ships, and let Q be the position of the lighthouse. The displacement of the lighthouse from the ith ship is the vector $\mathbf{d}_i$ joining P_i to Q. The displacement of the second ship from the first is the vector **d** joining P_1 to P_2. We have $\mathbf{d} + \mathbf{d}_2 = \mathbf{d}_1$ (Figure 1.2.12), and so $\mathbf{d} = \mathbf{d}_1 - \mathbf{d}_2$. That is, the displacement from one ship to the other is the difference between the displacements from the ships to the lighthouse.   ◆

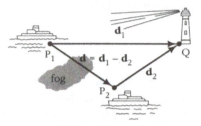

FIGURE 1.2.12. Vector methods can be used to locate objects.

We can also represent the velocity of a moving object as a vector. For the moment, we will consider only objects moving at uniform speed along straight lines. Suppose, for example, that a boat is steaming across a lake at 10 kilometers per hour (km/h) in the northeast direction. After 1 hour of travel, the displacement is $(10/\sqrt{2}, 10/\sqrt{2}) \approx (7.07, 7.07)$; see Figure 1.2.13.

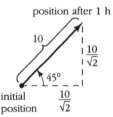

FIGURE 1.2.13. If an object moves northeast at 10 km/h, its velocity vector has components $(7.07, 7.07)$.

The vector whose components are $(10/\sqrt{2}, 10/\sqrt{2})$ is called the *velocity vector* of the boat. In general, if an object is moving uniformly along a straight line, *its **velocity vector** is the displacement vector from the position at any moment to the position 1 unit of time later.*

Displacement and Velocity

If an object has a (constant) velocity vector **v**, then in t units of time the resulting displacement vector of the object is $\mathbf{d} = t\mathbf{v}$; see Figure 1.2.14.

Returning to our boat on the lake, note that if a current appears, moving due eastward at 2 km/h, and the boat continues to point in the same direction with its engine running at the same rate, its displacement after 1 hour will have

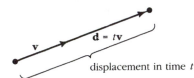

FIGURE 1.2.14. Displacement $=$ time $\times$ velocity.

components given by $(10/\sqrt{2}+2, 10/\sqrt{2})$; see Figure 1.2.15. The new velocity vector, therefore, has components $(10/\sqrt{2}+2, 10/\sqrt{2})$. We note that this is the sum of the original velocity vector $(10/\sqrt{2}, 10/\sqrt{2})$ of the boat and the velocity vector $(2, 0)$ of the current.

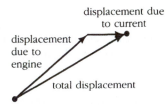

FIGURE 1.2.15. The total displacement is the sum of the displacements due to the engine and the current.

Similarly, consider a seagull that flies in calm air with velocity vector $\mathbf{v}$. If a wind comes up with velocity $\mathbf{w}$ and the seagull continues flying the same way, its actual velocity will be $\mathbf{v} + \mathbf{w}$. One can see the direction of the vector $\mathbf{v}$ because it points along the axis of the seagull. By comparing the direction of actual motion with the direction of $\mathbf{v}$, you can get an idea of the wind direction (see Figure 1.2.16).

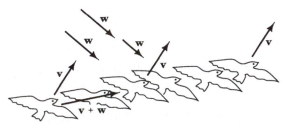

FIGURE 1.2.16. The velocity $\mathbf{w}$ of the wind can be estimated by comparing the "wingflap" velocity $\mathbf{v}$ with the actual velocity $\mathbf{v} + \mathbf{w}$.

Another example comes from medicine. An electrocardiograph detects the flow of electricity in the heart; both its magnitude and its direction are important. The net flow can be represented at every instant by a vector called the ***cardiac*** *vector*. The motion of this vector (see Figure 1.2.17) gives physicians useful information about the heart's function.

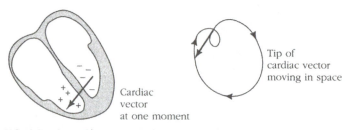

FIGURE 1.2.17. The magnitude and direction of electrical flow in the heart are indicated by the cardiac vector.

Example 12 A bird is flying in a straight line with velocity vector $10\mathbf{i} + 6\mathbf{j} + \mathbf{k}$ (in kilometers per hour). Suppose that (x, y) are its coordinates on the ground and z is its height above the ground.

(a) If the bird is at position $(1, 2, 3)$ at a certain moment, where is it 1 hour later? 1 minute later?

(b) How many seconds does it take the bird to climb 10 meters?

Solution

(a) The displacement vector from $(1, 2, 3)$ after 1 hour is $10\mathbf{i} + 6\mathbf{j} + \mathbf{k}$, so the new position is $(1, 2, 3) + (10, 6, 1) = (11, 8, 4)$. After 1 minute, the displacement vector from $(1, 2, 3)$ is

$$\frac{1}{60}(10\mathbf{i} + 6\mathbf{j} + \mathbf{k}) = \frac{1}{6}\mathbf{i} + \frac{1}{10}\mathbf{j} + \frac{1}{60}\mathbf{k},$$

and so the new position is

$$(1, 2, 3) + \left(\frac{1}{6}, \frac{1}{10}, \frac{1}{60}\right) = \left(\frac{7}{6}, \frac{21}{10}, \frac{181}{60}\right).$$

(b) After t seconds ($= t/3600$ hours), the displacement vector from $(1, 2, 3)$ is $(t/3600)(10\mathbf{i} + 6\mathbf{j} + \mathbf{k}) = (t/360)\mathbf{i} + (t/600)\mathbf{j} + (t/3600)\mathbf{k}$. The increase in altitude is the z-component $t/3600$. This will equal 10 m ($= (1/100)$km) when $t/3600 = 1/100$, that is, when $t = 36$ seconds.
◆

Example 13 Physical forces have magnitude and direction and may thus be represented by vectors. If several forces act at once on an object, the resultant force is represented by the sum of the individual force vectors. Suppose that forces

$\mathbf{i} + \mathbf{k}$ and $\mathbf{j} + \mathbf{k}$ are acting on a body. What third force must we impose to counteract the two, *i.e.,* to make the total force equal to zero?

Solution The force $\mathbf{F}$ should be chosen so that $(\mathbf{i} + \mathbf{k}) + (\mathbf{j} + \mathbf{k}) + \mathbf{F} = \mathbf{0}$; that is, $\mathbf{F} = -(\mathbf{i} + \mathbf{k}) - (\mathbf{j} + \mathbf{k}) = -\mathbf{i} - \mathbf{j} - 2\mathbf{k}$. (Recall that $\mathbf{0}$ is the *zero vector*, the vector whose components are all zero.) ◆

Real-World Problems vs. Made-Up Problems

You have probably wondered, after taking a year of one-variable calculus, and possibly other mathematics courses, how realistic are the problems and examples? You may have asked: "Will I really be able to solve some *significant* problems in physics, engineering, ecology, or biology just as easily?"

The answer to this question is somewhat negative, but there is a positive side too. The negative news is that truly realistic problems are normally (but not always) more complex than the ones presented in elementary texts, and one cannot solve them without seriously studying how the problem is modelled and bringing to bear more knowledge of the other field and of mathematics. To even understand such problems properly and to proceed to reliable solutions usually requires more time than one has available in the classroom. For example, consider the problem "How much solar energy is received in a day at a point on the earth (ignoring cloud cover) as a function of the day of the year and the latitude?" This is one of the "cleanest" but still realistic problems that we know of, yet it requires almost all the tools of calculus and multivariable calculus to solve, and it takes some time to understand. (The solution to this problem is given in the instructor's guide for this book — your instructor can provide it to you if you are interested.) Other sources, such as the UMAP Modules: Tools for Teaching, published by the Undergraduate Mathematics and Its Applications project, contain extended applications.

On the positive side, knowing multivariable and vector calculus well and the "made-up" or simplified types of problems presented, can form a basis of knowledge, upon which one can build, eventually enabling one to solve the often complex physical problems presented by the natural world.

Exercises for §1.2

In Exercises 1–4, compute $\|\mathbf{u}\|$, $\|\mathbf{v}\|$, and $\mathbf{u} \cdot \mathbf{v}$ for the given vectors in $\mathbb{R}^3$.

1. $\mathbf{u} = 15\mathbf{i} - 2\mathbf{j} + 4\mathbf{k}, \mathbf{v} = \pi\mathbf{i} + 3\mathbf{j} - \mathbf{k}$

2. $\mathbf{u} = 2\mathbf{j} - \mathbf{i}, \mathbf{v} = -\mathbf{j} + \mathbf{i}$

3. $\mathbf{u} = 2\mathbf{i} + 10\mathbf{j} - 12\mathbf{k}, \mathbf{v} = 3\mathbf{i} + 4\mathbf{k}$

4. $\mathbf{u} = \sqrt{3}\mathbf{i} + \pi\mathbf{j} + c\mathbf{k}, \mathbf{v} = 4\mathbf{i} - \mathbf{j} - \mathbf{k}$, where c is a constant.

Normalize the vectors given for Exercises 5–8.

5. The vector $\mathbf{u}$ in Exercise 1

6. The vector $\mathbf{v}$ in Exercise 1

7. The vector $\mathbf{u}$ in Exercise 2

8. The vector $\mathbf{v}$ in Exercise 2

In Exercises 9–12, find the angle between the given vectors. If necessary, express your answer in terms of $\cos^{-1}$.

9. The vectors in Exercise 1

10. The vectors in Exercise 2

11. The vectors in Exercise 3

12. The vectors in Exercise 4

In Exercises 13–16, verify the Cauchy–Schwarz inequality for the given pair of vectors.

13. The vectors in Exercise 1

14. The vectors in Exercise 2

15. The vectors in Exercise 3

16. The vectors in Exercise 4

17. Find two nonparallel vectors, both orthogonal to $(1, 1, 1)$.

18. What restrictions must be made on b so that the vector $2\mathbf{i} + b\mathbf{j}$ is orthogonal to (a) $-3\mathbf{i} + 2\mathbf{j} + \mathbf{k}$, (b) $\mathbf{k}$.

19. Tell without calculating whether $\|8\mathbf{i} - 12\mathbf{k}\| \cdot \|6\mathbf{j} + \mathbf{k}\| - |(8\mathbf{i} - 12\mathbf{k}) \cdot (6\mathbf{j} + \mathbf{k})|$ is equal to zero. Explain.

20. Find the line through $(3, 1, -2)$ that intersects and is perpendicular to the line $x = -1, y = -2 + t, z = -1 + t$. [Hint: If (x_0, y_0, z_0) is the point of intersection, find its coordinates.]

In Exercises 21–24, find the orthogonal projection of $\mathbf{u}$ onto $\mathbf{v}$.

21. $\mathbf{u} = -\mathbf{i} + \mathbf{j} + \mathbf{k}, \mathbf{v} = 2\mathbf{i} + \mathbf{j} - 3\mathbf{k}$

22. $\mathbf{u} = 2\mathbf{i} + \mathbf{j} - 3\mathbf{k}, \mathbf{v} = -\mathbf{i} + \mathbf{j} + \mathbf{k}$

23. $\mathbf{u} = 3\mathbf{i} + 4\mathbf{j} - 5\mathbf{k}, \mathbf{v} = \mathbf{i} + \mathbf{j} + \mathbf{k}$

24. $\mathbf{u} = \mathbf{i} + \mathbf{j} + \mathbf{k}, \mathbf{v} = 3\mathbf{i} + 4\mathbf{j} - 5\mathbf{k}$

25. Use the formula $(\mathbf{i} + \mathbf{j} + \mathbf{k}) \cdot \mathbf{i} = 1$ to find the angle between the diagonal of a cube and one of its edges. Sketch.

26.(a) If $\|\mathbf{u}\| = \|\mathbf{v}\|$, and $\mathbf{u}$ and $\mathbf{v}$ are not parallel, show that $\mathbf{u} + \mathbf{v}$ and $\mathbf{u} - \mathbf{v}$ are perpendicular.

   (b) Use the result of part (a) to prove that any triangle inscribed in a circle, with one side of the triangle as a diameter, is a right triangle.

27. A ship at position $(1, 0)$ on a nautical chart (with north in the positive y-direction) sights a rock at position $(2, 4)$. What is the vector joining the ship to the rock? What angle does the vector make with due north? (This is called the **bearing** of the rock from the ship.)

28. Suppose that the ship in Exercise 27 is pointing due north and traveling at a speed of 4 knots relative to the water. There is a current flowing due east at 1 knot. The units of the chart are nautical miles; 1 knot = 1 nautical mile per hour.

   (a) If there were no current, what vector $\mathbf{u}$ would represent the velocity of the ship relative to the sea bottom?

   (b) If the ship were just drifting with the current, what vector $\mathbf{v}$ would represent its velocity relative to the sea bottom?

   (c) What vector $\mathbf{w}$ represents the total velocity of the ship?

   (d) Where would the ship be after 1 hour?

   (e) Should the captain change course?

   (f) What if the rock were an iceberg?

29. The wind velocity $\mathbf{v}_1$ is 40 miles per hour (mi/h) from east to west while an airplane travels with air speed $\mathbf{v}_2$ of 100 mi/h due north. The speed of the airplane relative to the ground is the vector sum $\mathbf{v}_1 + \mathbf{v}_2$.

   (a) Find $\mathbf{v}_1 + \mathbf{v}_2$.

   (b) Draw a figure to scale.

30. An airplane is located at position $(3, 4, 5)$ at noon and traveling with velocity $400\mathbf{i} + 500\mathbf{j} - \mathbf{k}$ kilometers per hour. The pilot spots an airport at position $(23, 29, 0)$.

(a) At what time will the plane pass directly over the airport? (Assume that the earth is flat and that the vector $\mathbf{k}$ points straight up.)

(b) How high above the airport will the plane be when it passes?

31. A boat whose top speed in still water is 12 knots, points north and steams at full power. If there is an eastward current of 5 knots, what is the speed of the boat?

32. A 1-kilogram mass located at the origin is suspended by ropes attached to the points $(1, 1, 1)$ and $(-1, -1, 1)$. If the force of gravity is pointing in the direction of the vector $-\mathbf{k}$, what is the vector describing the force along each rope? [Hint: Use the symmetry of the problem. A 1-kilogram mass weighs 9.8 newtons (N).]

33. A force of 6 N (newtons) points to the upper right, making an angle of $\pi/4$ radians with the y axis. The force acts on an object that moves along the line segment from $(1, 2)$ to $(5, 4)$.

(a) Find a formula for the force vector $\mathbf{F}$.

(b) Find the angle θ between the displacement direction $\mathbf{D} = (5 - 1)\mathbf{i} + (4 - 2)\mathbf{j}$ and the force direction $\mathbf{F}$.

(c) The **work done** is $\mathbf{F} \cdot \mathbf{D}$, or equivalently, $\|\mathbf{F}\| \|\mathbf{D}\| \cos\theta$. Compute the work from both formulas and compare.

34. Imagine that you look to the side as you walk on a windless, rainy day. Now you stop walking.

(a) How does the (apparent) direction of the falling rain change?

(b) Explain the observation in (a) in terms of vectors.

(c) Suppose that you know your walking speed. How could you determine the speed at which the rain is falling?

Exercises 35–38 form a unit.

35. Suppose that $\mathbf{e}_1$ and $\mathbf{e}_2$ are perpendicular unit vectors in the plane, and let $\mathbf{v}$ be an arbitrary vector. Show that $\mathbf{v} = (\mathbf{v} \cdot \mathbf{e}_1)\mathbf{e}_1 + (\mathbf{v} \cdot \mathbf{e}_2)\mathbf{e}_2$. The numbers $\mathbf{v} \cdot \mathbf{e}_1$ and $\mathbf{v} \cdot \mathbf{e}_2$ are called the **components** of $\mathbf{v}$ in the directions of $\mathbf{e}_1$ and $\mathbf{e}_2$. This expression of $\mathbf{v}$ as a sum of vectors pointing in the directions of $\mathbf{e}_1$ and $\mathbf{e}_2$ is called the **orthogonal decomposition** of $\mathbf{v}$ relative to $\mathbf{e}_1$ and $\mathbf{e}_2$.

36. Consider the vectors $\mathbf{e}_1 = (1/\sqrt{2})(\mathbf{i} + \mathbf{j})$ and $\mathbf{e}_2 = (1/\sqrt{2})(\mathbf{i} - \mathbf{j})$ in the plane. Check that $\mathbf{e}_1$ and $\mathbf{e}_2$ are unit vectors perpendicular to each other, and express each of the following vectors in the form $\mathbf{v} = a_1\mathbf{e}_1 + a_2\mathbf{e}_2$ (that is, as a **linear combination** of $\mathbf{e}_1$ and $\mathbf{e}_2$):

(a) $\mathbf{v} = \mathbf{i}$

(b) $\mathbf{v} = \mathbf{j}$

(c) $\mathbf{v} = 2\mathbf{i} + \mathbf{j}$

(d) $\mathbf{v} = -2\mathbf{i} - \mathbf{j}$

37. Suppose that a force $\mathbf{F}$ (for example, gravity) is acting vertically downward on an object sitting on a plane that is inclined at an angle of 45° to the horizontal. Express this force as a sum of a force acting parallel to the plane and one acting perpendicular to it.

38. Suppose that an object moving in direction $\mathbf{i} + \mathbf{j}$ is acted on by a force given by the vector $\mathbf{F} = 2\mathbf{i} + \mathbf{j}$. Express this force as a sum of a force in the direction of motion and a force perpendicular to the direction of motion.

39. Show that if $\mathbf{a}$ is a nonzero vector and r is a nonzero scalar, then the orthogonal projection of a vector on $r\mathbf{a}$ is the same as its orthogonal projection on $\mathbf{a}$.

1.3
2 × 2 and
3 × 3 Matrices
and
Determinants

Vectors can be thought of as *lists* of numbers, whereas matrices are *arrays* of numbers. In this section we study some basic properties of matrices in preparation for the next section on the cross product of vectors.

A 2 × 2 ***matrix*** is an array

$$\begin{bmatrix} a_{11} & a_{12} \\ a_{21} & a_{22} \end{bmatrix}$$

of four scalars. For example,

$$\begin{bmatrix} 2 & 1 \\ 0 & 4 \end{bmatrix}, \begin{bmatrix} -1 & 0 \\ 1 & 1 \end{bmatrix}, \quad \text{and} \quad \begin{bmatrix} 13 & 7 \\ 6 & 11 \end{bmatrix}$$

are 2 × 2 matrices. The ***determinant***

$$\begin{vmatrix} a_{11} & a_{12} \\ a_{21} & a_{22} \end{vmatrix}$$

of such a matrix is the *number* defined by the equation

$$\begin{vmatrix} a_{11} & a_{12} \\ a_{21} & a_{22} \end{vmatrix} = a_{11}a_{22} - a_{12}a_{21}.$$

Example 1 $\begin{vmatrix} 1 & 1 \\ 1 & 1 \end{vmatrix} = 1 - 1 = 0; \quad \begin{vmatrix} 1 & 2 \\ 3 & 4 \end{vmatrix} = 4 - 6 = -2; \quad \begin{vmatrix} 5 & 6 \\ 7 & 8 \end{vmatrix} = 40 - 42 = -2.\blacklozenge$

A 3×3 matrix is an array of scalars

$$\begin{bmatrix} a_{11} & a_{12} & a_{13} \\ a_{21} & a_{22} & a_{23} \\ a_{31} & a_{32} & a_{33} \end{bmatrix}$$

where a_{ij} denotes the entry in the array that is in the ith row and the jth column. We define the **determinant** of a 3×3 matrix by the rule

$$\begin{vmatrix} a_{11} & a_{12} & a_{13} \\ a_{21} & a_{22} & a_{23} \\ a_{31} & a_{32} & a_{33} \end{vmatrix} = a_{11} \begin{vmatrix} a_{22} & a_{23} \\ a_{32} & a_{33} \end{vmatrix} - a_{12} \begin{vmatrix} a_{21} & a_{23} \\ a_{31} & a_{33} \end{vmatrix} + a_{13} \begin{vmatrix} a_{21} & a_{22} \\ a_{31} & a_{32} \end{vmatrix}.$$

Without some mnemonic device, this formula would be difficult to memorize. The rule to learn is that you move along the first row, multiplying a_{1j} by the determinant of the 2×2 matrix obtained by crossing out the first row and the jth column, and then you add these up, remembering to put a minus in front of the a_{12} term. For example, the determinant multiplied by the middle term of the preceding formula, namely,

$$\begin{vmatrix} a_{21} & a_{23} \\ a_{31} & a_{33} \end{vmatrix},$$

is obtained by crossing out the first row and the second column of the given 3×3 matrix. Schematically, the rule is illustrated in Figure 1.3.1.

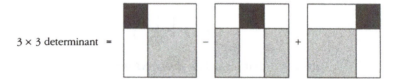

3 × 3 determinant =

FIGURE 1.3.1. One way to remember how to evaluate a 3 × 3 determinant.

Example 2

$$\begin{vmatrix} 1 & 0 & 0 \\ 0 & 1 & 0 \\ 0 & 0 & 1 \end{vmatrix} = 1 \begin{vmatrix} 1 & 0 \\ 0 & 1 \end{vmatrix} - 0 \begin{vmatrix} 0 & 0 \\ 0 & 1 \end{vmatrix} + 0 \begin{vmatrix} 0 & 1 \\ 0 & 0 \end{vmatrix} = 1.$$

$$\begin{vmatrix} 1 & 2 & 3 \\ 4 & 5 & 6 \\ 7 & 8 & 9 \end{vmatrix} = 1 \begin{vmatrix} 5 & 6 \\ 8 & 9 \end{vmatrix} - 2 \begin{vmatrix} 4 & 6 \\ 7 & 9 \end{vmatrix} + 3 \begin{vmatrix} 4 & 5 \\ 7 & 8 \end{vmatrix} = -3 + 12 - 9 = 0. \quad \blacklozenge$$

An important property of determinants is that interchanging two rows or two columns results in a change of sign. For 2×2 determinants, this is verified as follows: For rows, we have

$$\begin{vmatrix} a_{11} & a_{12} \\ a_{21} & a_{22} \end{vmatrix} = a_{11}a_{22} - a_{21}a_{12}$$

$$= -(a_{21}a_{12} - a_{11}a_{22}) = -\begin{vmatrix} a_{21} & a_{22} \\ a_{11} & a_{12} \end{vmatrix}$$

and for columns,

$$\begin{vmatrix} a_{11} & a_{12} \\ a_{21} & a_{22} \end{vmatrix} = -(a_{12}a_{21} - a_{11}a_{22}) = -\begin{vmatrix} a_{12} & a_{11} \\ a_{22} & a_{21} \end{vmatrix}.$$

Similarly, in the 3×3 case, *the interchange of any two rows or any two columns changes the sign of the determinant.*

We also note that *the determinant of a matrix with two identical rows is zero.* Simple observations like this will lead to improved methods for evaluating determinants.

The ***transpose*** of a matrix $\begin{bmatrix} a & b \\ c & d \end{bmatrix}$ is defined to be the matrix $\begin{bmatrix} a & c \\ b & d \end{bmatrix}$ obtained by reflection across the main (upper left to lower right) diagonal. For a 3×3 matrix $\begin{bmatrix} a & b & c \\ d & e & f \\ g & h & i \end{bmatrix}$, the transpose is $\begin{bmatrix} a & d & g \\ b & e & h \\ c & f & i \end{bmatrix}$.

Example 3   (a) Find the transpose of $\begin{bmatrix} 1 & 5 \\ -3 & 2 \end{bmatrix}$.

(b) Show that the determinant of a 2×2 matrix is equal to the determinant of its transpose: $\begin{vmatrix} a & b \\ c & d \end{vmatrix} = \begin{vmatrix} a & c \\ b & d \end{vmatrix}$.

(c) Check (b) for the matrix in (a).

Solution

(a) The transpose of $\begin{bmatrix} 1 & 5 \\ -3 & 2 \end{bmatrix}$ is $\begin{bmatrix} 1 & -3 \\ 5 & 2 \end{bmatrix}$.

(b) $\begin{vmatrix} a & b \\ c & d \end{vmatrix} = ad - bc$; $\begin{vmatrix} a & c \\ b & d \end{vmatrix} = ad - cb$. Since $bc = cb$, they are equal.

(c) $\begin{vmatrix} 1 & 5 \\ -3 & 2 \end{vmatrix} = 2 + 15 = 17$; also, $\begin{vmatrix} 1 & -3 \\ 5 & 2 \end{vmatrix} = 2 + 15 = 17$.   ♦

Another fundamental property of determinants is that *we can factor scalars out of any row or column*. For 2×2 determinants, this means that

$$\begin{vmatrix} \alpha a_{11} & a_{12} \\ \alpha a_{21} & a_{22} \end{vmatrix} = \begin{vmatrix} a_{11} & \alpha a_{12} \\ a_{21} & \alpha a_{22} \end{vmatrix} = \alpha \begin{vmatrix} a_{11} & a_{12} \\ a_{21} & a_{22} \end{vmatrix}$$

$$= \begin{vmatrix} \alpha a_{11} & \alpha a_{12} \\ a_{21} & a_{22} \end{vmatrix} = \begin{vmatrix} a_{11} & a_{12} \\ \alpha a_{21} & \alpha a_{22} \end{vmatrix}.$$

Similarly, for 3×3 determinants we have

$$\begin{vmatrix} \alpha a_{11} & \alpha a_{12} & \alpha a_{13} \\ a_{21} & a_{22} & a_{23} \\ a_{31} & a_{32} & a_{33} \end{vmatrix} = \alpha \begin{vmatrix} a_{11} & a_{12} & a_{13} \\ a_{21} & a_{22} & a_{23} \\ a_{31} & a_{32} & a_{33} \end{vmatrix} = \begin{vmatrix} a_{11} & \alpha a_{12} & a_{13} \\ a_{21} & \alpha a_{22} & a_{23} \\ a_{31} & \alpha a_{32} & a_{33} \end{vmatrix}$$

and so on. These results follow from the definitions. In particular, *if any row or column consists of zeros, then the value of the determinant is zero.*

A third fundamental fact about determinants is that *if we change a row (or column) by adding any multiple of another row (or column) to it, the value of the determinant remains the same.* For the 2×2 case this means that for any number k,

$$\begin{vmatrix} a_1 & a_2 \\ b_1 & b_2 \end{vmatrix} = \begin{vmatrix} a_1 + kb_1 & a_2 + kb_2 \\ b_1 & b_2 \end{vmatrix} = \begin{vmatrix} a_1 & a_2 \\ b_1 + ka_1 & b_2 + ka_2 \end{vmatrix}$$

$$= \begin{vmatrix} a_1 + ka_2 & a_2 \\ b_1 + kb_2 & b_2 \end{vmatrix} = \begin{vmatrix} a_1 & ka_1 + a_2 \\ b_1 & kb_1 + b_2 \end{vmatrix}.$$

For the 3×3 case,

$$\begin{vmatrix} a_1 & a_2 & a_3 \\ b_1 & b_2 & b_3 \\ c_1 & c_2 & c_3 \end{vmatrix} = \begin{vmatrix} a_1 + kb_1 & a_2 + kb_2 & a_3 + kb_3 \\ b_1 & b_2 & b_3 \\ c_1 & c_2 & c_3 \end{vmatrix} = \begin{vmatrix} a_1 + ka_2 & a_2 & a_3 \\ b_1 + kb_2 & b_2 & b_3 \\ c_1 + kc_2 & c_2 & c_3 \end{vmatrix}$$

and so on. Again, this property can be proved using the definition of the determinant (see Exercise 9).

Example 4 Evaluate

$$\begin{vmatrix} 2 & 1 & 1 \\ 3 & 1 & 1 \\ 4 & 0 & -1 \end{vmatrix}.$$

Solution Subtract the second row from the first:

$$\begin{vmatrix} 2 & 1 & 1 \\ 3 & 1 & 1 \\ 4 & 0 & -1 \end{vmatrix} = \begin{vmatrix} -1 & 0 & 0 \\ 3 & 1 & 1 \\ 4 & 0 & -1 \end{vmatrix} = -1 \begin{vmatrix} 1 & 1 \\ 0 & -1 \end{vmatrix} = 1. \quad \blacklozenge$$

Historical Note

Determinants seem to have been invented and first used by Gottfried Wilhelm Leibniz in 1693 in connection with solutions of linear equations. Colin Maclaurin and Gabriel Cramer developed their properties between 1729 and 1750; in particular, they showed that the solution of the system of equations

$$a_{11}x_1 + a_{12}x_2 + a_{13}x_3 = b_1,$$
$$a_{21}x_1 + a_{22}x_2 + a_{23}x_3 = b_2,$$
$$a_{31}x_1 + a_{32}x_2 + a_{33}x_3 = b_3,$$

is

$$x_1 = \frac{1}{\Delta}\begin{vmatrix} b_1 & a_{12} & a_{13} \\ b_2 & a_{22} & a_{23} \\ b_3 & a_{32} & a_{33} \end{vmatrix}, \quad x_2 = \frac{1}{\Delta}\begin{vmatrix} a_{11} & b_1 & a_{13} \\ a_{21} & b_2 & a_{23} \\ a_{31} & b_3 & a_{33} \end{vmatrix},$$

and

$$x_3 = \frac{1}{\Delta}\begin{vmatrix} a_{11} & a_{12} & b_1 \\ a_{21} & a_{22} & b_2 \\ a_{31} & a_{32} & b_3 \end{vmatrix},$$

where

$$\Delta = \begin{vmatrix} a_{11} & a_{12} & a_{13} \\ a_{21} & a_{22} & a_{23} \\ a_{31} & a_{32} & a_{33} \end{vmatrix}.$$

This is known as **Cramer's rule**. Gabriel Cramer (1704−1752) published this formula in his book *Introduction a l'Analyse des Lignes Courbes Algebriques* (1750); however, this result was probably already known to Colin Maclaurin. Later, Alexandre-Theophile Vandermonde (1772) and Augustin Cauchy (1812), treating determinants as a separate topic worthy of special attention, developed the field more systematically, with contributions by Laplace, Jacobi, and others. Formulas for volumes of parallelepipeds in terms of determinants are due to Joseph Louis Lagrange (1775). We shall study these in the next section. For the full history up to 1900, see T. Muir, *The Theory of Determinants in the Historical Order of Development*, reprinted by Dover, New York, 1960.

Example 5 Suppose that the rows **a**, **b**, and **c** of a 3 × 3 matrix satisfy

$$\mathbf{a} = \alpha\mathbf{b} + \beta\mathbf{c}; \quad i.e., \quad (a_1, a_2, a_3) = \alpha(b_1, b_2, b_3) + \beta(c_1, c_2, c_3).$$

(We say that **a** is a **linear combination** of **b** and **c**.) Show that

$$\begin{vmatrix} a_1 & a_2 & a_3 \\ b_1 & b_2 & b_3 \\ c_1 & c_2 & c_3 \end{vmatrix} = 0.$$

Solution   The determinant in question is

$$\begin{vmatrix} \alpha b_1 + \beta c_1 & \alpha b_2 + \beta c_2 & \alpha b_3 + \beta c_3 \\ b_1 & b_2 & b_3 \\ c_1 & c_2 & c_3 \end{vmatrix}.$$

Subtracting α times the second row and then β times the third row from the first row gives

$$\begin{vmatrix} 0 & 0 & 0 \\ b_1 & b_2 & b_3 \\ c_1 & c_2 & c_3 \end{vmatrix},$$

which is equal to 0.   ◆

The 2×2 determinants in the formula for the 3×3 determinant are called **minors**, and the formula itself is called the **expansion by minors of the first row**.

It turns out that a determinant can be evaluated by expanding in minors of *any* row or column. To do the expansion, multiply each entry in a given row or column by the 2×2 determinant obtained by crossing out the row and column of the given entry. Signs are assigned to the products according to the checkerboard pattern:

$$\begin{vmatrix} + & - & + \\ - & + & - \\ + & - & + \end{vmatrix}.$$

Example 6   Find the value of

$$\begin{vmatrix} 1 & 0 & 3 \\ 2 & 1 & -2 \\ 5 & 0 & 4 \end{vmatrix}.$$

Solution   It is efficient to expand along the row or down the column with the most zeros. Here we expand the determinant by minors of the second column; the only nonzero term is $\begin{vmatrix} 1 & 3 \\ 5 & 4 \end{vmatrix}(1) = -11.$ Since there is a + sign at the place in the checkerboard corresponding to the 1, -11 is the value of the determinant.   ◆

Exercises for §1.3

Evaluate the determinants in Exercises 1–8.

1. $\begin{vmatrix} 1 & 1 \\ -1 & 1 \end{vmatrix}$

2. $\begin{vmatrix} 1 & 1 \\ -1 & 0 \end{vmatrix}$

3. $\begin{vmatrix} 6 & 5 \\ 12 & 10 \end{vmatrix}$

4. $\begin{vmatrix} 0 & 0 \\ 3 & 17 \end{vmatrix}$

5. $\begin{vmatrix} 1 & 0 & 1 \\ 0 & 1 & 0 \\ 1 & 0 & 1 \end{vmatrix}$

6. $\begin{vmatrix} 1 & 1 & 0 \\ 0 & 1 & 1 \\ 0 & 1 & 1 \end{vmatrix}$

7. $\begin{vmatrix} 2 & -1 & 0 \\ 4 & 3 & 2 \\ 3 & 0 & 1 \end{vmatrix}$

8. $\begin{vmatrix} 1 & 1 & 1 \\ 2 & 2 & 2 \\ 3 & 3 & 3 \end{vmatrix}$

9. Show that adding three times the second row of a 3 × 3 matrix to the first row leaves the determinant unchanged.

10. Show that the determinant of any 3 × 3 matrix is equal to the determinant of its transpose.

11. Find a 2 × 2 matrix with determinant 27, all of whose entries are negative.

12. Find a 3 × 3 matrix with determinant 31, all of whose entries are negative.

13. Show that if

$$\begin{vmatrix} a & b \\ c & d \end{vmatrix} \neq 0,$$

then the solutions to the equations

$$ax + by = e,$$
$$cx + dy = f,$$

are given by the formulas

$$x = \frac{\begin{vmatrix} e & b \\ f & d \end{vmatrix}}{\begin{vmatrix} a & b \\ c & d \end{vmatrix}}, \quad y = \frac{\begin{vmatrix} a & e \\ c & f \end{vmatrix}}{\begin{vmatrix} a & b \\ c & d \end{vmatrix}}.$$

This result is an instance of *Cramer's rule*.

14. Verify Cramer's rule for 3×3 systems: suppose that the determinant

$$D = \begin{vmatrix} a_1 & b_1 & c_1 \\ a_2 & b_2 & c_2 \\ a_3 & b_3 & c_3 \end{vmatrix}$$

is unequal to zero. Then the solution of the equations

$$a_1 x + b_1 y + c_1 z = d_1,$$
$$a_2 x + b_2 y + c_2 z = d_2,$$
$$a_3 x + b_3 y + c_3 z = d_3,$$

is given by the formulas

$$x = \frac{\begin{vmatrix} d_1 & b_1 & c_1 \\ d_2 & b_2 & c_2 \\ d_3 & b_3 & c_3 \end{vmatrix}}{D}, \quad y = \frac{\begin{vmatrix} a_1 & d_1 & c_1 \\ a_2 & d_2 & c_2 \\ a_3 & d_3 & c_3 \end{vmatrix}}{D}, \quad z = \frac{\begin{vmatrix} a_1 & b_1 & d_1 \\ a_2 & b_2 & d_2 \\ a_3 & b_3 & d_3 \end{vmatrix}}{D}.$$

15. Use the 2×2 Cramer rule (Exercise 13) to solve the system of equations $4x + 3y = 2, 2x - 6y = 1$.

16. Use the 3×3 Cramer rule (Exercise 14) to solve the system of equations $-x + y = 14, 2x + y + z = 8, x + y + 5z = -1$.

1.4
The Cross Product and Planes

In §**1.2** we studied a *scalar* $\mathbf{a} \cdot \mathbf{b}$ obtained from two vectors $\mathbf{a}$ and $\mathbf{b}$. In this section we define a *vector* $\mathbf{a} \times \mathbf{b}$, called the *cross product* of $\mathbf{a}$ and $\mathbf{b}$. This new vector will have the pleasing geometric property that it is perpendicular to the plane spanned by $\mathbf{a}$ and $\mathbf{b}$. Our definition is based on determinants, a concept presented in the preceding section.

Definition of the Cross Product

Let $\mathbf{a} = a_1\mathbf{i} + a_2\mathbf{j} + a_3\mathbf{k}$ and $\mathbf{b} = b_1\mathbf{i} + b_2\mathbf{j} + b_3\mathbf{k}$ be vectors in $\mathbb{R}^3$. The **cross product** or **vector product** of $\mathbf{a}$ and $\mathbf{b}$, denoted $\mathbf{a} \times \mathbf{b}$, is defined to be the vector

$$\mathbf{a} \times \mathbf{b} = \begin{vmatrix} a_2 & a_3 \\ b_2 & b_3 \end{vmatrix} \mathbf{i} - \begin{vmatrix} a_1 & a_3 \\ b_1 & b_3 \end{vmatrix} \mathbf{j} + \begin{vmatrix} a_1 & a_2 \\ b_1 & b_2 \end{vmatrix} \mathbf{k},$$

or, symbolically,

$$\mathbf{a} \times \mathbf{b} = \begin{vmatrix} \mathbf{i} & \mathbf{j} & \mathbf{k} \\ a_1 & a_2 & a_3 \\ b_1 & b_2 & b_3 \end{vmatrix}.$$

Even though we only defined determinants for arrays of scalars, the formal expression involving *vectors* is a useful memory aid for the cross product.

Example 1 Find $(3\mathbf{i} - \mathbf{j} + \mathbf{k}) \times (\mathbf{i} + 2\mathbf{j} - \mathbf{k})$.

Solution

$$(3\mathbf{i} - \mathbf{j} + \mathbf{k}) \times (\mathbf{i} + 2\mathbf{j} - \mathbf{k}) = \begin{vmatrix} \mathbf{i} & \mathbf{j} & \mathbf{k} \\ 3 & -1 & 1 \\ 1 & 2 & -1 \end{vmatrix}$$

$$= \begin{vmatrix} -1 & 1 \\ 2 & -1 \end{vmatrix} \mathbf{i} - \begin{vmatrix} 3 & 1 \\ 1 & -1 \end{vmatrix} \mathbf{j} + \begin{vmatrix} 3 & -1 \\ 1 & 2 \end{vmatrix} \mathbf{k}$$

$$= -\mathbf{i} + 4\mathbf{j} + 7\mathbf{k}. \quad \blacklozenge$$

Certain algebraic properties of the cross product follow from the definition. If $\mathbf{a}, \mathbf{b}$, and $\mathbf{c}$ are vectors and α, β, and γ are scalars, then

(a) $\mathbf{a} \times \mathbf{b} = -(\mathbf{b} \times \mathbf{a})$;

(b) $\mathbf{a} \times (\beta\mathbf{b} + \gamma\mathbf{c}) = \beta(\mathbf{a} \times \mathbf{b}) + \gamma(\mathbf{a} \times \mathbf{c})$,
    $(\alpha\mathbf{a} + \beta\mathbf{b}) \times \mathbf{c} = \alpha(\mathbf{a} \times \mathbf{c}) + \beta(\mathbf{b} \times \mathbf{c})$.

Note that $\mathbf{a} \times \mathbf{a} = -(\mathbf{a} \times \mathbf{a})$, by property (a). Thus, $\mathbf{a} \times \mathbf{a} = \mathbf{0}$. In particular,

$$\mathbf{i} \times \mathbf{i} = \mathbf{0}, \quad \mathbf{j} \times \mathbf{j} = \mathbf{0}, \quad \mathbf{k} \times \mathbf{k} = \mathbf{0}.$$

Also,

$$\mathbf{i} \times \mathbf{j} = \mathbf{k}, \quad \mathbf{j} \times \mathbf{k} = \mathbf{i}, \quad \mathbf{k} \times \mathbf{i} = \mathbf{j},$$

which can be remembered by writing $\mathbf{i}, \mathbf{j}$, and $\mathbf{k}$ in a circle like this:

Example 2 (a) Compute $(3\mathbf{i} + 2\mathbf{j} - \mathbf{k}) \times (\mathbf{j} - \mathbf{k})$ by using the properties of the cross product.

(b) Find $\mathbf{i} \times (\mathbf{i} \times \mathbf{j})$ and $(\mathbf{i} \times \mathbf{i}) \times \mathbf{j}$. Are they equal?

Solution

(a) We use the products $\mathbf{i} \times \mathbf{j} = \mathbf{k}$, etc., and the algebraic rules as follows:

$$
\begin{aligned}
(3\mathbf{i} + 2\mathbf{j} - \mathbf{k}) \times (\mathbf{j} - \mathbf{k}) &= (3\mathbf{i} + 2\mathbf{j} - \mathbf{k}) \times \mathbf{j} - (3\mathbf{i} + 2\mathbf{j} - \mathbf{k}) \times \mathbf{k} \\
&= 3\mathbf{i} \times \mathbf{j} + 2\mathbf{j} \times \mathbf{j} - \mathbf{k} \times \mathbf{j} - 3\mathbf{i} \times \mathbf{k} \\
&\quad - 2\mathbf{j} \times \mathbf{k} + \mathbf{k} \times \mathbf{k} \\
&= 3\mathbf{k} + 0 + \mathbf{i} + 3\mathbf{j} - 2\mathbf{i} + 0 \\
&= -\mathbf{i} + 3\mathbf{j} + 3\mathbf{k}.
\end{aligned}
$$

This can be checked using the definition in the preceding box.

(b) We find that $\mathbf{i} \times (\mathbf{i} \times \mathbf{j}) = \mathbf{i} \times \mathbf{k} = -\mathbf{j}$, whereas $(\mathbf{i} \times \mathbf{i}) \times \mathbf{j} = \mathbf{0} \times \mathbf{j} = \mathbf{0}$, so the two expressions are not equal. This example means that the cross product is *not associative*, that is, one cannot move parentheses as in ordinary multiplication. ◆

To give a geometric interpretation of the cross product, we first introduce the triple product. Given three vectors $\mathbf{a}, \mathbf{b}$, and $\mathbf{c}$, the real number

$$(\mathbf{a} \times \mathbf{b}) \cdot \mathbf{c}$$

is called the ***triple product*** of $\mathbf{a}, \mathbf{b}$, and $\mathbf{c}$ (in that order). To obtain a formula for it, let $\mathbf{a} = a_1\mathbf{i} + a_2\mathbf{j} + a_3\mathbf{k}, \mathbf{b} = b_1\mathbf{i} + b_2\mathbf{j} + b_3\mathbf{k}$, and $\mathbf{c} = c_1\mathbf{i} + c_2\mathbf{j} + c_3\mathbf{k}$. Then

$$
\begin{aligned}
(\mathbf{a} \times \mathbf{b}) \cdot \mathbf{c} &= \left(\begin{vmatrix} a_2 & a_3 \\ b_2 & b_3 \end{vmatrix} \mathbf{i} - \begin{vmatrix} a_1 & a_3 \\ b_1 & b_3 \end{vmatrix} \mathbf{j} + \begin{vmatrix} a_1 & a_2 \\ b_1 & b_2 \end{vmatrix} \mathbf{k} \right) \\
&\quad \cdot (c_1\mathbf{i} + c_2\mathbf{j} + c_3\mathbf{k}) \\
&= \begin{vmatrix} a_2 & a_3 \\ b_2 & b_3 \end{vmatrix} c_1 - \begin{vmatrix} a_1 & a_3 \\ b_1 & b_3 \end{vmatrix} c_2 + \begin{vmatrix} a_1 & a_2 \\ b_1 & b_2 \end{vmatrix} c_3.
\end{aligned}
$$

This is the expansion by minors of the third row of the determinant, so

$$
(\mathbf{a} \times \mathbf{b}) \cdot \mathbf{c} = \begin{vmatrix} a_1 & a_2 & a_3 \\ b_1 & b_2 & b_3 \\ c_1 & c_2 & c_3 \end{vmatrix}.
$$

If $\mathbf{c}$ is a vector in the plane spanned by the vectors $\mathbf{a}$ and $\mathbf{b}$, then the third row in the determinant expression for $(\mathbf{a} \times \mathbf{b}) \cdot \mathbf{c}$ is a linear combination of the first and second rows and therefore $(\mathbf{a} \times \mathbf{b}) \cdot \mathbf{c} = 0$. (Compare Example 5 in §**1.3**.) In other words, *the vector* $\mathbf{a} \times \mathbf{b}$ *is orthogonal to any vector in the plane spanned by* $\mathbf{a}$ *and* $\mathbf{b}$, *in particular to both* $\mathbf{a}$ *and* $\mathbf{b}$.

Next, we calculate the length of $\mathbf{a} \times \mathbf{b}$. Note that

$$
\begin{aligned}
\|\mathbf{a} \times \mathbf{b}\|^2 &= \begin{vmatrix} a_2 & a_3 \\ b_2 & b_3 \end{vmatrix}^2 + \begin{vmatrix} a_1 & a_3 \\ b_1 & b_3 \end{vmatrix}^2 + \begin{vmatrix} a_1 & a_2 \\ b_1 & b_2 \end{vmatrix}^2 \\
&= (a_2 b_3 - a_3 b_2)^2 + (a_1 b_3 - b_1 a_3)^2 + (a_1 b_2 - b_1 a_2)^2.
\end{aligned}
$$

If we square the terms in the last expression, we can recollect them to give

$$(a_1^2 + a_2^2 + a_3^2)(b_1^2 + b_2^2 + b_3^2) - (a_1 b_1 + a_2 b_2 + a_3 b_3)^2,$$

which equals

$$\begin{aligned}
\|\mathbf{a}\|^2 \|\mathbf{b}\|^2 - (\mathbf{a} \cdot \mathbf{b})^2 &= \|\mathbf{a}\|^2 \|\mathbf{b}\|^2 - \|\mathbf{a}\|^2 \|\mathbf{b}\|^2 \cos^2\theta \\
&= \|\mathbf{a}\|^2 \|\mathbf{b}\|^2 \sin^2\theta
\end{aligned}$$

where θ is the angle between $\mathbf{a}$ and $\mathbf{b}, 0 \leq \theta \leq \pi$. Taking square roots and using $\sqrt{k^2} = |k|$, we find that $\|\mathbf{a} \times \mathbf{b}\| = \|\mathbf{a}\| \|\mathbf{b}\| |\sin\theta|$.

Combining our results, we conclude that $\mathbf{a} \times \mathbf{b}$ is a *vector perpendicular to the plane spanned by* $\mathbf{a}$ *and* $\mathbf{b}$ *with length* $\|\mathbf{a}\| \|\mathbf{b}\| |\sin\theta|$. We see from Figure 1.4.1 that this length is also the area of the parallelogram (with base $\|\mathbf{a}\|$ and height $\|\mathbf{b}\sin\theta\|$) spanned by $\mathbf{a}$ and $\mathbf{b}$. There are still two possible vectors that satisfy these conditions, because there are two choices of direction that are perpendicular (or normal) to the plane $\mathcal{P}$ spanned by $\mathbf{a}$ and $\mathbf{b}$. This is clear from Figure 1.4.1, which shows the two choices $\mathbf{n}_1$ and $-\mathbf{n}_1$ perpendicular to $\mathcal{P}$, with $\|\mathbf{n}_1\| = \|-\mathbf{n}_1\| = \|\mathbf{a}\| \|\mathbf{b}\| |\sin\theta|$.

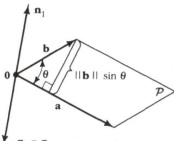

FIGURE 1.4.1. $\mathbf{n}_1$ and $\mathbf{n}_2$ are the two possible vectors orthogonal to both $\mathbf{a}$ and $\mathbf{b}$, and with norm $\|\mathbf{a}\| \|\mathbf{b}\| |\sin\theta|$.

Which vector represents $\mathbf{a} \times \mathbf{b}$, $\mathbf{n}_1$ or $-\mathbf{n}_1$? The answer is $\mathbf{n}_1$. Try a few cases such as $\mathbf{k} = \mathbf{i} \times \mathbf{j}$ to verify this. The following "right-hand rule" determines the direction of $\mathbf{a} \times \mathbf{b}$ in general. Take your right hand and place it so your fingers curl from $\mathbf{a}$ towards $\mathbf{b}$ through the *acute* angle θ, as in Figure 1.4.2. Then your thumb points in the direction of $\mathbf{a} \times \mathbf{b}$.

Example 3   Find the area of the parallelogram spanned by the vectors $\mathbf{a} = \mathbf{i} + 2\mathbf{j} + 3\mathbf{k}$ and $\mathbf{b} = -\mathbf{i} - \mathbf{k}$.

Solution   We calculate the cross product of $\mathbf{a}$ and $\mathbf{b}$ by applying the component or determinant formula, with $a_1 = 1, a_2 = 2, a_3 = 3, b_1 = -1, b_2 = 0, b_3 = -1$:

$$\mathbf{a} \times \mathbf{b} = [(2)(-1) - (3)(0)]\mathbf{i} + [(3)(-1) - (1)(-1)]\mathbf{j} + [(1)(0) - (2)(-1)]\mathbf{k}$$

$$= -2\mathbf{i} - 2\mathbf{j} + 2\mathbf{k}.$$

Thus the area is

$$\|\mathbf{a} \times \mathbf{b}\| = \sqrt{(-2)^2 + (-2)^2 + (2)^2} = 2\sqrt{3}. \quad \blacklozenge$$

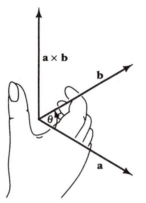

FIGURE 1.4.2. The right-hand rule for determining in which of the two possible directions **a** × **b** points.

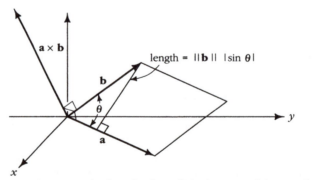

FIGURE 1.4.3. The length of **a** × **b** is the area of the parallelogram formed by **a** and **b**.

Given two vectors in space, the problem of finding a third vector that is orthogonal to both of them can be turned into a pair of linear equations if we use the dot product. The cross product, though, gives the solution directly, as in the following example.

Example 4   Find a unit vector orthogonal to the vectors $\mathbf{i} + \mathbf{j}$ and $\mathbf{j} + \mathbf{k}$.

Solution A vector perpendicular to both $\mathbf{i} + \mathbf{j}$ and $\mathbf{j} + \mathbf{k}$ is the vector

$$(\mathbf{i} + \mathbf{j}) \times (\mathbf{j} + \mathbf{k}) = \begin{vmatrix} \mathbf{i} & \mathbf{j} & \mathbf{k} \\ 1 & 1 & 0 \\ 0 & 1 & 1 \end{vmatrix} = \mathbf{i} - \mathbf{j} + \mathbf{k}.$$

Since $\|\mathbf{i} - \mathbf{j} + \mathbf{k}\| = \sqrt{3}$, the vector

$$\frac{1}{\sqrt{3}}(\mathbf{i} - \mathbf{j} + \mathbf{k})$$

is a unit vector perpendicular to $\mathbf{i} + \mathbf{j}$ and $\mathbf{j} + \mathbf{k}$. ◆

The Cross Product

Geometric definition: $\mathbf{a} \times \mathbf{b}$ is the vector such that:

1. $\|\mathbf{a} \times \mathbf{b}\| = \|\mathbf{a}\| \|\mathbf{b}\| \sin\theta$, the area of the parallelogram spanned by $\mathbf{a}$ and $\mathbf{b}$ (θ is the angle between $\mathbf{a}$ and $\mathbf{b}$; $0 \leq \theta \leq \pi$); see Figure 1.4.3.
2. $\mathbf{a} \times \mathbf{b}$ is perpendicular to $\mathbf{a}$ and $\mathbf{b}$, and the triple $(\mathbf{a}, \mathbf{b}, \mathbf{a} \times \mathbf{b})$ obeys the right-hand rule.

Component formula:

$$(a_1\mathbf{i} + a_2\mathbf{j} + a_3\mathbf{k}) \times (b_1\mathbf{i} + b_2\mathbf{j} + b_3\mathbf{k}) = \begin{vmatrix} \mathbf{i} & \mathbf{j} & \mathbf{k} \\ a_1 & a_2 & a_3 \\ b_1 & b_2 & b_3 \end{vmatrix}$$

$$= (a_2 b_3 - a_3 b_2)\mathbf{i} + (a_3 b_1 - a_1 b_3)\mathbf{j} + (a_1 b_2 - a_2 b_1)\mathbf{k}.$$

Algebraic rules:

1. $\mathbf{a} \times \mathbf{b} = \mathbf{0}$ if and only if $\mathbf{a}$ and $\mathbf{b}$ are parallel or $\mathbf{a}$ or $\mathbf{b}$ is zero.
2. $\mathbf{a} \times \mathbf{b} = -\mathbf{b} \times \mathbf{a}$.
3. $\mathbf{a} \times (\mathbf{b} + \mathbf{c}) = \mathbf{a} \times \mathbf{b} + \mathbf{a} \times \mathbf{c}$.
4. $(\mathbf{a} + \mathbf{b}) \times \mathbf{c} = \mathbf{a} \times \mathbf{c} + \mathbf{b} \times \mathbf{c}$.
5. $(\alpha\mathbf{a}) \times \mathbf{b} = \alpha(\mathbf{a} \times \mathbf{b})$.

Multiplication table:

	×	i	j	k
	i	0	**k**	−**j**
First	**j**	−**k**	0	**i**
factor	**k**	**j**	−**i**	0

Second factor

Example 5   Derive an identity from the formulas

$$\|\mathbf{u} \times \mathbf{v}\| = \|\mathbf{u}\| \, \|\mathbf{v}\| \sin\theta \quad \text{and} \quad \mathbf{u} \cdot \mathbf{v} = \|\mathbf{u}\| \, \|\mathbf{v}\| \cos\theta$$

by eliminating θ.

Solution   Seeing $\sin\theta$ and $\cos\theta$ multiplied by the same expression suggests squaring the two formulas and adding the results. We get

$$\|\mathbf{u} \times \mathbf{v}\|^2 + (\mathbf{u} \cdot \mathbf{v})^2 = \|\mathbf{u}\|^2 \|\mathbf{v}\|^2 (\sin^2\theta + \cos^2\theta) = \|\mathbf{u}\|^2 \|\mathbf{v}\|^2,$$

so

$$\|\mathbf{u} \times \mathbf{v}\|^2 = \|\mathbf{u}\|^2 \|\mathbf{v}\|^2 - (\mathbf{u} \cdot \mathbf{v})^2.$$

This formula is interesting because it establishes a link between the dot and cross products.   ◆

Using the cross product, we may obtain a basic geometric interpretation of 2×2 and 3×3 determinants. Let $\mathbf{a} = a_1\mathbf{i} + a_2\mathbf{j}$ and $\mathbf{b} = b_1\mathbf{i} + b_2\mathbf{j}$ be two vectors in the plane. If θ is the angle between $\mathbf{a}$ and $\mathbf{b}$, we have seen that $\|\mathbf{a} \times \mathbf{b}\| = \|\mathbf{a}\| \, \|\mathbf{b}\| \, |\sin\theta|$ is the area of the parallelogram with adjacent sides $\mathbf{a}$ and $\mathbf{b}$. The cross product as a determinant is

$$\mathbf{a} \times \mathbf{b} = \begin{vmatrix} \mathbf{i} & \mathbf{j} & \mathbf{k} \\ a_1 & a_2 & 0 \\ b_1 & b_2 & 0 \end{vmatrix} = \begin{vmatrix} a_1 & a_2 \\ b_1 & b_2 \end{vmatrix} \mathbf{k}.$$

Thus the area $\|\mathbf{a} \times \mathbf{b}\|$ is the absolute value of the determinant

$$\begin{vmatrix} a_1 & a_2 \\ b_1 & b_2 \end{vmatrix} = a_1 b_2 - a_2 b_1.$$

Geometry of 2 × 2 Determinants

The absolute value of the determinant $\begin{vmatrix} a_1 & a_2 \\ b_1 & b_2 \end{vmatrix}$ is the area of the parallelogram whose adjacent sides are the vectors $\mathbf{a} = a_1\mathbf{i} + a_2\mathbf{j}$ and $\mathbf{b} = b_1\mathbf{i} + b_2\mathbf{j}$. The sign of the determinant is $+$ when $\mathbf{a}, \mathbf{b}, \mathbf{k}$ form a right-handed triple; that is, the right-hand rule in Figure 1.4.2 holds (with $\mathbf{k}$ in place of $\mathbf{a} \times \mathbf{b}$).

There is an analogous interpretation of 3×3 determinants as volumes.

Geometry of 3 × 3 Determinants

The absolute value of the determinant

$$D = \begin{vmatrix} a_1 & a_2 & a_3 \\ b_1 & b_2 & b_3 \\ c_1 & c_2 & c_3 \end{vmatrix}$$

is the volume of the parallelepiped whose adjacent sides are the vectors

$$\mathbf{a} = a_1\mathbf{i} + a_2\mathbf{j} + a_3\mathbf{k}, \ \mathbf{b} = b_1\mathbf{i} + b_2\mathbf{j} + b_3\mathbf{k}, \ \text{and} \ \mathbf{c} = c_1\mathbf{i} + c_2\mathbf{j} + c_3\mathbf{k}.$$

To prove the statement in the box above, we follow Figure 1.4.4 and note that $\|\mathbf{a} \times \mathbf{b}\|$ is the area of the parallelogram with adjacent sides $\mathbf{a}$ and $\mathbf{b}$. Moreover, $(\mathbf{a} \times \mathbf{b}) \cdot \mathbf{c} = \|\mathbf{a} \times \mathbf{b}\| \, \|\mathbf{c}\| \cos\psi$, where ψ is the angle that $\mathbf{c}$ makes with the normal to the plane spanned by $\mathbf{a}$ and $\mathbf{b}$. Since the volume of the parallelepiped with adjacent sides $\mathbf{a}, \mathbf{b}$, and $\mathbf{c}$ is the product of the area of the base $\|\mathbf{a} \times \mathbf{b}\|$ and the altitude $\|\mathbf{c}\| \, |\cos\psi|$, it follows that the volume is $|(\mathbf{a} \times \mathbf{b}) \cdot \mathbf{c}|$. We saw earlier that $(\mathbf{a} \times \mathbf{b}) \cdot \mathbf{c} = D$, so the volume equals the absolute value of D.

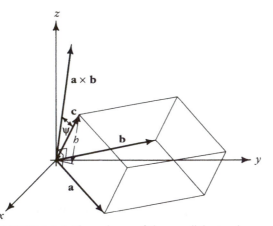

FIGURE 1.4.4. The volume of the parallelepiped spanned by $\mathbf{a}, \mathbf{b}, \mathbf{c}$ is the absolute value of the determinant of the 3 × 3 matrix with $\mathbf{a}, \mathbf{b}, \mathbf{c}$ as its rows.

Example 6   Find the volume of the parallelepiped spanned by the vectors $\mathbf{i} + 3\mathbf{k}, 2\mathbf{i} + \mathbf{j} - 2\mathbf{k}$, and $5\mathbf{i} + 4\mathbf{k}$.

Solution   The volume is the absolute value of

$$\begin{vmatrix} 1 & 0 & 3 \\ 2 & 1 & -2 \\ 5 & 0 & 4 \end{vmatrix}.$$

If we expand this determinant by minors of the second column, the only nonzero term is

$$\begin{vmatrix} 1 & 3 \\ 5 & 4 \end{vmatrix}(1) = -11.$$

so the volume equals 11.   ◆

To conclude this section, we use vector methods to determine the equation of a plane in space. Let $\mathcal{P}$ be a plane in space, $P_0 = (x_0, y_0, z_0)$ a point in the plane, and $\mathbf{n} = A\mathbf{i} + B\mathbf{j} + C\mathbf{k}$ a vector normal to the plane (see Figure 1.4.5).

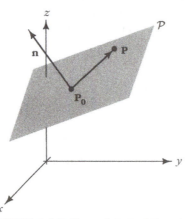

FIGURE 1.4.5. The points P of the plane through P_0 and perpendicular to $\mathbf{n}$ satisfy the equation $\overrightarrow{P_0P} \cdot \mathbf{n} = 0$.

Let $P = (x, y, z)$ be a point in $\mathbb{R}^3$. Then P lies on the plane $\mathcal{P}$ if and only if the vector $\overrightarrow{P_0P} = (x - x_0)\mathbf{i} + (y - y_0)\mathbf{j} + (z - z_0)\mathbf{k}$ is perpendicular to $\mathbf{n}$, *i.e.*, $\overrightarrow{P_0P} \cdot \mathbf{n} = 0$, or

$$(A\mathbf{i} + B\mathbf{j} + C\mathbf{k}) \cdot [(x - x_0)\mathbf{i} + (y - y_0)\mathbf{j} + (z - z_0)\mathbf{k}] = 0.$$

Hence

$$A(x - x_0) + B(y - y_0) + C(z - z_0) = 0.$$

Equation of a Plane in Space

The equation of the plane through (x_0, y_0, z_0) that has a normal vector $\mathbf{n} = A\mathbf{i} + B\mathbf{j} + C\mathbf{k}$ is

$$A(x - x_0) + B(y - y_0) + C(z - z_0) = 0$$

i.e.,

$$Ax + By + Cz + D = 0$$

where $D = -Ax_0 - By_0 - Cz_0$.

The four numbers A, B, C, and D are not determined uniquely by the plane $\mathcal{P}$. To see this, note that (x, y, z) satisfies the equation $Ax + By + Cz + D = 0$ if and only if it also satisfies the relation

$$(\lambda A)x + (\lambda B)y + (\lambda C)z + (\lambda D) = 0$$

for any constant $\lambda \neq 0$. If A, B, C, D and A', B', C', D' determine the same plane $\mathcal{P}$, then $A = \lambda A', B = \lambda B', C = \lambda C', D = \lambda D'$ for a scalar λ. We say that A, B, C, D are **determined by $\mathcal{P}$ up to a scalar multiple**. Conversely, given A, B, C, D and A', B', C', D', they determine the same plane only if $A = \lambda A', B = \lambda B', C = \lambda C', D = \lambda D'$ for some scalar λ.

Example 7   Determine an equation for the plane perpendicular to the vector $\mathbf{i} + \mathbf{j} + \mathbf{k}$ and containing the point $(1, 0, 0)$.

Solution   Using the general form $A(x - x_0) + B(y - y_0) + C(z - z_0) = 0$, the plane is $1(x - 1) + 1(y - 0) + 1(z - 0) = 0$; that is, $x + y + z = 1$. ◆

Example 8   Find an equation for the plane containing the points $(1, 1, 1), (2, 0, 0)$, and $(1, 1, 0)$.

Solution   *Method 1*. This is a "brute force" method which you can use if you have forgotten the vector methods. The equation for any plane is of the form $Ax + By + Cz + D = 0$. Since the points $(1, 1, 1), (2, 0, 0)$, and $(1, 1, 0)$ lie in the plane, we have

$$\begin{aligned} A + B + C + D &= 0, \\ 2A \quad\quad\quad\;\; + D &= 0, \\ A + B \quad\;\; + D &= 0. \end{aligned}$$

Proceeding by elimination, we reduce this system of equations to the form

$$2A + D = 0 \quad \text{(second equation)}$$
$$2B + D = 0 \quad \text{(2 × third − second)},$$
$$C = 0 \quad \text{(first − third)}.$$

Since the numbers $A, B, C,$ and D are determined only up to a scalar multiple, we can fix the value of one of them and then the others will be determined uniquely. If we let $D = -2$, then $A = +1, B = +1, C = 0$. Thus an equation of the plane that contains the given points is $x + y - 2 = 0$.

Method 2. Let $P = (1,1,1), Q = (2,0,0), R = (1,1,0)$. Any vector normal to the plane must be orthogonal to the vectors $\overrightarrow{QP}$ and $\overrightarrow{RP}$, which are parallel to the plane, since their endpoints lie on the plane. Thus, $\mathbf{n} = \overrightarrow{QP} \times \overrightarrow{RP}$ is normal to the plane. Computing the cross product, we have

$$\mathbf{n} = \begin{vmatrix} \mathbf{i} & \mathbf{j} & \mathbf{k} \\ -1 & 1 & 1 \\ 0 & 0 & 1 \end{vmatrix} = \mathbf{i} + \mathbf{j}.$$

Since the point $(2,0,0)$ lies on the plane, we conclude that the equation is $(x - 2) + (y - 0) - 0 \cdot (z - 0) = 0$, or $x + y - 2 = 0$. ◆

Let us now determine the distance from a point $E = (x_1, y_1, z_1)$ to the plane $\mathcal{P}$ with the equation $A(x - x_0) + B(y - y_0) + C(z - z_0) = Ax + By + Cz + D = 0$. To do so, consider the unit normal vector

$$\mathbf{n} = \frac{A\mathbf{i} + B\mathbf{j} + C\mathbf{k}}{\sqrt{A^2 + B^2 + C^2}},$$

which is a unit vector normal to the plane. Drop a perpendicular from E to the plane and construct the triangle REQ shown in Figure 1.4.6. The distance $d = \| \overrightarrow{EQ} \|$ is the length of the projection of $\mathbf{v} = \overrightarrow{RE}$ (the vector from R to E) onto $\mathbf{n}$; thus,

$$\begin{aligned} \text{distance} = |\mathbf{v} \cdot \mathbf{n}| &= |[(x_1 - x_0)\mathbf{i} + (y_1 - y_0)\mathbf{j} + (z_1 - z_0)\mathbf{k}] \cdot \mathbf{n}| \\ &= \frac{|A(x_1 - x_0) + B(y_1 - y_0) + C(z_1 - z_0)|}{\sqrt{A^2 + B^2 + C^2}}. \end{aligned}$$

If the plane is given in the form $Ax + By + Cz + D = 0$, then for any point (x_0, y_0, z_0) on it, $D = -(Ax_0 + By_0 + Cz_0)$. Substitution into the previous formula gives the following:

> ## Distance from a Point to a Plane
> The distance from (x_1, y_1, z_1) to the plane $Ax + By + Cz + D = 0$ is
> $$\text{distance} = \frac{|Ax_1 + By_1 + Cz_1 + D|}{\sqrt{A^2 + B^2 + C^2}}.$$

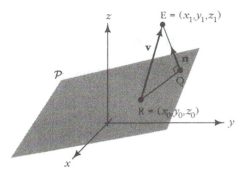

FIGURE 1.4.6. The geometry for determining the distance from the point E to the plane $\mathcal{P}$.

Example 9   Find the distance from $Q = (2, 0, -1)$ to the plane $3x - 2y + 8z + 1 = 0$.

Solution   We substitute into the formula in the preceding box the values $x_1 = 2, y_1 = 0, z_1 = -1$ (the point) and $A = 3, B = -2, C = 8, D = 1$ (the plane) to give

$$\text{distance} = \frac{|3 \cdot 2 + (-2) \cdot 0 + 8(-1) + 1|}{\sqrt{3^2 + (-2)^2 + 8^2}} = \frac{|-1|}{\sqrt{77}} = \frac{1}{\sqrt{77}}. \quad \blacklozenge$$

Exercises for §1.4

Calculate the cross products in Exercises 1–8.

1. $(\mathbf{i} - \mathbf{j} + \mathbf{k}) \times (\mathbf{j} - \mathbf{k})$

2. $\mathbf{i} \times (\mathbf{j} - \mathbf{k})$

3. $(\mathbf{i} + \mathbf{j}) \times [(\mathbf{k} - \mathbf{j}) + (3\mathbf{j} - 2\mathbf{i} + \mathbf{k})]$

4. $(a\mathbf{i} + \mathbf{j} - \mathbf{k}) \times \mathbf{i}$

5. $[(3\mathbf{i} + 2\mathbf{j}) \times 3\mathbf{j}] \times (2\mathbf{i} - \mathbf{j} + \mathbf{k})$

6. $(\mathbf{i} \times \mathbf{j}) \times (\mathbf{i} + \mathbf{j} + \mathbf{k})$

7. $(\mathbf{i} + 2\mathbf{j} + 3\mathbf{k}) \times (\mathbf{i} + 3\mathbf{k})$

8. $(\mathbf{i} + \mathbf{j} + \mathbf{k}) \times (\mathbf{i} + \mathbf{k})$

In Exercises 9–12, describe all unit vectors orthogonal to the given vectors.

9. $\mathbf{i}, \mathbf{j}$

10. $-5\mathbf{i} + 9\mathbf{j} - 4\mathbf{k}, 7\mathbf{i} + 8\mathbf{j} + 9\mathbf{k}$

11. $-5\mathbf{i} + 9\mathbf{j} - 4\mathbf{k}, 7\mathbf{i} + 8\mathbf{j} + 9\mathbf{k}, \mathbf{0}$

12. $2\mathbf{i} - 4\mathbf{j} + 3\mathbf{k}, -4\mathbf{i} + 8\mathbf{j} - 6\mathbf{k}$

In Exercises 13–16, find the length of the given vector.

13. $\mathbf{i} + 3\mathbf{j} - 2\mathbf{k}$

14. $\mathbf{i} + 3\mathbf{j} + 2\mathbf{k}$

15. $3\mathbf{i} + 4\mathbf{j} + 5\mathbf{k}$

16. $\mathbf{i} - 6\mathbf{k} + 2\mathbf{j}$

Find the area or volume of the figure described in Exercises 17–24.

17. The parallelogram spanned by $\mathbf{i} + 2\mathbf{j} + \mathbf{k}$ and $\mathbf{i} + \mathbf{j} + \mathbf{k}$

18. The parallelogram spanned by $\mathbf{i} - \mathbf{j}$ and $\mathbf{i} + \mathbf{j}$

19. The parallelogram spanned by $\mathbf{i}$ and $\mathbf{i} - 2\mathbf{j}$

20. The parallelogram spanned by $\mathbf{i} - \mathbf{j} - \mathbf{k}$ and $\mathbf{i} + \mathbf{j} + \mathbf{k}$

21. The triangle with vertices $(-1, 0, 0), (2, 0, 0)$, and $(0, -2, 3)$

22. The triangle with vertices $(0, 0, 0), (1, 1, 1)$, and $(0, -2, 3)$

23. The parallelepiped with sides $\mathbf{i}, 3\mathbf{j} - \mathbf{k}$, and $4\mathbf{i} + 2\mathbf{j} - \mathbf{k}$

24. The parallelepiped with sides $2\mathbf{i} + \mathbf{j} - \mathbf{k}, 5\mathbf{i} - 3\mathbf{k}$, and $\mathbf{i} - 2\mathbf{j} + \mathbf{k}$

In Exercises 25–32, find an equation for the plane that:

25. is perpendicular to $\mathbf{v} = (1, 1, 1)$ and passes through $(1, 0, 0)$.

26. is perpendicular to $\mathbf{v} = (1, 2, 3)$ and passes through $(1, 1, 1)$.

27. is perpendicular to the line $l(t) = (5, 0, 2)t + (3, -1, 1)$ and passes through $(5, -1, 0)$.

28. is perpendicular to the line $l(t) = (-1, -2, 3)t + (0, 7, 1)$ and passes through $(2, 4, -1)$.

29. passes through $(0, 0, 0), (2, 0, -1)$, and $(0, 4, -3)$.

30. passes through $(1, 2, 0), (0, 1, -2)$, and $(4, 0, 1)$.

31. passes through $(3, 2, -1)$ and $(1, -1, 2)$ and is parallel to the line $\mathbf{v} = (1, -1, 0) + t(3, 2, -2)$.

32. contains the line $\mathbf{v} = (-1, 1, 2) + t(3, 2, 4)$ and is perpendicular to the plane $2x + y - 3z + 4 = 0$.

In Exercises 33–36, find the distance between

33. the point $(2, 1, -1)$ and the plane $x + 2y + 2z + 5 = 0$.

34. the point $(1, 1, -5)$ and the plane $12x + 13y + 5z + 2 = 0$.

35. the point $(6, 1, 0)$ and the plane passing through the origin that is perpendicular to $\mathbf{i} + 2\mathbf{j} + \mathbf{k}$.

36. the point (a, b, c) and the plane passing through $(1, 0, 0), (0, 1, 0)$, and $(0, 0, 1)$.

37. If the coordinates of all the vertices of a parallelogram in the plane are integers, what must be true about the area of the parallelogram? Justify your answer.

38. Answer the question in Exercise 37 for a triangle.

39. Answer the question in Exercise 37 for a trapezoid.

40. Answer the question in Exercise 37 for a general quadrilateral.

41. Prove the "back-cab" identity $\mathbf{A} \times (\mathbf{B} \times \mathbf{C}) = \mathbf{B}(\mathbf{A} \cdot \mathbf{C}) - \mathbf{C}(\mathbf{A} \cdot \mathbf{B})$ in which the parentheses are "at the back" by using the component formula for the cross product. [When we write a scalar after a vector, like $\mathbf{B}(\mathbf{A} \cdot \mathbf{C})$, we mean the same thing as $(\mathbf{A} \cdot \mathbf{C})\mathbf{B}$.]

42. Use the back-cab identity (Exercise 41) to derive a formula for $\mathbf{A} \times (\mathbf{B} \times \mathbf{C}) - (\mathbf{A} \times \mathbf{B}) \times \mathbf{C}$. Is the cross product associative?

43. Given vectors $\mathbf{a}$ and $\mathbf{b}$, with $\mathbf{a} \neq \mathbf{0}$, do the equations $\mathbf{x} \times \mathbf{a} = \mathbf{b}$ and $\mathbf{x} \cdot \mathbf{a} = \|\mathbf{a}\|$ determine a unique vector $\mathbf{x}$? Argue both geometrically and algebraically.

44. Show that the plane that passes through the three points $A = (a_1, a_2, a_3)$, $B = (b_1, b_2, b_3)$, and $C = (c_1, c_2, c_3)$ consists of the points $P = (x, y, z)$ given by

$$\begin{vmatrix} a_1 - x & a_2 - y & a_3 - z \\ b_1 - x & b_2 - y & b_3 - z \\ c_1 - x & c_2 - y & c_3 - z \end{vmatrix} = 0.$$

(Hint: Write the determinant as a triple product.)

45. Find two vectors whose cross product is $\mathbf{i} + \mathbf{j} + \mathbf{k}$.

46. Find a vector whose cross product with $\mathbf{i} - \mathbf{j}$ is $\mathbf{i} + \mathbf{j}$.

1.5
n-Dimensional Euclidean Space

In §**1.1** we saw that a vector in Euclidean three space could be thought of in any of three ways:

(**i**) as a *triple* (x, y, z) where x, y, and z are real numbers;

(**ii**) as a *point* in space;

(**iii**) as a *directed line segment* in space.

The first of these points of view is easily extended from three to any number of dimensions. We define $\mathbb{R}^n$, where n is a positive integer (possibly greater than three), as the set of all ordered n-tuples $(x_1, x_2, \ldots, x_n)$, where the x_i are real numbers. For instance, $(1, \sqrt{5}, 2, 4) \in \mathbb{R}^4$.

The set $\mathbb{R}^n$ is known as ***Euclidean n-space***, and we may think of its elements $\mathbf{a} = (a_1, a_2, \ldots, a_n)$ as ***vectors*** or ***n-vectors***. By setting $n = 1, 2$, or 3, we recover the line, the plane, and three-dimensional space, respectively.

We launch our study of Euclidean n-space by introducing algebraic operations analogous to those introduced in §**1.1** for $\mathbb{R}^2$ and $\mathbb{R}^3$. Addition and scalar multiplication are defined as follows:

(**i**) $(a_1, a_2, \ldots, a_n) + (b_1, b_2, \ldots, b_n) = (a_1 + b_1, a_2 + b_2, \ldots, a_n + b_n)$; and

(**ii**) for any real number α,

$$\alpha(a_1, a_2, \ldots, a_n) = (\alpha a_1, \alpha a_2, \ldots, \alpha a_n).$$

The geometric significance of these operations for $\mathbb{R}^2$ and $\mathbb{R}^3$ was discussed in §**1.1**.

Note

What is the point of going from $\mathbb{R}, \mathbb{R}^2$, and $\mathbb{R}^3$, which seem comfortable and "real world," to $\mathbb{R}^n$? Our world is *three-dimensional*, not *n*-dimensional!

First, it *is true* that the bulk of this book is about $\mathbb{R}^2$ and $\mathbb{R}^3$. However, many of the ideas work in $\mathbb{R}^n$ with little extra effort, so why not do it? Second, the ideas really are *useful*! For instance, if you are studying a chemical reaction involving five chemicals, you will probably want to store and manipulate their concentrations as a 5-tuple; that is, an element of $\mathbb{R}^5$. The laws governing chemical reaction rates also demand that we do *calculus* in this five-dimensional space.

The n vectors in $\mathbb{R}^n$ defined by

$$\mathbf{e}_1 = (1,0,0,\ldots,0), \mathbf{e}_2 = (0,1,0,\ldots,0), \ldots, \mathbf{e}_n = (0,0,\ldots,0,1)$$

are called the **standard basis vectors** of $\mathbb{R}^n$, and they generalize the three mutually orthogonal unit vectors $\mathbf{i}, \mathbf{j}, \mathbf{k}$ of $\mathbb{R}^3$. Any vector $\mathbf{a} = (a_1, a_2, \ldots, a_n)$ can be written in terms of the $\mathbf{e}_i$'s as $\mathbf{a} = a_1 \mathbf{e}_1 + a_2 \mathbf{e}_2 + \cdots + a_n \mathbf{e}_n$.

For two vectors $\mathbf{a} = (a_1, a_2, a_3)$ and $\mathbf{b} = (b_1, b_2, b_3)$ in $\mathbb{R}^3$, we defined the **dot product** or **inner product** $\mathbf{a} \cdot \mathbf{b}$ to be the real number $\mathbf{a} \cdot \mathbf{b} = a_1 b_1 + a_2 b_2 + a_3 b_3$. This definition easily extends to $\mathbb{R}^n$; specifically, for $\mathbf{a} = (a_1, a_2, \ldots, a_n), \mathbf{b} = (b_1, b_2, \ldots, b_n)$, we define $\mathbf{a} \cdot \mathbf{b} = a_1 b_1 + a_2 b_2 + \cdots + a_n b_n$. We also define the **length** or **norm** of a vector $\mathbf{a}$ by the formula

$$\text{length of } \mathbf{a} = \|\mathbf{a}\| = \sqrt{\mathbf{a} \cdot \mathbf{a}} = \sqrt{a_1^2 + a_2^2 + \cdots + a_n^2}.$$

The algebraic properties (i)–(iv) stated at the beginning of §**1.2** are still valid for the dot product in $\mathbb{R}^n$. Each can be proven by a simple computation. Using these, we will prove algebraically the following inequality, which was proved by a geometric argument for $\mathbb{R}^3$ in §**1.2**.

Cauchy–Schwarz Inequality

For vectors $\mathbf{a}$ and $\mathbf{b}$ in $\mathbb{R}^n$, we have

$$|\mathbf{a} \cdot \mathbf{b}| \leq \|\mathbf{a}\| \, \|\mathbf{b}\|.$$

Proof If either $\mathbf{a}$ or $\mathbf{b}$ is zero, the inequality reduces to $0 \leq 0$, so we can assume that both are nonzero. Then

$$0 \leq \left\| \frac{\mathbf{a}}{\|\mathbf{a}\|} - \frac{\mathbf{b}}{\|\mathbf{b}\|} \right\|^2 = 1 - \frac{2\mathbf{a} \cdot \mathbf{b}}{\|\mathbf{a}\| \, \|\mathbf{b}\|} + 1.$$

Simplifying gives

$$\mathbf{a} \cdot \mathbf{b} \leq \|\mathbf{a}\| \, \|\mathbf{b}\|.$$

The same argument replacing the minus sign by a plus sign gives

$$-\mathbf{a} \cdot \mathbf{b} \leq \|\mathbf{a}\| \, \|\mathbf{b}\|.$$

The two together yield the stated inequality, since $|\mathbf{a} \cdot \mathbf{b}| = \pm \mathbf{a} \cdot \mathbf{b}$, depending on the sign of $\mathbf{a} \cdot \mathbf{b}$. ∎

If the vectors $\mathbf{a}$ and $\mathbf{b}$ in the Cauchy–Schwarz inequality are both nonzero, the quotient $\mathbf{a} \cdot \mathbf{b}/(\|\mathbf{a}\| \, \|\mathbf{b}\|)$ lies between -1 and 1. Any number in the interval $[-1, 1]$ is the cosine of a unique angle θ between 0 and π. By analogy with the two- and three-dimensional cases, we call

$$\theta = \cos^{-1}\left(\frac{\mathbf{a} \cdot \mathbf{b}}{\|\mathbf{a}\| \, \|\mathbf{b}\|} \right)$$

the *angle between the vectors* $\mathbf{a}$ and $\mathbf{b}$.

Note

The reasoning above is opposite to that in §**1.2**. That is because, in $\mathbb{R}^n$, we have no geometry to begin with, so we start with algebraic definitions and define the geometric notions in terms of them.

As in §**1.2**, we can derive the triangle inequality from the Cauchy–Schwarz inequality: $\mathbf{a} \cdot \mathbf{b} \leq |\mathbf{a} \cdot \mathbf{b}| \leq \|\mathbf{a}\| \, \|\mathbf{b}\|$, so that

$$\|\mathbf{a} + \mathbf{b}\|^2 = \|\mathbf{a}\|^2 + 2\mathbf{a} \cdot \mathbf{b} + \|\mathbf{b}\|^2 \leq \|\mathbf{a}\|^2 + 2\|\mathbf{a}\| \, \|\mathbf{b}\| + \|\mathbf{b}\|^2.$$

Hence, we get $\|\mathbf{a}+\mathbf{b}\|^2 \leq (\|\mathbf{a}\| + \|\mathbf{b}\|)^2$; taking square roots gives the following result:

Triangle Inequality

Let **a** and **b** be vectors in $\mathbb{R}^n$. Then

$$\|\mathbf{a} + \mathbf{b}\| \leq \|\mathbf{a}\| + \|\mathbf{b}\|.$$

If the Cauchy–Schwarz and triangle inequalities are written in terms of components, they look like this:

$$\left| \sum_{i=1}^{n} a_i b_i \right| \leq \sqrt{\sum_{i=1}^{n} a_i^2} \sqrt{\sum_{i=1}^{n} b_i^2};$$

and

$$\sqrt{\sum_{i=1}^{n} (a_i + b_i)^2} \leq \sqrt{\sum_{i=1}^{n} a_i^2} + \sqrt{\sum_{i=1}^{n} b_i^2}.$$

Example 1   Verify the Cauchy–Schwarz and triangle inequalities for $\mathbf{a} = (1, 2, 0, -1)$ and $\mathbf{b} = (-1, 1, 1, 0)$.

Solution

$$
\begin{aligned}
\|\mathbf{a}\| &= \sqrt{1^2 + 2^2 + 0^2 + (-1)^2} = \sqrt{6}, \\
\|\mathbf{b}\| &= \sqrt{(-1)^2 + 1^2 + 1^2 + 0^2} = \sqrt{3}, \\
\mathbf{a} \cdot \mathbf{b} &= 1(-1) + 2 \cdot 1 + 0 \cdot 1 + (-1)0 = 1, \\
\mathbf{a} + \mathbf{b} &= (0, 3, 1, -1), \\
\|\mathbf{a} + \mathbf{b}\| &= \sqrt{0^2 + 3^2 + 1^2 + (-1)^2} = \sqrt{11}.
\end{aligned}
$$

We compute $\mathbf{a} \cdot \mathbf{b} = 1$ and $\|\mathbf{a}\| \, \|\mathbf{b}\| = \sqrt{6}\sqrt{3} \approx 4.24$, which verifies the Cauchy–Schwarz inequality. Similarly, we can check the triangle inequality:

$$\|\mathbf{a} + \mathbf{b}\| = \sqrt{11} \approx 3.32,$$

while

$$\|\mathbf{a}\| + \|\mathbf{b}\| = \sqrt{6} + \sqrt{3} \approx 2.45 + 1.73 \approx 4.18,$$

which is larger.   ◆

By analogy with $\mathbb{R}^3$, we can define the notion of distance in $\mathbb{R}^n$; namely, if **a** and **b** are points in $\mathbb{R}^n$, the **_distance between a and b_** is defined to be $\|\mathbf{a} - \mathbf{b}\|$, or the length of the vector $\mathbf{a} - \mathbf{b}$. There is _no cross product_ defined on

$\mathbb{R}^n$ except for $n = 3$, because it is only in $\mathbb{R}^3$ that there is a unique direction perpendicular to two (linearly independent) vectors. It is only the dot product that has been generalized.

Generalizing 2×2 and 3×3 matrices (see §**1.3**), we can consider $m \times n$ matrices, that is, arrays of mn numbers:

$$
A = \begin{bmatrix}
a_{11} & a_{12} & \cdots & a_{1n} \\
a_{21} & a_{22} & \cdots & a_{2n} \\
\vdots & \vdots & & \vdots \\
a_{m1} & a_{m2} & \cdots & a_{mn}
\end{bmatrix}.
$$

Note that an $m \times n$ matrix has m rows and n columns. The entry a_{ij} goes in the ith row and the jth column. We shall also write A as $[a_{ij}]$. A **square matrix** is one for which $m = n$. We define addition and multiplication by a scalar componentwise, as we did for vectors. Given two $m \times n$ matrices A and B, we can add them to obtain a new $m \times n$ matrix $C = A + B$, whose ijth entry c_{ij} is the sum of a_{ij} and b_{ij}. It is clear that $A + B = B + A$. Similarly, the difference $D = A - B$ is the matrix whose entries are $d_{ij} = a_{ij} - b_{ij}$.

Example 2

(a) $\begin{bmatrix} 2 & 1 & 0 \\ 3 & 4 & 1 \end{bmatrix} + \begin{bmatrix} -1 & 1 & 3 \\ 0 & 0 & 7 \end{bmatrix} = \begin{bmatrix} 1 & 2 & 3 \\ 3 & 4 & 8 \end{bmatrix}.$

(b) $\begin{bmatrix} 1 & 2 \end{bmatrix} + \begin{bmatrix} 0 & -1 \end{bmatrix} = \begin{bmatrix} 1 & 1 \end{bmatrix}.$

(c) $\begin{bmatrix} 2 & 1 \\ 1 & 2 \end{bmatrix} - \begin{bmatrix} 1 & 0 \\ 0 & 1 \end{bmatrix} = \begin{bmatrix} 1 & 1 \\ 1 & 1 \end{bmatrix}.$ ◆

Given a scalar λ and an $m \times n$ matrix A, we can multiply A by λ to obtain a new $m \times n$ matrix $\lambda A = C$, whose ijth entry c_{ij} is the product λa_{ij}.

Example 3

$$
3 \begin{bmatrix} 1 & -1 & 2 \\ 0 & 1 & 5 \\ 1 & 0 & 3 \end{bmatrix} = \begin{bmatrix} 3 & -3 & 6 \\ 0 & 3 & 15 \\ 3 & 0 & 9 \end{bmatrix}. \quad ◆
$$

Next we turn to the most important operation, that of **matrix multiplication**. If $A = [a_{ij}]$ is an $m \times n$ matrix and $B = [b_{ij}]$ is an $n \times p$ matrix, then $AB = C$ is the $m \times p$ matrix whose ijth entry is

$$
c_{ij} = \sum_{k=1}^{n} a_{ik} b_{kj},
$$

which is the dot product of the ith row of A and the jth column of B (see Figure 1.5.1).

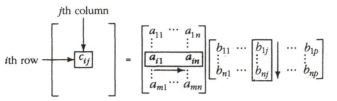

FIGURE 1.5.1. The computation of the (i, j)th entry in a matrix product.

Example 4 Let

$$A = \begin{bmatrix} 1 & 0 & 3 \\ 2 & 1 & 0 \\ 1 & 0 & 0 \end{bmatrix} \quad \text{and} \quad B = \begin{bmatrix} 0 & 1 & 0 \\ 1 & 0 & 0 \\ 0 & 1 & 1 \end{bmatrix}.$$

Then

$$AB = \begin{bmatrix} 0 & 4 & 3 \\ 1 & 2 & 0 \\ 0 & 1 & 0 \end{bmatrix} \quad \text{and} \quad BA = \begin{bmatrix} 2 & 1 & 0 \\ 1 & 0 & 3 \\ 3 & 1 & 0 \end{bmatrix}. \quad \blacklozenge$$

This example shows that matrix multiplication is **_not_** commutative. That is, in general,

$$AB \neq BA.$$

Note that for AB to be defined, the number of columns of A must equal the number of rows of B.

Example 5 Let

$$A = \begin{bmatrix} 2 & 0 & 1 \\ 1 & 1 & 2 \end{bmatrix} \quad \text{and} \quad B = \begin{bmatrix} 1 & 0 & 2 \\ 0 & 2 & 1 \\ 1 & 1 & 1 \end{bmatrix}.$$

Then

$$AB = \begin{bmatrix} 3 & 1 & 5 \\ 3 & 4 & 5 \end{bmatrix},$$

but BA is not defined.   ◆

Example 6   Let

$$A = \begin{bmatrix} 1 \\ 2 \\ 1 \\ 3 \end{bmatrix} \quad \text{and} \quad B = \begin{bmatrix} 2 & 2 & 1 & 2 \end{bmatrix}.$$

Then

$$AB = \begin{bmatrix} 2 & 2 & 1 & 2 \\ 4 & 4 & 2 & 4 \\ 2 & 2 & 1 & 2 \\ 6 & 6 & 3 & 6 \end{bmatrix} \quad \text{and} \quad BA = [13]. \quad ◆$$

If we have three matrices A, B, and C such that the products AB and BC are defined, then the products $(AB)C$ and $A(BC)$ will be defined and equal (that is, *matrix multiplication is associative*). It is legitimate, therefore, to drop the parenthesis and denote the product by ABC. We will accept this without proof.

Example 7   Let

$$A = \begin{bmatrix} 3 \\ 5 \end{bmatrix}, \quad B = \begin{bmatrix} 1 & 1 \end{bmatrix}, \quad \text{and} \quad C = \begin{bmatrix} 1 \\ 2 \end{bmatrix}.$$

Then

$$A(BC) = \begin{bmatrix} 3 \\ 5 \end{bmatrix} [3] = \begin{bmatrix} 9 \\ 15 \end{bmatrix}.$$

Also,

$$AB(C) = \begin{bmatrix} 3 & 3 \\ 5 & 5 \end{bmatrix} \begin{bmatrix} 1 \\ 2 \end{bmatrix} = \begin{bmatrix} 9 \\ 15 \end{bmatrix}$$

as well.   ◆

It is convenient when working with matrices to represent the vector $\mathbf{a} = (a_1, \ldots, a_n)$ in $\mathbb{R}^n$ by the $n \times 1$ *column matrix* $\begin{bmatrix} a_1 \\ a_2 \\ \vdots \\ a_n \end{bmatrix}$, which we also denote by $\mathbf{a}$.

Less frequently, one also encounters the $1 \times n$ **row matrix** $[a_1 \ldots a_n]$, which is called

the **transpose** of $\begin{bmatrix} a_1 \\ \vdots \\ a_n \end{bmatrix}$ and is denoted by $\mathbf{a}^T$. Note that an element of $\mathbb{R}^n$ (e.g.

$(4, 1, 3.2, 6)$) is written with parentheses and commas, while a row matrix (e.g. $[7\ 1\ 3\ 2]$) is written with square brackets and no commas. It is also convenient at this point to use the letters $\mathbf{x}, \mathbf{y}, \mathbf{z}, \ldots$ instead of $\mathbf{a}, \mathbf{b}, \mathbf{c}, \ldots$ to denote vectors and their components.

If $A = \begin{bmatrix} a_{11} & \cdots & a_{1n} \\ \vdots & & \vdots \\ a_{m1} & \cdots & a_{mn} \end{bmatrix}$ is a $m \times n$ matrix, and $\mathbf{x} = (x_1, \ldots, x_n)$ is in $\mathbb{R}^n$,

we can form the matrix product

$$Ax = \begin{bmatrix} a_{11} & \cdots & a_{1n} \\ \vdots & & \vdots \\ a_{m1} & \cdots & a_{mn} \end{bmatrix} \begin{bmatrix} x_1 \\ \vdots \\ x_n \end{bmatrix} = \begin{bmatrix} y_1 \\ \vdots \\ y_m \end{bmatrix}.$$

The resulting $m \times 1$ column matrix can be considered as a vector $(y_1, \ldots, y_m)$ in $\mathbb{R}^m$. In this way, the matrix A determines a function from $\mathbb{R}^n$ to $\mathbb{R}^m$ defined by $\mathbf{y} = A\mathbf{x}$. This function, sometimes indicated by the notation $\mathbf{x} \mapsto A\mathbf{x}$ to emphasize that $\mathbf{x}$ is transformed into $A\mathbf{x}$, is called a **linear transformation**, since it has the linearity properties

$$\begin{aligned} A(\mathbf{x} + \mathbf{y}) &= A\mathbf{x} + A\mathbf{y} \quad \text{for } \mathbf{x} \text{ and } \mathbf{y} \text{ in } \mathbb{R}^n \\ A(\alpha\mathbf{x}) &= \alpha(A\mathbf{x}) \quad \text{for } \mathbf{x} \text{ in } \mathbb{R}^n \text{ and } \alpha \text{ in } \mathbb{R}. \end{aligned}$$

In this book we will be using linear algebra very sparingly. If you have had a course in linear algebra it will only enrich your multivariable and vector calculus experience. If you have not had such a course there is no need to worry — we will provide everything that you will need for multivariable and vector calculus.

Example 8 If

$$A = \begin{bmatrix} 1 & 0 & 3 \\ -1 & 0 & 1 \\ 2 & 1 & 2 \\ -1 & 2 & 1 \end{bmatrix},$$

then the function $\mathbf{x} \mapsto A\mathbf{x}$ from $\mathbb{R}^3$ to $\mathbb{R}^4$ is defined by

$$
\begin{bmatrix} x_1 \\ x_2 \\ x_3 \end{bmatrix} \mapsto \begin{bmatrix} 1 & 0 & 3 \\ -1 & 0 & 1 \\ 2 & 1 & 2 \\ -1 & 2 & 1 \end{bmatrix} \begin{bmatrix} x_1 \\ x_2 \\ x_3 \end{bmatrix} = \begin{bmatrix} x_1 + 3x_3 \\ -x_1 + x_3 \\ 2x_1 + x_2 + 2x_3 \\ -x_1 + 2x_2 + x_3 \end{bmatrix}. \quad \blacklozenge
$$

Example 9 The following illustrates what happens to a specific vector when mapped by the 4×3 matrix of Example 8:

$$
A\mathbf{e}_2 = \begin{bmatrix} 1 & 0 & 3 \\ -1 & 0 & 1 \\ 2 & 1 & 2 \\ -1 & 2 & 1 \end{bmatrix} \begin{bmatrix} 0 \\ 1 \\ 0 \end{bmatrix} = \begin{bmatrix} 0 \\ 0 \\ 1 \\ 2 \end{bmatrix} = \text{second column of } A. \quad \blacklozenge
$$

An $n \times n$ matrix is said to be **invertible** if there is an $n \times n$ matrix B such that

$$
AB = BA = I_n,
$$

where

$$
I_n = \begin{bmatrix} 1 & 0 & 0 & \cdots & 0 \\ 0 & 1 & 0 & \cdots & 0 \\ 0 & 0 & 1 & \cdots & 0 \\ \vdots & \vdots & \vdots & & \vdots \\ 0 & 0 & 0 & \cdots & 1 \end{bmatrix}
$$

is the $n \times n$ identity matrix. The matrix I_n has the property that $I_n C = C I_n = C$ for any $n \times n$ matrix C. We denote B by A^{-1} and call A^{-1} the **inverse** of A. The inverse, when it exists, is unique.

Example 10 If

$$
A = \begin{bmatrix} 2 & 4 & 0 \\ 0 & 2 & 1 \\ 3 & 0 & 2 \end{bmatrix}, \quad \text{then} \quad A^{-1} = \frac{1}{20} \begin{bmatrix} 4 & -8 & 4 \\ 3 & 4 & -2 \\ -6 & 12 & 4 \end{bmatrix},
$$

since $AA^{-1} = I_3 = A^{-1}A$, as may be checked by matrix multiplication. $\blacklozenge$

If A is invertible, the equation $A\mathbf{x} = \mathbf{y}$ can be solved for the vector $\mathbf{x}$ by multiplying both sides by A^{-1} to obtain $\mathbf{x} = A^{-1}\mathbf{y}$.

In §**1.3** we defined the determinant of a 3×3 matrix. This can be generalized to $n \times n$ determinants. We illustrate here how to write the determinant of a

4×4 matrix in terms of the determinants of 3×3 matrices; by expansion along the first row:

$$\begin{vmatrix} a_{11} & a_{12} & a_{13} & a_{14} \\ a_{21} & a_{22} & a_{23} & a_{24} \\ a_{31} & a_{32} & a_{33} & a_{34} \\ a_{41} & a_{42} & a_{43} & a_{44} \end{vmatrix} = a_{11} \begin{vmatrix} a_{22} & a_{23} & a_{24} \\ a_{32} & a_{33} & a_{34} \\ a_{42} & a_{43} & a_{44} \end{vmatrix} - a_{12} \begin{vmatrix} a_{21} & a_{23} & a_{24} \\ a_{31} & a_{33} & a_{34} \\ a_{41} & a_{43} & a_{44} \end{vmatrix}$$

$$+ a_{13} \begin{vmatrix} a_{21} & a_{22} & a_{24} \\ a_{31} & a_{32} & a_{34} \\ a_{41} & a_{42} & a_{44} \end{vmatrix} - a_{14} \begin{vmatrix} a_{21} & a_{22} & a_{23} \\ a_{31} & a_{32} & a_{33} \\ a_{41} & a_{42} & a_{43} \end{vmatrix}$$

(note that the signs alternate $+, -, +, -, \ldots$). A similar formula holds if we expand along the first column, or in fact along any row or column. Continuing this way, one defines 5×5 determinants in terms of 4×4 determinants, and so on.

The basic properties of 3×3 determinants reviewed in §**1.3** remain valid for $n \times n$ determinants. In particular, we note the fact that if A is an $n \times n$ matrix and B is the matrix formed by adding a scalar multiple of the kth row (or column) of A to the lth row (or, respectively, column) of A, then the determinant of A is equal to the determinant of B. (This fact is used in Example 11 below.)

A basic theorem of linear algebra states that an *$n \times n$ matrix A is invertible if and only if the determinant of A is not zero.* Another basic property is that $\det(AB) = (\det A)(\det B)$. In this text we leave these assertions unproved.

Example 11   Let

$$A = \begin{bmatrix} 1 & 0 & 1 & 0 \\ 1 & 1 & 1 & 1 \\ 2 & 1 & 0 & 1 \\ 1 & 1 & 0 & 2 \end{bmatrix}.$$

Find $\det A$. Does A have an inverse?

Solution   Adding $(-1) \times$ first column to the third column and then expanding by minors of the first row, we get

$$\det A = \begin{vmatrix} 1 & 0 & 0 & 0 \\ 1 & 1 & 0 & 1 \\ 2 & 1 & -2 & 1 \\ 1 & 1 & -1 & 2 \end{vmatrix} = 1 \begin{vmatrix} 1 & 0 & 1 \\ 1 & -2 & 1 \\ 1 & -1 & 2 \end{vmatrix}.$$

Adding $(-1) \times$ first column to the third column of this 3×3 determinant gives

$$\det A = \begin{vmatrix} 1 & 0 & 0 \\ 1 & -2 & 0 \\ 1 & -1 & 1 \end{vmatrix} = \begin{vmatrix} -2 & 0 \\ -1 & 1 \end{vmatrix} = -2.$$

Thus, $\det A = -2 \neq 0$, and so A has an inverse.   ♦

Exercises for §1.5

Perform the calculations indicated in Exercises 1–4.

1. $(1, 4, 5, 6, 7) + (1, 2, 3, 4, 5) =$

2. $(1, 2, 3, \ldots, n) + (0, 1, 2, \ldots, n - 1) =$

3. $2(1, 2, 3, 4, 5) \cdot (5, 4, 3, 2, 1) =$

4. $4(5, 4, 3, 2) \cdot 6(8, 4, 1, 7) =$

Verify the Cauchy–Schwarz inequality and the triangle inequality for the vectors given in Exercises 5–8.

5. $\mathbf{a} = (2, 0, -1), \mathbf{b} = (4, 0, -2)$

6. $\mathbf{a} = (1, 0, 2, 6), \mathbf{b} = (3, 8, 4, 1)$

7. $\mathbf{i} + \mathbf{j} + \mathbf{k}, \mathbf{i} + \mathbf{k}$

8. $2\mathbf{i} + \mathbf{j}, 3\mathbf{i} + 4\mathbf{j}$

Perform the calculations indicated in Exercises 9–12.

9. $\begin{bmatrix} 1 & 2 & 3 \\ 4 & 5 & 6 \\ 7 & 8 & 9 \end{bmatrix} + \begin{bmatrix} 4 & 7 & 3 \\ 8 & 2 & 1 \\ 0 & 6 & 6 \end{bmatrix} =$

10. $\begin{bmatrix} 0 & 6 & 3 \\ 2 & 9 & 8 \\ 1 & 3 & 3 \\ 2 & 7 & 6 \end{bmatrix} + \begin{bmatrix} 2 & 1 & 7 \\ 6 & 6 & 6 \\ 4 & 4 & 4 \\ 9 & 8 & 1 \end{bmatrix} =$

11. $\begin{bmatrix} 2 & 3 & 4 \\ 7 & 7 & 7 \\ 1 & 1 & 1 \end{bmatrix} - \begin{bmatrix} 2 & 3 & 1 \\ 1 & 1 & 2 \\ 3 & 2 & 3 \end{bmatrix} =$

12. $6 \begin{bmatrix} 1 & 2 & 3 \\ 4 & 5 & 6 \\ 7 & 8 & 9 \end{bmatrix} =$

In Exercises 13–20, find the matrix product or explain why it is not defined.

13. $\begin{bmatrix} 1 & 2 & 3 \end{bmatrix} \begin{bmatrix} 4 \\ 5 \\ 6 \end{bmatrix}$

14. $\begin{bmatrix} \frac{1}{4} & \frac{1}{2} & \frac{1}{4} \end{bmatrix} \begin{bmatrix} 1 \\ 2 \\ 1 \end{bmatrix}$

15. $\begin{bmatrix} 0 & 1 \\ 1 & 0 \end{bmatrix} \begin{bmatrix} a & b \\ c & d \end{bmatrix}$

16. $\begin{bmatrix} 1 \\ 2 \\ 3 \end{bmatrix} \begin{bmatrix} 4 \\ 5 \\ 6 \end{bmatrix}$

17. $\begin{bmatrix} 1 & 2 \\ 3 & 4 \\ 5 & 6 \end{bmatrix} \begin{bmatrix} 0 & 0 & 0 \\ 3 & 2 & 1 \end{bmatrix}$

18. $\begin{bmatrix} 1 & 0 & 0 \\ 0 & 1 & 0 \\ 0 & 0 & 1 \end{bmatrix} \begin{bmatrix} a & b & c \\ d & e & f \\ g & h & i \end{bmatrix}$

19. $\begin{bmatrix} 0 & 1 \\ 2 & 3 \end{bmatrix} \begin{bmatrix} 1 \\ 2 \\ 3 \end{bmatrix}$

20. $\left(\begin{bmatrix} 1 & 0 \\ 2 & 3 \end{bmatrix} \begin{bmatrix} 2 & 4 \\ 1 & -1 \end{bmatrix} \right) \begin{bmatrix} 1 & 1 \\ 0 & 1 \end{bmatrix}$

In Exercises 21–24, for the given *A*, (a) define the mapping $\mathbf{x} \mapsto A\mathbf{x}$ as was done in Example 8, and (b) calculate $A\mathbf{a}$.

21. $A = \begin{bmatrix} 1 & 2 & 3 \\ 4 & 5 & 6 \\ 7 & 8 & 9 \end{bmatrix}, \quad \mathbf{a} = \begin{bmatrix} 1 \\ 2 \\ 3 \end{bmatrix}$

22. $A = \begin{bmatrix} 0 & 1 & 0 \\ 1 & 0 & 1 \\ 0 & 1 & 0 \end{bmatrix}, \quad \mathbf{a} = \begin{bmatrix} 4 \\ 9 \\ 8 \end{bmatrix}$

23. $A = \begin{bmatrix} 4 & 5 \\ 9 & 0 \\ 1 & 1 \\ 7 & 3 \end{bmatrix}, \quad \mathbf{a} = \begin{bmatrix} 7 \\ 9 \end{bmatrix}$

24. $A = \begin{bmatrix} 4 & 4 & 4 & 0 \\ 3 & 5 & 5 & 7 \end{bmatrix}, \quad \mathbf{a} = \begin{bmatrix} 7 \\ 9 \\ 0 \\ 1 \end{bmatrix}$

In Exercises 25–28, determine whether the given matrix has an inverse.

25. $\begin{bmatrix} 1 & 2 \\ 0 & 1 \end{bmatrix}$

26. $\begin{bmatrix} 1 & 1 \\ 1 & 1 \end{bmatrix}$

27. $\begin{bmatrix} 0 & 4 & 5 \\ 7 & 0 & 6 \\ 8 & 9 & 0 \end{bmatrix}$

28. $\begin{bmatrix} 1 & 2 & 3 \\ 4 & 5 & 6 \\ 7 & 8 & 9 \end{bmatrix}$

29. Let $I = \begin{bmatrix} 1 & 0 \\ 0 & 1 \end{bmatrix}$ and $A = \begin{bmatrix} 1 & 2 \\ 0 & 1 \end{bmatrix}$. Find a matrix B such that $AB = I$. Check that $BA = I$.

30. Verify that if $ad - bc \neq 0$, the inverse of $\begin{bmatrix} a & b \\ c & d \end{bmatrix}$ is $\dfrac{1}{ad - bc} \begin{bmatrix} d & -b \\ -c & a \end{bmatrix}$.

31. Assuming $\det(AB) = (\det A)(\det B)$, verify that $(\det A)(\det A^{-1}) = 1$ and conclude that if A has an inverse, then $\det A \neq 0$.

32. Show that, if A, B and C are $n \times n$ matrices such that $AB = I_n$ and $CA = I_n$, then $B = C$.

33. Define the **transpose** A^T of an $m \times n$ matrix A as follows, A^T is the $n \times m$ matrix whose ijth entry is the jith entry of A. For instance

$$\begin{bmatrix} a & b \\ c & d \end{bmatrix}^T = \begin{bmatrix} a & c \\ b & d \end{bmatrix}, \qquad \begin{bmatrix} a & b \\ d & e \\ g & h \end{bmatrix}^T = \begin{bmatrix} a & d & g \\ b & e & h \end{bmatrix}.$$

In §**1.3** we saw that, if A is a 2×2 or 3×3 matrix, then $\det(A^T) = \det A$. Use this to prove the same fact for 4×4 matrices.

34. Let B be the $m \times 1$ column matrix $\begin{bmatrix} \frac{1}{m} \\ \frac{1}{m} \\ \vdots \\ \frac{1}{m} \end{bmatrix}$. If $A = [a_1 \ldots a_m]$ is any row matrix, what is AB?

35. Prove that if A is a 4×4 matrix, then

(a) if B is a matrix obtained from a 4×4 matrix A by multiplying any row or column by a scalar λ, then $\det B = \lambda \det A$; and

(b) $\det(\lambda A) = \lambda^4 \det A$.

In Exercises 36–38, $A, B,$ and C denote $n \times n$ matrices.

36. Is $\det(A + B) = \det A + \det B$? Give a proof or counterexample.

37. Does $(A + B)(A - B) = A^2 - B^2$?

38. Assuming the law $\det(AB) = (\det A)(\det B)$, prove that $\det(ABC) = (\det A)(\det B)(\det C)$.

1.6
Curves in the Plane and in Space

One often thinks of a curve as a line drawn on paper, such as a straight line, a circle, or a sine curve. It is useful to think of a curve mathematically as the set of values of a function that maps an interval of real numbers into the plane or space. We shall call such a map a ***path***. We usually denote a path by **c**. The image of the path then corresponds to a line we see on paper (see Figure 1.6.1). Often we write t for the independent variable and imagine it to be *time*, so that $\mathbf{c}(t)$ is the position at time t of a moving particle, which traces out a curve as t varies.

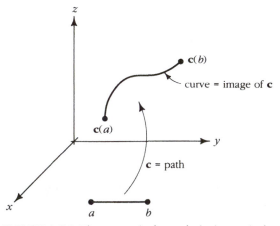

FIGURE 1.6.1. The map **c** is the path; its image is the curve we "see."

Paths and Curves

A **path** in $\mathbb{R}^n$ is a map $\mathbf{c} : [a, b] \to \mathbb{R}^n$; it is a **path in the plane** if $n = 2$ and a **path in space** if $n = 3$. The collection of points $\mathbf{c}(t)$ as t varies in $[a, b]$ is called a **curve**, and $\mathbf{c}(a)$ and $\mathbf{c}(b)$ are its **endpoints**. The path $\mathbf{c}(t)$ is said to **parametrize** this curve.

If $\mathbf{c}$ is a path in $\mathbb{R}^3$, we can write $\mathbf{c}(t) = (x(t), y(t), z(t))$ and we call $x(t), y(t)$, and $z(t)$ the **component functions** of $\mathbf{c}$. We form component functions similarly in $\mathbb{R}^2$ or, generally, in $\mathbb{R}^n$.

Strictly speaking, we should distinguish between $\mathbf{c}(t)$ as a *point* in space and as a *vector* based at the origin. However, not doing so should not cause confusion.

Example 1   The straight line L in $\mathbb{R}^3$ through the point (x_0, y_0, z_0) in the direction of vector $\mathbf{v}$ is the image of the path

$$\mathbf{c}(t) = (x_0, y_0, z_0) + t\mathbf{v}$$

for $t \in \mathbb{R}$ (see Figure 1.6.2). Thus, our notion of curve includes straight lines as special cases.   ◆

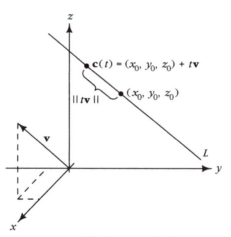

FIGURE 1.6.2. L is the straight line in space through (x_0, y_0, z_0) and in direction $\mathbf{v}$; its equation is $\mathbf{c}(t) = (x_0, y_0, z_0) + t\mathbf{v}$.

Example 2   The unit circle $x^2 + y^2 = 1$ in the plane is the image of the path

$$\mathbf{c} : \mathbb{R} \to \mathbb{R}^2, \qquad \mathbf{c}(t) = (\cos t, \sin t)$$

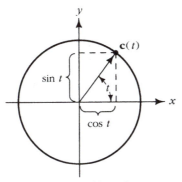

FIGURE 1.6.3. $\mathbf{c}(t) = (\cos t, \sin t)$ is a path whose image is the unit circle.

(see Figure 1.6.3). ◆

Example 3   The path $\mathbf{c}(t) = (t, t^2)$ traces out a parabolic arc. This curve coincides with the graph $f(x) = x^2$ (see Figure 1.6.4). ◆

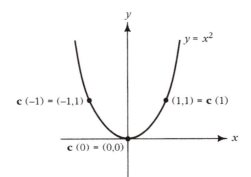

FIGURE 1.6.4. The image of $\mathbf{c}(t) = (t, t^2)$ is the parabola $y = x^2$.

Example 4   A wheel of radius R rolls to the right along a straight line at speed v. Use vector methods to find the path $\mathbf{c}(t)$ of the point on the wheel that initially lies at a distance r below the center.

Solution   We place the wheel in the xy plane with its center initially at $(0, R)$, so that the position of the center at time t is given by the path $\mathbf{C}(t) = (vt, R)$. (Refer to Figure 1.6.5.)

The position of the point $\mathbf{c}(t)$ *relative to the center* is given by the vector $\mathbf{d}(t) = \mathbf{c}(t) - \mathbf{C}(t)$ that has the initial value $-r\mathbf{j}$ and rotates in the *clockwise* direction. The rate of rotation is such that the wheel makes a full rotation after the center has moved a distance $2\pi R$ (equal to the circumference of the wheel). This takes a time $2\pi R/v$, so the angular velocity $d\theta/dt$ of the wheel is v/R.

Since the rotation is clockwise, the vector function $\mathbf{d}(t)$ is of the form

$$\mathbf{d}(t) = r\left(\cos\left[-\frac{v}{R}t + \theta\right]\mathbf{i} + \sin\left[-\frac{v}{R}t + \theta\right]\mathbf{j}\right)$$

for some initial angle θ. Since $\mathbf{d}(0) = -r\mathbf{j}$, we have $\cos\theta = 0$ and $\sin\theta = -1$, so $\theta = -\pi/2$, and hence

$$\mathbf{d}(t) = r\left(\cos\left[-\frac{v}{R}t - \frac{\pi}{2}\right]\mathbf{i} + \sin\left[-\frac{v}{R}t - \frac{\pi}{2}\right]\mathbf{j}\right).$$

Using $\cos(\varphi - \pi/2) = \sin\varphi$ and $\sin(\varphi - \pi/2) = -\cos\varphi$, we get

$$\mathbf{d}(t) = r\left(-\sin\frac{vt}{R}\mathbf{i} - \cos\frac{vt}{R}\mathbf{j}\right).$$

Finally, the path $\mathbf{c}(t)$ is given by adding the components of the vector function $\mathbf{d}(t)$ to the coordinates of the path $\mathbf{C}(t)$: the result is

$$\mathbf{c}(t) = \left(vt - r\sin\frac{vt}{R}, R - r\cos\frac{vt}{R}\right).$$

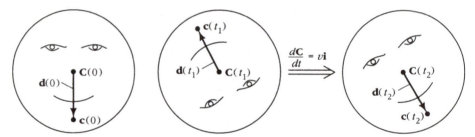

FIGURE 1.6.5. The vector $\mathbf{d}(t)$ points from the wheel's center, $\mathbf{C}(t)$, to the position $\mathbf{c}(t)$ of a point on the wheel and rotates in the clockwise direction while the wheel moves to the right.

In the special case $v = R = r = 1$, we get $\mathbf{c}(t) = (t - \sin t, 1 - \cos t)$. The image of this curve is shown in Figure 1.6.6; it is called a **cycloid.** ◆

Note

This is a good point to dig out your single-variable calculus text and review differentiation techniques. Target the definition of the derivative and the chain rule for special attention.

The definition below and the definitions in Chapter 2 use the limit concept as well. At this point you do not need to know or review the technical definition of limit. However, you should recall the basic idea of this concept, and, for Chapter 2, review a few properties (the limit of a sum is the sum of the limits, etc.).

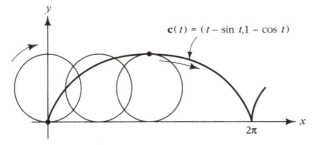

FIGURE 1.6.6. The path traced by a point moving on a rolling circle is called a cycloid.

Historical Note

The cycloid is traced out by a circle rolling along a *straight line*; when it rolls on a *circle*, we get an **epicycle**, as in Figure 1.6.7. Since ancient times the circle and sphere have been considered to be the perfect forms in geometry. For the Greeks they were the symbols of the ultimate symmetry of the divine. What forms could be better suited to describe the immutable and eternal motion of the planets than circular motion? From the Earth, these motions appear to be quite complicated. The motion of the Sun and the Moon can be roughly described as circular with constant speed. The orbits of the other planets seem much more complicated, because as the planets go through one revolution they reverse direction for a time, then reverse again to go forward all with changing speed, something akin to epicycles. In the third century, B.C., Apollonius of Perga suggested that the celestial orbits could be explained by means of a combination of circular motions like epicycles. This idea was to become the most important astronomical theory of the next 2000 years. It was ultimately replaced by the Copernicus–Kepler–Newton theory, which we will discuss later.

The French mathematician Blaise Pascal also studied the cycloid in 1649 as a way of distracting himself at a time when he was suffering from a painful toothache. When the pain disappeared, he took it as a sign that God was not displeased with his thoughts. Pascal's results stimulated other mathematicians to investigate this curve, and subsequently numerous remarkable properties were found. One of these was found by the Dutchman Christian Huygens, who used it in the construction of a "perfect" pendulum clock.

If we think of $\mathbf{c}(t)$ as the path traced out by a particle and as t as the time, it is reasonable to define the velocity vector as follows.

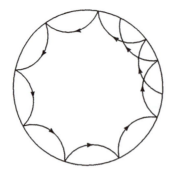

FIGURE 1.6.7. An epicycle.

Velocity Vector

The **velocity** of a path $\mathbf{c}(t)$ is defined by

$$\mathbf{c}'(t) = \lim_{h \to 0} \frac{\mathbf{c}(t+h) - \mathbf{c}(t)}{h}.$$

We normally draw the vector $\mathbf{c}'(t)$ with its tail at the point $\mathbf{c}(t)$. The **speed** of the path $\mathbf{c}(t)$ is $s = \|\mathbf{c}'(t)\|$, the length of the velocity vector. If $\mathbf{c}(t) = (x(t), y(t))$ in $\mathbb{R}^2$, then

$$\mathbf{c}'(t) = (x'(t), y'(t)) = x'(t)\mathbf{i} + y'(t)\mathbf{j}$$

and if $\mathbf{c}(t) = (x(t), y(t), z(t))$ in $\mathbb{R}^3$, then

$$\mathbf{c}'(t) = (x'(t), y'(t), z'(t)) = x'(t)\mathbf{i} + y'(t)\mathbf{j} + z'(t)\mathbf{k}.$$

The theory of limits is discussed in Chapter 2, but the reader should be familiar with the idea from one-variable calculus. Here, $x'(t)$ is the ordinary one-variable derivative dx/dt. If we accept limits of vectors interpreted componentwise, the formulas for the velocity vector come from the definition of the derivative. However, the limit can be interpreted in the sense of vectors as well. In Figure 1.6.8, we see that $[\mathbf{c}(t+h) - \mathbf{c}(t)]/h$ becomes tangent to the curve as $h \to 0$.

Tangent Vector

The velocity $\mathbf{c}'(t)$ is a vector tangent to the path $\mathbf{c}(t)$.

Example 5   Compute the tangent vector to the curve $\mathbf{c}(t) = (t, t^2, e^t)$ at $t = 0$.

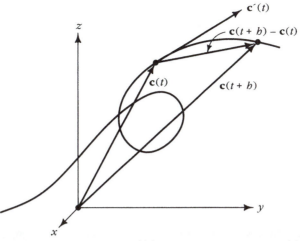

FIGURE 1.6.8. The vector $\mathbf{c}'(t)$ is tangent to the curve $\mathbf{c}(t)$.

Solution   Here $\mathbf{c}'(t) = (1, 2t, e^t)$, and so at $t = 0$ we obtain the tangent vector $(1, 0, 1)$.   ◆

Example 6   Describe the path $\mathbf{c}(t) = (\cos t, \sin t, t)$. Find the velocity vector at the point on the curve where $t = \pi/2$.

Solution   For a given t, the point $(\cos t, \sin t, 0)$ lies on the circle $x^2 + y^2 = 1$ in the xy plane. Therefore the point $(\cos t, \sin t, t)$ lies t units above the point $(\cos t, \sin t, 0)$ if t is positive and $-t$ units below $(\cos t, \sin t, 0)$ if t is negative. As t increases, $(\cos t, \sin t, t)$ wraps around the cylinder $x^2 + y^2 = 1$ with the z-coordinate increasing. The curve thus describes what is called a **helix**, which is depicted in Figure 1.6.9. At $t = \pi/2, \mathbf{c}'(\pi/2) = (-\sin \pi/2, \cos \pi/2, 1) = (-1, 0, 1) = -\mathbf{i} + \mathbf{k}$.   ◆

Example 7   The cycloidal path of a particle on the edge of a wheel of radius R with velocity v is given by $\mathbf{c}(t) = (vt - R \sin (vt/R), R - R \cos (vt/R))$. (See Example 4.) Find the velocity $\mathbf{c}'(t)$ of the particle as a function of t. When is the velocity zero? Is the velocity vector ever vertical?

Solution

$$
\begin{aligned}
\mathbf{c}'(t) &= \left(\frac{d}{dt} \left(vt - R \sin \frac{vt}{R} \right), \frac{d}{dt} \left(R - R \cos \frac{vt}{R} \right) \right) \\
&= \left(v - v \cos \frac{vt}{R}, v \sin \frac{vt}{R} \right).
\end{aligned}
$$

In vector notation, $\mathbf{c}'(t) = (v - v \cos (vt/R))\mathbf{i} + (v \sin (vt/R))\mathbf{j}$. The component in the direction of $\mathbf{i}$ is $v(1 - \cos (vt/R))$, which is zero whenever vt/R is an integer multiple of 2π. For such values of t, $\sin (vt/R)$ is zero as well, so

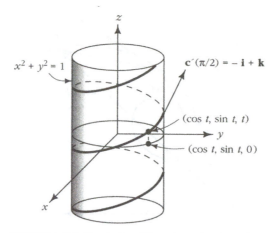

FIGURE 1.6.9. The helix $c(t) = (\cos t,\ \sin t, t)$ wraps around the cylinder $x^2 + y^2 = 1$.

the only times at which the velocity is zero are when $t = 2\pi nR/v$ for some integer n. At such times, $\mathbf{c}(t) = (2\pi nR, 0)$, so the moving point is touching the ground. These moments occur at time intervals of $2\pi R/v$ (more frequently for small wheels, as well as for rapidly rolling ones).

The velocity vector is never vertical, since the horizontal component vanishes only when the vertical one does as well. ◆

Figure 1.6.10 shows some velocity vectors superimposed on the cycloidal path of Figure 1.6.6.

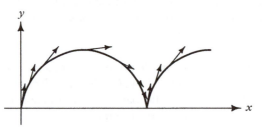

FIGURE 1.6.10. Velocity vectors for the path traced by a point on the rim of a rolling wheel.

The tangent line to a path at a point is the line through the point in the direction of the tangent vector. Using the point–direction form of the equation of a line, we obtain the parametric equation for the tangent line.

Tangent Line to a Path

If $\mathbf{c}(t)$ is a path, its **tangent line** at the point $\mathbf{c}(t_0)$ is

$$l(t) = \mathbf{c}(t_0) + (t - t_0)\mathbf{c}'(t_0).$$

For convenience we have written the equation so l goes through $\mathbf{c}(t_0)$ at $t = t_0$ (rather than $t = 0$). See Figure 1.6.11.

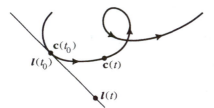

FIGURE 1.6.11. The tangent line to a path.

Example 8   A path in $\mathbb{R}^3$ goes through the point $(3, 6, 5)$ at $t = 0$ with tangent vector $\mathbf{i} - \mathbf{j}$. Find the equation of the tangent line.

Solution   The equation of the tangent line is

$$l(t) = (3, 6, 5) + t(\mathbf{i} - \mathbf{j}) = (3, 6, 5) + t(1, -1, 0) = (3 + t, 6 - t, 5).$$

In coordinates (x, y, z), the tangent line is $x = 3 + t, y = 6 - t, z = 5$.   ◆

Physically, we can interpret motion along the tangent line as the path that a point on a curve would follow if it were set free at a certain moment.

Example 9   Suppose that a particle follows the path $\mathbf{c}(t) = (e^t, e^{-t}, \cos t)$ until it flies off on a tangent at $t = 1$. Where is it at $t = 3$?

Solution   The velocity vector is $(e^t, -e^{-t}, -\sin t)$, which at $t = 1$ is the vector $(e, -1/e, -\sin 1)$. The particle is at $(e, 1/e, \cos 1)$ at $t = 1$. The tangent line is $l(t) = (e, 1/e, \cos 1) + (t - 1)(e, -1/e, -\sin 1)$. At $t = 3$, the position on this line is

$$l(3) = \left(e, \frac{1}{e}, \cos 1\right) + 2\left(e, -\frac{1}{e}, -\sin 1\right) = \left(3e, -\frac{1}{e}, \cos 1 - 2\sin 1\right).$$ ◆

Exercises for §1.6

Sketch the curves in Exercises 1–4.

1. $x = \sin t, y = 4\cos t, 0 \le t \le 2\pi$

2. $x = 2\sin t, y = 4\cos t, 0 \le t \le 2\pi$

3. $\mathbf{c}(t) = (2t - 1, t + 2, t)$

4. $\mathbf{c}(t) = (-t, 2t, 1/t), 1 \le t \le 3$

In Exercises 5–8, determine the velocity vector of the given path.

5. $\mathbf{c}(t) = 6t\mathbf{i} + 3t^2\mathbf{j} + t^3\mathbf{k}$

6. $\mathbf{c}(t) = (\sin 3t)\mathbf{i} + (\cos 3t)\mathbf{j} + 2t^{3/2}\mathbf{k}$

7. $\mathbf{r}(t) = (\cos^2 t, 3t - t^3, t)$

8. $\mathbf{r}(t) = (4e^t, 6t^4, \cos t)$

In Exercises 9–12, compute the tangent vector to the curve.

9. $\mathbf{c}(t) = (e^t, \cos t)$

10. $\mathbf{c}(t) = (3t^2, t^3)$

11. $\mathbf{c}(t) = (t\sin t, 4t)$

12. $\mathbf{c}(t) = (t^2, e^2)$

13. When is the velocity vector of a point on the rim of a rolling wheel *horizontal*? What is the speed at this point?

14. If the position of a particle in space is $(6t, 3t^2, t^3)$ at time t, what is its velocity vector at $t = 0$?

In Exercises 15 and 16, determine the equation of the tangent line to the given path at the specified value of t.

15. $(\sin 3t, \cos 3t, 2t^{5/2}); t = 1$

16. $(\cos^2 t, 3t - t^3, t); t = 0$

In Exercises 17–20, suppose that a particle following the given path $\mathbf{c}(t)$ flies off on a tangent at $t = t_0$. Compute the position of the particle at the given time t_1.

17. $\mathbf{c}(t) = (t^2, t^3 - 4t, 0), t_0 = 2, t_1 = 3$

18. $\mathbf{c}(t) = (e^t, e^{-t}, \cos t), t_0 = 1, t_1 = 2$

19. $\mathbf{c}(t)$ as in Exercise 8, $t_0 = 0, t_1 = 1$

20. $\mathbf{c}(t) = (\sin e^t, t, 4 - t^3), t_0 = 1, t_1 = 2$

Review Exercises for Chapter 1

Complete the calculations in Exercises 1–4.

1. $(3, 2) + (-1, 6) =$

2. $(1, 2, 3) + 2(-1, -2, 7) =$

3. $(3\mathbf{i} + 2\mathbf{j}) + (8\mathbf{i} - \mathbf{j} - \mathbf{k}) =$

4. $(8\mathbf{i} + 3\mathbf{j} - \mathbf{k}) - 6(\mathbf{i} - \mathbf{j} - \mathbf{k}) =$

5.(a) Draw the vector $\mathbf{v}$ joining $(-2, 0)$ to $(4, 6)$ and find the components of $\mathbf{v}$.

  (b) Add $\mathbf{v}$ to the vector joining $(-2, 0)$ to $(1, 1)$.

6.(a) Draw the vector $\mathbf{v}$ joining $(1, 1)$ and $(-3, 6)$ and find the components of $\mathbf{v}$.

  (b) Add $\mathbf{v}$ to the vector joining $(1, 1)$ to $(5, 2)$.

Write an equation or set of equations, to describe the lines given in Exercises 7–10.

7. The line through $(1, 1, 2)$ and $(2, 2, 3)$.

8. The line through $(0, 0, -1)$ and $(1, 1, 3)$.

9. The line through $(1, 1, 1)$ in the direction of $\mathbf{i} - \mathbf{j} - \mathbf{k}$.

10. The line through $(1, -1, 2)$ in the direction of $\mathbf{i} + \mathbf{j} + \mathbf{k}$.

11. Describe the set of all lines through the origin in space which make an angle of $\pi/3$ with the x axis.

12. Consider the set of all points P in space such that the vector from $\mathbf{0}$ to P has length 2 and makes an angle of $45°$ with $\mathbf{i} + \mathbf{j}$.

13. Thales' theorem states that the angle θ in Figure 1.R.1 (a) is $\pi/2$. Prove this using the vectors **a** and **b** shown in Figure 1.R.1 (b).

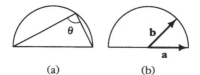

(a)   (b)

FIGURE 1.R.1. Every triangle inscribed in a semicircle is a right triangle.

14. Show that the midpoint of the hypotenuse of a right triangle is equidistant from all three vertices.

15. A bird is headed northeast with speed 40 kilometers per hour. A wind from the north at 15 kilometers per hour begins to blow, but the bird continues to head northeast (*i.e.*, its "beak heading" is northeast) and flies at the same rate relative to the air. Find the speed of the bird relative to the earth's surface.

16. An airplane flying in a straight line at 500 miles per hour for 12 minutes moves 35 miles north and 93.65 miles east. How much does its altitude change? Can you determine whether the airplane is climbing or descending? (Ignore the curvature of the earth.)

17. The work W done in moving an object from $(0,0)$ to $(7,2)$ subject to a force **F** is $W = \mathbf{F} \cdot \mathbf{r}$ where **r** is the vector with head at $(7,2)$ and tail at $(0,0)$. The units are feet and pounds.

    (a) Suppose that the force $\mathbf{F} = 10 \cos \theta \mathbf{i} + 10 \sin \theta \mathbf{j}$. Find W in terms of θ.

    (b) Suppose that the force **F** has magnitude 6 pounds and makes an angle of $\pi/6$ radian with the horizontal, pointing right. Find W in feet-lbs.

18. A 4-kilogram mass located at the origin is suspended by ropes attached to the points $(1,0,1)$ and $(-1,0,1)$. If the force of gravity is pointing in the direction of the vector $-\mathbf{k}$, what is the vector describing the force along each rope? [Hint: Use the symmetry of the problem. A 1-kilogram mass weighs 9.8 newtons (N).]

In Exercises 19–22, compute $\mathbf{v} \cdot \mathbf{w}$.

19. $\mathbf{v} = -\mathbf{i} + \mathbf{j}, \mathbf{w} = \mathbf{k}$

20. $\mathbf{v} = \mathbf{i} + 2\mathbf{j} - \mathbf{k}, \mathbf{w} = 3\mathbf{i} + \mathbf{j}$

21. $\mathbf{v} = -2\mathbf{i} - \mathbf{j} + \mathbf{k}, \mathbf{w} = 3\mathbf{i} + 2\mathbf{j} - 2\mathbf{k}$

22. $\mathbf{v} = 8\mathbf{i} + 3\mathbf{j} - \mathbf{k}, \mathbf{w} = \mathbf{i} - \mathbf{j} - \mathbf{k}$

Find the cosine of the angles between the vectors given in Exercises 23–24.

23. $\mathbf{v} = \mathbf{i} + \mathbf{j}, \mathbf{w} = \mathbf{k}$

24. $\mathbf{v} = -2\mathbf{i} - \mathbf{j} + \mathbf{k}, \mathbf{w} = 3\mathbf{i} + 2\mathbf{j} - 2\mathbf{k}$

25. Suppose that $\mathbf{v} \cdot \mathbf{w} = 0$ for all vectors $\mathbf{w}$. Show that $\mathbf{v} = \mathbf{0}$. (Note: This is not the same as showing that $\mathbf{0} \cdot \mathbf{w} = 0$.)

26. Suppose that $\mathbf{u} \cdot \mathbf{w} = \mathbf{v} \cdot \mathbf{w}$ for all vectors $\mathbf{w}$. Show that $\mathbf{u} = \mathbf{v}$.

Evaluate the determinants in Exercises 27–30.

27. $\begin{vmatrix} 2 & -1 \\ 0 & 1 \end{vmatrix}$

28. $\begin{vmatrix} 0 & 1 \\ -1 & 0 \end{vmatrix}$

29. $\begin{vmatrix} 1 & 0 & 0 \\ 0 & -1 & 1 \\ 0 & 1 & 1 \end{vmatrix}$

30. $\begin{vmatrix} 6 & 2 & -3 \\ 2 & 2 & 3 \\ 4 & 8 & -1 \end{vmatrix}$

31. Show that $\begin{vmatrix} 66 & 628 & 246 \\ 88 & 435 & 24 \\ 2 & -1 & 1 \end{vmatrix} = \begin{vmatrix} 68 & 627 & 247 \\ 86 & 436 & 23 \\ 2 & -1 & 1 \end{vmatrix}$.

32. Show that $\begin{vmatrix} n & n+1 & n+2 \\ n+3 & n+4 & n+5 \\ n+6 & n+7 & n+8 \end{vmatrix}$
has the same value no matter what n is. What is this value?

Evaluate the products in Exercises 33–36.

33. $(8\mathbf{i} + 3\mathbf{j} - \mathbf{k}) \times (\mathbf{i} - \mathbf{j} - \mathbf{k})$

34. $(\mathbf{i} - \mathbf{j} - \mathbf{k}) \times (8\mathbf{i} + 3\mathbf{j} \cdot \mathbf{k})$

35. $[(2\mathbf{i} - \mathbf{j}) \times (3\mathbf{i} + \mathbf{j})] \cdot (2\mathbf{j} + \mathbf{k})$

36. $(\mathbf{i} + \mathbf{j}) \times (\mathbf{i} - \mathbf{j})$

37. If the triple product $(\mathbf{v} \times \mathbf{j}) \cdot \mathbf{k}$ is zero, what can you say about the vector $\mathbf{v}$?

38. Suppose that $\mathbf{a} \times \mathbf{b} = \mathbf{a}' \times \mathbf{b}$ for all $\mathbf{b}$. Is it true that $\mathbf{a} = \mathbf{a}'$?

In Exercises 39–42, find the area or volume of the figure described.

39. The parallelogram spanned by $3\mathbf{i} - 2\mathbf{j} + \mathbf{k}$ and $8\mathbf{i} - \mathbf{k}$.

40. The parallelogram spanned by $2\mathbf{i} - \mathbf{j}$ and $3\mathbf{i} - 2\mathbf{j}$.

41. The triangle with vertices $(1, 2), (0, 1), (-1, 1)$.

42. The parallelepiped spanned by the vectors $(1, 0, 1), (1, 1, 1)$, and $(-3, 2, 0)$.

43. Let $\mathbf{a}$ and $\mathbf{b}$ be two vectors in the plane, $\mathbf{a} = (a_1, a_2), \mathbf{b} = (b_1, b_2)$, and let λ be a real number. Show that the area of the parallelogram determined by $\mathbf{a}$ and $\mathbf{b} + \lambda\mathbf{a}$ is the same as that determined by $\mathbf{a}$ and $\mathbf{b}$. Sketch. Relate this result to a known property of determinants.

44. The volume of a *tetrahedron* with concurrent edges $\mathbf{a}, \mathbf{b}$, and $\mathbf{c}$ is given by $V = (1/6)\mathbf{a} \cdot (\mathbf{b} \times \mathbf{c})$. See Figure 1.R.2.

   (a) Express the volume as a determinant.

   (b) Evaluate V when $\mathbf{a} = \mathbf{i} + \mathbf{j} + \mathbf{k}, \mathbf{b} = \mathbf{i} - \mathbf{j} + \mathbf{k}, \mathbf{c} = \mathbf{i} + \mathbf{j}$.

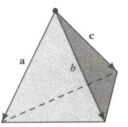

FIGURE 1.R.2. A tetrahedron with edges $\mathbf{a}, \mathbf{b}$ and $\mathbf{c}$.

In Exercises 45–48, find a unit vector with the given property.

45. Orthogonal to $3\mathbf{i} + 2\mathbf{k}$ and $\mathbf{j} - \mathbf{k}$.

46. Orthogonal to the plane $x - 6y + z = 12$.

47. Parallel to both the planes $8x + y + z = 1$ and $x - y - z = 0$.

48. Orthogonal to the line $x = 2t - 1, y = -t - 1, z = t + 2$, and the vector $\mathbf{i} - \mathbf{j}$.

In Exercises 49–52, write an equation or set of equations to describe each of the following geometric figures.

49. The plane through $(1, 1, 2), (2, 2, 3)$, and $(0, 0, 0)$.

50. The plane through $(1, 2, -1)$ that is parallel to both $\mathbf{i} - \mathbf{j} + 2\mathbf{k}$ and $\mathbf{i} - 3\mathbf{k}$.

51. The line orthogonal to the plane in Exercise 50 and passing through $(0, 0, 3)$.

52. The line orthogonal to the plane spanned by $\mathbf{i} + \mathbf{j}$ and $3\mathbf{k}$ passing through $(2, 3, 1)$.

Find the products in Exercises 53–56, or else explain why they are not defined.

53. $\begin{bmatrix} 1 & 2 & 4 \end{bmatrix} \begin{bmatrix} 2 \\ -1 \\ 1 \end{bmatrix}$

54. $\begin{bmatrix} 0 & -1 \\ 1 & 0 \end{bmatrix} \begin{bmatrix} 1 & 0 \\ 0 & -1 \end{bmatrix}$

55. $\begin{bmatrix} 1 & 2 & 3 & 4 & 5 \\ 6 & 7 & 8 & 9 & 10 \end{bmatrix} \begin{bmatrix} 1 \\ -1 \\ 1 \\ -1 \\ 1 \end{bmatrix}$

56. $\begin{bmatrix} 1 & 2 & 3 & 4 \\ 5 & 6 & 7 & 8 \end{bmatrix} \begin{bmatrix} 1 & 2 \\ 3 & 4 \\ 5 & 6 \end{bmatrix}$

57. As a particular case of the associative law of matrix multiplication, note that for two $n \times n$ matrices A and B, and a vector $\mathbf{x}$ in $\mathbb{R}^n$,

$$(AB)\mathbf{x} = A(B\mathbf{x}).$$

What does this equality imply about the relationship between the composition of the functions $f(\mathbf{x}) = B\mathbf{x}, g(\mathbf{y}) = A\mathbf{y}$ and matrix multiplication?

58. If a particle with mass m moves with velocity $\mathbf{v}$, its *momentum* is $\mathbf{p} = m\mathbf{v}$. In a game of marbles, a marble with mass 2 grams is shot with velocity 2 meters per second, hits two marbles with mass 1 gram each, and comes to a dead halt. One of the marbles flies off with a velocity of 3 meters per second at an angle of $45°$ to the incident direction of the larger marble as in Figure 1.R.3. Assuming that the total momentum before and after the

collision is the same (law of conservation of momentum), at what angle and speed does the second marble move?

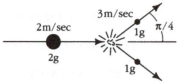

FIGURE 1.R.3. Momentum and marbles.

59.(a) Using vector methods, show that the distance between two nonparallel lines l_1 and l_2 is given by

$$d = \frac{|(\mathbf{v}_2 - \mathbf{v}_1) \cdot (\mathbf{a}_1 \times \mathbf{a}_2)|}{\|\mathbf{a}_1 \times \mathbf{a}_2\|},$$

where $\mathbf{v}_1, \mathbf{v}_2$ are any points on l_1 and l_2, respectively, and $\mathbf{a}_1$ and $\mathbf{a}_2$ are the directions of l_1 and l_2. [Hint: Consider the plane through l_2 which is parallel to l_1. Show that $(\mathbf{a}_1 \times \mathbf{a}_2)/\|\mathbf{a}_1 \times \mathbf{a}_2\|$ is a unit normal for this plane; now project $\mathbf{v}_2 - \mathbf{v}_1$ onto this normal direction.]

(b) Find the distance between the line l_1 determined by the points $(-1, -1, 1)$ and $(0, 0, 0)$ and the line l_2 determined by the points $(0, -2, 0)$ and $(2, 0, 5)$.

60. Show that two planes given by the equations $Ax + By + Cz + D_1 = 0$ and $Ax + By + Cz + D_2 = 0$ are parallel, and that the distance between two such planes is

$$\frac{|D_1 - D_2|}{\sqrt{A^2 + B^2 + C^2}}.$$

61. In 3-space, show that the vector $\mathbf{v} = \|\mathbf{a}\|\mathbf{b} + \|\mathbf{b}\|\mathbf{a}$ bisects the angle between $\mathbf{a}$ and $\mathbf{b}$.

62. Show that if all the vertices of a rectangle in space have coordinates that are multiples of 3, then the area of the rectangle is a multiple of 9.

63. Find the solution of the equation $x + 2y + 3z = 4$ that is closest to the origin.

64. Show that the cross product satisfies the *Jacobi identity*:

$$\mathbf{a} \times (\mathbf{b} \times \mathbf{c}) + \mathbf{b} \times (\mathbf{c} \times \mathbf{a}) + \mathbf{c} \times (\mathbf{a} \times \mathbf{b}) = 0.$$

In Exercises 65–68, at the indicated time t_0, compute the tangent line to the curve.

65. $\mathbf{c}(t) = (t^3 + 1, e^{-t}, \cos(\pi t/2)); t_0 = 1$

66. $\mathbf{c}(t) = (t^2 - 1, \, \cos(t^2), t^4); t_0 = \sqrt{\pi}$

67. $\mathbf{c}(t) = (e^t, \, \sin t, \, \cos t); t_0 = 0$

68. $\mathbf{c}(t) = \left[\dfrac{t^2}{1 + t^2} \right] \mathbf{i} + t\mathbf{j} + \mathbf{k}; t_0 = 2$

69. Sketch the graph of the curve $\mathbf{c}(t) = (\sin 2t, \, \cos 2t, 3t)$ for $-\pi \leq t \leq \pi$.

70. Find the tangent line to the curve in Exercise 69 at $t = \pi/4$.

In Exercises 71 and 72, suppose that a particle following the given path flies off on a tangent at $t = t_0$. Compute the position of the particle at the given time t_1.

71. $\mathbf{c}(t) = (t, t^2, t \cos t); t_0 = \pi; t_1 = 2\pi$

72. $\mathbf{c}(t) = 3t^2\mathbf{i} - (\sin t)\mathbf{j} - e^t\mathbf{k}; t_0 = 1/2; t_1 = 1$

2

Differentiation

I turn away with fright and horror from the lamentable evil of functions which do not have derivatives.

Charles Hermite *(in a letter to Thomas Jan Stieltjes)*

This chapter extends differential calculus from one to several variables. As in one-variable calculus, we relate this calculus to the graphs of functions, so the chapter opens with a study of graphs. We develop the notion of partial derivative using ideas of one-variable calculus and then consider the total derivative as a matrix of partial derivatives. This approach is especially convenient for expressing the chain rule for functions of several variables. We apply differential calculus to find equations of tangent planes to graphs and then, using a vector operation called the gradient, to calculate tangent lines to level curves and tangent planes to level surfaces.

2.1 Graphs and Level Surfaces

We launch our investigation of real-valued functions by developing methods for visualizing them. In particular, we shall introduce the notions of graph, level curve, and level surface for such functions.

Let f be a function whose domain is a subset U of $\mathbb{R}^n$ and whose range is contained in $\mathbb{R}^m$. By this we mean that to each point $\mathbf{x} = (x_1, \ldots, x_n)$ in U, f assigns a value $f(\mathbf{x})$, an m-tuple in $\mathbb{R}^m$. Such functions f are called $\mathbb{R}^m$-*valued functions* or, if $m = 1$, *real-valued functions*, or *scalar-valued functions*. For example, the scalar-valued function $f(x, y, z) = (x^2 + y^2 + z^2)^{-3/2}$ maps the set U consisting of all $(x, y, z) \neq (0, 0, 0)$ in $\mathbb{R}^3$ ($n = 3$ in this case) to $\mathbb{R}$ ($m = 1$).

We write $f : U \subset \mathbb{R}^n \to \mathbb{R}^m$ to signify that the subset U of $\mathbb{R}^n$ is the domain of f and $\mathbb{R}^m$ is the target. When we wish to stress that the domain is in $\mathbb{R}^n$, we call such a function f a *function of n variables* or, if $n > 1$, *a function of several variables*.

Functions of several variables are not just mathematical abstractions, but arise naturally in problems studied in all the sciences. For example:

i. to specify the temperature T in a region U of space requires a function $T : U \subset \mathbb{R}^3 \to \mathbb{R}$ ($n = 3, m = 1$), where $T(x, y, z)$ is the temperature at the point (x, y, z);

ii. to specify the velocity for the steady flow of a fluid (or gas) moving in space requires a map $\mathbf{V} : \mathbb{R}^3 \to \mathbb{R}^3$, where $\mathbf{V}(x, y, z)$ is the velocity vector of the fluid at the point (x, y, z) (see Figure 2.1.1) (The domain U consists of those (x, y, z) for which the velocity is defined.);

iii. to specify the reaction rate of a solution consisting of six reacting chemicals A, B, C, D, E, F in proportions x, y, z, w, u, v, respectively, requires a map $\sigma : U \subset \mathbb{R}^6 \to \mathbb{R}$, where $\sigma(x, y, z, w, u, v)$ gives the rate when the chemicals are in the indicated proportions;

iv. to specify the "cardiac vector" (the vector giving the magnitude and direction of electric current flow in the heart) as it depends on time requires a map $\mathbf{c} : \mathbb{R} \to \mathbb{R}^3$.

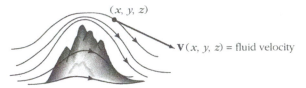

FIGURE 2.1.1. A fluid in motion defines a vector field **V** giving the velocity of the fluid particles at each point in space.

For $f : U \subset \mathbb{R} \to \mathbb{R}$, the **graph** of f is the subset of $\mathbb{R}^2$ consisting of all points $(x, f(x))$ in the plane, for x in U. This subset is normally thought of as a curve in $\mathbb{R}^2$.

To visualize a scalar function of n variables, we consider its graph in $(n+1)$-dimensional space, defined as follows:

The Graph of a Function

If $f : U \subset \mathbb{R}^n \to \mathbb{R}$ is a function of n variables, its **graph** consists of the set of points $(x_1, \ldots, x_n, f(x_1, \ldots, x_n))$ in $\mathbb{R}^{n+1}$ for $(x_1, \ldots, x_n)$ in U.

For $n = 2$, the **graph** of a function $f(x, y)$ of two variables is the surface consisting of all points (x, y, z) in space such that (x, y) is in the domain of the function and $z = f(x, y)$. See Figure 2.1.2.

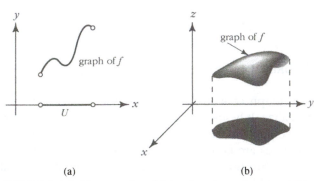

FIGURE 2.1.2. The graphs of (a) a function of one variable and (b) a function of two variables.

Example 1 Sketch the graph of (a) $f(x, y) = x - y + 2$ and (b) $f(x, y) = 3x$.

Solution (a) We recognize $z = x - y + 2$ (that is, $x - y - z + 2 = 0$) as the equation of a plane. The normal to the plane is $(1, -1, -1)$, and the plane meets the axes at $(-2, 0, 0)$, $(0, 2, 0)$, and $(0, 0, 2)$. From this information we sketch the graph of f in Figure 2.1.3.

(b) The graph of $f(x, y) = 3x$ is the plane $z = 3x$. It contains the y axis and is shown in Figure 2.1.4. ◆

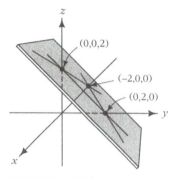

FIGURE 2.1.3. The graph $z = x - y + 2$ is a plane.

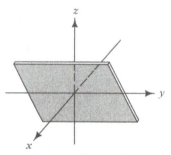

FIGURE 2.1.4. The graph $z = 3x$.

The graph of a function of three variables is a set in $\mathbb{R}^4$, which is hard to visualize. It helps in this case to look instead at **level sets**, that is, sets obtained by setting the function equal to a constant. Level sets for functions of two variables are familiar to hikers and weather watchers. For instance, level sets of the function $h(x, y)$, which is the height of a hill at position (x, y), are obtained by setting $h(x, y) = c$ for various constants c, as in Figure 2.1.5. (In this case, the level sets are also called *level curves* or *level contours*.)

Similarly, isotherms on a weather map, which give curves of constant temperature, are level curves of the function $T(x, y)$ that gives the temperature at ground level at latitude x and longitude y, as in Figure 2.1.6.

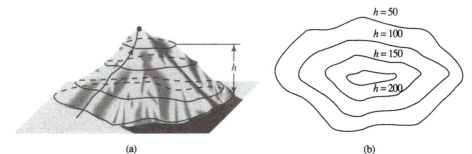

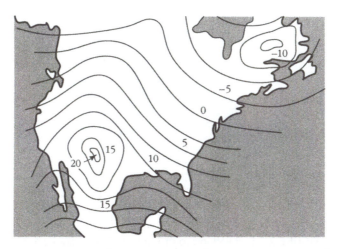

(a)   (b)

FIGURE 2.1.5. Contour plot of a hill.

FIGURE 2.1.6. Isotherms are curves of constant temperature (in degrees Celsius).

Level Curves

Let f be a function of two variables and let c be a constant. The set of all (x, y) in the plane such that $f(x, y) = c$ is called a **level curve** of f (with value c).

Example 2 The function $f(x, y) = x + y + 2$ has as its graph the inclined plane $z = x + y + 2$. This plane intersects the xy plane ($z = 0$) in the line $y = -x - 2$ and the z axis at the point $(0, 0, 2)$. For any value $c \in \mathbb{R}$, the level curve of value c is the straight line $y = -x + (c - 2)$. We indicate a few of the level curves of the function in Figure 2.1.7, which is a contour map of the function f. The graph of f is shown in Figure 2.1.8. ♦

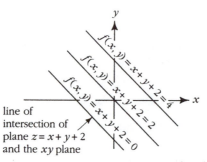

FIGURE 2.1.7. The level curves of $f(x,y) = x+y+2$ show the behavior of this function.

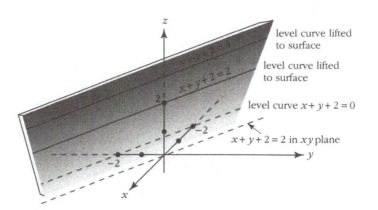

FIGURE 2.1.8. Relationship of level curves of Figure 2.1.7 to the graph of the function $f(x,y) = x+y+2$, which is the plane $z = x+y+2$.

Example 3 (a) Use level curves to sketch the graph of $f(x,y) = x^2 + y^2$ (this graph is called a ***paraboloid of revolution***).

(b) Sketch the surface $z = x^2 + y^2 - 4x - 6y + 13$.

Solution (a) Setting $z = $ constant, we get the circle $x^2 + y^2 = c$. Taking $c = 1^2, 2^2, 3^2, 4^2$, we get circles of radius $1, 2, 3$, and 4. These are placed on the planes $z = 1^2 = 1, z = 2^2 = 4, z = 3^2 = 9$, and $z = 4^2 = 16$ to give the graph shown in Figure 2.1.9.

If we set $x = 0$, we obtain the parabola $z = y^2$; if we set $y = 0$, we obtain the parabola $z = x^2$. The graph is rotationally symmetric about the z axis since z depends only on $r = \sqrt{x^2 + y^2}$; i.e., $z = r^2$.

(b) Completing the square, we write $z = x^2 + y^2 - 4x - 6y + 13$ as

$$\begin{aligned} z &= x^2 - 4x + \quad y^2 - 6y \quad + 13 \\ &= x^2 - 4x + 4 + y^2 - 6y + 9 + 13 - 13 \\ &= (x-2)^2 + (y-3)^2. \end{aligned}$$

The level surface for value c is thus the circle $(x-2)^2 + (y-3)^2 = c$ with center $(2,3)$ and radius $\sqrt{c}$. Comparing this result with (a) we find that the surface is again a paraboloid of revolution, with its axis shifted to the line $x = 2, y = 3$. (See Figure 2.1.10.) ◆

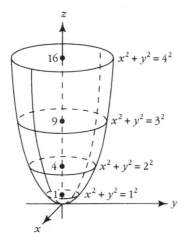

FIGURE 2.1.9. The sections of the graph $z = x^2 + y^2$ by planes $z = c$ are circles.

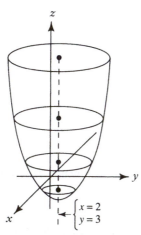

FIGURE 2.1.10. The graph $z = x^2 + y^2 - 4x - 6y + 13$ is a shifted paraboloid of revolution.

For a function of three variables, the set obtained by letting f be a constant is called a ***level surface***.

Level Surfaces

Let f be a function of three variables and let c be a constant. The set of all points (x, y, z) in space such that $f(x, y, z) = c$ is called a **level surface** of f (with value c).

For a function of n variables a **level set** $f(x_1, \ldots, x_n) = c$ is a subset of $\mathbb{R}^n$ (a level curve if $n = 2$, a level surface if $n = 3$). For $n = 4, 5, \ldots$, this set is hard to visualize, so we concentrate on the cases $n = 2, 3$.

Example 4   Let $f(x, y, z) = x - y + z + 2$. Sketch the level surfaces with values $1, 2, 3$. (That is, with $c = 1, 2, 3$ in the preceding box.)

Solution   In each case we set $f(x, y, z) = c$:

$$
\begin{aligned}
c = 1: & \quad x - y + z + 2 = 1 \quad (i.e., \; x - y + z + 1 = 0), \\
c = 2: & \quad x - y + z + 2 = 2 \quad (i.e., \; x - y + z = 0), \\
c = 3: & \quad x - y + z + 2 = 3 \quad (i.e., \; x - y + z - 1 = 0).
\end{aligned}
$$

These surfaces are parallel planes and are sketched in Figure 2.1.11.   ◆

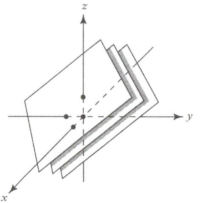

FIGURE 2.1.11. Three level surfaces of the function $f(x, y, z) = x - y + z + 2$.

Example 5   Sketch the level surface of $f(x, y, z) = x^2 + y^2 + z^2 - 8$ with value 1.

Solution   The surface $x^2 + y^2 + z^2 - 8 = 1$ (that is, $x^2 + y^2 + z^2 = 9$) is the set of points (x, y, z) whose distance from the origin is $\sqrt{9} = 3$; it is a sphere with radius 3 and center at the origin. (See Figure 2.1.12.)   ◆

Instead of taking horizontal slices through the graph of a function of two variables, which gives us level curves, we can also take slices by *vertical* planes.

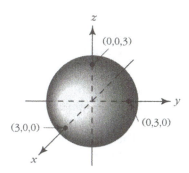

FIGURE 2.1.12. The level surface of $x^2 + y^2 + z^2 - 8$ with value 1 is a sphere of radius 3.

This technique is called the ***method of sections***. We summarize this and some other techniques in the following box.

Plotting Surfaces

To plot a surface given by an equation in x, y, and z:

1. Note any symmetries.

2. See if any variables x, y, or z are missing from the equation. If so, the surface is a "cylinder" parallel to the axis of the missing variable, and its cross section is the curve described by other variables.

3. If the surface is a graph $z = f(x, y)$, find the level curves $f(x, y) = c$ for various convenient values of c and draw these curves on the planes $z = c$. Smoothly join these curves with a surface in space. Draw the curves obtained by setting $x = 0$ and $y = 0$ or other convenient values to help clarify the picture.

4. If the surface has the form $F(x, y, z) = c$, then either:

(a) solve for one of the variables in terms of the other two and use step 3 if it is convenient to do so, or

(b) set x equal to various constant values to obtain curves in y and z; draw these curves in the corresponding $x =$ constant planes. Repeat with $y =$ constant or $z =$ constant or both. Fill in the curves obtained with a surface.

The steps in the box above are explained in more detail as they are carried out in the following examples.

Note

These steps are meant to be general guidelines to help you do examples. It is better to understand them than to try to memorize them. As you gain experience, you will also gain flexibility in using the ideas efficiently. Note that in item 2, "cylinder" is used in a more general sense than a "can." Both the surfaces in Figures 2.1.13 and 2.1.14 are cylinders, but only the second is shaped like a can (an infinitely long one). If the equation had been $x^2 + 4y^2 = 25$, we would get an elliptical cylinder.

Example 6   Sketch the surfaces in xyz space given by (a) $z = -y^2$ and (b) $x^2 + y^2 = 25$.

Solution   (a) *Step 1.* The equation $z = -y^2$ is unchanged when y is replaced by $-y$. This shows that the surface is symmetric when reflected across the xz plane.

Step 2. The variable x is missing from the equation. Hence, all sections $x = $ constant must look the same; they are copies of the parabola $z = -y^2$. Thus we draw the parabola $z = -y^2$ in the yz plane and extend it parallel to the x axis as shown in Figure 2.1.13. The surface is called a ***parabolic cylinder***. Its symmetry under the reflection in the xz plane is visible in the figure. Steps 3 and 4 are not needed for this example.

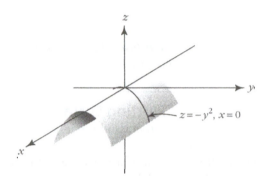

FIGURE 2.1.13. The surface $z = -y^2$ is a parabolic cylinder.

(b) *Step 1.* The surface is symmetric under reflections in the yz and xz planes.

Step 2. In this case, the variable z does not occur in the equation. This means that if we take any point (x, y) in the xy plane such that $x^2 + y^2 = 25$ (a circle of radius 5), then (x, y, z) will be on the surface for *every* z. Therefore, the surface is a right circular cylinder, as shown in Figure 2.1.14. Step 3 does not apply here, and Step 4 is not necessary.   ♦

Example 7   The graph of the quadratic function $f(x, y) = x^2 - y^2$ is called a ***hyperbolic paraboloid***, or ***saddle***, centered at the origin. Sketch the graph.

Solution   *Step 1.* The equation $z = x^2 - y^2$ of the graph of f is unchanged if we change the sign of either x or y. As a result, the graph will be symmetric across the yz and xz planes.

Step 2. No variables are missing from the equation.

Step 3. We draw some level curves. These level curves are given by $x^2 - y^2 = c$. Consider the values $c = 0, \pm 1$, and ± 4. For $c = 0$, we have $y^2 = x^2$, or $y = \pm x$,

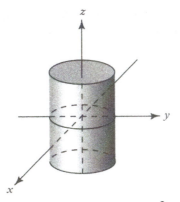

FIGURE 2.1.14. The surface $x^2 + y^2 = 25$ is a right circular cylinder.

so that this level consists of two straight lines through the origin. For $c = 1$, the level curve is $x^2 - y^2 = 1$, which is a hyperbola that passes vertically through the x axis at the points $(\pm 1, 0)$ (see Figure 2.1.15).

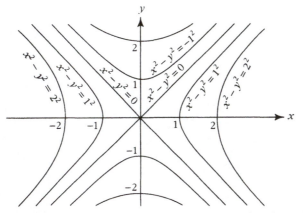

FIGURE 2.1.15. Level curves for the function $f(x, y) = x^2 - y^2$.

Similarly, for $c = 4$, the level curve is defined by $x^2 - y^2 = 4$ (or $y = \pm\sqrt{x^2 - 4}$), the hyperbola passing vertically through the x axis at $(\pm 2, 0)$. For $c = -1$, we obtain the curve $x^2 - y^2 = -1$, that is, $x = \pm\sqrt{y^2 - 1}$, the hyperbola passing horizontally through the y axis at $(0, \pm 1)$. And for $c = -4$, the hyperbola through $(0, \pm 2)$ is obtained. These level curves are shown in Figure 2.1.15. Since it is not easy to visualize the graph of f from these data alone, we shall use step 4 and compute two vertical sections. The section $y = 0$ gives $z = x^2$, which is a parabola opening upward. For the yz plane, we set $x = 0$, giving $z = -y^2$, which is a parabola opening downward. The graph may now be visualized by lifting the level curves to the appropriate heights and smoothing out the resulting surface. Their placement is aided by computing the parabolic sections. This procedure generates the hyperbolic saddle indicated in Figure 2.1.16. Compare this with the computer-generated graphs in Figure 2.1.17 (note that the orientation of the axes has been changed).   ♦

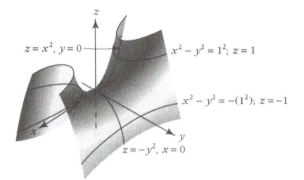

FIGURE 2.1.16. The graph $z = x^2 - y^2$ is a hyperbolic paraboloid, or "saddle."

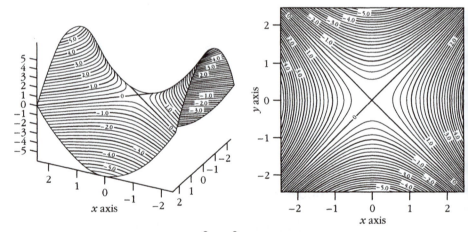

FIGURE 2.1.17. The graphs of $z = x^2 - y^2$ and its level curves.

We have seen how to graph several quadric functions. More generally, a **quadric surface** is a three-dimensional figure defined by a quadratic equation in three variables:

$$ax^2 + by^2 + cz^2 + dxy + exz + fyz + gx + hy + kz + m = 0.$$

The quadric surfaces are the three-dimensional analogues of the conic sections (circles, parabolas, ellipses, and hyperbolas).

Example 8   Particular conic sections can degenerate to points or lines. Similarly, some quadric surfaces can degenerate to points, lines, or planes. Match the sample equations in three-dimensional space to the appropriate descriptions:

$$
\begin{array}{llll}
\text{(a)} & x^2 + 3y^2 + z^2 = 0 & \text{(1)} & \text{No points at all} \\
\text{(b)} & z^2 = 0 & \text{(2)} & \text{A single point} \\
\text{(c)} & x^2 + y^2 = 0 & \text{(3)} & \text{A line} \\
\text{(d)} & x^2 + y^2 + z^2 + 1 = 0 & \text{(4)} & \text{One plane} \\
\text{(e)} & x^2 - y^2 = 0 & \text{(5)} & \text{Two planes}
\end{array}
$$

Solution   Equation (a) matches (2) since only $(0, 0, 0)$ satisfies the equation; (b) matches (4) since this is the plane $z = 0$; (c) matches (3) since this is the z axis, where $x = 0$ and $y = 0$; (d) matches (1) since a nonnegative number added to 1 can never be zero; (e) matches (5) since the equation $x^2 - y^2 = 0$ is equivalent to the two equations $x + y = 0$ or $x - y = 0$, which define two planes.   ◆

Example 9   The surface defined by an equation of the form $x^2/a^2 + y^2/b^2 - z^2/c^2 = -1$ is called a ***hyperboloid of two sheets***. Sketch the surface $x^2 + 4y^2 - z^2 = -4$.

Solution   The section of the surface in the plane $z = c$ has the equation $x^2 + 4y^2 = c^2 - 4$. This is an ellipse when $|c| > 2$, a point when $c = \pm 2$, and is empty when $|c| < 2$. The section in the xz plane is the hyperbola $x^2 - z^2 = -4$, and the section in the yz plane is the hyperbola $4y^2 - z^2 = -4$. The surface is symmetric with respect to each of the coordinate planes. A sketch is given in Figure 2.1.18.   ◆

Example 10   The surface defined by an equation of the form $x^2/a^2 + y^2/b^2 + z^2/c^2 = 1$ is called an ***ellipsoid***. Sketch the surface $x^2/9 + y^2/16 + z^2 = 1$.

Solution   First, let z be constant. Then we get $x^2/9 + y^2/16 = 1 - z^2$. This is an ellipse centered at the origin if $-1 < z < 1$. If $z = 1$, we just get the point $(0, 0, 1)$. Likewise, $(0, 0, -1)$ is on the surface. If $|z| > 1$ there are no (x, y) satisfying the equation.

Setting $x = $ constant or $y = $ constant, we also get ellipses. We must have $|x| \leq 3$ and $|y| \leq 4$. The surface, shaped a little bit like a stepped-on football, is easiest to draw if the intersections with the three coordinate planes are drawn first. (See Figure 2.1.19.)   ◆

Example 11   The surface defined by an equation of the form $x^2/a^2 + y^2/b^2 - z^2/c^2 = 1$ is called a ***hyperboloid of one sheet***. Sketch the surface $x^2 + y^2 - z^2 = 4$.

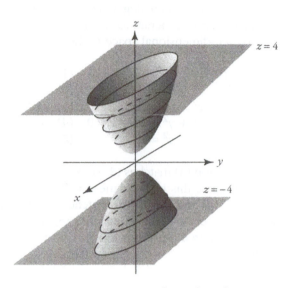

FIGURE 2.1.18. The surface $x^2 + 4y^2 - z^2 = -4$ is a hyperboloid of two sheets (shown with some of its sections in planes of the form $z = $ constant).

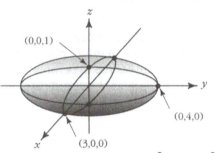

FIGURE 2.1.19. The surface $(x^2/9) + (y^2/16) + z^2 = 1$ is an ellipsoid.

Solution   If z is a constant, then $x^2 + y^2 = 4 + z^2$ is a circle. Thus, in any plane parallel to the xy plane, we get a circle. Our job of drawing the surface is simplified if we note right away that the surface is rotationally symmetric about the z axis (since z depends only on $r^2 = x^2 + y^2$). Thus we can draw the curve traced by the surface in the yz plane (or xz plane) and revolve it about the z axis. Setting $x = 0$, we get $y^2 - z^2 = 4$, a hyperbola. Hence we get the surface shown in Figure 2.1.20, a one-sheeted hyperboloid. Since this surface is symmetric about the z axis, it is also called a hyperboloid of revolution.   ◆

The hyperboloid of one sheet has the property that it is **ruled**: that is, the surface is composed of straight lines. It is therefore easy to make with string models or straight boards and is useful in architecture. (See Figure 2.1.21.)

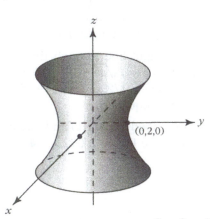

FIGURE 2.1.20. The surface $x^2 + y^2 - z^2 = 4$ is a one-sheeted hyperboloid of revolution.

FIGURE 2.1.21. One can make a hyperboloid with a wire frame and string.

Example 12 Consider the equation $x^2 + y^2 - z^2 = 0$.

(a) What are the horizontal cross sections for $z = \pm 1, \pm 2, \pm 3$?

(b) What are the vertical cross sections for $x = 0$ or $y = 0$?

(c) Show that any straight line through the origin making a 45° angle with the z axis lies in the surface.

(d) Sketch this surface.

Solution

(a) Rewriting the equation as $x^2 + y^2 = z^2$ shows that the horizontal cross sections are circles centered around the z axis with radius $|z|$. Therefore, for $z = \pm 1, \pm 2$, and ± 3, the cross sections are circles of radius 1, 2, and 3.

(b) When $x = 0$, the equation is $y^2 - z^2 = 0$ or $y = \pm z$, the graph of which is two straight lines. When $y = 0$, the equation is $x^2 - z^2 = 0$ or $x = \pm z$, again giving two straight lines.

(c) Any point on a straight line through the origin making a 45° angle with the z axis satisfies $|z|/\sqrt{x^2 + y^2 + z^2} = \cos 45° = 1/\sqrt{2}$. Squaring gives $z^2/(x^2 + y^2 + z^2) = 1/2$, or $x^2 + y^2 + z^2 = 2z^2$, or $x^2 + y^2 - z^2 = 0$, which is the original equation.

(d) Draw a line as described in part (c) and rotate it around the z axis. The surface is a **cone**. (See Figure 2.1.22.) ◆

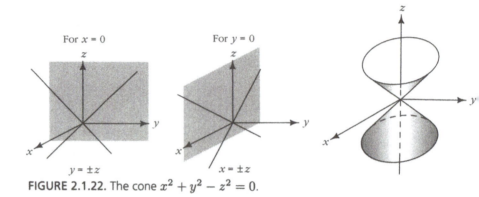

FIGURE 2.1.22. The cone $x^2 + y^2 - z^2 = 0$.

Since complicated functions can be difficult to draw "by hand," the computer is a useful aid. For example, Figure 2.1.23 shows the graph of the function

$$z = (x^2 + 3y^2)e^{1-(x^2+y^2)}.$$

Figure 2.1.24 shows the level curves of this function in the xy plane. Study these pictures to help develop your powers of three-dimensional visualization; attempt to reconstruct the graph in your mind by looking at the level curves. In Chapter 3 we will see how calculus can help us understand features of graphs like this.

Exercises for §2.1

In Exercises 1–4, sketch the given plane.

1. $f(x, y) = x - y + 2$

2. $f(x, y) = 1 - x - y$

3. $z = x + 2$

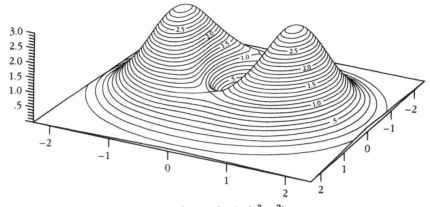

FIGURE 2.1.23. Graph of $z = (x^2 + 3y^2)e^{1-(x^2+y^2)}$.

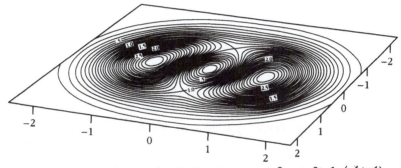

FIGURE 2.1.24. Level curves for the function $z = (x^2 + 3y^2)e^{1-(x^4+y^4)}$.

4. $z = 2x - y + 1$

In Exercises 5–8, sketch the level curves for $c = -1, 0, 1, 2$ of the given function.

5. The plane in Exercise 1

6. The plane in Exercise 2

7. $f(x, y) = x^2 + y^2 + 1$

8. $f(x, y) = 2 - x^2 - y^2$

In Exercises 9–12, sketch the indicated level curves.

9. $f(x, y) = x^2 + 4y^2, c = 0, 1, 4, 9$

10. $f(x, y) = (100 - x^2 - y^2), c = 0, 2, 4, 6, 8, 10$

11. $f(x, y) = x/y, c = -3, -2, -1, 0, 1, 2, 3$

12. $f(x, y) = x^2 + xy, c = -3, -2, -1, 0, 1, 2, 3$

In Exercises 13–16, sketch some level surfaces of the given function.

13. $f(x, y, z) = -x^2 - y^2 - z^2$

14. $f(x, y, z) = 4x^2 + y^2 + 9z^2$

15. $f(x, y, z) = x^2 + y^2$

16. $f(x, y, z) = 4x^2 + 3y^2$

In Exercises 17 and 18 sketch the graph of each function by computing some level curves and sections.

17. $f(x, y) = |y|$

18. $f(x, y) = xy$

In Exercises 19 and 20 sketch the indicated level sets of the given function.

19. $f(x, y, z) = xy + yz, c = 0$

20. $f(x, y, z) = xy + z^2, c = 0$

In Exercises 21–28, sketch or describe the surface in $\mathbb{R}^3$ defined by the given equation.

21. $4x^2 + y^2 = 16$

22. $y^2 + z^2 = 4$

23. $z = x^2$

24. $\dfrac{x^2}{9} + \dfrac{y^2}{12} + \dfrac{z^2}{9} = 1$

25. $4x^2 - 3y^2 + 2z^2 = 0$

26. $\dfrac{y^2}{9} + \dfrac{z^2}{4} = 1 + \dfrac{x^2}{16}$

27. $z = \dfrac{y^2}{4} - \dfrac{x^2}{9}$

28. $x^2 + y^2 + z^2 + 4x - by + 9z - b = 0$, where b is a constant.

2.2
Partial
Derivatives
and
Continuity

In §**2.1** we considered a few methods for graphing functions. By these methods alone it may be impossible to grasp even the general features of a complicated function. From one-variable calculus we know that the derivative can greatly aid us in this task. For example, the derivative enables us to locate maxima and minima and to find slopes.

Building on the concepts of one-variable calculus, we expect that a continuous function is one that has no "breaks" in its graph, whereas a differentiable function from $\mathbb{R}^2$ to $\mathbb{R}$ is one with a well-defined plane tangent to the graph at each point. Thus, there must not be any sharp folds or peaks in the graph (see Figures 2.2.1 and 2.2.2). In other words, the graph must be *smooth*.

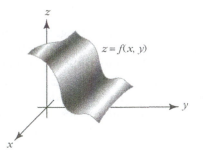

FIGURE 2.2.1. A smooth graph.

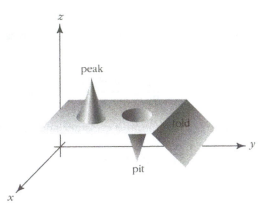

FIGURE 2.2.2. This graph is not smooth.

To determine by computation whether the graph of a function of several variables is smooth, we need a definition of what we mean by "$f(x_1, \ldots, x_n)$ is differentiable at $\mathbf{x} = (x_1, \ldots, x_n)$." Toward that end, we introduce in this section the notion of the ***partial derivative***.

Let us begin with the "bottom line": the calculation of partial derivatives.

Partial Differentiation

If f is a function of several variables, to calculate the **partial derivative** with respect to a certain variable, treat the remaining variables as constants and differentiate as usual by using the rules of one-variable calculus.

If $z = f(x, y)$ is a function of two variables, the two partial derivatives are denoted $\dfrac{\partial f}{\partial x}$ and $\dfrac{\partial f}{\partial y}$, or $\dfrac{\partial z}{\partial x}$ and $\dfrac{\partial z}{\partial y}$.

If $u = f(x, y, z)$ is a function of three variables, the partial derivatives are denoted $\dfrac{\partial f}{\partial x}, \dfrac{\partial f}{\partial y}$, and $\dfrac{\partial f}{\partial z}$, or $\dfrac{\partial u}{\partial x}, \dfrac{\partial u}{\partial y}$, and $\dfrac{\partial u}{\partial z}$. The symbol ∂, a modification of d, is called "del."

The notion of partial derivative really belongs to the calculus of functions of one variable. When one fixes the value of a variable, one is converting a function of two variables to a function of one variable. For example, if $f(x, y) = x^2y + y^3$ and we fix the value of y at $y = 2$, then $f(x, 2) = 2x^2 + 8$ is now a function of x alone. In computations of the partial derivatives of $f(x, y)$, as we shall presently see, one does not literally replace y (or x) by a constant, rather one just *thinks* of y as constant when computing $\partial f / \partial x$, and x as constant when computing $\partial f / \partial y$.

Example 1 If $f(x, y) = x^2y + y^3$, find $\partial f / \partial x$ and $\partial f / \partial y$.

Solution To find $\partial f / \partial x$ we think of y as a constant and differentiate with respect to x. This yields

$$\frac{\partial f}{\partial x} = \frac{d(x^2y + y^3)}{dx} = 2xy.$$

Similarly, to find $\partial f / \partial y$ we hold x constant and differentiate with respect to y:

$$\frac{\partial f}{\partial y} = \frac{d(x^2y + y^3)}{dy} = x^2 + 3y^2. \quad \blacklozenge$$

To indicate that a partial derivative is to be evaluated at a particular point (x_0, y_0), we write

$$\frac{\partial f}{\partial x}(x_0, y_0) \quad \text{or} \quad \frac{\partial f}{\partial x}\bigg|_{x=x_0, y=y_0} \quad \text{or} \quad \frac{\partial f}{\partial x}\bigg|_{x_0, y_0}.$$

We will also use some other notations for partial derivatives. Instead of $\partial f / \partial x$ (or $\partial f / \partial y$), we may write f_x (or f_y). If $z = f(x, y)$, we may replace the name of the function f by the dependent variable z and write $\partial z / \partial x$ or z_x for f_x.

Example 2   If $z = \cos xy + x \cos y = f(x, y)$, find the partial derivatives

$$\frac{\partial z}{\partial x}(x_0, y_0) \quad \text{and} \quad f_y\left(2, \frac{\pi}{2}\right).$$

Solution   First we fix y and differentiate with respect to x, giving

$$\frac{\partial z}{\partial x}(x, y) = \frac{d(\cos xy + x \cos y)}{dx} = -y \sin xy + \cos y.$$

Evaluating at (x_0, y_0) gives

$$\frac{\partial z}{\partial x}(x_0, y_0) = -y_0 \sin x_0 y_0 + \cos y_0.$$

Similarly, we fix x and differentiate with respect to y to obtain

$$f_y(x, y) = \frac{d(\cos xy + x \cos y)}{dy} = -x \sin xy - x \sin y.$$

At $(2, \pi/2)$, we get $f_y(2, \pi/2) = -2 \sin \pi - 2 \sin(\pi/2) = -2.$ ◆

The computation of partial derivatives of functions $f(x_1, \ldots, x_n)$ of n variables goes in a similar way. To find $\partial f / \partial x_1$, think of all variables except x_1 as constants and differentiate with respect to x_1, etc..

Example 3   If $f(x_1, x_2, x_3, x_4) = \sin(x_1 x_3) - \cos(x_2 x_4)$, find $\partial f / \partial x_3$.

Solution   Here $\partial f / \partial x_3 = x_1 \cos(x_1 x_3)$ by the chain rule.   ◆

As with the derivative of a function of one variable, there is a geometric interpretation of the partial derivative as the slope of a tangent line. Consider the graph of a function $f(x, y)$. Fix a point (x_0, y_0) and take planes through (x_0, y_0) parallel to the xz and yz planes, respectively. The intersection of these planes with the graph $z = f(x, y)$ results in two curves that intersect at $(x_0, y_0, f(x_0, y_0))$. In the plane parallel to the xz plane, the variable y is constant $(y = y_0)$. The equation of the curve of intersection in this plane is $z = f(x, y_0)$, and the slope of the tangent line at $x = x_0$ is

$$\left. \frac{d}{dx} f(x, y_0) \right|_{x=x_0} = \frac{\partial f}{\partial x}(x_0, y_0);$$

that is, the partial derivative of f with respect to x evaluated at the point (x_0, y_0). (See Figure 2.2.3.)

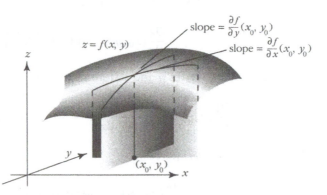

FIGURE 2.2.3. The partial derivative as the slope of a tangent line.

Similarly the slope of the tangent line to the curve of intersection in the plane parallel to the yz plane at (x_0, y_0) is $\partial f / \partial y(x_0, y_0)$.

In terms of limits, these partial derivatives can be written as

$$\frac{\partial f}{\partial x}(x_0, y_0) = \lim_{h \to 0} \frac{f(x_0 + h, y_0) - f(x_0, y_0)}{h}$$

and

$$\frac{\partial f}{\partial y}(x_0, y_0) = \lim_{k \to 0} \frac{f(x_0, y_0 + k) - f(x_0, y_0)}{k}.$$

The following example shows how partial derivatives, like derivatives of functions of one variable, may also be interpreted as rates of change.

Example 4 The temperature (in degrees Celsius) near Dawson Creek at noon on April 14, 1901, was given by $T = -(0.0003)x^2y + (0.9307)y$, where x and y are the latitude and longitude (in degrees), respectively. At what rate would the temperature be changing if one were to go directly north? (The latitude and longitude of Dawson Creek are $x = 55.7°$ and $y = 120.2°$, respectively.)

Solution   Proceeding directly north means increasing the latitude x and, since there is no east–west movement, y stays constant. Thus we calculate

$$\frac{\partial T}{\partial x} = -(0.0003) \cdot 2xy = -(0.0003) \cdot 2 \cdot (55.7) \cdot (120.2) \approx -4.017.$$

Thus the temperature drops as we proceed north from Dawson Creek, at the instantaneous rate of 4.017°C per degree of latitude .   ◆

A physical interpretation of partial derivatives is illustrated by the example of wave motion. Consider water in motion in a narrow tank, as illustrated in Figure 2.2.4. We will assume that the motion of the water is gentle enough so that, at any instant of time, the height z of the water above the bottom of the tank is a function of the position y measured along the long direction of the tank (this means that there are no "breaking waves"), and that the height of water is constant along the short direction of the tank. Since the water is in motion, the height z depends on the time as well as on y, so we may write $z = f(t, y)$; the domain of the function f consists of all pairs (t, y) such that t lies in the interval of time relevant for the experiment, and $a \leq y \leq b$, where a and b mark the ends of the tank.

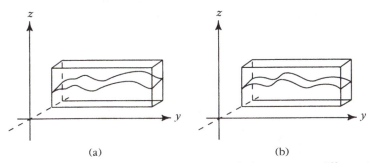

(a)   (b)

FIGURE 2.2.4. Moving water in a narrow tank shown at two different instants of time.

We can graph the entire function f as a surface in (t, y, z) space lying over the strip $a \leq y \leq b$ (see Figure 2.2.5); the section of this surface cut by a plane of the form $t = t_0$ is a curve that shows the configuration of the water at the moment t_0 (such as each of the "snapshots" in Figure 2.2.4). This curve is the graph of a function of *one* variable, $z = g(y)$, where g is defined by $g(y) = f(t_0, y)$. If we take the derivative of the function g at a point y_0 in (a, b), we get a number $g'(y_0)$ which represents the slope of the water's surface at the time t_0 and at the location y_0. (See Figure 2.2.6.) It could be observed as the slope of a small stick parallel to the sides of the tank floating on the water at that time and position.

Here is a summary of the steps we took:

1. Fix t at the value t_0.

2. Differentiate the resulting function of y.

3. Set y equal to y_0.

The number $g'(y_0)$ is the partial derivative of f with respect to y at the point y_0 at time t_0.

We can also define the partial derivative of f with respect to t at (t_0, y_0), the derivative is obtained by:

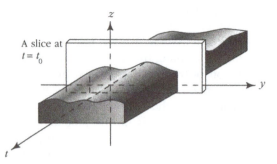

FIGURE 2.2.5. The motion of the water is depicted by a graph in (t, y, z) space; sections by planes of the form $t = t_0$ show the configuration of the water at various instants of time.

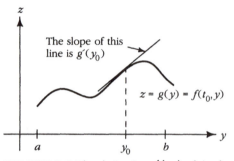

FIGURE 2.2.6. The derivative $g'(y_0)$ of the function $g(y) = f(t_0, y)$ represents the slope of the water's surface at time t_0 and position y_0.

1. Fixing y at the value y_0.

2. Differentiating the resulting function of t.

3. Setting t equal to t_0.

The result is $f_t(t_0, y_0)$. In the first step, we obtain the function $h(t) = f(t, y_0)$, which represents the vertical motion of the water's surface observed at the fixed position y_0. The derivative with respect to t is, therefore, the *vertical velocity* of the surface at the position y_0. It could be observed as the vertical velocity of a cork floating on the water at that position. Finally, setting t equal to t_0 merely involves observing the velocity at the specific time t_0.

This concludes our introduction to partial derivatives. The second main topic of this section is continuity. As in one-variable calculus, continuity depends on the limit concept (this time limits are taken in several variables at once), so we begin by defining limits, starting with functions of two variables.

Limits—Intuitive Approach

Let $f(x, y)$ be defined for all (x, y) *near* (x_0, y_0), but f need not be defined *at* (x_0, y_0). We say $f(x, y)$ ***has a limit*** l as (x, y) approaches (x_0, y_0) and write

$$\lim_{(x,y)\to(x_0,y_0)} f(x, y) = l$$

provided $f(x, y)$ becomes arbitrarily close to l when (x, y) is close enough to (x_0, y_0).

Although this definition seems straightforward, flaws are revealed when one attempts to implement the definition. For instance, what *exactly* does "becomes arbitrarily close" mean? We will clarify the intuitive definition below, but for calculating examples and getting on with the theory of calculus, it is sufficient.

In one-variable calculus one learns rules of limits, such as the sum and product rules. Similar rules apply in multivariable calculus. For example, the ***sum*** and ***product rules*** are:

$$\lim_{(x,y)\to(x_0,y_0)} [f(x, y) + g(x, y)] = \lim_{(x,y)\to(x_0,y_0)} f(x, y) + \lim_{(x,y)\to(x_0,y_0)} g(x, y)$$

and

$$\lim_{(x,y)\to(x_0,y_0)} [f(x, y)g(x, y)] = \left[\lim_{(x,y)\to(x_0,y_0)} f(x, y) \right] \left[\lim_{(x,y)\to(x_0,y_0)} g(x, y) \right],$$

respectively. Also, some limits, such as

$$\lim_{(x,y)\to(x_0,y_0)} x = x_0$$

are obvious. We shall accept these rules here, though, strictly speaking, they should be proved using the precise definition given later in this section.

The following three examples demonstrate the use of the limit laws for functions of two variables. We also give tables of values to illustrate the behavior of the functions.

Example 5   Let $f(x, y) = x^2 + y^2 + 2$. Compute $\lim_{(x,y)\to(0,0)} f(x, y)$.

Solution   Here f is the sum of three functions x^2, y^2, and 2. The limit of a sum is the sum of the limits, and the limit of a product is the product of the limits. Using these and the fact that $\lim_{(x,y)\to(x_0,y_0)} x = x_0$, we obtain

$$\lim_{(x,y)\to(x_0,y_0)} x^2 = \left(\lim_{(x,y)\to(x_0,y_0)} x \right) \left(\lim_{(x,y)\to(x_0,y_0)} x \right) = x_0^2$$

and, using the same reasoning, $\lim_{(x,y)\to(x_0,y_0)} y^2 = y_0^2$. Consequently,

$$\lim_{(x,y)\to(0,0)} f(x,y) = 0^2 + 0^2 + 2 = 2. \blacklozenge$$

A table of values for this example is:

	0.1	2.01	2.01000001	2.010001	2.0101	2.02
y	0.01	2.0001	2.00010001	2.000101	2.0002	2.0101
	0.001	2.000001	2.00000101	2.000002	2.000101	2.010001
	0.0001	2.00000001	2.00000002	2.00000101	2.00010001	2.01000001
	0		2.00000001	2.000001	2.0001	2.01
		0	0.0001	0.001	0.01	0.1

$$x$$

This table shows values of $f(x,y) = x^2 + y^2 + 2$ for (x,y) in the first quadrant. One sees clearly that the values approach 2 as x and y approach zero.

Example 6 Find $\displaystyle\lim_{(x,y)\to(0,0)} \frac{x^3 + 2x^2 + xy^2 + 2y^2}{x^2 + y^2}$.

Solution   The function is defined for all $(x,y) \neq (0,0)$. The numerator and denominator vanish when $(x,y) = (0,0)$. The numerator may be factored as $(x^2 + y^2)(x + 2)$, so

$$\lim_{(x,y)\to(0,0)} \frac{(x^2 + y^2)(x + 2)}{x^2 + y^2} = \lim_{(x,y)\to(0,0)} (x + 2) = 0 + 2 = 2. \quad \blacklozenge$$

Example 7 Show that the following limit does not exist:

$$\lim_{(x,y)\to(0,0)} \frac{\partial}{\partial x} \sqrt{x^2 + y^2}.$$

Solution   We have $(\partial/\partial x)\sqrt{x^2 + y^2} = x/\sqrt{x^2 + y^2}$ if $(x,y) \neq (0,0)$. Thus

$$\lim_{(x,y)\to(0,0)} \frac{\partial}{\partial x} \sqrt{x^2 + y^2} = \lim_{(x,y)\to(0,0)} \frac{x}{\sqrt{x^2 + y^2}}.$$

If we approach $(0,0)$ on the y axis, that is, along points $(0,y)$, we get zero. Thus the limit, if it exists, is zero. On the other hand, if we approach $(0,0)$ along the positive x axis, we have $y = 0$ and $x > 0$; then $x/\sqrt{x^2 + y^2} = 1$ because $x/\sqrt{x^2} = 1$, so the limit is 1. Since we obtain different answers in different directions, the limit cannot exist. $\blacklozenge$

A table of values for this example is as follows:

	0.1	0	0.00099	0.0099	0.0995	0.707
	0.01	0	0.0099	0.0995	0.707	0.995
y	0.001	0	0.0995	0.707	0.995	0.99995
	0.0001	0	0.707	0.995	0.99995	0.9999995
	0	1	1	1	1	
		0	0.0001	0.001	0.01	0.1

$$x$$

This table shows values of $f(x,y) = x/\sqrt{x^2 + y^2}$ for (x,y) in the first quadrant. The value of the function depends on the ratio of y to x and varies all the way from 0 to 1, no matter how close (x,y) comes to $(0,0)$. [With some hindsight, we may notice that $f(x,y)$ can be rewritten as $1/\sqrt{1 + (y/x)^2}$ when x is positive.]

The preceding example illustrates an important fact about limits: when the limit $\lim_{(x,y)\to(x_0,y_0)} f(x,y) = L$, the value of $f(x,y)$ must approach L when (x,y) approaches (x_0, y_0) in *any* direction.

In particular:

Double and Single Limits in Two Variables

If $\lim_{(x,y)\to(x_0,y_0)} f(x,y) = L$, then the two one-variable limits $\lim_{x\to x_0} f(x, y_0)$ and $\lim_{y\to y_0} f(x_0, y)$ *both* must equal L.

Note

The technical definition of limit comes next. It took mathematicians themselves several hundred years to develop this idea. Even with this accumulated experience it is natural to expect that it would take most people several readings and some struggling to master the subject of limits.

Write $d((x,y),(x_0,y_0)) = \sqrt{(x - x_0)^2 + (y - y_0)^2}$ for the distance between (x,y) and (x_0, y_0), with a similar notation $d((x,y,z),(x_0,y_0,z_0))$ in space. The *disk* $D_r(x_0, y_0)$ of radius r centered at (x_0, y_0) is, by definition, the set of all (x,y) such that $d((x,y),(x_0,y_0)) < r$, as shown in Figure 2.2.7. The limit concept now can be defined by the same ε, δ technique as in one-variable calculus.

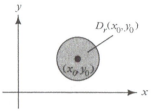

FIGURE 2.2.7. The disk $D_r(x_0, y_0)$ consists of the shaded region (excluding the solid circle).

The ε, δ Definition of Limit

Suppose that f is defined on a set A such that every disk about (x_0, y_0) intersects A in at least one point other than (x_0, y_0) . We write

$$\lim_{(x,y)\to(x_0,y_0)} f(x, y) = l$$

if, for every $\varepsilon > 0$, there is a $\delta > 0$ such that $|f(x, y) - l| < \varepsilon$ whenever $0 < d((x, y), (x_0, y_0)) < \delta$. A similar definition is made for functions of three or n variables and for vector valued functions. Note that this definition does not require that (x_0, y_0) be in A.

The ε, δ definition of a limit may be rephrased as follows: for every $\varepsilon > 0$, there is a $\delta > 0$ such that $|f(x, y) - l| < \varepsilon$ if $(x, y) \neq (x_0, y_0)$ and lies in $D_\delta(x_0, y_0)$ and in A.

Example 8   Prove the "obvious" limit, $\lim_{(x,y)\to(x_0,y_0)} x = x_0$, using ε's and δ's.

Solution   Let $\varepsilon > 0$ be given and let $f(x, y) = x$ and $l = x_0$. Here A is $\mathbb{R}^2$, the set of all pairs of real numbers. We must find a $\delta > 0$ such that $|f(x, y) - l| < \varepsilon$ whenever $d((x, y), (x_0, y_0)) < \delta$, that is, such that $|x - x_0| < \varepsilon$ whenever $\sqrt{(x - x_0)^2 + (y - y_0)^2} < \delta$. However, note that

$$|x - x_0| = \sqrt{(x - x_0)^2} \leq \sqrt{(x - x_0)^2 + (y - y_0)^2},$$

so if we choose $\delta = \varepsilon$, then $d((x, y), (x_0, y_0)) < \delta$ will imply $|x - x_0| < \varepsilon$. ◆

Example 9   Prove that

$$\lim_{(x,y)\to(0,0)} \frac{2x^2 y}{x^2 + y^2} = 0.$$

Solution   We want to show, that $2x^2y/(x^2 + y^2)$ is small (close to 0) when x and y are small but both not zero. For this purpose, we notice the inequality[1]

$$\left|\frac{2x^2y}{x^2 + y^2}\right| \le \left|\frac{2x^2y}{x^2}\right| = |2y|.$$

Given $\varepsilon > 0$, if we choose $\delta = \frac{1}{2}\varepsilon$, then $0 < \|(x,y) - (0,0)\| = \sqrt{x^2 + y^2} < \delta$ implies $|y| < \delta$. With this choice

$$\left|\frac{2x^2y}{x^2 + y^2} - 0\right| < \varepsilon,$$

so that the definition of limit is fulfilled. This is confirmed by the (computer-generated) graph in Figure 2.2.8.   ◆

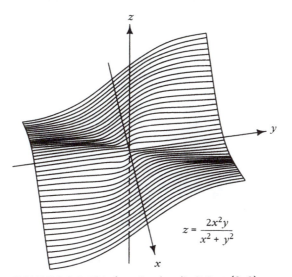

FIGURE 2.2.8. This function has limit 0 at $(0,0)$.

We can base the concept of continuity on that of limits, just as one does in one-variable calculus.

---

[1]This inequality $|2x^2y/(x^2 + y^2)| \le |2y|$ is derived assuming $x \ne 0$, but it is clearly valid if $x = 0$ and $y \ne 0$, as well.

Definition of Continuity

Let f be defined for all points *near* (x_0, y_0) including at (x_0, y_0). We say that f is **continuous** at (x_0, y_0) if

$$\lim_{(x,y)\to(x_0,y_0)} f(x,y) = f(x_0, y_0).$$

There is a similar definition for functions of several variables and vector valued functions.

Since the condition $\lim_{\mathbf{x}\to\mathbf{x}_0} f(\mathbf{x}) = f(\mathbf{x}_0)$ means that $f(\mathbf{x})$ is close to $f(\mathbf{x}_0)$ when $\mathbf{x}$ is close to $\mathbf{x}_0$, we see that our definition does indeed correspond to the idea that the graph of f be unbroken. Let us, for ease of visualization, continue to deal with real-valued functions, say $f : \mathbb{R}^2 \to \mathbb{R}$. In this case we can visualize f by drawing its graph, which consists of all points (x, y, z) in $\mathbb{R}^3$ with $z = f(x, y)$. The continuity of f thus means that its graph has no "breaks" in it (see Figure 2.2.9).

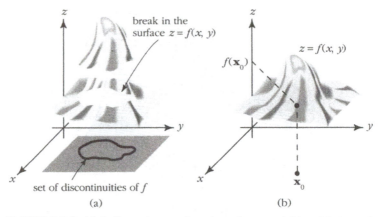

FIGURE 2.2.9. (a) A discontinuous function of two variables. (b) A continuous function.

Example 10 Show that the function

$$f(x,y) = \begin{cases} \dfrac{x^2 y}{x^2 + y^2} & \text{if } (x,y) \neq (0,0), \\ 0, & \text{if } (x,y) = (0,0), \end{cases}$$

is continuous at $(0,0)$.

Solution By Example 9, $\lim_{(x,y)\to(0,0)} f(x,y) = 0 = f(0,0)$, so the conditions of continuity at $(0,0)$ are fulfilled and f is continuous at this point. ◆

A basic operation on functions, called composition, is useful for a variety of purposes, as the reader knows from the chain rule of one-variable calculus.

Composition

Let g map A to B and f map B to C, as in Figure 2.2.10. The ***composition*** $f \circ g$ maps A to C and is defined by

$$(f \circ g)(\mathbf{x}) = f(g(\mathbf{x})).$$

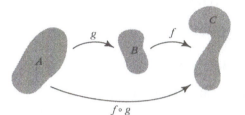

FIGURE 2.2.10. The composition of f and g.

For example, $\sin(x^2 + y^2)$ is the composition of $g(x, y) = x^2 + y^2$ with $f(u) = \sin u$. As in one-variable calculus it is convenient to let another letter like u denote the intermediate variable; here $u = x^2 + y^2$.

Compositions are important because they show how complicated functions are "built up" from simpler functions. Compositions of continuous functions turn out to be continuous and the composition of differentiable functions is also differentiable. The chain rule (see the shaded box in §**2.4**) shows how to differentiate such functions.

Here is the result for continuity, stated precisely.

Continuity and Composition

Let $g : A \subset \mathbb{R}^n \to \mathbb{R}^m$ and let $f : B \subset \mathbb{R}^m \to \mathbb{R}^p$. Suppose that $g(A) \subset B$, so that $f \circ g$ is defined on A. If g is continuous at $\mathbf{x}_0 \in A$ and f is continuous at $g(\mathbf{x}_0)$, then $f \circ g$ is continuous at $\mathbf{x}_0$.

We omit the formal proof, but here is the idea: we must show that as $\mathbf{x}$ gets close to $\mathbf{x}_0$, $f(g(\mathbf{x}))$ gets close to $f(g(\mathbf{x}_0))$. But as $\mathbf{x}$ gets close to $\mathbf{x}_0$, $g(\mathbf{x})$ gets close to $g(\mathbf{x}_0)$ (by continuity of g at $\mathbf{x}_0$); and as $g(\mathbf{x})$ gets close to $g(\mathbf{x}_0)$, $f(g(\mathbf{x}))$ gets close to $f(g(\mathbf{x}_0))$ (by continuity of f at $g(\mathbf{x}_0)$).

Example 11   Let $f(x, y, z) = (x^2 + y^2 + z^2)^{30} + \sin z^3$. Show that f is continuous.

Solution   Since f is the sum of the two functions $(x^2 + y^2 + z^2)^{30}$ and $\sin z^3$, it suffices to show that each is continuous. The first is the composite of $g(x, y, z) = (x^2 + y^2 + z^2)$ with $f(u) = u^{30}$, and the second is the composite of $h(x, y, z) = z^3$ with $k(u) = \sin u$, and so we have continuity by the preceding box. ◆

Exercises for §2.2

Compute $\partial f/\partial x$ and $\partial f/\partial y$ for the functions in Exercises 1–2, and compute $\partial z/\partial x$ and $\partial z/\partial y$ in Exercises 3–4; evaluate at the indicated points.

1. $f(x, y) = xy$; $(1, 1)$

2. $f(x, y) = \dfrac{x}{y}$; $(1, 1)$

3. $z = 3x^2 + 2y^2$; $(1, 2)$

4. $z = \sin(x^2 - 3xy)$; $(\sqrt{\pi}, \sqrt{\pi})$

Compute f_x, f_y, and f_z for the functions in Exercises 5–8, and evaluate them at the indicated points.

5. $f(x, y, z) = xyz$; $(1, 1, 1)$

6. $f(x, y, z) = \sqrt{x^2 + y^2 + z^2}$; $(3, 0, 4)$

7. $f(x, y, z) = \cos(xy^2) + e^{3xyz}$; $(\pi, 1, 1)$

8. $f(x, y, z) = x^{yz}$; $(1, 1, 0)$

Find the partial derivatives $\partial u/\partial x, \partial u/\partial y, \partial u/\partial z$ in Exercises 9–12.

9. $u = e^{xyz}(xy + xz + yz)$

10. $u = \sin(xy^2 z^3)$

11. $u = e^x \cos(yz^2)$

12. $u = (xy^3 + c^z)/(x^3y - c^z)$; c is a constant.

In Exercises 13–16, compute the indicated partial derivatives.

13. $\dfrac{\partial}{\partial s} e^{stu^2}$

14. $\dfrac{\partial}{\partial r}\left(\dfrac{1}{3}\pi r^2 h\right)$

15. $\dfrac{\partial}{\partial \lambda} \left(\dfrac{\cos \lambda \mu}{1 + \lambda^2 + \mu^2} \right)$

16. $\dfrac{\partial}{\partial a}(bcd)$

17. In the situation of Example 4, how fast is the temperature changing if we proceed directly west? (The longitude y is increasing as we go west.)

18. If three resistors with resistances R_1, R_2, and R_3 are connected in parallel, the total electrical resistance is determined by the equation

$$\frac{1}{R} = \frac{1}{R_1} + \frac{1}{R_2} + \frac{1}{R_3}.$$

   (a) What is $\partial R / \partial R_1$?

   (b) Suppose that R_1, R_2, and R_3 are variables whose values are currently 100, 200, and 300 ohms, respectively. How fast is R changing with respect to R_1?

In Exercises 19–24, compute the given limits if they exist (do not attempt a precise justification).

19. $\displaystyle\lim_{(x,y)\to(0,0)} \frac{4x^2 + 3y^2 + x^3y^3}{x^2 + y^2 + x^4y^4}$

20. $\displaystyle\lim_{(x,y)\to(0,0)} \frac{x^2y + y^3}{x^2 + y^2}$

21. $\displaystyle\lim_{(x,y)\to(0,1)} e^x y$

22. $\displaystyle\lim_{(x,y)\to(0,0)} \frac{e^{xy}}{x + 1}$

23. $\displaystyle\lim_{(x,y)\to(0,1)} e^{xy} \cos(\pi xy)$

24. $\displaystyle\lim_{(x,y)\to(1,1)} e^x \cos(\pi y)$

25. Prove that there is a number $\delta > 0$ such that whenever $|a| < \delta$, we have $|a^3 + 3a^2 + a| < 1/100$.

26. Prove that there is a number $\delta > 0$ such that if $x^2 + y^2 < \delta^2$, then

$$|x^2 + y^2 + 3xy + 180xy^5| < \frac{1}{10,000}.$$

27. Suppose that $\mathbf{x}$ and $\mathbf{y}$ are in $\mathbb{R}^n$ and $\mathbf{x} \neq \mathbf{y}$. Show that there is a continuous function $f : \mathbb{R}^n \to \mathbb{R}$ with $f(\mathbf{x}) = 1, f(\mathbf{y}) = 0$, and $0 \leq f(\mathbf{z}) \leq 1$ for every $\mathbf{z}$ in $\mathbb{R}^n$.

28. Let $f : A \subset \mathbb{R}^n \to \mathbb{R}$ be given and let $\mathbf{x}_0$ be such that f is defined for all x near $x_0, x \neq x_0$. We say that $\lim_{x \to x_0} f(\mathbf{x}) = \infty$ if for every $N > 0$ there is a $\delta > 0$ such that $0 < \| \mathbf{x} - \mathbf{x}_0 \| < \delta$ implies $f(\mathbf{x}) > N$.

  (a) Prove that $\lim_{x \to 1} (x - 1)^{-2} = \infty$.

  (b) Prove that $\lim_{x \to 0} (1/|x|) = \infty$. Is it true that $\lim_{x \to 0} 1/x = \infty$?

  (c) Prove that $\lim_{(x,y) \to (0,0)} 1/(x^2 + y^2) = \infty$.

29. Show that $f(x, y) = ye^x + \sin x + (xy)^4$ is continuous.

30. Show that $f(x, y) = \dfrac{x + y}{x - y}$ is continuous at (1,2).

31. Find
$$\lim_{\Delta y \to 0} \frac{3 + (x + y + \Delta y)^2 z - (3 + (x + y)^2 z)}{\Delta y}.$$

32. Can $\dfrac{\sin (x + y)}{x + y}$ be made continuous by suitably defining it at (0,0)?

33. Can $\dfrac{xy}{x^2 + y^2}$ be made continuous by suitably defining it at (0,0)?

34.(a) Use L'Hôpital's rule from one-variable calculus to calculate the limit
$$\lim_{x \to 0} \frac{\sin 2x - 2x}{x^3}.$$

  (b) Does $\displaystyle\lim_{(x,y) \to (0,0)} \dfrac{\sin 2x - 2x + y}{x^3 + y}$ exist?

2.3 Differentiability, the Derivative Matrix, and Tangent Planes

Now that we have some experience with partial derivatives, we can assemble them to form a matrix called the *derivative matrix*. This matrix, which is a useful computational tool, will also allow us to make a clear statement of the chain rule in terms of matrix multiplication.

In this section, we also study the related notions of tangent planes and linear approximations. To use these concepts, we must require that our functions have a property called *differentiability*, which is more restrictive than the existence of partial derivatives.

Why do we need a more restrictive definition of differentiability for functions of more than one variable? Why not say that a function is differentiable if its partial derivatives exist? There is one basic reason: For functions of one variable, differentiable functions are continuous, and the chain rule holds for the composition of differentiable functions. We must choose our definition so that these and other fundamental properties hold for functions of several variables as well.

In Example 4 below we shall see that a function of two variables may have both partial derivatives exist at a point but still not be continuous at that point. Thus *if* we took differentiability to mean the existence of partial derivatives, then differentiability would not imply continuity.

Note

The correct definition of differentiability for functions of two and more variables took mathematicians many years to establish. Don't be surprised, then, if it takes you some time to understand it.

We introduce the definition of differentiability by computing the equation of the tangent plane to the graph $z = f(x, y)$ at a point (x_0, y_0). See Figure 2.3.1.

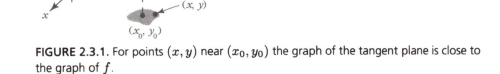

FIGURE 2.3.1. For points (x, y) near (x_0, y_0) the graph of the tangent plane is close to the graph of f.

We begin with the fact that a (nonvertical) plane passing through the point (x_0, y_0, z_0) is the graph of a linear function:

$$z = z_0 + a(x - x_0) + b(y - y_0) = g(x, y).$$

Since (x_0, y_0, z_0) lies on the graph of f, $z_0 = f(x_0, y_0)$ as in the figure. We need to determine the constants a and b. In Figure 2.2.3 we learned that the partial derivatives represent the slopes of curves in the graph. Therefore, from Figure 2.3.1, if the plane is tangent to the graph of f, then it is reasonable that the slopes of corresponding curves in the plane and the graph coincide at (x_0, y_0). This requirement gives

$$\frac{\partial g}{\partial x}(x_0, y_0) = a = \frac{\partial f}{\partial x}(x_0, y_0) \quad \text{and} \quad \frac{\partial g}{\partial y}(x_0, y_0) = b = \frac{\partial f}{\partial y}(x_0, y_0).$$

Hence we arrive at:

Tangent Plane to a Graph

The tangent plane to the graph $z = f(x, y)$ at the point (x_0, y_0) has the normal

$$\left(-\frac{\partial f}{\partial x}(x_0, y_0), -\frac{\partial f}{\partial y}(x_0, y_0), 1 \right)$$

and the equation

$$z = f(x_0, y_0) + \left[\frac{\partial f}{\partial x}(x_0, y_0) \right] (x - x_0) + \left[\frac{\partial f}{\partial y}(x_0, y_0) \right] (y - y_0).$$

Example 1   Find the plane tangent to the graph $z = x^2 + y^4 + e^{xy}$ at the point $(1, 0, 2)$.

Solution   We use the formula in the box above, with $x_0 = 1, y_0 = 0$, and $z_0 = f(x_0, y_0) = 2$. The partial derivatives are

$$\frac{\partial f}{\partial x} = 2x + ye^{xy} \quad \text{and} \quad \frac{\partial f}{\partial y} = 4y^3 + xe^{xy}.$$

At $(1, 0, 2)$, they are 2 and 1, respectively. Thus the tangent plane is

$$z = 2(x - 1) + 1(y - 0) + 2, \quad i.e., \quad z = 2x + y. \quad \blacklozenge$$

Example 2   Find the equation of the plane tangent to the hemisphere $z = \sqrt{1 - x^2 - y^2}$ at a point (x_0, y_0). Interpret your result geometrically.

Solution   Letting $f(x, y) = \sqrt{1 - x^2 - y^2}$, we calculate that

$$\frac{\partial f}{\partial x} = f_x = -\frac{x}{\sqrt{1 - x^2 - y^2}} \quad \text{and} \quad \frac{\partial f}{\partial y} = f_y = -\frac{y}{\sqrt{1 - x^2 - y^2}}.$$

The equation of the tangent plane at (x_0, y_0, z_0) is therefore

$$z = \sqrt{1 - x_0^2 - y_0^2} - \frac{x_0}{\sqrt{1 - x_0^2 - y_0^2}} (x - x_0) - \frac{y_0}{\sqrt{1 - x_0^2 - y_0^2}} (y - y_0),$$

or

$$z = z_0 - \frac{x_0}{z_0}(x - x_0) - \frac{y_0}{z_0}(y - y_0).$$

A normal vector to this plane is $(x_0/z_0)\mathbf{i} + (y_0/z_0)\mathbf{j} + \mathbf{k}$. Multiplying by z_0, we find that another normal vector is $x_0\mathbf{i} + y_0\mathbf{j} + z_0\mathbf{k}$. Thus we have recovered the geometric property that the tangent plane at a point P of a sphere is perpendicular to the vector from the center of the sphere to P. ◆

For functions of one variable, the tangent line is the graph of a "good approximation" to the function. To recall the argument behind this statement, we begin with the definition of the derivative:

$$\lim_{\Delta x \to 0} \frac{f(x_0 + \Delta x) - f(x_0)}{\Delta x} = f'(x_0).$$

Let $x = x_0 + \Delta x$ and rewrite this as

$$\lim_{x \to x_0} \frac{f(x) - f(x_0)}{x - x_0} = f'(x_0).$$

Using the "trivial limit" $\lim_{x \to x_0} f'(x_0) = f'(x_0)$, and the difference rule, we can rewrite the preceding equation as

$$\lim_{x \to x_0} \left[\frac{f(x) - f(x_0)}{x - x_0} - f'(x_0) \right] = f'(x_0) - f'(x_0) = 0$$

or

$$\lim_{x \to x_0} \frac{f(x) - f(x_0) - f'(x_0)(x - x_0)}{x - x_0} = 0.$$

Thus the tangent line l through $(x_0, f(x_0))$ with slope $f'(x_0)$ is close to f in the sense that the difference between $f(x)$ and $l(x) = f(x_0) + f'(x_0)(x - x_0)$ goes to zero *even* when divided by $x - x_0$ as x goes to x_0. This is the notation of a "good approximation" that we will adapt to functions of several variables for our definition of differentiability, with the tangent line replaced by the tangent plane.

Differentiability

Let $f : \mathbb{R}^2 \to \mathbb{R}$. We say f is **differentiable at** (x_0, y_0), if $\partial f / \partial x$ and $\partial f / \partial y$ exist at (x_0, y_0) and if

$$\frac{f(x,y) - f(x_0, y_0) - \left[\frac{\partial f}{\partial x}(x_0, y_0)\right](x - x_0) - \left[\frac{\partial f}{\partial y}(x_0, y_0)\right](y - y_0)}{\|(x,y) - (x_0, y_0)\|} \to 0$$

as $(x,y) \to (x_0, y_0)$. This equation expresses what we mean by saying that

$$f(x_0, y_0) + \left[\frac{\partial f}{\partial x}(x_0, y_0)\right](x - x_0) + \left[\frac{\partial f}{\partial y}(x_0, y_0)\right](y - y_0)$$

is a **good approximation** to the function f near (x_0, y_0).

Shortly, we will give simple tests that determine whether or not a given function is differentiable.

Sometimes one says that the function

$$l(x,y) = f(x_0, y_0) + f_x(x_0, y_0)(x - x_0) + f_y(x_0, y_0)(y - y_0)$$

is the *linear approximation* to f near (x_0, y_0). (This is common "abuse of language" in calculus—l is really a linear function plus a constant—but we will not fuss with the distinction.) Likewise, for a function of three variables, the linear approximation at (x_0, y_0, z_0) to $f(x,y,z)$ is the function:

$$\begin{aligned} l(x,y,z) = \ & f(x_0, y_0, z_0) + f_x(x_0, y_0, z_0)(x - x_0) \\ & + f_y(x_0, y_0, z_0)(y - y_0) + f_z(x_0, y_0, z_0)(z - z_0). \end{aligned}$$

Just as in one-variable calculus, we can use the linear approximation to do approximate numerical computations. We will see some specific numerical examples in §**3.2**.

We now rewrite the definition of differentiability of $f(x,y)$ in matrix notation. To do so, we first recall from §**1.5** that, to multiply elements of $\mathbb{R}^2$ on the left by matrices, we represent the elements of $\mathbb{R}^2$ as 2×1 *column* matrices. We let $\mathbf{D}f(x_0, y_0)$ stand for the *row* matrix

$$\left[\frac{\partial f}{\partial x}(x_0, y_0) \quad \frac{\partial f}{\partial y}(x_0, y_0)\right],$$

so that the definition of differentiability asserts that

$$f(x_0, y_0) \quad + \quad \mathbf{D}f(x_0, y_0)\begin{bmatrix} x - x_0 \\ y - y_0 \end{bmatrix}$$

$$= f(x_0, y_0) + \left[\frac{\partial f}{\partial x}(x_0, y_0) \right] (x - x_0) + \left[\frac{\partial f}{\partial y}(x_0, y_0) \right] (y - y_0)$$

is a good approximation to f near (x_0, y_0). [As above, "good" is taken in the sense that this expression differs from $f(x, y)$ by something small times $\sqrt{(x - x_0)^2 + (y - y_0)^2}$.]

Now we are ready for the general definition of differentiability of maps from $\mathbb{R}^n$ to $\mathbb{R}^m$. Let f map a set U in $\mathbb{R}^n$ to $\mathbb{R}^m$ and $\mathbf{x}_0$ be in U. Write $f(\mathbf{x}) = (f_1(\mathbf{x}), f_2(\mathbf{x}), \ldots, f_m(\mathbf{x}))$ where each f_j maps U to $\mathbb{R}$, so the f_j are the "components" of f. Assume that U contains some disk about $\mathbf{x}_0$; that is, there is a number $r > 0$ such that all $\mathbf{x}$ with $\|\mathbf{x} - \mathbf{x}_0\| < r$ lie in U.

The Derivative

The **derivative matrix** of f at $\mathbf{x}_0$ is the matrix $\mathbf{D}f(\mathbf{x}_0)$ whose ijth entry is $\partial f_i / \partial x_j$ evaluated at $\mathbf{x}_0$. We say that f is **differentiable** at $\mathbf{x}_0$ if the partial derivatives of each f_i exist at $\mathbf{x}_0$ and if

$$\lim_{\mathbf{x} \to \mathbf{x}_0} \frac{\|f(\mathbf{x}) - f(\mathbf{x}_0) - \mathbf{D}f(\mathbf{x}_0)(\mathbf{x} - \mathbf{x}_0)\|}{\|\mathbf{x} - \mathbf{x}_0\|} = 0.$$

Again, $\mathbf{x} - \mathbf{x}_0$ is considered as a column matrix so it can be multiplied by $\mathbf{D}f(\mathbf{x}_0)$.

If $m = 1$, so $f : U \subset \mathbb{R}^n \to \mathbb{R}$,

$$\mathbf{D}f(\mathbf{x}_0) = \left[\frac{\partial f}{\partial x_1}(\mathbf{x}_0) \cdots \frac{\partial f}{\partial x_n}(\mathbf{x}_0) \right].$$

For the general case, where f maps a subset of $\mathbb{R}^n$ to $\mathbb{R}^m$, $f(x_1, \ldots, x_n) = (f_1(x_1, \ldots, x_n), \ldots, f_m(x_1, \ldots, x_n))$ the derivative matrix is the $m \times n$ matrix

$$\mathbf{D}f(\mathbf{x}_0) = \begin{bmatrix} \dfrac{\partial f_1}{\partial x_1} & \cdots & \dfrac{\partial f_1}{\partial x_n} \\ \vdots & & \vdots \\ \dfrac{\partial f_m}{\partial x_1} & \cdots & \dfrac{\partial f_m}{\partial x_n} \end{bmatrix},$$

where $\partial f_i / \partial x_j$ is evaluated at $\mathbf{x}_0$. Sometimes we want to emphasize $\mathbf{x}_0$ as a variable and so drop the subscript 0. In that case, $\mathbf{D}f(\mathbf{x})$ is the derivative matrix at $\mathbf{x}$. The derivative matrix is sometimes called the *Jacobian matrix*.

Example 3   Calculate the matrices of partial derivatives for (a) $f(x, y) = (e^{x+y} + y, y^2 x)$, (b) $f(x, y) = (x^2 + \cos y, y e^x)$ and (c) $f(x, y, z) = (z e^x, -y e^z)$.

Solution   (a) Here $f : \mathbb{R}^2 \to \mathbb{R}^2$ is defined by $f_1(x, y) = e^{x+y} + y$ and $f_2(x, y) = y^2 x$. Hence $\mathbf{D}f(x, y)$ is the 2×2 matrix

$$\mathbf{D}f(x, y) = \begin{bmatrix} e^{x+y} & e^{x+y} + 1 \\ y^2 & 2xy \end{bmatrix}.$$

(b) We have

$$\mathbf{D}f(x, y) = \begin{bmatrix} 2x & -\sin y \\ y e^x & e^x \end{bmatrix}.$$

(c) Here

$$\mathbf{D}f(x, y, z) = \begin{bmatrix} z e^x & 0 & e^x \\ 0 & -e^z & -y e^z \end{bmatrix}. \quad \blacklozenge$$

In one-variable calculus we learned that a differentiable function is continuous. The proof is simple:

$$\begin{aligned} \lim_{x \to x_0} f(x) &= \lim_{x \to x_0} [(f(x) - f(x_0)) + f(x_0)] \\ &= \left\{ \lim_{x \to x_0} [f(x) - f(x_0)] \right\} + f(x_0) \\ &= \left\{ \lim_{x \to x_0} \left[\frac{f(x) - f(x_0)}{x - x_0} \right] \cdot (x - x_0) \right\} + f(x_0) \\ &= f'(x_0) \cdot 0 + f(x_0) = f(x_0). \end{aligned}$$

This argument, appropriately modified (details are left to the interested reader), works in multivariable calculus as well.

Differentiability and Continuity
If $f : U \subset \mathbb{R}^n \to \mathbb{R}^m$ is differentiable at $\mathbf{x}_0$, then it is continuous at $\mathbf{x}_0$.

The next example shows that the existence of partial derivatives at a point is not enough to insure that a function is continuous.

Example 4   Let $f : \mathbb{R}^2 \to \mathbb{R}$ be defined by

$$f(x, y) = \begin{cases} 1, & \text{if } x = 0 \text{ or if } y = 0, \\ 0, & \text{otherwise.} \end{cases}$$

Since f is constant on the x and y axes, where it equals 1,

$$\frac{\partial f}{\partial x}(0,0) = 0 \quad \text{and} \quad \frac{\partial f}{\partial y}(0,0) = 0.$$

But f is not continuous at $(0,0)$, because $\lim_{(x,y)\to(0,0)} f(x,y)$ does not exist since there are points arbitrarily close to $(0,0)$, where the function takes both the values 0 and 1. ♦

Since partial derivatives are based on one-variable calculus, it is usually easy to tell when the partial derivatives of a function exist. On the other hand, to determine if a function is differentiable, we examine the approximation condition, which may seem difficult to verify. Fortunately there is a simple criterion, given in the following box, that tells us when a function is differentiable.

Condition for Differentiability

Let $f : U \subset \mathbb{R}^n \to \mathbb{R}^m$. Suppose that the partial derivatives $\partial f_i / \partial x_j$ of f all exist and are continuous in a neighborhood of a point $\mathbf{x}$ in U. Then f is differentiable at $\mathbf{x}$.

$$\begin{array}{c} \text{Definition} \\ \text{of derivative} \\ \downarrow \end{array}$$

$$\begin{array}{ccc} \text{Partials exist and} & \Longrightarrow \quad \text{Differentiable} \quad \Longrightarrow & \text{Partials exist} \\ \text{are continuous} & & \end{array}$$

We will omit the proof of this result. A function whose partial derivatives exist and are continuous is said to be **continuously differentiable**, of *class C^1*, or **smooth**. Thus this result says that *any C^1 function is differentiable.*

Example 5   Let

$$f(x,y) = \frac{\cos x + e^{xy}}{x^2 + y^2}.$$

Show that f is differentiable at all points $(x,y) \neq (0,0)$.

Solution   Observe that the partial derivatives

$$\frac{\partial f}{\partial x} = \frac{(x^2 + y^2)(ye^{xy} - \sin x) - 2x(\cos x + e^{xy})}{(x^2 + y^2)^2}$$

$$\frac{\partial f}{\partial y} = \frac{(x^2 + y^2)xe^{xy} - 2y(\cos x + e^{xy})}{(x^2 + y^2)^2}$$

are continuous except when $x = 0$ and $y = 0$, so this function is C^1 and hence differentiable at all points other than $(0,0)$. (It is not defined at $(0,0)$.)   ◆

Exercises for §2.3

In Exercises 1–8, find the plane tangent to the given graph at the indicated point.

1. $z = x^3 + y^3 - 6xy$; $(1, 2, -3)$

2. $z = (\cos x)(\cos y)$; $(0, \pi/2, 0)$

3. $z = (\cos x)(\sin y)$; $(0, \pi/2, 1)$

4. $z = 1/xy$; $(1, 1, 1)$

5. $f(x, y) = x^2 + 2y^3 + 1$; $(1, 1, 4)$

6. $f(x, y) = x^2 + 2y^3 + 1$; $(-1, -1, 0)$

7. $f(x, y) = x - y + 2$; $(x_0, y_0, f(x_0, y_0))$ where $(x_0, y_0) = (1, 1)$

8. $f(x, y) = x^2 + 4y^2$; $(x_0, y_0, f(x_0, y_0))$ where $(x_0, y_0) = (2, -1)$

In Exercises 9–12, find the linear approximation of the given mapping at the indicated point.

9. $f(x, y) = (e^x, \sin xy)$; $(1, 3)$

10. $f(x, y, z) = (x - y, y + z)$; $(1, 0, 1)$

11. $f(x, y, z) = (x + e^z + y, yx^2)$; $(1, 1, 0)$

12. $f(x, y) = (xe^y + \cos y, x, x + e^y)$; $(1, 0)$

In Exercises 13–16, compute the derivative matrix $\mathbf{D}f(\mathbf{x}_0)$ of the given function.

13. The function in Exercise 9

14. The function in Exercise 10

15. The function in Exercise 11

16. The function in Exercise 12

In Exercises 17–20, show that each of the following functions is differentiable at each point in its domain. Determine which of the functions are C^1.

17. $f(x, y) = \dfrac{2xy}{(x^2 + y^2)^2}$

18. $f(x, y) = \dfrac{x}{y} + \dfrac{y}{x}$

19. $f(r, \theta) = \dfrac{1}{2} r \sin 2\theta, r > 0$

20. $f(x, y) = \dfrac{xy}{\sqrt{x^2 + y^2}}$

2.4 The Chain Rule

One of the most important formulas in one-variable calculus is the chain rule, which tells us how to differentiate a "function of a function." For example, to differentiate the function $y = \sin(x^2)$ with respect to x, introduce the intermediate variable $u = x^2$, so that $y = \sin u$. The chain rule states that

$$\frac{dy}{dx} = \frac{dy}{du} \frac{du}{dx}.$$

In our example, $dy/du = \cos u$ and $du/dx = 2x$, so the chain rule gives $dy/dx = \cos u \cdot 2x$. Substituting x^2 for u in the last equation gives the final result $dy/dx = 2x \cos(x^2)$.

We also recall the chain rule in "function" notation. If f and g are differentiable functions from $\mathbb{R}$ to $\mathbb{R}$, then their *composition* is the function $h(x) = f(g(x))$, and the chain rule states that $h'(x) = f'(g(x))g'(x)$.

For functions of several variables, the chain rule is a bit more complicated. First we consider the case where z is a function of x and y, and x and y are functions of t; we can then regard z as a function of t. In this case the chain rule states that

$$\frac{dz}{dt} = \frac{\partial z}{\partial x} \frac{dx}{dt} + \frac{\partial z}{\partial y} \frac{dy}{dt}.$$

Example 1 Verify the chain rule for $z = xy, x = t^3$ and $y = t^2$.

Solution Here $z = t^5$, so $dz/dt = 5t^4$. On the other hand,

$$\frac{\partial z}{\partial x} \frac{dx}{dt} + \frac{\partial z}{\partial y} \frac{dy}{dt} = y \cdot 3t^2 + x \cdot 2t = t^2 \cdot 3t^2 + t^3 \cdot 2t = 5t^4,$$

so the chain rule checks. ◆

The Chain Rule

Let $z = f(x, y)$ have continuous partial derivatives and let x and y be differentiable functions of t. Then

$$\frac{dz}{dt} = \frac{\partial z}{\partial x}\frac{dx}{dt} + \frac{\partial z}{\partial y}\frac{dy}{dt}.$$

For three intermediate variables, if u depends on x, y, and z, and x, y, and z depend on t, then

$$\frac{du}{dt} = \frac{\partial u}{\partial x}\frac{dx}{dt} + \frac{\partial u}{\partial y}\frac{dy}{dt} + \frac{\partial u}{\partial z}\frac{dz}{dt}.$$

In "function" notation, if $u = f(x, y, z), x = g(t), y = h(t)$, and $z = k(t)$, the chain rule reads

$$\frac{du}{dt} = f_x g' + f_y h' + f_z k'.$$

Example 2   Suppose that a duck is swimming in the circle $x = \cos t, y = \sin t$, while the water temperature is given by the formula $T = x^2 e^y - xy^3$. Find dT/dt : (a) by the chain rule; (b) by expressing T in terms of t and differentiating.

Solution   (a) $\partial T/\partial x = 2xe^y - y^3; \partial T/\partial y = x^2 e^y - 3xy^2; dx/dt = -\sin t; dy/dt = \cos t$. By the chain rule, $dT/dt = (\partial T/\partial x)(dx/dt) + (\partial T/\partial y)(dy/dt)$, so

$$
\begin{aligned}
\frac{dT}{dt} &= (2xe^y - y^3)(-\sin t) + (x^2 e^y - 3xy^2)\cos t \\
&= (2\cos t e^{\sin t} - \sin^3 t)(-\sin t) + (\cos^2 t e^{\sin t} - 3\cos t \sin^2 t)\cos t \\
&= -2\cos t \sin t e^{\sin t} + \sin^4 t + \cos^3 t e^{\sin t} - 3\cos^2 t \sin^2 t.
\end{aligned}
$$

(b) Substituting for x and y in the formula for T gives

$$T = \cos^2 t e^{\sin t} - \cos t \sin^3 t,$$

and differentiating this gives

$$
\begin{aligned}
\frac{dT}{dt} &= 2\cos t(-\sin t)e^{\sin t} + \cos^2 t e^{\sin t}\cos t \\
&\quad + \sin t \sin^3 t - (\cos t)3\sin^2 t \cos t \\
&= -2\cos t \sin t e^{\sin t} + \cos^3 t e^{\sin t} + \sin^4 t - 3\cos^2 t \sin^2 t,
\end{aligned}
$$

which is the same as the answer in part (a).   ◆

An intuitive argument for the chain rule is based on linear approximations. If the position of a swimming duck changes from the point (x, y) to the point $(x + \Delta x, y + \Delta y)$, the temperature change ΔT is given approximately by $(\partial T/\partial x)\Delta x + (\partial T/\partial y)\Delta y$. On the other hand, the linear approximation for functions of one variable gives $\Delta x \approx (dx/dt)\Delta t$ and $\Delta y \approx (dy/dt)\Delta t$. Substituting,

$$\Delta T \approx \frac{\partial T}{\partial x}\frac{dx}{dt}\Delta t + \frac{\partial T}{\partial y}\frac{dy}{dt}\Delta t.$$

Hence

$$\frac{\Delta T}{\Delta t} \approx \frac{\partial T}{\partial x}\frac{dx}{dt} + \frac{\partial T}{\partial y}\frac{dy}{dt}.$$

As $\Delta t \to 0$, the approximations become more and more accurate and the ratio $\Delta T/\Delta t$ approaches dT/dt, so the preceding approximation formula becomes the chain rule. The argument for three variables is similar.

Example 3   Verify the chain rule for $u = xe^{yz}$ and $(x, y, z) = (e^t, t, \sin t)$.

Solution   Substituting the formulas for $x, y,$ and z in the formula for u gives

$$u = e^t \cdot e^{t\sin t} = e^{t(1+\sin t)},$$

so

$$\frac{du}{dt} = [t\cos t + (1 + \sin t)]\, e^{t(1+\sin t)}.$$

The chain rule says that this should equal

$$\frac{\partial u}{\partial x}\frac{dx}{dt} + \frac{\partial u}{\partial y}\frac{dy}{dt} + \frac{\partial u}{\partial z}\frac{dz}{dt} = e^{yz}e^t + xze^{yz}\cdot 1 + xye^{yz}\cos t$$

$$= e^{t\sin t}e^t + e^t\sin t\, e^{t\sin t} + e^t\cdot t\cdot e^{t\sin t}\cos t$$

$$= e^{t(1+\sin t)}(1 + \sin t + t\cos t),$$

which it does.   ◆

Note

Just as in one-variable calculus, it is important to be flexible about the names of variables. In the following example, the independent variable will be called θ instead of t.

Example 4   Let $f(x, y)$ be a function of x and y and let $x = \cos \theta$, and $y = \sin \theta$. Calculate $df/d\theta$.

Solution   By the chain rule,

$$\frac{df}{d\theta} = \frac{\partial f}{\partial x}\frac{\partial x}{\partial \theta} + \frac{\partial f}{\partial y}\frac{\partial y}{\partial \theta}; \quad i.e., \quad \frac{df}{d\theta} = -\sin \theta\frac{\partial f}{\partial x} + \cos \theta\frac{\partial f}{\partial y}. \quad ◆$$

We now apply the chain rule to the problem of finding tangent lines to curves on surfaces. Let us recall the concept of a tangent vector to a curve from §1.6.

Curves and Tangents

A **path** in $\mathbb{R}^3$ is a mapping of an interval of real numbers to $\mathbb{R}^3$. It is written

$$\mathbf{c}(t) = (g(t), h(t), k(t)) = g(t)\mathbf{i} + h(t)\mathbf{j} + k(t)\mathbf{k}.$$

The *tangent* or *velocity vector* to the curve $\mathbf{c}(t)$ at $\mathbf{c}(t_0)$ is given by

$$\mathbf{c}'(t_0) = g'(t_0)\mathbf{i} + h'(t_0)\mathbf{j} + k'(t_0)\mathbf{k}.$$

The following box tells us how to construct curves on those surfaces which are the graphs of functions.

Curves on Graphs

If $(g(t), h(t))$ is a curve in the plane and $f(x, y)$ is a differentiable function of two variables, then the curve

$$\mathbf{c}(t) = (g(t), h(t), f(g(t), h(t)))$$

lies on the graph $z = f(x, y)$.

Example 5 If $(g(t), h(t))$ is a curve in the unit disk $x^2 + y^2 \le 1$ in the xy plane, then $\mathbf{c}(t) = (g(t), h(t), \sqrt{1 - g(t)^2 - h(t)^2})$ lies on the upper hemisphere $z = \sqrt{1 - x^2 - y^2}$ (see Figure 2.4.1). ◆

Tangents to Curves on Surfaces

If $\mathbf{c}(t) = (g(t), h(t), f(g(t), h(t)))$ lies on the surface $z = f(x, y)$, then the tangent to this curve at the point $\mathbf{c}(t_0)$ is

$$\mathbf{c}'(t_0) = g'(t_0)\mathbf{i} + h'(t_0)\mathbf{j} + \left(\frac{\partial f}{\partial x} g'(t_0) + \frac{\partial f}{\partial y} h'(t_0) \right) \mathbf{k}$$

where the partial derivatives $\partial f / \partial x$ and $\partial f / \partial y$ are evaluated at $(g(t_0), h(t_0))$. This vector lies in the tangent plane to the surface. (See Figure 2.4.2.)

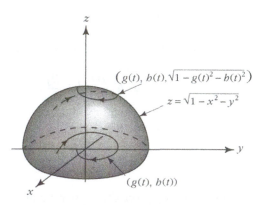

FIGURE 2.4.1. If $(g(t), h(t))$ is a curve in the disk, then the curve $(g(t), h(t), \sqrt{1 - g(t)^2 - h(t)^2})$ is a curve on the upper hemisphere.

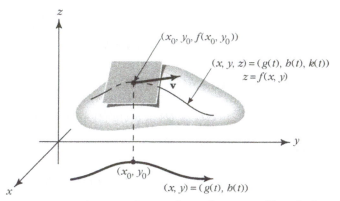

FIGURE 2.4.2. If a curve lies on the surface $z = f(x, y)$, then the tangent line (with direction vector $\mathbf{v}$) to the curve lies in the tangent plane of the surface.

To establish the statements in the box above, write the curve as $\mathbf{c}(t) = (g(t), h(t), k(t))$ where $k(t) = f(g(t), h(t))$. From the definition of the derivative,

$$\mathbf{c}'(t) = g'(t)\mathbf{i} + h'(t)\mathbf{j} + k'(t)\mathbf{k}.$$

Notice that $\mathbf{D}c(t)$ as a column matrix has the same components as $\mathbf{c}'(t)$. By the chain rule,

$$k'(t) = \frac{\partial f}{\partial x}g'(t) + \frac{\partial f}{\partial y}h'(t)$$

as we claimed.

To show that $\mathbf{c}'(t)$ lies in the direction of the tangent plane, we will show that it is perpendicular to the normal. By the first box in §2.3, a normal is $(-\partial f/\partial x, -\partial f/\partial y, 1)$, whereas we have just seen that

$$\mathbf{c}' = \left(g', h', \frac{\partial f}{\partial x}g' + \frac{\partial f}{\partial y}h' \right).$$

The dot product of these two vectors is zero.

Example 6   Show that for any curve $\mathbf{c}(t)$ on the upper hemisphere of Example 5, the tangent vector $\mathbf{c}'(t)$ is perpendicular to $\mathbf{c}(t)$.

Solution

$$\begin{aligned} \mathbf{c}'(t) &= g'(t)\mathbf{i} + h'(t)\mathbf{j} + \left(\frac{\partial f}{\partial x}g'(t) + \frac{\partial f}{\partial y}h'(t) \right)\mathbf{k} \\ &= g'(t)\mathbf{i} + h'(t)\mathbf{j} + k'(t)\mathbf{k}. \end{aligned}$$

Since $\mathbf{c}(t) = g(t)\mathbf{i} + h(t)\mathbf{j} + k(t)\mathbf{k}$,

$$\mathbf{c}(t) \cdot \mathbf{c}'(t) = g(t)g'(t) + h(t)h'(t) + k(t)k'(t).$$

But

$$\|\mathbf{c}(t)\|^2 = 1 = g(t)^2 + h(t)^2 + k(t)^2$$

since $\mathbf{c}(t)$ lies on the unit sphere. Differentiating this equation gives

$$g(t)g'(t) + h(t)h'(t) + k(t)k'(t) = 0 = \mathbf{c}(t) \cdot \mathbf{c}'(t).$$

You should compare this with Example 2 in §2.3.   ◆

We now move on to more general forms of the chain rule.

The Chain Rule for Two Intermediate and Two Independent Variables

Let $f : \mathbb{R}^2 \to \mathbb{R}$ and $g : \mathbb{R}^2 \to \mathbb{R}^2$ be differentiable. Write f as a function of the variables u and v and write $g(x, y) = (u(x, y), v(x, y))$. Define $h : \mathbb{R}^2 \to \mathbb{R}$ by setting

$$h(x, y) = f(u(x, y), v(x, y)).$$

Then

$$
\begin{bmatrix} \dfrac{\partial h}{\partial x} & \dfrac{\partial h}{\partial y} \end{bmatrix} = \begin{bmatrix} \dfrac{\partial f}{\partial u} & \dfrac{\partial f}{\partial v} \end{bmatrix} \begin{bmatrix} \dfrac{\partial u}{\partial x} & \dfrac{\partial u}{\partial y} \\[2mm] \dfrac{\partial v}{\partial x} & \dfrac{\partial v}{\partial y} \end{bmatrix}.
$$

To justify this matrix form of the chain rule, we note that multiplying out this matrix product gives the two equations

$$\frac{\partial h}{\partial x} = \frac{\partial f}{\partial u}\frac{\partial u}{\partial x} + \frac{\partial f}{\partial v}\frac{\partial v}{\partial x}$$

$$\frac{\partial h}{\partial y} = \frac{\partial f}{\partial u}\frac{\partial u}{\partial y} + \frac{\partial f}{\partial v}\frac{\partial v}{\partial y}.$$

The first equation above follows from the chain rule in the box on page 134 by considering the composition $f(u(x, y), v(x, y)) = h(x, y)$ as a function of x only, holding y fixed. Application of this chain rule holding x fixed gives the second equation.

Example 7 Verify the chain rule for

$$f(u, v) = uv \quad \text{and} \quad g(x, y) = (x^2 - y^2, x^2 + y^2) = (u(x, y), v(x, y)).$$

Solution Here $h(x, y) = (x^2 - y^2)(x^2 + y^2) = x^4 - y^4$ and so

$$\frac{\partial h}{\partial x} = 4x^3 \quad \text{and} \quad \frac{\partial h}{\partial y} = -4y^3.$$

However, by the chain rule

$$\frac{\partial h}{\partial x} = \frac{\partial f}{\partial u}\frac{\partial u}{\partial x} + \frac{\partial f}{\partial v}\frac{\partial v}{\partial x} = v(2x) + u(2x)$$
$$= (x^2 + y^2)(2x) + (x^2 - y^2)(2x) = 4x^3.$$

Similarly

$$\frac{\partial h}{\partial y} = \frac{\partial f}{\partial u}\frac{\partial u}{\partial y} + \frac{\partial f}{\partial v}\frac{\partial v}{\partial y} = v(-2y) + u(2y)$$
$$= (x^2 + y^2)(-2y) + (x^2 - y^2)(2y) = -4y^3,$$

so the chain rule is verified.   ◆

The chain rule also generalizes to the following case by an analogous argument.

The Chain Rule for Three Intermediate and Two Independent Variables

Let $f : \mathbb{R}^3 \to \mathbb{R}$ and $g : \mathbb{R}^2 \to \mathbb{R}^3$ be differentiable. Write f as a function of the variables $u, v,$ and w, and $g(x, y) = (u(x, y), v(x, y), w(x, y))$. Define $h : \mathbb{R}^2 \to \mathbb{R}$ by setting

$$h(x, y) = f(u(x, y), v(x, y), w(x, y)).$$

Then

$$\begin{bmatrix} \dfrac{\partial h}{\partial x} & \dfrac{\partial h}{\partial y} \end{bmatrix} = \begin{bmatrix} \dfrac{\partial f}{\partial u} & \dfrac{\partial f}{\partial v} & \dfrac{\partial f}{\partial w} \end{bmatrix} \begin{bmatrix} \dfrac{\partial u}{\partial x} & \dfrac{\partial u}{\partial y} \\[2mm] \dfrac{\partial v}{\partial x} & \dfrac{\partial v}{\partial y} \\[2mm] \dfrac{\partial w}{\partial x} & \dfrac{\partial w}{\partial y} \end{bmatrix}.$$

Similar reasoning also gives

The Chain Rule for Three Intermediate and Three Independent Variables

Let $f : \mathbb{R}^3 \to \mathbb{R}$ and let $g : \mathbb{R}^3 \to \mathbb{R}^3$ be differentiable. Write $g(x, y, z) = (u(x, y, z), v(x, y, z), w(x, y, z))$ and define $h : \mathbb{R}^3 \to \mathbb{R}$ by setting

$$h(x, y, z) = f(u(x, y, z), v(x, y, z), w(x, y, z)).$$

Then

$$\begin{bmatrix} \dfrac{\partial h}{\partial x} & \dfrac{\partial h}{\partial y} & \dfrac{\partial h}{\partial z} \end{bmatrix} = \begin{bmatrix} \dfrac{\partial f}{\partial u} & \dfrac{\partial f}{\partial v} & \dfrac{\partial f}{\partial w} \end{bmatrix} \begin{bmatrix} \dfrac{\partial u}{\partial x} & \dfrac{\partial u}{\partial y} & \dfrac{\partial u}{\partial z} \\[2mm] \dfrac{\partial v}{\partial x} & \dfrac{\partial v}{\partial y} & \dfrac{\partial v}{\partial z} \\[2mm] \dfrac{\partial w}{\partial x} & \dfrac{\partial w}{\partial y} & \dfrac{\partial w}{\partial z} \end{bmatrix}.$$

Example 8   Verify the chain rule for $\partial h/\partial x$ in the form of the preceding box for

$$f(u, v, w) = u^2 + v^2 - w,$$

where

$$u(x, y, z) = x^2 y, \quad v(x, y, z) = y^2, \quad w(x, y, z) = e^{-xz}.$$

Solution   Here

$$
\begin{aligned}
h(x, y, z) &= f(u(x, y, z), v(x, y, z), w(x, y, z)) \\
&= (x^2 y)^2 + y^4 - e^{-xz} = x^4 y^2 + y^4 - e^{-xz}.
\end{aligned}
$$

Differentiating directly,

$$\frac{\partial h}{\partial x} = 4x^3 y^2 + ze^{-xz}.$$

On the other hand, using the chain rule

$$
\begin{aligned}
\frac{\partial h}{\partial x} &= \frac{\partial f}{\partial u}\frac{\partial u}{\partial x} + \frac{\partial f}{\partial v}\frac{\partial v}{\partial x} + \frac{\partial f}{\partial w}\frac{\partial w}{\partial x} \\
&= 2u(2xy) + 2v \cdot 0 + (-1)(-ze^{-xz}) = (2x^2 y)(2xy) + ze^{-xz},
\end{aligned}
$$

which is the same as above.   ◆

Now we are ready for a general statement of the chain rule. This form compactly summarizes *all* the previous cases in *one* rule. It also covers the case where there are several dependent variables.[2]

---

[2]This result is in a shaded box, indicating that it is especially important. Five years after you have taken multivariable calculus, the chain rule is one of those results that should have stuck with you.

> ### The Chain Rule—General Case
>
> Let $g(x_1,\ldots,x_n) = (g_1(x_1,\ldots,x_n),\ldots,g_m(x_1,\ldots,x_n))$ be m functions of n variables and $f(u_1,\ldots,u_m) = (f_1(u_1,\ldots,u_m), f_2(u_1,\ldots,u_m),\ldots, f_p(u_1,\ldots,u_m))$ be p functions of m variables. Assume that f and g are differentiable. Then $f \circ g$ is differentiable and
>
> $$\mathbf{D}(f \circ g)(\mathbf{x}_0) = \mathbf{D}f(g(\mathbf{x}_0)) \cdot \mathbf{D}g(\mathbf{x}_0)$$
>
> *i.e.*, if $h = f \circ g$, and $u = g(\mathbf{x})$,
>
> $$\begin{bmatrix} \dfrac{\partial h_1}{\partial x_1} & \cdots & \dfrac{\partial h_1}{\partial x_n} \\[2mm] \dfrac{\partial h_2}{\partial x_1} & \cdots & \dfrac{\partial h_2}{\partial x_n} \\ \vdots & & \vdots \\ \dfrac{\partial h_p}{\partial x_1} & \cdots & \dfrac{\partial h_p}{\partial x_n} \end{bmatrix} = \begin{bmatrix} \dfrac{\partial f_1}{\partial u_1} & \cdots & \dfrac{\partial f_1}{\partial u_m} \\[2mm] \vdots & & \vdots \\ \dfrac{\partial f_p}{\partial u_1} & \cdots & \dfrac{\partial f_p}{\partial u_m} \end{bmatrix} \begin{bmatrix} \dfrac{\partial g_1}{\partial x_1} & \cdots & \dfrac{\partial g_1}{\partial x_n} \\[2mm] \vdots & & \vdots \\ \dfrac{\partial g_m}{\partial x_1} & \cdots & \dfrac{\partial g_m}{\partial x_n} \end{bmatrix}.$$

Again, the proof is similar to the cases discussed earlier.

Example 9 Given $g(x, y) = (x^2+1, y^2)$ and $f(u, v) = (u+v, u, v^2)$, compute the derivative matrix of $f \circ g$ at $(x, y) = (1, 1)$ using the chain rule.

Solution Compute the derivative matrices of f and g:

$$\mathbf{D}f(u, v) = \begin{bmatrix} \dfrac{\partial f_1}{\partial u} & \dfrac{\partial f_1}{\partial v} \\[2mm] \dfrac{\partial f_2}{\partial u} & \dfrac{\partial f_2}{\partial v} \\[2mm] \dfrac{\partial f_3}{\partial u} & \dfrac{\partial f_3}{\partial v} \end{bmatrix} = \begin{bmatrix} 1 & 1 \\ 1 & 0 \\ 0 & 2v \end{bmatrix} \quad \text{and} \quad \mathbf{D}g(x, y) = \begin{bmatrix} 2x & 0 \\ 0 & 2y \end{bmatrix}.$$

When $(x, y) = (1, 1)$, we get $g(1, 1) = (2, 1)$. Hence

$$\mathbf{D}(f \circ g)(1, 1) = \mathbf{D}f(2, 1)\mathbf{D}g(1, 1) = \begin{bmatrix} 1 & 1 \\ 1 & 0 \\ 0 & 2 \end{bmatrix} \begin{bmatrix} 2 & 0 \\ 0 & 2 \end{bmatrix} = \begin{bmatrix} 2 & 2 \\ 2 & 0 \\ 0 & 4 \end{bmatrix}. \quad \blacklozenge$$

Some other general properties of derivative matrices follow. The proofs are similar to their one dimensional counterparts, so they are omitted.

Additional Derivative Rules

(i) Constant Multiple Rule. Let $f : U \subset \mathbb{R}^n \to \mathbb{R}^m$ be differentiable at $\mathbf{x}_0$ and let c be a real number. Then $h(\mathbf{x}) = cf(\mathbf{x})$ is differentiable at $\mathbf{x}_0$ and

$$\mathbf{D}h(\mathbf{x}_0) = c\mathbf{D}f(\mathbf{x}_0) \qquad \text{(equality of matrices).}$$

(ii) Sum Rule. Let $f : U \subset \mathbb{R}^n \to \mathbb{R}^m$ and $g : U \subset \mathbb{R}^n \to \mathbb{R}^m$ be differentiable at $\mathbf{x}_0$. Then $h(\mathbf{x}) = f(\mathbf{x}) + g(\mathbf{x})$ is differentiable at $\mathbf{x}_0$ and

$$\mathbf{D}h(\mathbf{x}_0) = \mathbf{D}f(\mathbf{x}_0) + \mathbf{D}g(\mathbf{x}_0) \qquad \text{(sum of matrices).}$$

(iii) Product Rule. Let $f : U \subset \mathbb{R}^n \to \mathbb{R}$ and $g : U \subset \mathbb{R}^n \to \mathbb{R}$ be differentiable at $\mathbf{x}_0$ and let $h(\mathbf{x}) = g(\mathbf{x})f(\mathbf{x})$. Then $h : U \subset \mathbb{R}^n \to \mathbb{R}$ is differentiable at $\mathbf{x}_0$ and

$$\mathbf{D}h(\mathbf{x}_0) = g(\mathbf{x}_0)\mathbf{D}f(\mathbf{x}_0) + f(\mathbf{x}_0)\mathbf{D}g(\mathbf{x}_0).$$

(Note that each side of this equation is a $1 \times n$ matrix.)

(iv) Quotient Rule. With the same hypotheses as in rule (iii), let $h(\mathbf{x}) = f(\mathbf{x})/g(\mathbf{x})$ and suppose that g is never zero on U. Then h is differentiable at $\mathbf{x}_0$ and

$$\mathbf{D}h(\mathbf{x}_0) = \frac{g(\mathbf{x}_0)\mathbf{D}f(\mathbf{x}_0) - f(\mathbf{x}_0)\mathbf{D}g(\mathbf{x}_0)}{[g(\mathbf{x}_0)]^2}.$$

Example 10   Verify the formula for $\mathbf{D}h$ in rule (iv) of the preceding box with $f(x,y,z) = x^2 + y^2 + z^2$ and $g(x,y,z) = x^2 + 1$.

Solution   Here

$$h(x,y,z) = \frac{x^2 + y^2 + z^2}{x^2 + 1},$$

so that by direct differentiation

$$
\begin{aligned}
\mathbf{D}h(x,y,z) &= \begin{bmatrix} \dfrac{\partial h}{\partial x} & \dfrac{\partial h}{\partial y} & \dfrac{\partial h}{\partial z} \end{bmatrix} \\[2mm]
&= \begin{bmatrix} \dfrac{(x^2+1)2x - (x^2+y^2+z^2)2x}{(x^2+1)^2} & \dfrac{2y}{x^2+1} & \dfrac{2z}{x^2+1} \end{bmatrix} \\[2mm]
&= \begin{bmatrix} \dfrac{2x(1-y^2-z^2)}{(x^2+1)^2} & \dfrac{2y}{x^2+1} & \dfrac{2z}{x^2+1} \end{bmatrix}.
\end{aligned}
$$

By rule (iv), we get

$$\mathbf{D}h = \frac{g\mathbf{D}f - f\mathbf{D}g}{g^2} = \frac{(x^2+1)[2x, 2y, 2z] - (x^2 + y^2 + z^2)[2x, 0, 0]}{(x^2+1)^2},$$

which agrees with what we obtained directly. ◆

Exercises for §2.4

In Exercises 1–4, verify the chain rule by doing the problem in two ways.

1. Suppose that a duck is swimming in a straight line $x = 3 + 8t, y = 3 - 2t$, while the water temperature is given by the formula $T = x^2 \cos y - y^2 \sin x$. Find dT/dt.

2. Suppose that a duck is swimming along the curve $x = (3 + t)^2, y = 2t^2$, while the water temperature is given by the formula $T = e^x(y^2 + x^2)$. Find dT/dt.

3. Find $\dfrac{du}{dt}$ if $u = \dfrac{x}{y} + \dfrac{y}{z} + \dfrac{z}{x}$, and $x = e^t, y = e^{t^2}, z = e^{t^3}$.

4. Find du/dt if $u = \sin(xy)$, and $x = t^2 + t, y = t^3$.

5. Let $z = \sqrt{x^2 + y^2} + 2xy^2$, where x and y are functions of u. Find an expression for dz/du.

6. If $u = \sin(a + \cos b)$, where a and b are functions of t, what is du/dt?

7. Suppose that a function is given in terms of rectangular coordinates by $u = f(x, y, z)$. If $x = r \cos \theta \sin \phi, y = r \sin \theta \sin \phi, z = r \cos \phi$, express $\partial u/\partial r, \partial u/\partial \theta$, and $\partial u/\partial \phi$ in terms of $\partial u/\partial x, \partial u/\partial y$, and $\partial u/\partial z$.

8. Suppose that x, y, z are as in Exercise 7 and $u = x^2 + y^2 + z^2$. Find $\partial u/\partial r, \partial u/\partial \theta$ and $\partial u/\partial \phi$.

In Exercises 9–12, find the tangent vector to the curve $\mathbf{c}(t) = (g(t), h(t), f(g(t), h(t)))$ on the surface $z = f(x, y)$ when $t = 1$, where

9. $g(t) = t, h(t) = t^2, z = x^2 - y^2$

10. $g(t) = t, h(t) = \cos t, z = e^{xy}$

11. $g(t) = t^3 + 1, h(t) = t^2, z = xy$

12. $g(t) = e^t, h(t) = e^{-t}, z = x/y$

In Exercises 13 and 14, compute the derivative matrices using the chain rule.

13. $z = u^2 + v^2; u = 2x + 7, v = 3x + y + 7$

14. $z = \sin u \cos v; u = 3x^2 - 2y, v = x - 3y$

In Exercises 15–18, find the derivative matrix of the composition of the two functions and evaluate at the given point.

15. $z = f(x, y); x = u \sin v, y = e^{uv}$; at $(u, v) = (0, 1)$

16. $w = f(x, y, z); x = 3 \sin t^2, y = s^2 t^2 u^2, z = s(\sin u)$; at $(s, t, u) = (\pi/3, 1, \pi/3)$

17. $w = f(u, v); u = xyz, v = x + y + z$; at $(3, 3, 3)$

18. $z = f(x, y); x = r \cos \theta, y = r \sin \theta$, at $(r, \theta) = (r, \pi/6)$

In Exercises 19 and 20, (a) compute the derivative matrices; (b) express (u, v) in terms of (t, s) and calculate $\mathbf{D}(u, v)$; and (c) verify that the chain rule holds.

19. $x = t + s, y = t - s; u = x^2 + y, v = x^2 - y^2$

20. $x = t^2 - s^2, y = ts; u = x, v = -y$

In Exercises 21 and 22, calculate $f \circ g$ and $\mathbf{D}(f \circ g)$ at the given point.

21. $f(u, v) = (\tan(u - 1) - e^v, u^2 - v^2)$ and $g(x, y) = (e^{x-y}, x - y); (1, 1)$

22. $f(u, v, w) = (e^{u-w}, \cos(v + u) + \sin(u + v + w))$ and
    $g(x, y) = (e^x, \cos(y - x), e^{-y}); (0, 0)$

23. Show that applying the chain rule to $z = x/y$ (where x and y are arbitrary functions of t) gives the quotient rule for functions of one variable.

24. Suppose that the temperature at the point (x, y, z) in space is $T(x, y, z) = x^2 + y^2 + z^2$. Let a particle follow the right circular helix $\sigma(t) = (\cos t, \sin t, t)$ and let $T(t)$ be its temperature at time t.

   (a) What is $T'(t)$?

   (b) Find an approximate value for the temperature at $t = (\pi/2) + 0.01$.

25. What is wrong with the following argument?
    Suppose that $w = f(x, y, z), z = g(x, y)$ and $\partial z/\partial x \neq 0$. By the chain rule,

    $$\frac{\partial w}{\partial x} = \frac{\partial w}{\partial x}\frac{\partial x}{\partial x} + \frac{\partial w}{\partial y}\frac{\partial y}{\partial x} + \frac{\partial w}{\partial z}\frac{\partial z}{\partial x} = \frac{\partial w}{\partial x} + \frac{\partial w}{\partial z}\frac{\partial z}{\partial x}.$$

    Hence

    $$0 = \frac{\partial w}{\partial z}\frac{\partial z}{\partial x},$$

    and thus $\partial w/\partial z = 0$, so w is independent of z.

26. Suppose that f is a differentiable function of one variable and that a function $u = g(x, y)$ is defined by

$$u = g(x, y) = xyf\left(\frac{x + y}{xy}\right).$$

Show that u satisfies a (partial) differential equation of the form

$$x^2\frac{\partial u}{\partial x} - y^2\frac{\partial u}{\partial y} = G(x, y)u$$

and find the function $G(x, y)$.

2.5
Gradients and Directional Derivatives

This section introduces the gradient, the first of several basic operators used in vector calculus. This operation, applied to real-valued functions, will be used to obtain information about their graphs and level sets.

The Gradient

If $f(x, y, z)$ is a real-valued function of three variables, its **gradient**, which is denoted ∇f or grad f, is defined by

$$\nabla f = \frac{\partial f}{\partial x}\mathbf{i} + \frac{\partial f}{\partial y}\mathbf{j} + \frac{\partial f}{\partial z}\mathbf{k}.$$

For $f(x, y)$ the gradient is

$$\nabla f = \frac{\partial f}{\partial x}\mathbf{i} + \frac{\partial f}{\partial y}\mathbf{j}.$$

Notice that ∇f is the same as the derivative $\mathbf{D}f$, except that it is regarded as a vector rather than as a row matrix.

Example 1   Find ∇f if $f(x, y, z) = xy - z^2$.

Solution   Substituting the partial derivatives of f into the formula for the gradient of f, we find $\nabla f(x, y, z) = y\mathbf{i} + x\mathbf{j} - 2z\mathbf{k}$.   ◆

Example 2   Find ∇f for the function $f(x, y, z) = e^{xy} - x\cos(yz^2)$.

Solution   Here $f_x(x, y, z) = ye^{xy} - \cos(yz^2)$, $f_y(x, y, z) = xe^{xy} + xz^2\sin(yz^2)$, and $f_z(x, y, z) = 2xyz\sin(yz^2)$, so

$$\begin{aligned}\nabla f(x, y, z) &= \left[ye^{xy} - \cos(yz^2)\right]\mathbf{i} + \left[xe^{xy} + xz^2\sin(yz^2)\right]\mathbf{j} \\ &\quad + \left[2xyz\sin(yz^2)\right]\mathbf{k}.\end{aligned}$$ ◆

Notice that the vector $\nabla f(x, y, z)$ is a function of the point (x, y, z) in space where the partial derivatives are evaluated. A rule $\mathbf{F}$ that assigns a vector $\mathbf{F}(x, y, z)$ in space to each point (x, y, z) of some domain in space is called a ***vector field***. Thus, for a given function f, *the gradient* ∇f *is a vector field*. Similarly, a vector field in the xy plane is a rule $\mathbf{F}$ which assigns to each point (x, y) a vector $\mathbf{F}(x, y)$ in the plane.

We may sketch a vector field $\mathbf{F}(x, y)$ in the plane by choosing several values for (x, y), evaluating $\mathbf{F}(x, y)$ at each point, and drawing the vector $\mathbf{F}(x, y)$ with its tail at the point (x, y). The same thing may be done for vector fields in space, although they are more difficult to visualize.

Example 3   Sketch the vector field ∇f, where $f(x, y) = x^2 + y^2$.

Solution $\nabla f(x, y) = 2x\mathbf{i} + 2y\mathbf{j} = 2(x\mathbf{i} + y\mathbf{j})$. To sketch ∇f as a vector field we draw, for each $(x, y), 2(x\mathbf{i} + y\mathbf{j})$ as a vector based at (x, y). Doing this for several points (x, y) we obtain a sketch as in Figure 2.5.1.   ◆

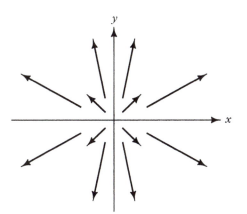

FIGURE 2.5.1. The vector field $2x\mathbf{i} + 2y\mathbf{j}$.

Example 4   Sketch the vector field ∇f if $f(x, y) = \frac{1}{4}(x^2 - y^2)$.

Solution   Here $\nabla f(x, y) = \frac{1}{2}(x\mathbf{i} - y\mathbf{j})$. The vector $\frac{1}{2}(x\mathbf{i} - y\mathbf{j})$ is obtained from the vector $x\mathbf{i} + y\mathbf{j}$ by reflection about the horizontal direction and shrinking by a factor $\frac{1}{2}$. Doing this for several points (x, y) and basing the vectors $\frac{1}{2}(x\mathbf{i} - y\mathbf{j})$ at (x, y) instead of the origin, we obtain the sketch in Figure 2.5.2.   ◆

In a number of situations arising later in the book, the vector $\mathbf{r}$ from the origin to a point (x, y, z) plays a basic role. The next example illustrates its use.

Example 5   Let $\mathbf{r} = x\mathbf{i} + y\mathbf{j} + z\mathbf{k}$ and $r = \|\mathbf{r}\| = \sqrt{x^2 + y^2 + z^2}$. (Note that r is a scalar and $\mathbf{r}$ is a vector.) Show that

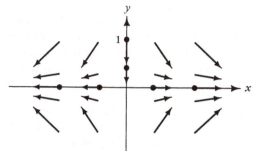

FIGURE 2.5.2. The vector field $(x\mathbf{i} - y\mathbf{j})/2$.

(a) $\nabla r = \dfrac{\mathbf{r}}{r}$

(b) $\nabla \left(\dfrac{1}{r} \right) = -\dfrac{\mathbf{r}}{r^3}, \quad r \neq 0.$

What is $\|\nabla(1/r)\|$?

Solution   (a) By the definition of the gradient,

$$
\begin{aligned}
\nabla r &= \frac{\partial r}{\partial x}\mathbf{i} + \frac{\partial r}{\partial y}\mathbf{j} + \frac{\partial r}{\partial z}\mathbf{k} \\
&= \left(\frac{x}{\sqrt{x^2 + y^2 + z^2}}, \frac{y}{\sqrt{x^2 + y^2 + z^2}}, \frac{z}{\sqrt{x^2 + y^2 + z^2}} \right) = \frac{\mathbf{r}}{r},
\end{aligned}
$$

where $\mathbf{r}$ is the point (x, y, z). Thus ∇r is the unit vector in the direction of (x, y, z).

(b)

$$
\nabla \left(\frac{1}{r} \right) = \frac{\partial}{\partial x}\left(\frac{1}{r} \right)\mathbf{i} + \frac{\partial}{\partial y}\left(\frac{1}{r} \right)\mathbf{j} + \frac{\partial}{\partial z}\left(\frac{1}{r} \right)\mathbf{k}.
$$

The partial derivatives are

$$
\frac{\partial}{\partial x}\left(\frac{1}{r} \right) = \frac{\partial}{\partial x}\left(\frac{1}{\sqrt{x^2 + y^2 + z^2}} \right) = -\frac{x}{(x^2 + y^2 + z^2)^{3/2}} = -\frac{x}{r^3},
$$

and, similarly,

$$
\frac{\partial}{\partial y}\left(\frac{1}{r} \right) = -\frac{y}{r^3}, \quad \frac{\partial}{\partial z}\left(\frac{1}{r} \right) = -\frac{z}{r^3}.
$$

Thus,

$$\nabla\left(\frac{1}{r}\right) = -\frac{x}{r^3}\mathbf{i} - \frac{y}{r^3}\mathbf{j} - \frac{z}{r^3}\mathbf{k} = -\frac{1}{r^3}(x\mathbf{i} + y\mathbf{j} + z\mathbf{k}) = -\frac{\mathbf{r}}{r^3},$$

as required. Finally,

$$\left\|\nabla\left(\frac{1}{r}\right)\right\| = \left|-\frac{1}{r^3}\right|\|\mathbf{r}\| = \frac{r}{r^3} = \frac{1}{r^2} = \frac{1}{x^2 + y^2 + z^2}. \quad \blacklozenge$$

For scalar-valued functions of two or three variables, the gradient provides an elegant formulation of the chain rule:

The Chain Rule and Gradients

Let f be a function of two (or three) variables, $\mathbf{c}(t)$ a curve in the plane (or in 3-space), and $h(t) = f(\mathbf{c}(t))$ the composite function. Then

$$\frac{d}{dt}h(t) = h'(t) = \nabla f(\mathbf{c}(t)) \cdot \mathbf{c}'(t).$$

Example 6   Verify the chain rule for $u = f(x, y, z) = xy - z^2$ and $\mathbf{c}(t) = (\sin t, \cos t, e^t)$.

Solution   We compute

$$\frac{\partial f}{\partial x} = y, \quad \frac{\partial f}{\partial y} = x, \quad \text{and} \quad \frac{\partial f}{\partial z} = -2z.$$

Therefore $\nabla f(x, y, z) = y\mathbf{i} + x\mathbf{j} - 2z\mathbf{k}$. Substituting $\mathbf{c}(t) = (\sin t, \cos t, e^t)$ gives

$$\nabla f(\mathbf{c}(t)) = \cos t\,\mathbf{i} + \sin t\,\mathbf{j} - 2e^t\mathbf{k}.$$

Differentiating $\mathbf{c}(t), \mathbf{c}'(t) = \cos t\,\mathbf{i} - \sin t\,\mathbf{j} + e^t\mathbf{k}$. Letting $u(t) = f(\mathbf{c}(t))$, the chain rule gives

$$\frac{du}{dt} = \nabla f(\mathbf{c}(t)) \cdot \mathbf{c}'(t) = \cos^2 t - \sin^2 t - 2e^{2t}.$$

To verify this directly we compute $u(t) = f(\mathbf{c}(t)) = xy - z^2$ where $x = \sin t, y = \cos t, z = e^t$ and so

$$u(t) = \sin t \cos t - e^{2t}.$$

Thus

$$\frac{du}{dt} = \cos^2 t - \sin^2 t - 2e^{2t}.$$

Thus the chain rule is verified in this case.   ◆

Example 7   Suppose that f takes the value 2 at all points on a path $\mathbf{c}(t)$. What can you say about the two vectors $\nabla f(\mathbf{c}(t))$ and $\mathbf{c}'(t)$?

Solution   If $f(\mathbf{c}(t))$ is always equal to 2, the derivative $d/dt\,f(\mathbf{c}(t))$ is zero. By the chain rule, $0 = \nabla f(\mathbf{c}(t)) \cdot \mathbf{c}'(t)$, so the gradient vector and the velocity vector are perpendicular at all points on $\mathbf{c}(t)$.   ◆

Let f be a real-valued function of two or three variables, and consider the function h from $\mathbb{R}$ to $\mathbb{R}$ defined by $h(t) = f(\mathbf{x} + t\mathbf{v})$. The set of points of the form $\mathbf{x} + t\mathbf{v}$ as t varies is the straight line L through the point $\mathbf{x}$ in the direction of the vector $\mathbf{v}$. The function $h(t) = f(\mathbf{x} + t\mathbf{v})$ represents the function f restricted to the line L. We may ask: How fast are the values of f changing along the line L at the point $\mathbf{x}$? Since the rate of change of a function is a derivative, we would answer that the value we seek is the derivative of the function $h(t)$ at $t = 0$. If we set $\mathbf{c}(t) = \mathbf{x} + t\mathbf{v}$, then applying the chain rule we find that

$$\frac{dh}{dt} = \nabla f(\mathbf{c}(t)) \cdot \mathbf{c}'(t).$$

At $t = 0, \mathbf{c}(0) = \mathbf{x}$ and $\mathbf{c}'(t) = \mathbf{v}$, so

$$\left.\frac{dh}{dt}\right|_{t=0} = \nabla f(\mathbf{x}) \cdot \mathbf{v}.$$

If $\mathbf{v}$ is a unit vector ($\|\mathbf{v}\| = 1$), then $\mathbf{v}$ specifies a direction in the plane or space, and we call $\nabla f(\mathbf{x}) \cdot \mathbf{v}$ the *directional derivative* of f at the point $\mathbf{x}$ in the direction $\mathbf{v}$.

Now we may ask an additional question. Since the directional derivative is the rate at which f is changing in the direction $\mathbf{v}$ at the point $\mathbf{x}$, in what direction is f changing the fastest at the point $\mathbf{x}$? For example, if you are on a mountain without tent or supplies, and nightfall is coming, you may want to move in the direction which decreases your altitude the fastest.

We can solve this problem as follows. By the chain rule in the preceding box, we are seeking a unit vector $\mathbf{v}$ so that

$$\nabla f(\mathbf{x}) \cdot \mathbf{v}$$

is as large as possible. Write the dot product as follows (see §1.2):

$$\nabla f(\mathbf{x}) \cdot \mathbf{v} = \|\nabla f(\mathbf{x})\| \, \|\mathbf{v}\| \cos \theta = \|\nabla f(\mathbf{x})\| \cos \theta$$

where θ is the angle between the vectors $\nabla f(\mathbf{x})$ and $\mathbf{v}$.

Since $-1 \le \cos \theta \le 1$, the maximum value is achieved when $\cos \theta = 1$, *i.e.*, when $\theta = 0$. This means that the vectors $\nabla f(\mathbf{x})$ and $\mathbf{v}$ point in the *same* direction. Thus, $\mathbf{v} = \lambda \nabla f(\mathbf{x})$ where λ is a positive constant. Since $\|\mathbf{v}\| = 1 = \|\lambda \nabla f(\mathbf{x})\| = \lambda \|\nabla f(\mathbf{x})\|$ we see that $\lambda = 1/\|\nabla f(\mathbf{x})\|$ and thus

$$\mathbf{v} = \frac{\nabla f(\mathbf{x})}{\|\nabla f(\mathbf{x})\|}.$$

Hence $\nabla f(\mathbf{x})/\|\nabla f(\mathbf{x})\|$ is the direction in which f is *increasing* the fastest. Likewise $-\nabla f(\mathbf{x})/\|\nabla f(\mathbf{x})\|$, corresponding to $\theta = \pi$, is the direction in which f is *decreasing* the fastest.

The following box summarizes our findings.

Gradients and Directional Derivatives

The **directional derivative** at $\mathbf{x}$ in the direction of a unit vector $\mathbf{v}$ is the rate of change of f along the straight line through $\mathbf{x}$ in the direction $\mathbf{v}$; *i.e.*, along $\mathbf{c}(t) = \mathbf{x} + t\mathbf{v}$. The directional derivative at $\mathbf{x}$ in the direction $\mathbf{v}$ equals $\nabla f(\mathbf{x}) \cdot \mathbf{v}$. For $\mathbf{x}$ fixed, the directional derivative is the greatest when

$$\mathbf{v} = \frac{\nabla f(\mathbf{x})}{\|\nabla f(\mathbf{x})\|}$$

and the least when

$$\mathbf{v} = -\frac{\nabla f(\mathbf{x})}{\|\nabla f(\mathbf{x})\|}.$$

This formula holds for functions of both two and three variables.

Example 8   Let $f(x, y) = x^2 - y^2$. In what direction from $(0, 1)$ should one proceed to increase f the fastest? Illustrate your answer with a sketch.

Solution   The required direction is that of the vector

$$
\begin{aligned}
\nabla f(0, 1) &= \frac{\partial f}{\partial x}\mathbf{i} + \frac{\partial f}{\partial y}\mathbf{j} \quad \text{at} \quad (0, 1) \\
&= 2x\mathbf{i} - 2y\mathbf{j} \quad \text{at} \quad (0, 1) \\
&= -2\mathbf{j}.
\end{aligned}
$$

Thus one should head toward the origin along the y axis. The graph of f, sketched in Figure 2.5.3, illustrates this.   ◆

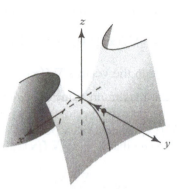

FIGURE 2.5.3. Starting from $(0, 1)$, moving along the y axis makes the graph rise the steepest.

Note

Why do we choose **v** to be a *unit* vector? The reason is that then $\mathbf{c}(t) = \mathbf{x} + t\,\mathbf{v}$ has unit speed along L and so $\nabla f \cdot \mathbf{v}$ can be interpreted as the rate of change of f with respect to *distance* along L. This corresponds to what we mean by saying "the rate of change of f in direction **v**." It may help to imagine how the situation changes if, in walking down a mountain, you change direction *and* change your speed.

Note that in Example 8 and the examples below we do not bother to normalize the vector. This is permitted as long as we just speak about the *direction*. If the actual rate of change of f is asked for, then the magnitude of **v** in $\nabla f(\mathbf{x}) \cdot \mathbf{v}$ is important.

In the next examples we apply these ideas to functions of three variables.

Example 9   Let $u = f(x, y, z) = (\sin xy)e^{-z^2}$. In what direction from $(1, \pi, 0)$ should one proceed to increase f most rapidly?

Solution   We compute the gradient:

$$
\begin{aligned}
\nabla f &= \frac{\partial u}{\partial x}\mathbf{i} + \frac{\partial u}{\partial y}\mathbf{j} + \frac{\partial u}{\partial z}\mathbf{k} \\
&= y\cos(xy)e^{-z^2}\mathbf{i} + x\cos(xy)e^{-z^2}\mathbf{j} + (-2z\sin xy)e^{-z^2}\mathbf{k}.
\end{aligned}
$$

At $(1, \pi, 0)$ this becomes

$$\pi\cos(\pi)\mathbf{i} + \cos(\pi)\mathbf{j} = -\pi\mathbf{i} - \mathbf{j}.$$

Thus one should proceed in the direction of the vector $-\pi\mathbf{i} - \mathbf{j}$.   ◆

Example 10　Captain Astro is drifting in space near the sunny side of Mercury and notices that the hull of her ship is beginning to melt. The temperature in her vicinity is given by $T = e^{-x} + e^{-2y} + e^{3z}$. If she is at $(1, 1, 1)$, in what direction should she proceed in order to *cool* fastest?

Solution　In order to cool the fastest, the captain should proceed in the direction in which T is decreasing the fastest; that is, in the direction $-\nabla T(1, 1, 1)$. Since

$$\nabla T = \frac{\partial T}{\partial x}\mathbf{i} + \frac{\partial T}{\partial y}\mathbf{j} + \frac{\partial T}{\partial z}\mathbf{k} = -e^{-x}\mathbf{i} - 2e^{-2y}\mathbf{j} + 3e^{3z}\mathbf{k},$$

$$-\nabla T(1, 1, 1) = e^{-1}\mathbf{i} + 2e^{-2}\mathbf{j} - 3e^{3}\mathbf{k}$$

is the direction required.　◆

The tangent plane to a graph $z = f(x, y)$ was defined as the graph of the linear approximation to f. We found (§2.4) that the tangent plane at a point could also be characterized as the plane containing the tangent *lines* to all curves on the surface through the given point. For a general surface, we take this as a definition of the tangent plane.

Tangent Plane to a Surface

Let S be a surface in space and $\mathbf{r}_0$ a point on S. If there is a plane that contains the tangent lines at $\mathbf{r}_0$ to all curves through $\mathbf{r}_0$ in S, then this plane is called the **tangent plane** to S at $\mathbf{r}_0$. A normal to the tangent plane is sometimes said to be **perpendicular to S**.

Here is how to find the tangent plane to a level surface.

Gradients and Tangent Planes

Let $\mathbf{r}_0$ lie on the level surface S defined by $f(x, y, z) = b$, and suppose that $\nabla f(\mathbf{r}_0) \neq \mathbf{0}$. Then $\nabla f(\mathbf{r}_0)$ is normal to the tangent plane to S at $\mathbf{r}_0$. (See Figure 2.5.4.)

To prove this assertion, first observe that $f(\mathbf{c}(t)) = b$ if the curve $\mathbf{c}(t)$ lies in S, with $\mathbf{c}(t_0) = \mathbf{r}_0$. Hence

$$\frac{d}{dt} f(\mathbf{c}(t)) = 0.$$

By the chain rule in terms of gradients, this gives

$$\nabla f(\mathbf{c}(t)) \cdot \mathbf{c}'(t) = 0.$$

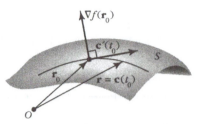

FIGURE 2.5.4. The gradient of f at $\mathbf{r}_0$ is perpendicular to the tangent vector of any curve in the level surface.

Setting $t = t_0$, we have $\nabla f(\mathbf{r}_0) \cdot \mathbf{c}'(t_0) = 0$ for every $\mathbf{c}$ in S, so $\nabla f(\mathbf{r}_0)$ is normal to the tangent plane. [We required $\nabla f(\mathbf{r}_0) \neq \mathbf{0}$ so that there would be a well-defined plane orthogonal to $\nabla f(\mathbf{r}_0)$.]

Example 11   Let $u = f(x, y, z) = x^2 + y^2 - z^2$. Find $\nabla f(0, 0, 1)$. Plot this on the level surface $f(x, y, z) = -1$.

Solution   We have

$$\nabla f = \frac{\partial u}{\partial x}\mathbf{i} + \frac{\partial u}{\partial y}\mathbf{j} + \frac{\partial u}{\partial z}\mathbf{k} = 2x\mathbf{i} + 2y\mathbf{j} - 2z\mathbf{k}.$$

At $(0, 0, 1)$, $\nabla f(0, 0, 1) = -2\mathbf{k}$.

The level surface $x^2 + y^2 - z^2 = -1$ is a hyperboloid of two sheets. If we plot $\nabla f(0, 0, 1)$ on it (Figure 2.5.5), we see that it is indeed perpendicular to the surface.   ◆

Example 12   Find a unit normal to the surface $\sin(xy) = e^z$ at $(1, \pi/2, 0)$.

Solution   Let $f(x, y, z) = \sin(xy) - e^z$, so the surface is $f(x, y, z) = 0$. A normal is $\nabla f = y\cos(xy)\mathbf{i} + x\cos(xy)\mathbf{j} - e^z\mathbf{k}$. At $(1, \pi/2, 0)$, we get $\nabla f = -\mathbf{k}$. Thus $-\mathbf{k}$ (or $\mathbf{k}$) is the required unit normal. (It already has length 1, so there is no need to normalize.)   ◆

Example 13   The gravitational force exerted on a mass m at (x, y, z) by a mass M at the origin is, by Newton's law of gravitation,

$$\mathbf{F} = -\frac{GMm}{r^3}\mathbf{r}, \quad \text{where} \quad \mathbf{r} = x\mathbf{i} + y\mathbf{j} + z\mathbf{k} \quad \text{and} \quad r = \|\mathbf{r}\|,$$

where G is a constant. Write $\mathbf{F}$ as the negative of the gradient of a function V (called the **gravitational potential**) and verify that $\mathbf{F}$ is orthogonal to the level surfaces of V.

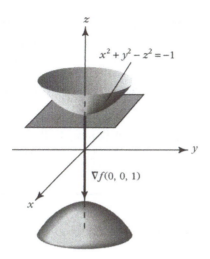

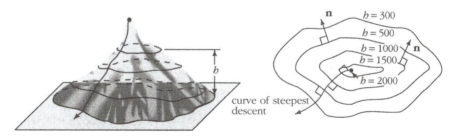

FIGURE 2.5.5. The vector $\nabla f(0,0,1)$ is perpendicular to the level surface of f.

Solution   By Example 5 in this section, $\nabla(1/r) = -(\mathbf{r}/r^3)$. Therefore we can choose $V = -GMm/r$ to give $\mathbf{F} = -\nabla V$. Note that on each level surface of V, r is a constant, so the surface is a sphere. Since $\mathbf{F}$ is a positive multiple of $-\mathbf{r}$, $\mathbf{F}$ points toward the origin and thus is orthogonal to these spheres.   ◆

If we combine the fact that ∇f is the direction in which f is *increasing* the fastest ($-\nabla f$ is the direction in which f is *decreasing* the fastest) and the fact that ∇f is orthogonal to the level surfaces (or curves) of f, we see that *the direction in which the function f is increasing or decreasing most rapidly is perpendicular to the level surfaces (or curves) of f.* For example, to get down most directly from the top of a hill, one should proceed in a direction perpendicular to the level contours. (See Figure 2.5.6.)

FIGURE 2.5.6. The curve of steepest descent is perpendicular to the level curves. (a) Steepest descent of a hill. (b) Contour map of hill 2000 feet high.

The gradient also enables us to compute the equation of the tangent plane to the level surface S at $\mathbf{r}_0$. Indeed, $\nabla f(\mathbf{r}_0)$ will be a normal to this plane, which passes through $\mathbf{r}_0$. Therefore its equation can be read off as follows:

> ## Equation of the Tangent Plane to a Level Surface
> The normal to the tangent plane at $\mathbf{r}_0 = (x_0, y_0, z_0)$ of a level surface $f(x, y, z) = b$ is $\nabla f(\mathbf{r}_0)$. The equation of this plane is
>
> $$f_x(x_0, y_0, z_0)(x - x_0) + f_y(x_0, y_0, z_0)(y - y_0) + f_z(x_0, y_0, z_0)(z - z_0) = 0.$$
>
> The equation of the tangent line at (x_0, y_0) to the curve $f(x, y) = c$ is
>
> $$f_x(x_0, y_0)(x - x_0) + f_y(x_0, y_0)(y - y_0) = 0.$$

Example 14 Compute the equation of the plane tangent to the surface $3xy + z^2 = 4$ at the point $(1, 1, 1)$.

Solution Here $f(x, y, z) = 3xy + z^2$ and $\nabla f = (3y, 3x, 2z)$, which at $(1, 1, 1)$ is the vector $3\mathbf{i} + 3\mathbf{j} + 2\mathbf{k}$. Thus the tangent plane is

$$3(x - 1) + 3(y - 1) + 2(z - 1) = 0 \quad \text{or} \quad 3x + 3y + 2z = 8. \quad \blacklozenge$$

Example 15 (a) Find a unit normal to the ellipsoid $x^2 + 2y^2 + 3z^2 = 10$ at each of the points $(\sqrt{10}, 0, 0), (-\sqrt{10}, 0, 0), (1, 0, \sqrt{3})$, and $(-1, 0, -\sqrt{3})$.

(b) Do the vectors you have found point to the inside or outside of the ellipsoid?

(c) Give equations to the tangent planes to the surface at the two points of the surface with $x_0 = y_0 = 1$.

Solution

(a) Letting $f(x, y, z) = x^2 + 2y^2 + 3z^2$, we find $\nabla f(x, y, z) = (2x, 4y, 6z)$. At $(\sqrt{10}, 0, 0)$, a unit normal to the ellipsoid is

$$\frac{\nabla f(\sqrt{10}, 0, 0)}{\|\nabla f(\sqrt{10}, 0, 0)\|} = \frac{(2\sqrt{10}, 0, 0)}{\left((2\sqrt{10})^2 + 0^2 + 0^2\right)^{1/2}} = (1, 0, 0).$$

At $(-\sqrt{10}, 0, 0)$, it is $(-1, 0, 0)$. At $(1, 0, \sqrt{3})$, it is

$$\frac{\nabla f(1, 0, \sqrt{3})}{\|\nabla f(1, 0, \sqrt{3})\|} = \left(\frac{1}{\sqrt{28}}, 0, \frac{3\sqrt{3}}{\sqrt{28}}\right),$$

and at $(-1, 0, -\sqrt{3})$ it is $(-1/\sqrt{28}, 0, -3\sqrt{3}/\sqrt{28})$.

(b) The vectors are all pointing to the outside of the ellipsoid.

(c) The two points are $(1, 1, \sqrt{7/3})$, and $(1, 1, -\sqrt{7/3})$. Evaluating the gradient, $\nabla f(1, 1, \sqrt{7/3}) = (2, 4, 2\sqrt{21})$ and $\nabla f(1, 1, -\sqrt{7/3})$ $= (2, 4, -2\sqrt{21})$, so the tangent planes to the surface at the points $(1, 1, \sqrt{7/3})$ and $(1, 1, -\sqrt{7/3})$ are given by $2(x-1) + 4(y-1) + 2\sqrt{21}(z - \sqrt{7/3}) = 0$ and $2(x-1) + 4(y-1) - 2\sqrt{21}(z + \sqrt{7/3}) = 0$, respectively.   ◆

Example 16   Find the equation of the tangent line to $xy = 6$ at $x = 1, y = 6$.

Solution   With $f(x, y) = xy$, we have $f_x(x, y) = y$ and $f_y(x, y) = x$. Then $f_x(1, 6) = 6$ and $f_y(1, 6) = 1$, so from the preceding box, the equation of the tangent line through $(1, 6)$ is

$$6(x - 1) + 1(y - 6) = 0 \quad \text{or} \quad y = -6x + 12. \quad ◆$$

In the next example we check that the equation given in §**2.3** for the tangent plane to a graph is consistent with the one given here.

Example 17   Let $z = g(x, y)$. The graph of g may be defined as the level surface $f(x, y, z) = 0$, where $f(x, y, z) = z - g(x, y)$. Compute the gradient of f and verify that it is perpendicular to the tangent plane of the graph $z = g(x, y)$ as defined in §**2.3**.

Solution   With $f(x, y, z) = z - g(x, y)$,

$$\begin{aligned} \nabla f(x, y, z) &= f_x(x, y, z)\mathbf{i} + f_y(x, y, z)\mathbf{j} + f_z(x, y, z)\mathbf{k} \\ &= -g_x(x, y)\mathbf{i} - g_y(x, y)\mathbf{j} + \mathbf{k}. \end{aligned}$$

This is the normal to the tangent plane at (x, y) to the graph of g.   ◆

Exercises for §2.5

In Exercises 1–8, compute the gradient of the given function.

1. $f(x, y) = \sqrt{x^2 + y^2}$

2. $f(x, y) = xe^{x^2 + y^2}$

3. $f(x, y, z) = xy + yz + xz$

4. $f(x, y, z) = x + y^2 + z^3$

5. $f(x, y, z) = xy^2 + yz^2 + zx^2$

6. $f(x, y) = \log\left(\sqrt{x^2 + y^2}\right)$

7. $f(x, y) = (x^2 + y^2)\log\sqrt{x^2 + y^2}$

8. $f(x,y,z) = \dfrac{1}{r^3}$ $(r \neq 0)$

In Exercises 9–12, sketch the gradient vector field of the given function.

9. $f(x,y) = \dfrac{x^2}{8} + \dfrac{y^2}{12} + 6$

10. $f(x,y) = \dfrac{x^2}{8} - \dfrac{y^2}{12}$

11. $f(x,y) = x - y$

12. $f(x,y) = x + y^2$

In Exercises 13–16, verify the chain rule for the given function and curve.

13. $f(x,y) = x^2 + xy; \mathbf{c}(t) = (e^t, \cos t)$

14. $f(x,y) = e^{xy}; \mathbf{c}(t) = (6t, 3t^2)$

15. $f(x,y,z) = \sqrt{x^2 + y^2 + z^2}; \mathbf{c}(t) = (\sin t, \cos t, t)$

16. $f(x,y,z) = xy + yz + xz; \mathbf{c}(t) = (t, t, t)$

In Exercises 17–24, compute the directional derivative of each function at the given point in the direction of the unit vector *parallel* to the given vector $\mathbf{v}$.

17. $f(x,y) = x^2 + y^2 - 3xy^3; (x_0, y_0) = (1, 2); \mathbf{v} = (1/2, \sqrt{3}/2)$

18. $f(x,y) = e^x \cos y; (x_0, y_0) = (0, \pi/4); \mathbf{v} = (\mathbf{i} + 3\mathbf{j})/\sqrt{10}$

19. $f(x,y,z) = x^2 - 2xy + 3z^2; (x_0, y_0, z_0) = (1, 1, 2); \mathbf{v} = (\mathbf{i} + \mathbf{j} - \mathbf{k})/\sqrt{3}$

20. $f(x,y,z) = e^{(x^2+y^2+z^2)}; (x_0, y_0, z_0) = (1, 10, 100); \mathbf{v} = (1, -1, -1)/\sqrt{3}$

21. $f(x,y) = x^y; (x_0, y_0) = (e, c); \mathbf{v} = 5\mathbf{i} + 12\mathbf{j}$

22. $f(x,y,z) = e^x + yz; (x_0, y_0, z_0) = (1, 1, 1); \mathbf{v} = (1, -1, 1)$

23. $f(x,y,z) = xyz; (x_0, y_0, z_0) = (1, 0, 1); \mathbf{v} = (1, 0, -1)$

24. $f(x,y,z) = 1/(x^2 + y^2 + z^2); (x_0, y_0, z_0) = (2, 3, 1); \mathbf{v} = (\mathbf{j} - 2\mathbf{k} + \mathbf{i})$

In Exercises 25 and 26, determine the direction in which the function is increasing and decreasing fastest at $(1, 1, 1)$.

25. $f(x,y,z) = ze^x \sin y$

26. $f(x,y,z) = e^{xz} \sin y - e^{-xz} \cos y$

In Exercises 27 and 28, find $\nabla f(0, 0, 1)$ and plot it on the level surface of f that passes through $(0, 0, 1)$.

27. $f(x, y, z) = x^2 + y^2 + z^2$

28. $f(x, y, z) = z - x + y$

In Exercises 29–32, find a unit normal to the given surface at the given point.

29. $xyz = 8$ at $(1, 1, 8)$

30. $x^2 y^2 + y - z + 1 = 0$ at $(0, 0, 1)$

31. $\cos(xy) = e^z - 2$ at $(1, \pi, 0)$

32. $e^{xyz} = e$ at $(1, 1, 1)$

In Exercises 33–36, find the equation for the tangent plane to the surface at the indicated point.

33. $x^2 + 2y^2 + 3z^2 = 10$ at $(1, \sqrt{3}, 1)$

34. $xyz^2 = 1$ at $(1, 1, 1)$

35. $x^2 + 2y^2 + 3xz = 10$ at $(1, 2, 1/3)$

36. $y^2 - x^2 = 3$ at $(1, 2, 8)$

In Exercises 37–40, find the equation for the tangent line to the curve at the indicated point.

37. $x^2 + 2y^2 = 3$ at $(1, 1)$

38. $xy = 17$ at $\left(x_0, \dfrac{17}{x_0}\right)$

39. $\cos(x + y) = 0$ at $x = \pi/2, y = 0$

40. $e^{xy} = 2$ at $(1, \log 2)$

41. Suppose that $f(\mathbf{c}(t))$ attains a minimum at the time t_0. What can you say about the angle between $\nabla f(\mathbf{c}(t_0))$ and $\mathbf{c}'(t_0)$?

42. The height h of the Hawaiian volcano Mauna Loa is (roughly) described by the function $h(x, y) = 2.59 - 0.00024y^2 - 0.00065x^2$, where h is the height above sea level in miles and x and y measure east–west and north–south distances in miles from the top of the mountain.
At $(x, y) = (-2, -4)$:

(a) How fast is the height increasing in the direction $(1,1)$ (that is, northeastward)? Express your answer in miles of height per mile of horizontal distance travelled.

(b) In what direction is the steepest upward path?

43. Captain Astro is once again in trouble near the sunny side of Mercury. She is at location $(1,1,1)$, and the temperature of the ship's hull when she is at location (x,y,z) will be given by $T(x,y,z) = e^{-x^2-2y^2-3z^2}$, where x, y, and z are measured in meters.

(a) In what direction should she proceed in order to decrease the temperature most rapidly?

(b) If the ship travels at e^8 meters per second, how fast will the temperature decrease if she proceeds in that direction?

(c) Unfortunately, the metal of the hull will crack if cooled at a rate greater than $\sqrt{14}e^2$ degrees per second. Describe the set of possible directions in which she may proceed to bring the temperature down at no more than that rate.

44. What rate of change does $\nabla f(x,y,z) \cdot (-\mathbf{j})$ represent?

45. A function $f(x,y)$ has, at the point $(1,3)$, directional derivatives of $+2$ in the direction toward $(2,3)$ and -2 in the direction toward $(1,4)$. Determine the gradient vector at $(1,3)$ and compute the directional derivative in the direction toward $(3,6)$.

46. Coulomb's law states that, in suitable units, the electric force on a charge q at (x,y,z) produced by a charge Q at the origin is $\mathbf{F} = Qq\mathbf{r}/r^3$. Find V so that $\mathbf{F} = -\nabla V$ and verify that $\mathbf{F}$ is orthogonal to the level surfaces of V.

2.6
Implicit
Differentiation

Many important mathematical functions are defined implicitly, an example being the inverse trigonometric function, $y = \arcsin x$, also written as $y = \sin^{-1} x$, defined implicitly by $x = \sin y$. Interestingly, partial derivatives play an important role in the process of implicit as well as explicit differentiation.

Suppose that $y = f(x)$ and that x and y satisfy the relation $x^3 + xy^2 + 8x \sin y = 0$. From one-variable calculus we say that this defines y "implicitly" as a function of x. The main point is that although we may not be able to find y as a function of x explicitly, we *can* still calculate dy/dx in terms of y and x. In this case, using the chain rule for functions of one variable to differentiate with respect to x gives

$$3x^2 + y^2 + 2xy\frac{dy}{dx} + 8\sin y + 8x\cos y\frac{dy}{dx} = 0,$$

which we can solve for dy/dx to obtain

$$\frac{dy}{dx} = -\frac{3x^2 + y^2 + 8\sin y}{2xy + 8x\cos y}.$$

A general analysis, replacing the specific relation $x^3 + xy^2 + 8x\sin y = 0$ by a general relation of the form $F(x, y) = 0$ uses partial derivatives. We differentiate $F(x, y) = 0$ with respect to x using the chain rule:

$$\frac{\partial F}{\partial x}\frac{dx}{dx} + \frac{\partial F}{\partial y}\frac{dy}{dx} = 0,$$

or, since dx/dx is equal to 1,

$$\frac{\partial F}{\partial x} + \frac{\partial F}{\partial y}\frac{dy}{dx} = 0.$$

Solving for dy/dx gives the following result:

Implicit Differentiation and Partial Derivatives

If $y = f(x)$ is a function satisfying the relation $z = F(x, y) = 0$, then

$$\frac{dy}{dx} = -\frac{\partial z/\partial x}{\partial z/\partial y},$$

i.e.,

$$\frac{dy}{dx} = f'(x) = -\frac{F_x(x, f(x))}{F_y(x, f(x))}.$$

Example 1 Suppose that y is defined implicitly in terms of x by $e^{x-y} + x^2 - y = 1$. Find dy/dx at $x = 0, y = 0$ using the formula in the preceding box.

Solution Here $z = F(x, y) = e^{x-y} + x^2 - y - 1$, so

$$\frac{\partial z}{\partial x} = e^{x-y} + 2x \quad \text{and} \quad \frac{\partial z}{\partial x}\bigg|_{\substack{x=0 \\ y=0}} = 1.$$

Likewise

$$\frac{\partial z}{\partial y} = -e^{x-y} - 1 \quad \text{and} \quad \frac{\partial z}{\partial y}\bigg|_{\substack{x=0 \\ y=0}} = -2.$$

Therefore

$$-\frac{\partial z/\partial x}{\partial z/\partial y} = \frac{1}{2}$$

and so $dy/dx = 1/2$.   ◆

The formula

$$\frac{dy}{dx} = -\frac{\partial z/\partial x}{\partial z/\partial y}$$

makes sense as long as $\partial z/\partial y \neq 0$. In fact there is a result stated below, called the *implicit function theorem*, which guarantees that $F(x,y) = 0$ does indeed define y as a function of x, provided that $\partial z/\partial y \neq 0$.

Example 2   Discuss what happens to y as a function of x if $\partial z/\partial y = 0$ for the example $x - y^3 = 0$.

Solution   The equation $z = F(x,y) = x - y^3 = 0$ implicitly defines the function $y = f(x) = \sqrt[3]{x}$. We have $\partial z/\partial x = 1$ and $\partial z/\partial y = -3y^2$, so $\partial z/\partial y$ vanishes when $y = 0$; this is just the point on the graph $y = \sqrt[3]{x}$ where the cube-root function is not differentiable and the tangent line becomes vertical. ◆

It may be that at points (x_0, y_0) on $F(x,y) = 0$ where $F_y = 0, y$ is *not* defined implicitly as a function of x. For example, consider the function $F(x,y) = x^2 + y^2 - 1$ and the equation $F(x,y) = 0$. The set of points in the xy plane satisfying this equation is the unit circle $x^2 + y^2 = 1$. (Figure 2.6.1.)

At the point $(1,0)$, $\partial F/\partial y = 2y = 0$. Near the point $(1,0)$, the unit circle cannot be described as a graph of some function $y = f(x)$ since if it could, there would have to be two y-values for each x-value, contradicting the definition of a function.

The **implicit function theorem** states that *if $\partial z/\partial y(x_0, y_0) \neq 0$, then in some disk about (x_0, y_0) the set of points (x,y) satisfying $F(x,y) = 0$ is the graph of a function $y = f(x)$ with $dy/dx = -F_x/F_y$.*

We can also find the tangent planes to surfaces defined implicitly rather than explicitly. Consider an expression of the form

$$F(x, y, z) = 0;$$

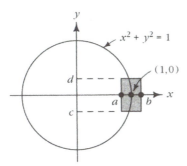

FIGURE 2.6.1. The nonsolvability of $x^2 + y^2 = 1$ for y as an function of x near $(1, 0)$.

for example $x^2 + y^2 + \sin z - 1 = 0$. When does this define z implicitly as a function of x and y; *i.e.*, when is there a function $f(x,y)$ such that $F(x, y, z) = 0$ is also described by $z = f(x, y)$? Moreover, what is the tangent plane to the surface S described by the graph of $z = f(x,y)$?

In the two-variable case a normal vector to the level curve $f(x, y) = 0$ at (x_0, y_0) is given by $\nabla f(x_0, y_0)$. In the three-variable case a normal vector to the surface $F(x, y, z) = 0$ at a point (x_0, y_0, z_0) is given by

$$\nabla F(x_0, y_0, z_0) = \frac{\partial F}{\partial x}(x_0, y_0, z_0)\mathbf{i} + \frac{\partial F}{\partial y}(x_0, y_0, z_0)\mathbf{j} + \frac{\partial F}{\partial z}(x_0, y_0, z_0)\mathbf{k}.$$

Another version of the implicit function theorem states that *provided $(\partial F/\partial z)(x_0, y_0, z_0) \neq 0$, then near (x_0, y_0, z_0) the level surface S is the graph of some differentiable function $z = f(x, y)$;* *i.e.*, we can solve the equation

$$F(x, y, z) = 0$$

as $z = f(x, y)$ for all (x, y, z) "near" (x_0, y_0, z_0). The tangent plane to the level surface S at (x_0, y_0, z_0) is the plane through (x_0, y_0, z_0) that is perpendicular to $\nabla F(x_0, y_0, z_0)$. From §1.4, the equation of the plane that has a normal $A\mathbf{i} + B\mathbf{j} + C\mathbf{k}$ and that passes through (x_0, y_0, z_0) is given by

$$A(x - x_0) + B(y - y_0) + C(z - z_0) = 0.$$

Therefore the equation of the plane tangent to the level surface $F(x, y, z) = 0$ at (x_0, y_0, z_0) is given by

$$F_x(x - x_0) + F_y(y - y_0) + F_z(z - z_0) = 0$$

where the partial derivatives F_x, F_y, F_z are evaluated at (x_0, y_0, z_0).

We summarize this as follows.

Level Surfaces and the Implicit Function Theorem

Consider a level surface S described by $F(x, y, z) = 0$, with F a C^1 function. Let (x_0, y_0, z_0) be a point on S satisfying

$$\frac{\partial F}{\partial z}(x_0, y_0, z_0) \neq 0.$$

Then near (x_0, y_0, z_0), S can be described as a graph of a C^1 function $z = f(x, y)$. The equation of the tangent plane at (x_0, y_0, z_0) is given by

$$F_x(x - x_0) + F_y(y - y_0) + F_z(z - z_0) = 0,$$

all partial derivatives of F being evaluated at (x_0, y_0, z_0).

Example 3   Near what points may the surface

$$x^3 + 3y^2 + 8xz^2 - 3z^3y = 1$$

be represented as a graph of a differentiable function $z = k(x, y)$?

Solution   Here we take $F(x, y, z) = x^3 + 3y^2 + 8xz^2 - 3z^3y - 1$ and attempt to solve $F(x, y, z) = 0$ for z as a function of (x, y). By the implicit function theorem, this may be done near a point (x_0, y_0, z_0) if $(\partial F/\partial z)(x_0, y_0, z_0) \neq 0$, that is, if

$$z_0(16x_0 - 9z_0y_0) \neq 0,$$

which means, in turn,

$$z_0 \neq 0 \quad \text{and} \quad 16x_0 \neq 9z_0y_0. \quad \blacklozenge$$

Example 4   Find the equation of the tangent plane to the level surface

$$F(x, y, z) = x^2 + y^2 + \sin z - 1 = 0$$

at $(1, 0, 0)$.

Solution   The partial derivatives are $F_x = 2x, F_y = 2y$ and $F_z = \cos z$, so $F_x(1,0,0) = 2, F_y(1,0,0) = 0$, and $F_z(1,0,0) = 1$. Therefore, the equation of the tangent plane is

$$2(x-1) + z = 0. \quad \blacklozenge$$

Exercises for §2.6

In Exercises 1–4, suppose that y is defined implicitly in terms of x by the given equation. Find dy/dx.

1. $x^2 + 2y^2 = 3$

2. $x^2 - y^2 = 7$

3. $x/y = 10$

4. $y - \sin x^3 + x^2 - y^2 = 1$

In Exercises 5–8, find dy/dx at the indicated point.

5. $3x^2 + y^2 - e^x = 0; x = 0, y = 1$

6. $x^2 + y^4 = 2; x = 1, y = 1$

7. $\cos(x + y) = x + 1/2; x = 0, y = \pi/3$

8. $\cos(xy) = 1/2; x = 1, y = \pi/3$

In Exercises 9–12, determine near what points the surface $F(x, y, z) = 0$ can be described as the graph of a differentiable function $z = f(x, y)$.

9. $x^2 + xyz + 2z^3 = 0$

10. $z^2 x^3 + xy^2 - 4 = 0$

11. $x^3 + y^3 + z^3 = 1$

12. $x = y^2 z^2$

In Exercises 13–16, find the equation of the tangent plane to the level surface $F(x, y, z) = 0$ for the given function F.

13. F as in Exercise 9 at the point $(1, -1, -1)$

14. F as in Exercise 10 at the point $(1, 0, 2)$

15. F as in Exercise 11 at the point $(0, 0, 1)$

16. F as in Exercise 12 at the point $(1, 1, 1)$

17.(a) Check *directly* where we can solve the equation $F(x, y) = y^2 + y + 3x + 1 = 0$ for y in terms of x. Calculate dy/dx.

   (b) Check that your answer in part (a) agrees with the answer you expect from implicit differentiation.

18.(a) Show that the curve $x^2 - y^2 = c$, for *any* value of c, satisfies the differential equation $dy/dx = x/y$.

   (b) Draw in a few of the curves $x^2 - y^2 = c$, say for $c = \pm 1$. At several points (x, y) along each of these curves, draw a short segment of slope x/y. Check that these segments appear to be tangent to the curve. What happens when $y = 0$? What happens when $c = 0$?

19. Suppose that a particle is ejected from the surface $x^2 + y^2 - z^2 = -1$ at the point $(1, 1, \sqrt{3})$ in a direction normal to the surface at time $t = 0$ with a speed of 10 units per second. When and where does it cross the xy plane?

20. Let y be a function of x satisfying $F(x, y, x + y) = 0$, where $F(x, y, z)$ is a given function. Find a formula for dy/dx.

21. Suppose that $z = f(x, y)$ is defined implicitly by an equation of the form $F(x, y, z) = 0$. Find formulas for the partial derivatives f_x and f_y in terms of F_x, F_y, and F_z.

Review Exercises for Chapter 2

In Exercises 1–4, draw some level curves (in the xy plane) for the given function f and specified values of c.

1. $f(x, y) = 4 - 3x + 2y, c = 0, 1, 2, 3$

2. $f(x, y) = (x^2 + y^2)^{1/2}, c = 0, 1, 2, 3, 4, 5$

3. $f(x, y) = x^2 + y^2, c = 0, 1, 2$

4. $f(x, y) = x^2 + 2y^2 + 1, c = -10, -1, 0, 1, 2, 10$

In Exercises 5–8, describe the level surfaces $f(x, y, z) = c$ for the given function f and values c.

5. $f(x, y, z) = 2x^2 + y^2 + z^2, c = 0, c = 1$

6. $f(x, y, z) = x^2, c = 0, c = 1, c = 2$

7. $f(x, y, z) = xyz, c = 0$

8. $f(x, y, z) = x^2 + y^2 - z^2, c = 0$

In Exercises 9–12, sketch or describe the surfaces in $\mathbb{R}^3$ defined by the given equations.

9. $\dfrac{x}{4} = \dfrac{y^2}{4} + \dfrac{z^2}{4}$

10. $x^2 + y^2 - 2x = 0$

11. $z^2 = y^2 + 4$

12. $x = y^2/25 - z^2/9$

Calculate all first partial derivatives for the functions in Exercises 13–16.

13. $u = g(x, y) = \dfrac{\sin(\pi x)}{1 + y^2}$

14. $u = f(x, z) = \dfrac{x}{1 + \cos(2z)}$

15. $u = k(x, z) = xz^2 - \cos(xz^3)$

16. $u = n(x, y, z) = x^{yz}$

17. Find $\left. \dfrac{\partial}{\partial x} e^{x - \cos(xy)} \right|_{x=1, y=0}$.

18. Find $\left. \dfrac{\partial}{\partial s} \exp(rs^3 - r^3 s) \right|_{r=1, s=1}$.

19. Find $f_x(1, 0)$ if $f(x, y) = \cos(x + e^{xy})$.

20. Find $f_s(-1, 2)$ if $f(r, s) = (r + s^2)/(1 - r^2 - s^2)$.

Find the limits in Exercises 21–24, if they exist.

21. $\lim_{(x,y)\to(0,0)} (x^2 - 2xy + 4)$

22. $\lim_{(x,y)\to(0,0)} (x^3 - y^3 + 15)$

23. $\lim_{(x,y)\to(0,0)} \left(\dfrac{x^3 - y^3}{x^2 + y^2} \right)$

24. $\lim\limits_{(x,y)\to(0,0)} \dfrac{xy + x^3 + x - 2}{\sqrt{x^2 + y^2}}$

In Exercises 25–28, find the equation of the tangent plane to the given surface at the indicated point.

25. $z = x^2 + y^2; x = 1, y = 1$

26. $z = x \sin y; x = 2, y = \pi/4$

27. $z = e^{xy}; x = 0, y = 0$

28. $z = \sqrt{x^2 + y^2}; x = 3, y = 4$

In Exercises 29–32, compute the derivative matrix $\mathbf{D}f(x)$ of the given mapping.

29. $f(x, y) = (x^2 y, e^{-xy})$

30. $f(x) = (x, x)$

31. $f(x, y, z) = e^x + e^y + e^z$

32. $f(x, y, z) = (x, y, z)$

In Exercises 33–36, use the chain rule to find $\partial z/\partial x$ and $\partial z/\partial y$. Verify your results using direct substitution and then differentiating.

33. $z = \dfrac{u^2 + v^2}{u^2 - v^2}, u = e^{-x-y}, v = e^{xy}$

34. $z = uv, u = x + y, v = x - y$

35. $z = e^{uv}, u = x + y, v = x - y$

36. $z = \sin(uv), u = x^2, v = x^2 + y^2$

In Exercises 37–40, compute the matrices $\mathbf{D}G(1, 1)$ and $\mathbf{D}F(G(1, 1))$ and use matrix multiplication to compute $\partial z/\partial x$ at $(1, 1)$, where $z(x, y) = F(G(x, y))$. Verify that your answers agree with Exercises 33–36.

37. $F(u, v) = \dfrac{u^2 + v^2}{u^2 - v^2}, G(x, y) = (e^{-x-y}, e^{xy})$

38. $F(u, v) = uv, G(x, y) = (x + y, x - y)$

39. $F(u, v) = e^{uv}, G(x, y) = (x + y, x - y)$

40. $F(u, v) = \sin(uv), G(x, y) = (x^2, x^2 + y^2)$

In Exercises 41–44, compute the gradient of the given function.

41. $f(x, y, z) = xe^z + y \cos x$

42. $f(x, y, z) = (x + y + z)^{10}$

43. $f(x, y, z) = (x^2 + y)/z$

44. $f(x, y) = xy \sin \left[\dfrac{1}{x^2 + y^2} \right]$ if $(x, y) \neq (0, 0)$, $f(0, 0) = 0$

In Exercises 45 and 46, sketch the gradient vector field of the given function.

45. $f(x, y) = xy$

46. $f(x, y) = e^{xy}$

In Exercises 47 and 48, verify the chain rule for the given function and curve.

47. $f(x, y) = x \sin y, \mathbf{c}(t) = (t^2, t^3)$

48. $f(x, y) = e^{xy}(x - y), \mathbf{c}(t) = (t, 1/t)$

In Exercises 49 and 50, at the given point calculate (a) the directional derivative of the function in the direction $\mathbf{d}$ and (b) the direction in which the function is increasing most rapidly.

49. $f(x, y) = \sin(x^3 - 2y^3); \mathbf{d} = \dfrac{1}{\sqrt{2}}(\mathbf{i} - \mathbf{j}); (1, -1)$

50. $f(x, y) = 17x^y; \mathbf{d} = \dfrac{1}{\sqrt{2}}(\mathbf{i} + \mathbf{j}); (1, 1)$

51. A bug finds itself in a toxic environment. The toxicity level is given by $T(x, y) = 2x^2 - 4y^2$. The bug is at $(-1, 2)$. In what direction should it move to lower the toxicity fastest?

52. Find the direction in which the function $w = x^2 + xy$ increases most rapidly at the point $(-1, 1)$. What is the magnitude of ∇w in this direction? Interpret this magnitude geometrically.

In Exercises 53–56, find the equation of the plane tangent to the given surface at the indicated point.

53. $z = x^3 + 2y; (1, 1, 3)$

54. $z = \cos(x^2 + y^2); (0, 0, 1)$

55. $x^2 + y^2 + z^2 = 1; \left(\dfrac{1}{\sqrt{3}}, \dfrac{1}{\sqrt{3}}, \dfrac{1}{\sqrt{3}} \right)$

56. $x^3 + y^3 + z^3 = 3; (1,1,1)$

In Exercises 57 and 58, suppose that x and y are related by the equation given. Find dy/dx at the indicated point.

57. $x + \cos y = 1; (1, \pi/2)$

58. $x^4 + y^4 = 17; (-1, 2)$

In Exercises 59 and 60, suppose that $x = f(t)$ and $y = g(t)$ satisfy the given relation. Relate dx/dt and dy/dt.

59. $x^2 + xy - y^2 = 1$

60. $\cos(x - y) = 1/2$

61. Show that $x^3 - \sin y + y^4 + 7 = 8$ defines $y = f(x)$ as a function of x near the point $(1, 0)$; calculate $f'(1)$.

62. Show that $\sin(zx) - 2zy = 0$ defines a function $z = f(x, y)$ near $(1, 1, 0)$. Calculate $f_x(1, 1)$ and $f_y(1, 1)$.

63. A circular cone of sand is gradually collapsing. At a certain moment, the cone has a height of 10 meters and a base radius of 3 meters. If the height of the cone is decreasing at a rate of 1 meter per hour, how is the radius changing, assuming that the volume remains constant?

64. If $z = f(x - y)/y$, show that $z + y(\partial z/\partial x) + y(\partial z/\partial y) = 0$.

65. The displacement of a certain violin string placed on the x axis is given by $u = \sin(x - 6t) + \sin(x + 6t)$. Calculate the velocity of the string at $x = 1$ when the time t is $1/3$.

66. The height of water in a fish tank is $h(x, y, t) = 10 + \sin[(x^2 + 1)t] - \cos[3(x^2 + y^2 + 1)t]$ where $h(x, y, t)$ is the height of the water above the location (x, y) at the bottom of the tank at time t. Find the vertical velocity of the water's surface at $(1, 1)$ when $t = \pi/6$ and the equation of the tangent plane to the water's surface at that point.

3

Higher Derivatives and Extrema

*Because the shape of the whole universe is most perfect, and, in fact
designed by the wisest creator, nothing in all of the world will occur
in which no maximum or minimum rule is somehow shining forth.*

Leonhard Euler

In one-variable calculus, we search for the local maximum or minimum points of a differentiable function $f(x)$ by looking for critical points; that is, points x_0 for which $f'(x_0) = 0$. At each such point we check the sign of the second derivative $f''(x_0)$. If $f''(x_0) < 0$, $f(x_0)$ is a local maximum of f; if $f''(x_0) > 0$, $f(x_0)$ is a local minimum of f; if $f''(x_0) = 0$, the test fails. There is also a *first* derivative test: We search for absolute maxima and minima on a closed interval by finding the maxima and minima of f among the critical points and endpoints.

This chapter extends these methods to functions of several variables. We begin in §**3.1** with a discussion of higher order derivatives, followed by Taylor's theorem in §**3.2**. Maxima and minima are defined in §**3.3**, and the critical point test for locating them is presented. Tests using second derivatives are contained in §**3.4**.

In §**3.5** we study the problem of maximizing a real-valued function subject to supplementary conditions, also referred to as constraints and of maximizing a function defined on a bounded region of space. For example, we might wish to maximize $f(x, y, z)$ among those (x, y, z) constrained to lie on the unit sphere, $x^2 + y^2 + z^2 = 1$.

3.1
Higher Order Partial Derivatives

Since the partial derivatives of a function are new functions, we can also take their partial derivatives to obtain the **second partial derivatives** of f. Repeating this process, we obtain partial derivatives of third and higher order. As in one-variable calculus, the second derivatives are especially important since they determine which way the graphs "bend." Second derivatives also occur in many physical laws.

Functions $f(x, y)$ of two variables have two (first) partial derivatives f_x and f_y, and four second partial derivatives

$$(f_x)_x, \quad (f_x)_y, \quad (f_y)_x, \quad \text{and} \quad (f_y)_y.$$

The notation $(f_x)_x$ means the derivative of f_x with respect to x, $(f_x)_y$ the derivative of f_x with respect to y, etc. In $\partial/\partial x$ notation, these can be written

$$\frac{\partial}{\partial x}\left(\frac{\partial f}{\partial x}\right), \quad \frac{\partial}{\partial y}\left(\frac{\partial f}{\partial x}\right), \quad \frac{\partial}{\partial x}\left(\frac{\partial f}{\partial y}\right), \quad \text{and} \quad \frac{\partial}{\partial y}\left(\frac{\partial f}{\partial y}\right)$$

or if $z = f(x, y)$, we also write

$$\frac{\partial}{\partial x}\left(\frac{\partial z}{\partial x}\right), \quad \frac{\partial}{\partial y}\left(\frac{\partial z}{\partial x}\right), \quad \frac{\partial}{\partial x}\left(\frac{\partial z}{\partial y}\right), \quad \text{and} \quad \frac{\partial}{\partial y}\left(\frac{\partial z}{\partial y}\right).$$

Removing the parentheses, we can simplify this notation by using f_{xx}, f_{xy}, f_{yx} and f_{yy},

$$\frac{\partial^2 f}{\partial x^2}, \quad \frac{\partial^2 f}{\partial y \partial x}, \quad \frac{\partial^2 f}{\partial x \partial y}, \quad \text{and} \quad \frac{\partial^2 f}{\partial y^2}$$

or

$$\frac{\partial^2 z}{\partial x^2}, \quad \frac{\partial^2 z}{\partial y \partial x}, \quad \frac{\partial^2 z}{\partial x \partial y}, \quad \text{and} \quad \frac{\partial^2 z}{\partial y^2}.$$

Example 1   Find all second partial derivatives of $f(x, y) = x \log y$.

Solution

$$f_x = \log y, \quad f_y = \frac{x}{y}$$

$$f_{xx} = \frac{\partial^2 f}{\partial x^2} = 0, \quad f_{yy} = \frac{\partial^2 f}{\partial y^2} = -\frac{x}{y^2}$$

$$f_{xy} = \frac{\partial^2 f}{\partial y \partial x} = \frac{1}{y}, \quad f_{yx} = \frac{\partial^2 f}{\partial x \partial y} = \frac{1}{y}. \quad \blacklozenge$$

Example 2   Find all second partial derivatives of $f(x, y) = \sin(x^2 y)$.

Solution

$$f_x = \frac{\partial f}{\partial x} = 2xy \cos(x^2 y), \quad f_y = x^2 \cos(x^2 y)$$

$$f_{xx} = \frac{\partial^2 f}{\partial x^2} = 2y \cos(x^2 y) - 4x^2 y^2 \sin(x^2 y),$$

$$f_{yy} = \frac{\partial^2 f}{\partial y^2} = -x^4 \sin(x^2 y)$$

$$f_{xy} = \frac{\partial^2 f}{\partial y \partial x} = 2x \cos(x^2 y) - 2x^3 y \sin(x^2 y),$$

$$f_{yx} = \frac{\partial^2 f}{\partial x \partial y} = 2x \cos(x^2 y) - 2x^3 y \sin(x^2 y). \quad \blacklozenge$$

Although the *calculations* that produce f_{xy} and f_{yx} in Examples 1 and 2 are quite different, the *results* are identical. These are not coincidences, but instances of a general fact called the *equality of mixed partial derivatives*, the central result of this section. Leonhard Euler proved this result in 1734 in connection with his studies of the motion of fluids.

It is no exaggeration to say that the entire structure of vector analysis is crucially dependent on the equality of mixed partial derivatives.

Equality of Mixed Partial Derivatives

If $u = f(x, y)$ has continuous second partial derivatives, then the mixed partial derivatives are equal; that is,

$$f_{xy} = f_{yx}$$

or

$$\frac{\partial^2 f}{\partial y \partial x} = \frac{\partial^2 f}{\partial x \partial y}.$$

Proof We consider the following expression

$$S(\Delta x, \Delta y) = f(x_0 + \Delta x, y_0 + \Delta y) - f(x_0 + \Delta x, y_0) - f(x_0, y_0 + \Delta y) + f(x_0, y_0)$$

Holding y_0 and Δy fixed, define

$$g(x) = f(x, y_0 + \Delta y) - f(x, y_0),$$

so that $S(\Delta x, \Delta y) = g(x_0 + \Delta x) - g(x_0)$, *which expresses S as a difference of differences.* By the mean value theorem for functions of one variable, this difference equals $g'(\bar{x})\Delta x$ for some $\bar{x}$ between x_0 and $x_0 + \Delta x$. Hence

$$S(\Delta x, \Delta y) = \left[\frac{\partial f}{\partial x}(\bar{x}, y_0 + \Delta y) - \frac{\partial f}{\partial x}(\bar{x}, y_0) \right] \Delta x.$$

Applying the mean value theorem for the y variable this time gives

$$S(\Delta x, \Delta y) = \frac{\partial^2 f}{\partial y \partial x}(\bar{x}, \bar{y})(\Delta x \Delta y).$$

Since $\partial^2 f / \partial y \partial x$ is continuous, it follows that

$$\frac{\partial^2 f}{\partial y \partial x}(x_0, y_0) = \lim_{(\Delta x, \Delta y) \to (0,0)} \frac{1}{\Delta x \Delta y}[S(\Delta x, \Delta y)].$$

By exchanging the roles of x and y, one shows that $\partial^2 f / \partial x \partial y$ is given by the *same* limit formula. ■

There is a purely algebraic counterpart to this proof which may help you to understand it. Place four numbers $A, B, C,$ and D at the vertices of a rectangle, as in Figure 3.1.1. We can write the difference of differences (a *second* difference) in two ways:

$$(D - B) - (C - A) = (D - C) - (B - A)$$

(vertical difference, then horizontal) = (horizontal difference, then vertical).

Thus,

$$(f(x_0 + \Delta x, y_0 + \Delta y) - f(x_0 + \Delta x, y_0)) - (f(x_0, y_0 + \Delta y) - f(x_0, y_0))$$
$$= (f(x_0 + \Delta x, y_0 + \Delta y) - f(x_0, y_0 + \Delta y)) - (f(x_0 + \Delta x, y_0) - f(x_0, y_0)).$$

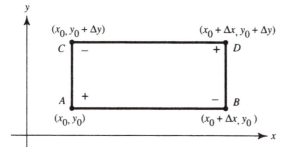

FIGURE 3.1.1. The algebra behind the equality of mixed partials, writing the difference of differences in two ways.

The equality of mixed partial derivatives is valid as well for functions of more than two variables. Differentiating $f(x_1, \ldots, x_n)$ by x_i and then by x_j gives $\dfrac{\partial}{\partial x_j}\left(\dfrac{\partial f}{\partial x_i}\right)$, which is also written without parentheses as

$$\frac{\partial^2 f}{\partial x_j \partial x_i}.$$

Example 3   Let $f(x, y, z) = e^{xy} + z\cos x$. Then

$$\frac{\partial f}{\partial x} = ye^{xy} - z\sin x, \quad \frac{\partial f}{\partial y} = xe^{xy}, \quad \frac{\partial f}{\partial z} = \cos x,$$

$$\frac{\partial^2 f}{\partial z \partial x} = -\sin x, \quad \frac{\partial^2 f}{\partial x \partial z} = -\sin x, \quad \text{etc.} \quad \blacklozenge$$

Equality of Mixed Partial Derivatives

If $f(x_1, \ldots, x_n)$ has continuous first and second partial derivatives, then the mixed partial derivatives are equal:

$$f_{x_i x_j} = f_{x_j x_i}$$

$$\frac{\partial^2 f}{\partial x_j \partial x_i} = \frac{\partial^2 f}{\partial x_i \partial x_j}.$$

After we know that mixed second partial derivatives are equal, then the equality of higher order derivatives follows as well under the assumption of continuity of the partial derivatives up to the required order. For example, if $f(x, y, z)$ is a function of x, y, z, then $\partial^3 f / \partial x \partial y \partial z$ represents the derivative of f first with respect to z, then with respect to y, and then with respect to x.

Example 4   Prove that

$$\frac{\partial^3 f}{\partial x \partial y \partial z} = \frac{\partial^3 f}{\partial y \partial z \partial x}.$$

Solution   By equality of mixed partials,

$$
\begin{aligned}
\frac{\partial^3 f}{\partial y \partial z \partial x} &= \frac{\partial}{\partial y}\left(\frac{\partial^2 f}{\partial z \partial x}\right) = \frac{\partial}{\partial y}\left(\frac{\partial^2 f}{\partial x \partial z}\right) \\
&= \frac{\partial^2}{\partial y \partial x}\left(\frac{\partial f}{\partial z}\right) = \frac{\partial^2}{\partial x \partial y}\left(\frac{\partial f}{\partial z}\right) = \frac{\partial^3 f}{\partial x \partial y \partial z}. \quad \blacklozenge
\end{aligned}
$$

We end this section with a discussion of the role that equations involving partial derivatives (partial differential equations) have played in the development of science.

Historical Note: Some Partial Differential Equations

Philosophy [nature] is written in that great book which ever is before our eyes—I mean the universe—but we cannot understand it if we do not first learn the language and grasp the symbols in which it is written. The book is written in mathematical language, and the symbols are triangles, circles, and other geometrical figures, without whose help it is impossible to comprehend a single word of it; without which one wanders in vain through a dark labyrinth. —Galileo

This quotation illustrates the belief, already popular in Galileo's time, that the workings of nature could be reduced to mathematics. In the latter part of the seventeenth century this thinking was dramatically reinforced when Newton used his law of gravitation and the new calculus to derive Kepler's three laws of celestial motion (see §**4.1**). This philosophy had a substantial impact on mathematics, and many mathematicians sought to "mathematize" nature. The extent to which mathematics pervades the physical sciences today (and, to an increasing amount, economics and the social and life sciences) is a testament to the success of these endeavors. In addition, attempts to mathematize nature have often led to new discoveries in mathematics itself.

Many of the laws of nature were described in terms of either ordinary differential equations (equations involving the derivatives of functions of one variable alone, such as $\mathbf{F} = md^2\mathbf{x}/dt^2$, where $\mathbf{F}$ is given by Newton's law of gravitation) or partial differential equations, that is, equations involving partial derivatives of functions. To give the reader some historical perspective and offer motivation for studying partial derivatives, we present a brief description of three of the most famous partial differential equations: the heat equation, the potential equation (or Laplace's equation), and the wave equation.

The Heat Equation In the early part of the nineteenth century, the French mathematician Joseph Fourier (1768–1830) took up the study of heat. Heat flow had obvious applications to both industrial and scientific problems. A better understanding of it would, for example, make possible more efficient smelting of metals and would enable scientists to determine the temperature of a body, given the temperature at its boundary, and thereby to approximate the temperature of the earth's interior. (For a more domestic application, see "On cooking a roast," M.S. Klamkin, *SIAM Review* 3 (1961), 167–169.)

Let a homogeneous body be represented by some region $\mathcal{B}$ in 3-space (Figure 3.1.2). Let $T(x, y, z, t)$ denote the temperature of the body at the point (x, y, z) at time t. Fourier proved, on the basis of energy principles, that T must satisfy the **heat equation**,

$$k\left(\frac{\partial^2 T}{\partial x^2} + \frac{\partial^2 T}{\partial y^2} + \frac{\partial^2 T}{\partial z^2}\right) = \frac{\partial T}{\partial t},$$

where k is a constant whose value depends on the conductivity of the material comprising the body.

Fourier used this equation to solve problems in heat conduction. In fact, his investigations into the solutions of the heat equation led him to the discovery of *Fourier series*, a way of decomposing arbitrary functions into sines and cosines.

The Potential Equation According to Example 13 of §**2.5**, the gravitational potential for a mass m at a point (x, y, z) caused by a point mass M situated at the origin is $V = -GmM/r$, where $r = \sqrt{x^2 + y^2 + z^2}$. The potential V satisfies **Laplace's equation**:

$$\frac{\partial^2 V}{\partial x^2} + \frac{\partial^2 V}{\partial y^2} + \frac{\partial^2 V}{\partial z^2} = 0$$

everywhere except at the origin. Pierre Simon de Laplace (1749–1827) studied this equation in his work on the gravitational attraction of extended masses. He gave arguments (later shown to be incorrect) that his equation held for any body and any point, whether inside or outside that body. Laplace was not the first person to write down "Laplace's equation." It appeared for the first time in one of Euler's major papers in 1752, "Principles of the Motions of Fluids," in which he derived the potential equation in connection with the motion of (incompressible) fluids. Euler remarked that he had no idea how to solve the equation. Poisson later showed that if (x, y, z) lies *inside* an attracting body, then V satisfies **Poisson's equation**

$$\frac{\partial^2 V}{\partial x^2} + \frac{\partial^2 V}{\partial y^2} + \frac{\partial^2 V}{\partial z^2} = -4\pi\rho$$

where ρ is the mass density of the attracting body. He was also the first to point out the importance of this equation for problems involving electric fields. Laplace's and Poisson's equations are fundamental to many fields besides fluid mechanics, gravitational fields, and electrostatic fields. For example they are useful for studying soap films and liquid crystals; see *Mathematics and Optimal Form* by S. Hildebrandt and A. Tromba, Scientific American Books, New York, 1985.

The Wave Equation The linear wave equation in space has the form

$$\frac{\partial^2 f}{\partial x^2} + \frac{\partial^2 f}{\partial y^2} + \frac{\partial^2 f}{\partial z^2} = c^2 \frac{\partial^2 f}{\partial t^2}.$$

The one-dimensional wave equation

$$\frac{\partial^2 f}{\partial x^2} = c^2 \frac{\partial^2 f}{\partial t^2}$$

was derived in 1727 by John Bernoulli and several years later by Jean Le Rond d'Alembert in the study of a vibrating string (such as a violin string). The wave equation is useful in the study of sound and water waves, vibrating bodies, and the propagation of electromagnetic radiation.

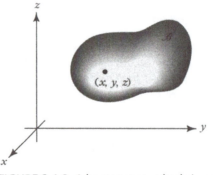

FIGURE 3.1.2. A homogeneous body in space.

Example 5   Does $u(x, y) = x^3 - 3xy^2$ satisfy Laplace's equation?

Solution   We calculate

$$\frac{\partial u}{\partial x} = 3x^2 - 3y^2 \quad \text{and so} \quad \frac{\partial^2 u}{\partial x^2} = 6x.$$

Also,

$$\frac{\partial u}{\partial y} = -6xy \quad \text{and so} \quad \frac{\partial^2 u}{\partial y^2} = -6x.$$

Thus, $\dfrac{\partial^2 u}{\partial x^2} + \dfrac{\partial^2 u}{\partial y^2} = 0$, so Laplace's equation is satisfied.   ◆

We conclude with the use of higher order partial derivatives in studying a partial differential equation.

Note

The calculation in the next example uses the hyperbolic secant function. You should familiarize yourself with hyperbolic functions before proceeding, by consulting your one-variable calculus text. Talk to your instructor if you have not covered them before.

Example 6   The partial differential equation $u_t + u_{xxx} + u u_x = 0$, called the **Korteweg–de Vries equation** (or KdV equation, for short), describes the motion of water waves in a shallow channel.

(a) Show that for any positive constant c, the function

$$u(x,t) = 3c \operatorname{sech}^2 \left[\frac{1}{2}(x - ct)\sqrt{c}\right]$$

is a solution of the Korteweg–de Vries equation. (This solution represents a traveling "hump" of water in the channel and is called a **soliton**. Solitons were first observed by J. Scott Russell around 1840 in barge canals near Edinburgh.)

(b) How do the shape and speed of the soliton depend on c?

Solution

(a) We compute u_t, u_x, u_{xx} and u_{xxx} using the chain rule and the differentiation formula $(d/dx)\operatorname{sech} x = -\operatorname{sech} x \tanh x$ from one-variable calculus. Letting $\alpha = (x - ct)\sqrt{c}/2, u = 3c \operatorname{sech}^2 \alpha$, so that

$$\begin{aligned} u_t &= 6c \operatorname{sech} \alpha \frac{\partial}{\partial t} \operatorname{sech} \alpha = -6c \operatorname{sech}^2 \alpha \tanh \alpha \frac{\partial \alpha}{\partial t} \\ &= 3c^{5/2} \operatorname{sech}^2 \alpha \tanh \alpha = c^{3/2} u \tanh \alpha. \end{aligned}$$

Also,

$$\begin{aligned} u_x &= -6c \operatorname{sech}^2 \alpha \tanh \alpha \frac{\partial \alpha}{\partial x} \\ &= -3c^{3/2} \operatorname{sech}^2 \alpha \tanh \alpha = -\sqrt{c} u \tanh \alpha, \end{aligned}$$

and so $u_t = -cu_x$. Next,

$$u_{xx} = -\sqrt{c}\left[u_x \tanh\alpha + u\left(\operatorname{sech}^2\alpha\frac{\sqrt{c}}{2}\right)\right]$$

$$= -\sqrt{c}(\tanh\alpha)u_x - \frac{u^2}{6}$$

$$= c(\tanh^2\alpha)u - \frac{u^2}{6} = c(1 - \operatorname{sech}^2\alpha)u - \frac{u^2}{6}$$

$$= cu - \frac{u^2}{3} - \frac{u^2}{6} = cu - \frac{u^2}{2}.$$

Thus, $u_{xxx} = cu_x - uu_x$. Hence,

$$u_t + u_{xxx} + uu_x = u_t + cu_x = 0.$$

(b) The speed of the soliton is c, since $u(x+ct,t) = u(x,0)$. The soliton is higher and thinner when c is larger. Its shape at time $t = 10$ is shown in Figure 3.1.3.   ◆

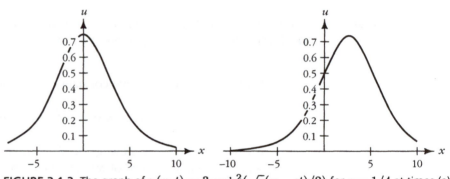

FIGURE 3.1.3. The graph of $u(x,t) = 3\operatorname{sech}^2(\sqrt{c}(x-ct)/2)$ for $c = 1/4$ at times (a) $t = 0$ and (b) $t = 10$.

Exercises for §3.1

For the given functions in Exercises 1–4, calculate each of the second partial derivatives

$$\frac{\partial^2 f}{\partial x^2}, \quad \frac{\partial^2 f}{\partial x\partial y}, \quad \frac{\partial^2 f}{\partial y\partial x}, \quad \text{and} \quad \frac{\partial^2 f}{\partial y^2},$$

and verify the equality of mixed partials.

1. $f(x,y) = 2xy/(x^2 + y^2)^2$, $(x,y) \neq (0,0)$

2. $f(x,y) = \cos(x^2 y^2)$

3. $f(x,y) = \dfrac{1}{\cos^2 x + e^{-y}}$

4. $f(x, y, z) = e^z + \dfrac{1}{x} + xe^{-y}, x \neq 0$

5. Let $f(x, y, z) = x^2 y + xy^2 + yz^2$. Find $f_{xy}, f_{yz}, f_{zx},$ and f_{xyz}.

6. Let $z = x^4 y^3 - x^8 + y^4$.

   (a) Compute $\dfrac{\partial^3 z}{\partial x \partial y \partial y}, \dfrac{\partial^3 z}{\partial y \partial x \partial y},$ and $\dfrac{\partial^3 z}{\partial y \partial y \partial x}.$

   (b) Compute $\dfrac{\partial^3 z}{\partial y \partial x \partial x}, \dfrac{\partial^3 z}{\partial x \partial y \partial x},$ and $\dfrac{\partial^3 z}{\partial x \partial x \partial y}.$

7. Show that $\dfrac{\partial^3 f}{\partial x \partial y \partial z} = \dfrac{\partial^3 f}{\partial x \partial z \partial y}.$

8. Show that $\dfrac{\partial^3 f}{\partial x^2 \partial y} = \dfrac{\partial^3 f}{\partial x \partial y \partial x}.$

9. Verify that

$$\frac{\partial^3 f}{\partial x \partial y \partial z} = \frac{\partial^3 f}{\partial z \partial y \partial x}$$

   for $f(x, y, z) = ze^{xy} + yz^3 x^2$.

10. Verify that $f_{xzw} = f_{zwx}$ for $f(x, y, z, w) = e^{xyz} \sin(xw)$.

11. Does $u(x, y) = x^2 - y^2$ satisfy Laplace's equation? Explain.

12. Does $f(x, y) = e^x \sin y$ satisfy Laplace's equation? Explain.

13. Let f and g be differentiable functions of one variable. Set $\phi = f(x - t) + g(x + t)$.

   (a) Prove that ϕ satisfies the wave equation: $\partial^2 \phi / \partial t^2 = \partial^2 \phi / \partial x^2$.

   (b) Sketch the graph of ϕ against t and x if $f(x) = x^2$ and $g(x) = 0$.

14. Let $w = f(x, y)$ be a function of two variables, and let $x = u + v, y = u - v$. Show that

$$\frac{\partial^2 w}{\partial u \partial v} = \frac{\partial^2 w}{\partial x^2} - \frac{\partial^2 w}{\partial y^2}.$$

15. Let $f(x, y, z) = e^{x+y} \sin z$, and let $x = g(s, t), y = h(s, t), z = k(s, t)$, and $m(s, t) = f(g(s, t), h(s, t), k(s, t))$. Find a formula for m_{st} using the chain rule, and verify that your answer is symmetric in s and t.

16.(a) Show that the function $g(x, t) = 2 + e^{-t} \sin x$ satisfies the heat equation: $g_t = g_{xx}$. [Here $g(x, t)$ represents the temperature in a metal rod at position x and time t.]

(b) Sketch the graph of g for $t \geq 0$. (Hint: Look at sections by the planes $t = 0, t = 1$, and $t = 2$.)

(c) What happens to $g(x, t)$ as $t \to \infty$? Interpret this limit in terms of the behavior of heat in the rod.

17. Let

$$f(x, y) = \begin{cases} xy(x^2 - y^2)/(x^2 + y^2), & (x, y) \neq (0, 0), \\ 0, & (x, y) = (0, 0). \end{cases}$$

(See Figure 3.1.4.)

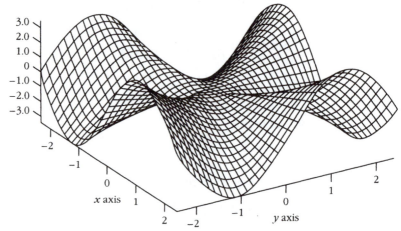

FIGURE 3.1.4. Computer-generated graph of the function in Exercise 17.

(a) If $(x, y) \neq (0, 0)$, calculate $\partial f / \partial x$ and $\partial f / \partial y$.

(b) Show that $(\partial f / \partial x)(0, 0) = 0 = (\partial f / \partial y)(0, 0)$.

(c) Show that $(\partial^2 f / \partial x \partial y)(0, 0) = 1, (\partial^2 f / \partial y \partial x)(0, 0) = -1$.

(d) What went wrong? Why are the mixed partials not equal?

3.2
Taylor's
Theorem

An important application of second and higher order partial derivatives is to the higher order approximation of functions, via Taylor's theorem.

For functions of one variable $y = f(x)$ with continuous derivatives of order $k + 1$, **Taylor's theorem** asserts that *with the point a held fixed*,

$$f(x) = f(a) + f'(a)(x - a) + \frac{f''(a)}{2!}(x - a)^2 + \cdots + \frac{f^{(k)}(a)}{k!}(x - a)^k + R_k$$

where $f^{(k)}$ denotes the kth derivative and

$$R_k = \int_a^x \frac{(x - t)^k}{k!} f^{(k+1)}(t) dt$$

is the remainder. For x near a, this error R_k is small "to order k" in the sense that

$$\frac{R_k}{(x-a)^k} \to 0 \quad \text{as} \quad x \to a.$$

In other words, R_k is small compared with the (already small) quantity $(x-a)^k$.

This result is proved, as in one-variable calculus texts, as follows. We begin with the fundamental theorem of calculus:

$$f(x) - f(a) = \int_a^x f'(t)dt.$$

Then the right hand side is integrated by parts using the formula

$$\int u\,dv = uv - \int v\,du$$

with $u = f'(t)$ and $v = x - t$. The result is

$$
\begin{aligned}
\int_a^x f'(t)dt &= -\int_a^x u\,dv = -\left(uv \,\Big|_a^x - \int_a^x v\,du \right) \\
&= f'(a)(x-a) + \int_a^x (x-t)f''(t)dt.
\end{aligned}
$$

Thus we have proved the first order form of Taylor's theorem (the theorem with $k = 1$):

$$f(x) = f(a) + f'(a)(x-a) + \int_a^x (x-t)f''(t)dt.$$

Note that the first two terms on the right-hand side equal the first two terms in the Taylor series of f. If we integrate by parts again with

$$u = f''(t) \quad \text{and} \quad v = \frac{(x-t)^2}{2},$$

we get

$$
\begin{aligned}
\int_a^x (x-t)f''(t)dt &= -\int_a^x u\,dv = -uv\,\big|_a^x + \int_a^x v\,du \\
&= \frac{f''(a)}{2}(x-a)^2 + \int_a^x \frac{(x-t)^2}{2} f'''(t)dt;
\end{aligned}
$$

so, substituting this into the first order form of Taylor's theorem gives

$$f(x) = f(a) + f'(a)(x-a) + \frac{f''(a)}{2}(x-a)^2 + \int_a^x \frac{(x-t)^2}{2} f'''(t)dt.$$

Repeating the procedure we obtain the general result claimed.

Our first goal in this section is to obtain a second order Taylor theorem for functions of *two* variables. We already know a first order version of this theorem: *If $z = f(x, y)$ is differentiable at (x_0, y_0), then*

$$f(x, y) = f(x_0, y_0) + (x - x_0)\left\{\frac{\partial f}{\partial x}(x_0, y_0)\right\} + (y - y_0)\left\{\frac{\partial f}{\partial y}(x_0, y_0)\right\} + R_1,$$

where $R_1/\|\mathbf{h}\| \to 0$ as $\|\mathbf{h}\| \to 0$, and where $\mathbf{h} = (x - x_0, y - y_0)$. Recall that

$$z = f(x_0, y_0) + (x - x_0)\left\{\frac{\partial f}{\partial x}(x_0, y_0)\right\} + (y - y_0)\left\{\frac{\partial f}{\partial y}(x_0, y_0)\right\}$$

is the equation of the tangent plane to the graph $z = f(x, y)$ at the point (x_0, y_0). The right-hand side is called the **linear** or **first order approximation** to f. Thus this version of Taylor's theorem says that the graph of the tangent plane is close to the graph of $z = f(x, y)$ for points (x, y) "near" (x_0, y_0), where "close" means that the difference in the z-values is R_1 where R_1 is so small that not only does $R_1 \to 0$ as (x, y) gets nearer and nearer to (x_0, y_0) but

$$\frac{R_1}{\sqrt{(x - x_0)^2 + (y - y_0)^2}} \to 0 \quad \text{as} \quad \sqrt{(x - x_0)^2 + (y - y_0)^2} \to 0$$

(see Figure 3.2.1). This is a stronger statement than just requiring that $R_1 \to 0$ as $\sqrt{(x - x_0)^2 + (y - y_0)^2} \to 0$; it says that R_1 is "much smaller" than the quantity $\sqrt{(x - x_0)^2 + (y - y_0)^2}$.

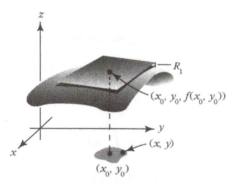

FIGURE 3.2.1. The difference between the graph of the tangent plane at (x_0, y_0) and the graph $z = f(x, y)$ gets very small as (x, y) approaches (x_0, y_0).

The second order Taylor formula is as follows:

A Second Order Taylor Formula

Let $z = f(x, y)$ have continuous partial derivatives up to second order (*i.e.,* all first and second order partial derivatives are continuous). Then

$$f(x, y) = f(x_0, y_0) + (x - x_0)\left\{\frac{\partial f}{\partial x}(x_0, y_0)\right\} + (y - y_0)\left\{\frac{\partial f}{\partial y}(x_0, y_0)\right\}$$

$$+ \frac{1}{2}(x - x_0)^2\left\{\frac{\partial^2 f}{\partial x^2}(x_0, y_0)\right\}$$

$$+ (x - x_0)(y - y_0)\left\{\frac{\partial^2 f}{\partial x \partial y}(x_0, y_0)\right\}$$

$$+ \frac{1}{2}(y - y_0)^2\left\{\frac{\partial^2 f}{\partial y^2}(x_0, y_0)\right\} + R_2$$

where $R_2/\|\mathbf{h}\|^2 \to 0$ as $\|\mathbf{h}\| \to 0$, where $\mathbf{h} = (x - x_0, y - y_0)$.

The polynomial on the right hand side, with R_2 deleted, is called the **quadratic** or **second order approximation** to f at (x_0, y_0).

Method of Proof We consider the function

$$g(s) = f(sx + (1 - s)x_0, sy + (1 - s)y_0)$$

defined for $0 \le s \le 1$. By Taylor's theorem for one variable applied to g on the interval $[0, 1]$, we get

$$g(1) = g(0) + g'(0) + \frac{1}{2}g''(0) + \int_0^1 \frac{(1 - t)^2}{2}g'''(t)dt.$$

Now $g(1) = f(x, y), g(0) = f(x_0, y_0)$ and, by the chain rule,

$$g'(0) = \frac{\partial f}{\partial x}(x_0, y_0)(x - x_0) + \frac{\partial f}{\partial y}(x_0, y_0)(y - y_0).$$

If one calculates $g''(0)$ in a similar way, one finds the second order terms in the Taylor formula for $f(x, y)$. The third derivative $g'''(t)$ contains expressions *cubic* in $\mathbf{h} = (x - x_0, y - y_0)$, so the remainder

$$R_2 = \int_0^1 \frac{(1 - t)^2}{2}g'''(t)dt$$

satisfies $R_2/\|\mathbf{h}\|^2 \to 0$ as $\|\mathbf{h}\| \to 0$. ∎

In the first order Taylor formula, the error R_1 satisfies $R_1/\|\mathbf{h}\| \to 0$ as $\mathbf{h} \to \mathbf{0}$ and in the second order Taylor formula, $R_2/\|\mathbf{h}\|^2 \to 0$ as $\|\mathbf{h}\| \to 0$. Thus R_1 is much smaller than $\|\mathbf{h}\|$ but R_2 is much smaller than $\|\mathbf{h}\|^2$. Since for $\mathbf{h}$ small $\|\mathbf{h}\|^2$ is smaller than $\|\mathbf{h}\|$ [for instance $(0.01)^2 = 0.0001$ is much smaller than 0.01], this says that R_2 goes to zero even "faster" than R_1, as $\|\mathbf{h}\| \to 0$, *i.e.*, as (x, y) gets closer to (x_0, y_0). This usually happens because R_1 is less than $\|\mathbf{h}\|^2$ times something fixed and R_2 is less than $\|\mathbf{h}\|^3$ times something fixed.

Example 1   Compute the second order Taylor formula for $f(x, y) = e^x \cos y$ about the point $x_0 = 0, y_0 = 0$.

Solution   Here

$$f(0,0) = 1, \quad \frac{\partial f}{\partial x}(0,0) = 1, \quad \frac{\partial f}{\partial y}(0,0) = 0,$$

$$\frac{\partial^2 f}{\partial x^2}(0,0) = 1, \quad \frac{\partial^2 f}{\partial y^2}(0,0) = -1, \quad \frac{\partial^2 f}{\partial x \partial y}(0,0) = 0.$$

and so

$$f(x, y) = 1 + x + \frac{1}{2}x^2 - \frac{1}{2}y^2 + R_2. \quad \blacklozenge$$

Example 2   Compute the second order Taylor formula for $f(x, y) = \sin(x + 2y)$, about the point $(x_0, y_0) = (0, 0)$.

Solution   Notice that

$$f(0,0) = 0,$$

$$\frac{\partial f}{\partial x}(0,0) = \cos(0 + 2 \cdot 0) = 1, \quad \frac{\partial f}{\partial y}(0,0) = 2\cos(0 + 2 \cdot 0) = 2,$$

$$\frac{\partial^2 f}{\partial x^2}(0,0) = 0, \quad \frac{\partial^2 f}{\partial y^2}(0,0) = 0, \quad \frac{\partial^2 f}{\partial x \partial y}(0,0) = 0.$$

Thus since $x_0 = 0, y_0 = 0$

$$f(x, y) = x + 2y + R_2$$

where $R_2/\|\mathbf{h}\|^2 \to 0$ as $\|\mathbf{h}\| \to 0$, where $\mathbf{h} = (x, y)$.

Although the expression $x + 2y$ does not contain any quadratic terms, it is still a second order Taylor formula because the coefficients of the quadratic terms are zero and $R_2/\|\mathbf{h}\|^2 \to 0$ as $\|\mathbf{h}\| \to 0$, where $\mathbf{h} = (x, y)$. ◆

Example 3   Find the first and second order Taylor approximations to $f(x, y) = \sin(xy)$ at the point $(x_0, y_0) = (1, \pi/2)$.

Solution   Here

$$
\begin{aligned}
f(x_0, y_0) &= \sin(x_0 y_0) = \sin(\pi/2) = 1 \\
f_x(x_0, y_0) &= y_0 \cos(x_0 y_0) = \frac{\pi}{2} \cos(\pi/2) = 0 \\
f_y(x_0, y_0) &= x_0 \cos(x_0 y_0) = \cos(\pi/2) = 0 \\
f_{xx}(x_0, y_0) &= -y_0^2 \sin(x_0 y_0) = -\frac{\pi^2}{4} \sin(\pi/2) = -\pi^2/4 \\
f_{xy}(x_0, y_0) &= \cos(x_0 y_0) - x_0 y_0 \sin(x_0 y_0) = -\frac{\pi}{2} \sin(\pi/2) = -\pi/2 \\
f_{yy}(x_0, y_0) &= -x_0^2 \sin(x_0 y_0) = -\sin(\pi/2) = -1.
\end{aligned}
$$

Thus the linear approximation is (see Fig. 3.2.2)

$$
\begin{aligned}
l(x, y) &= f(x_0, y_0) + f_x(x_0, y_0)(x - x_0) + f_y(x_0, y_0)(y - y_0) \\
&= 1 + 0 + 0 = 1
\end{aligned}
$$

and the second order approximation is (from the box on page 185)

$$
\begin{aligned}
g(x, y) &= 1 + 0 + 0 + \frac{1}{2}\left(-\frac{\pi^2}{4}\right)(x - 1)^2 + \left(-\frac{\pi}{2}\right)(x - 1)\left(y - \frac{\pi}{2}\right) \\
&\quad + \frac{1}{2}(-1)\left(\frac{y - \pi}{2}\right)^2 \\
&= 1 - \frac{\pi^2}{8}(x - 1)^2 - \frac{\pi}{2}(x - 1)\left(y - \frac{\pi}{2}\right) - \frac{1}{2}\left(y - \frac{\pi}{2}\right)^2.
\end{aligned}
$$

◆

Example 4   Find linear and quadratic approximations to $(3.98 - 1)^2/(5.97 - 3)^2$. Compare with the exact value.

Solution   Let $f(x, y) = (x - 1)^2/(y - 3)^2$. The desired expression is close to $f(4, 6) = 1$. To find the approximations, we differentiate:

$$
f_x = \frac{2(x - 1)}{(y - 3)^2}, \quad f_y = \frac{-2(x - 1)^2}{(y - 3)^3}
$$

$$
f_{xy} = f_{yx} = \frac{-4(x - 1)}{(y - 3)^3}, \quad f_{xx} = \frac{2}{(y - 3)^2}, \quad f_{yy} = \frac{6(x - 1)^2}{(y - 3)^4}.
$$

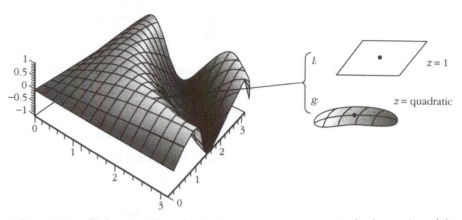

FIGURE 3.2.2. The linear and quadratic approximations to $z = \sin(xy)$ near $(1, \pi/2)$.

At the point of approximation, we have

$$f_x(4,6) = \frac{2}{3}, \; f_y = -\frac{2}{3}, \; f_{xy} = f_{yx} = -\frac{4}{9}, \; f_{xx} = \frac{2}{9}, \; f_{yy} = \frac{2}{3}.$$

The linear approximation is then

$$1 + \frac{2}{3}(-0.02) - \frac{2}{3}(-0.03) = 1.00666.$$

The quadratic approximation is

$$1 \; + \; \frac{2}{3}(-0.02) - \frac{2}{3}(-0.03) + \frac{2}{9}\frac{(-0.02)^2}{2} - \frac{4}{9}(-0.02)(-0.03) + \frac{2}{3}\frac{(-0.03)^2}{2}$$
$$= \; 1.00674.$$

The "exact" value is 1.00675.   ◆

Just as in one-variable calculus, we can form higher order Taylor approximations to functions of two or more variables. A monomial $x^k y^l$ is said to have
order $k + l$, and the "term of order n" consists of all the monomials of order n,
with appropriate numerical coefficients. Although we have not formally stated
how to find higher order Taylor approximations, the following example shows
how it can be done in a special case.

Example 5 Write the first four terms of the Taylor series for the function

$$f(x,y) = \frac{1}{1 - x - y^2}$$

around $(x, y) = (0, 0)$.

Solution   Using the formula

$$\frac{1}{1-r} = 1 + r + r^2 + r^3 \ldots,$$

valid for $|r| < 1$, for the sum of a geometric series, we can write

$$f(x,y) = \frac{1}{1 - (x + y^2)} = 1 + (x + y^2) + (x + y^2)^2 + (x + y^2)^3 + \ldots,$$

for $|x + y^2| < 1$. In this form, the terms are not arranged in the correct order, though. We expand

$$f(x,y) = 1 + x + y^2 + x^2 + 2xy^2 + y^4 + x^3 + 3x^2y^2 + 3xy^4 + y^6 + \ldots,$$

where all the terms in "..." are of order 4 or higher. Collecting according to the order, we have

$$f(x,y) = 1 + (x) + (y^2 + x^2) + (2xy^2 + x^3) + \ldots. \quad \blacklozenge$$

Exercises for §3.2

In each of Exercises 1–6, determine the second order Taylor formula for the given function about the given point (x_0, y_0).

1. $f(x,y) = (x + y)^2, x_0 = 0, y_0 = 0$

2. $f(x,y) = 1/(x^2 + y^2 + 1), x_0 = 0, y_0 = 0$

3. $f(x,y) = e^{x+y}, x_0 = 0, y_0 = 0$

4. $f(x,y) = e^{-x^2-y^2} \cos(xy), x_0 = 0, y_0 = 0$

5. $f(x,y) = \sin(xy) + \cos(xy), x_0 = 0, y_0 = 0$

6. $f(x,y) = e^{(x-1)^2} \cos y, x_0 = 1, y_0 = 0$

Find the linear and quadratic approximations for each of the quantities in Exercises 7–10, using the first and second order approximations. Compare your answers to the exact values.

7. $(1.01)^2(1 - \sqrt{1.98})$ [Hint: $1.96 = (1.4)^2$.]

8. $\tan\left(\dfrac{\pi + 0.01}{3.97}\right)$

9. $(0.99)^3 + (2.01)^3 - 6(0.99)(2.01)$

10. $(0.98) \sin\left(\dfrac{0.98\pi}{1.03}\right)$

Write the first four terms in the Taylor expansions of the functions in Exercises 11 and 12 about $(0,0)$.

11. $f(x,y) = \dfrac{x}{1-x^2-y^3}$

12. $f(x,y) = \dfrac{xy}{1-x-y}$

3.3
Maxima and Minima

Among the most basic geometric features of a function's graph are those points where the function attains its greatest and least values—its extrema. In this section we derive a method for determining these points. The method reveals local extrema as well—points where the function attains a maximum or minimum value relative to nearby points.

Definition of Maxima and Minima

Let $f(x,y)$ be a function of two variables. We say that (x_0, y_0) is a **local minimum point** for f if there is a disk (of positive radius) about (x_0, y_0) such that $f(x,y) \geq f(x_0, y_0)$ for all (x,y) in the disk.

Similarly, if $f(x,y) \leq f(x_0, y_0)$ for all (x,y) in some disk (of positive radius) about (x_0, y_0), we call (x_0, y_0) a **local maximum point** for f. A point that is either a local maximum or minimum point is called a **local extremum**.

Historical Note

Throughout history, people have searched for laws to describe the phenomena of the physical world. However, no general principle encompassing all phenomena was proposed until 1744, when the French scientist Pierre Louis Moreau de Maupertuis put forth his grand scheme of the universe.

The "metaphysical principle" of Maupertuis is the assumption that nature always operates with the greatest possible economy. In short, physical laws are a consequence of a principle of "economy of means" ; nature always acts in such a way as to minimize some quantity which Maupertuis called the "action." For example, physical systems often try to "rearrange themselves" to have a minimum energy, such as a ball rolling from a mountain peak to a valley, or the primordial irregular Earth assuming a more nearly spherical shape. As another example, the spherical shape of soap bubbles is connected with the fact that spheres are the surfaces of least area containing a fixed volume.

Minimization principles provide a philosophical cornerstone of many aspects of physics, engineering, and mathematics. Thus, describing many physical phenomena amounts to finding the maxima and minima of some scalar function. The Swiss mathematician Leonhard Euler provided much of the mathematical groundwork for the theory of maxima and minima of scalar quantities. For more information and history, consult *Mathematics and Optimal Form* by S. Hildebrandt and A. J. Tromba, Scientific American Books, N.Y. (1985).

We may also define **global** or **absolute** maximum and minimum points to be those at which a function attains the greatest and least values for all points in its domain. The definitions for functions of n variables are similar. See Figure 3.3.1.

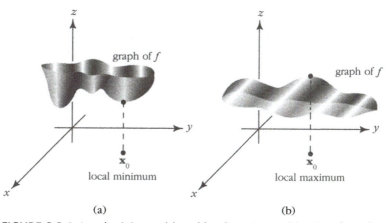

(a)   (b)

FIGURE 3.3.1. Local minimum (a) and local maximum (b) points for a function of two variables.

Example 1   Refer to Figure 3.3.2, a computer-drawn graph of $z = 2(x^2+y^2)e^{-x^2-y^2}$. Where are the maximum and minimum points?

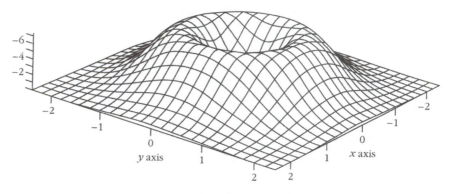

FIGURE 3.3.2. The volcano: $z = 2(x^2 + y^2)\exp{(-x^2 - y^2)}$.

Solution   There is a local (in fact, global) minimum at the volcano's center $(0,0)$, where $z = 0$. There are maximum points all around the crater's rim (the circle $x^2 + y^2 = 1$).   ◆

The search for local maxima or minima in one-variable calculus usually begins by looking for critical points; that is, those points where the derivative is zero. Geometrically these are the points where the tangent line is horizontal. The same is true for two variables, but in this case a critical point is where *both* partial derivatives are zero. Geometrically these are the points where the tangent plane to the graph is horizontal (Figure 3.3.1).

Critical Points

A point (x_0, y_0) is a ***critical point*** of $f(x, y)$ if

$$f_x(x_0, y_0) = f_y(x_0, y_0) = 0.$$

Thus, the tangent plane to the graph $z = f(x, y)$ at (x_0, y_0) is horizontal.

Just as in one-variable calculus, any maximum or minimum is a critical point:

First Derivative Test

If (x_0, y_0) is a local extremum of f and the partial derivatives of f exist at (x_0, y_0), then (x_0, y_0) is a critical point.

Let us prove this in the case of a local minimum; the proof for a local maximum is essentially the same.

By assumption, there is a disk of radius r about (x_0, y_0) on which $f(x, y) \geq f(x_0, y_0)$. In particular, if $|x - x_0| < r$, then $f(x, y_0) \geq f(x_0, y_0)$, so the function $g(x) = f(x, y_0)$ has a local minimum at x_0. By the first derivative test of one-variable calculus, $g'(x_0) = 0$; but $g'(x_0)$ is just $f_x(x_0, y_0)$. Similarly, the function $h(y) = f(x_0, y)$ has a local minimum at y_0, so $f_y(x_0, y_0) = 0$, and hence (x_0, y_0) is a critical point for f.

In summary, *when looking for local maxima and minima for $z = f(x, y)$, we begin by looking for the critical points*. Note that at a critical point, the gradient, $\nabla f = f_x \mathbf{i} + f_y \mathbf{j}$ is zero.

Example 2   Find all the critical points of $z = x^2 y + y^2 x$.

Solution   Differentiating, we obtain

$$\frac{\partial z}{\partial x} = 2xy + y^2, \qquad \frac{\partial z}{\partial y} = x^2 + 2yx.$$

Equating the partial derivatives to zero yields

$$2xy + y^2 = 0, \qquad 2xy + x^2 = 0.$$

Subtracting, we obtain $x^2 = y^2$. Thus $x = \pm y$. Substituting $x = +y$ in the first equation above, we find that

$$2y^2 + y^2 = 3y^2 = 0,$$

so that $y = 0$ and thus $x = 0$. If $x = -y$, then

$$-2y^2 + y^2 = -y^2 = 0,$$

and so $y = 0$ and therefore $x = 0$. Hence the only critical point is $(0,0)$. For $x = y$, $z = 2x^3$, which is both positive and negative for x near zero. Thus $(0,0)$ is not a relative extremum. ◆

Example 3   Verify that the critical points of the function in Example 1 occur at $(0,0)$ and on the circle $x^2 + y^2 = 1$.

Solution   Since $z = 2(x^2 + y^2)e^{-x^2-y^2}$, we have

$$
\begin{aligned}
\frac{\partial z}{\partial x} &= 4x(e^{-x^2-y^2}) + 2(x^2+y^2)e^{-x^2-y^2}(-2x) \\
&= e^{-x^2-y^2}[4x - 4x(x^2+y^2)] \\
&= 4x(e^{-x^2-y^2})(1 - x^2 - y^2)
\end{aligned}
$$

and

$$\frac{\partial z}{\partial y} = 4y(e^{-x^2-y^2})(1 - x^2 - y^2).$$

These vanish when $x = y = 0$ or when $x^2 + y^2 = 1$. ◆

In one-variable calculus, critical points need not be local maxima or minima; they can be inflection points. In several variables there is not only a critical point that behaves like an inflection point, but an entirely new type of critical point called a **saddle** point, like the one depicted in Figure 3.3.3. Near this point, the function takes larger or smaller values depending on the direction in which one travels away from the point.

Example 4   Let $z = x^2 - y^2$. Show that $(0,0)$ is a critical point. Is it a local maximum or minimum?

Solution   The partial derivatives $\partial z/\partial x = 2x$ and $\partial z/\partial y = -2y$ vanish at $(0,0)$, so the origin is a critical point. It is neither a local maximum nor minimum since $f(x,y) = x^2 - y^2$ is zero at $(0,0)$ and can be either positive (on the x axis) or negative (on the y axis) arbitrarily near the origin. This is also clear from the graph (see Figure 3.3.4), which shows a saddle point at $(0,0)$. ◆

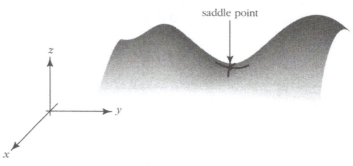

FIGURE 3.3.3. A saddle point.

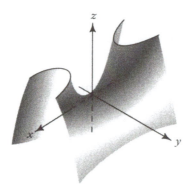

FIGURE 3.3.4. The graph $z = x^2 - y^2$ is a hyperbolic paraboloid, or "saddle."

In one-variable calculus, finding critical points is just the first step in locating the *absolute* maximum or minimum of a function (see Figure 3.3.5). We recall an important test in the following box.

Absolute Maxima and Minima
on Closed Intervals

Given a differentiable function $y = f(x)$, to find the absolute minimum and maximum values of f for x in $[a, b]$, we:

1. find all critical points $x_1, x_2, \ldots, x_n$ in the open interval (a, b);

2. consider the values $f(a), f(x_1), f(x_2), \ldots, f(x_n), f(b)$ and pick the largest (this will be the absolute maximum) and the smallest (this will be the absolute minimum).

Thus the problem of finding absolute maxima and minima is reduced via differential calculus to picking the largest and smallest values from a finite list of numbers.

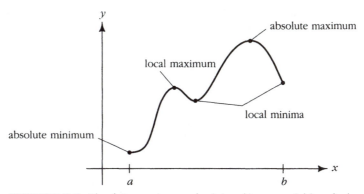

FIGURE 3.3.5. Absolute maxima and minima in one-variable calculus.

In §**3.5**, we will consider generalizations of the technique above to find absolute maxima and minima for functions of two and three variables.

On domains other than closed intervals, more care is necessary. For instance, the function $f(x) = 1/x, 0 < x \leq 1$ has no maximum on the half open interval $(0, 1]$. (See Figure 3.3.6.)

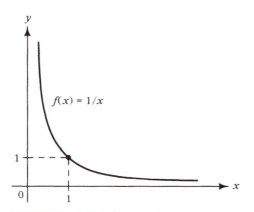

FIGURE 3.3.6. This function has no maximum on $(0, 1]$.

Another example of this type is the function $f(x) = x^2 + (1/x^2)$ defined on all of $\mathbb{R}$ except $x = 0$. (See Figure 3.3.7). Here $f(x)$ is *not* defined on a closed interval but on all of $\mathbb{R}$ with one exceptional point; namely, $x = 0$. We would like to locate the *minimum* value of such a function. Since $f(x)$ gets larger and larger as $x \to +\infty$ or $x \to -\infty$ or as x gets closer and closer to the exceptional point $x = 0$, it is reasonable (and true!) that *to locate the minimum value of f we need only consider the critical points of f and pick the smallest value.* For $f(x) = x^2 + (1/x^2)$, we set the derivative equal to zero:

$$f'(x) = 2x - \frac{2}{x^3} = 0.$$

Thus $x^4 - 1 = 0$, so $x = \pm 1$ are the two critical points and $f(\pm 1) = 2$ is the minimum value for f.

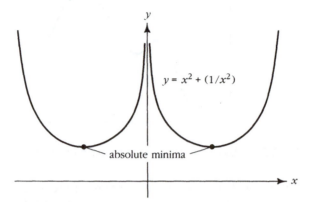

FIGURE 3.3.7. This function has two absolute minima and no maxima.

Similar care is needed for functions of two or more variables, as the next examples illustrate.

Example 5
(a) Find the minimum distance from the origin to a point on the plane $x + 3y - z = 6$.

(b) Find the minimum distance from the point $(1, 2, 0)$ to the cone $z^2 = x^2 + y^2$.

Solution

(a) To find the point closest to the origin, we minimize the distance $d = \sqrt{x^2 + y^2 + z^2}$, where $z = x + 3y - 6$. It is equivalent but simpler to minimize $f(x, y) = d^2 = x^2 + y^2 + (x + 3y - 6)^2$. The critical points are given by

$$\frac{\partial f}{\partial x} = 0, \quad i.e., \quad 2x + 2(x + 3y - 6) = 0$$

and

$$\frac{\partial f}{\partial y} = 0, \quad i.e., \quad 2y + 6(x + 3y - 6) = 0.$$

Solving these equations gives $y = 18/11, x = 6/11$. Since $f(x, y)$ becomes large if x or y gets large, it must attain a minimum somewhere, and the critical point that we just found is the only place it can be. At this point, $z = x + 3y - 6 = -6/11$, and so the minimum distance is $d = \sqrt{x^2 + y^2 + z^2} = 6\sqrt{11}/11$. (See Figure 3.3.8.)

(b) We minimize the square of the distance:

$$d^2 = (x - 1)^2 + (y - 2)^2 + z^2.$$

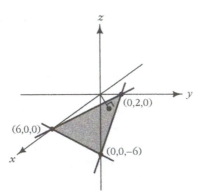

FIGURE 3.3.8. The point nearest to the origin on the plane $z = x + 3y - 6$ is $(6/11, 18/11, -6/11)$.

Substituting $z^2 = x^2 + y^2$, we have the problem of minimizing

$$f(x,y) = (x-1)^2 + (y-2)^2 + x^2 + y^2$$
$$= 2x^2 + 2y^2 - 2x - 4y + 5.$$

Now

$$f_x(x,y) = 4x - 2 \quad \text{and} \quad f_y(x,y) = 4y - 4.$$

Thus the critical point, obtained by setting these equal to zero, is $x = 1/2, y = 1$. Since $f(x,y)$ gets large as x or y gets large, this is the minimum point and so the minimum distance is

$$d = \sqrt{(1/2-1)^2 + (1-2)^2 + (1/2)^2 + 1}$$
$$= \sqrt{1/4 + 1 + 1/4 + 1} = \sqrt{5/2} \approx 1.581. \quad \blacklozenge$$

Example 6   A rectangular box, open at the top, is to hold 256 cubic centimeters of sand. Find the dimensions for which the surface area (bottom and four sides) is minimized.

Solution   Let x and y be the lengths of the sides of the base. Since the volume of the box is to be 256, the height must be $256/xy$. Two of the sides have area $x(256/xy)$, two sides have area $y(256/xy)$, and the base has area xy, so the total surface area is

$$f(x,y) = 2x\left(\frac{256}{xy}\right) + 2y\left(\frac{256}{xy}\right) + xy = \frac{512}{y} + \frac{512}{x} + xy.$$

The only relevant values of (x,y) are in the first quadrant ($x > 0$ and $y > 0$). As (x,y) approaches infinity or the boundary of the quadrant (the positive x

and y half-axes), $f(x, y)$ becomes large, so the minimum of f is attained at a critical point. The critical points are given by

$$0 = \frac{\partial f}{\partial x} = -\frac{512}{x^2} + y, \qquad 0 = \frac{\partial f}{\partial y} = -\frac{512}{y^2} + x.$$

The first equation gives $y = 512/x^2$; substituting this into the second equation gives $0 = -512(x^2/512)^2 + x = -x^4/512 + x$. Discarding the extraneous root $x = 0$, we have $x^3/512 = 1$, or $x = \sqrt[3]{512} = 8$. Thus $y = 512/x^2 = 8$, and the height is $256/xy = 4$, so the optimal box has a base which is 8 cm square and is half as high as it is wide. ◆

Example 7   The graph of the function $g(x, y) = 1/xy$ is the surface S in $\mathbb{R}^3$ shown in Figure 3.3.9. Find the points on S that are closest to the origin $(0, 0, 0)$.

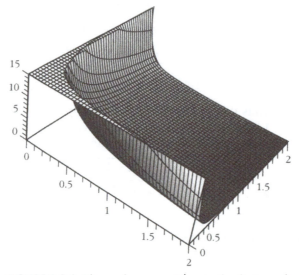

FIGURE 3.3.9. The surface $z = 1/xy$ in the first quadrant.

Solution   Each point on S is of the form $(x, y, 1/xy)$. The distance from this point to the origin is

$$d(x, y) = \sqrt{x^2 + y^2 + \frac{1}{x^2y^2}}.$$

It is easier to work with the square of d, so let $f(x, y) = x^2 + y^2 + (1/x^2y^2)$, which will have the same minimum point. Notice that $f(x,y)$ becomes very large as x and y get larger and larger or as (x,y) approaches the x or y axis where

f is not defined, so f must attain a minimum at some critical point. The critical points are determined by:

$$\frac{\partial f}{\partial x} = 2x - \frac{2}{x^3 y^2} = 0,$$

$$\frac{\partial f}{\partial y} = 2y - \frac{2}{y^3 x^2} = 0,$$

that is, $x^4 y^2 - 1 = 0$ and $x^2 y^4 - 1 = 0$. From the first equation we get $y^2 = 1/x^4$, and, substituting this into the second equation, we obtain

$$1 = \frac{x^2}{x^8} = \frac{1}{x^6}.$$

Thus $x = \pm 1$ and $y = \pm 1$, and it therefore follows that f has four critical points, namely, $(1, 1), (1, -1), (-1, 1)$, and $(-1, -1)$. Note that f has the value 3 for all these points, so they are all minimum points. Thus, the points on the surface closest to the point $(0, 0, 0)$ are $(1, 1, 1)$, $(1, -1, -1)$, $(-1, 1, -1)$ and $(-1, -1, 1)$, and the minimum distance is $\sqrt{3}$. ◆

Exercises for §3.3

Find the critical points of the functions in Exercises 1–8.

1. $f(x, y) = x^2 - y^2 + xy + x - y$

2. $f(x, y) = x^2 + y^2 - xy + y$

3. $f(x, y) = x^2 + y^2 + 2xy - 10$

4. $f(x, y) = x^2 + y^2 + 3xy - 3y$

5. $f(x, y) = e^{1 + x^2 - y^2}$

6. $f(x, y) = x^2 - 3xy + 5x - 2y + 6y^2 + 8$

7. $f(x, y) = 3x^2 + 2xy + 2x + y^2 + y + 4$

8. $f(x, y) = \sin(x^2 + y^2)$

9. Refer to Figure 3.3.10, a graph of the function $z = (x^3 - 3x)/(1 + y^2)$. Where are the local maximum and minimum points?

10. Refer to Figure 3.3.11, a graph of the function $z = \sin(\pi x)/(1 + y^2)$. Where are the maximum and minimum points?

11. Minimize the distance to the origin from the plane $x - y + 2z = 3$.

12. Find the distance from the plane given by $x + 2y + 3z - 10 = 0$:
    (a) To the origin. (b) To the point $(1, 1, 1)$.

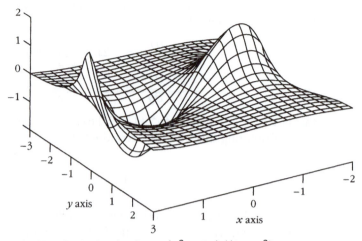

FIGURE 3.3.10. Graph of $z = (x^3 - 3x)/(1 + y^2)$.

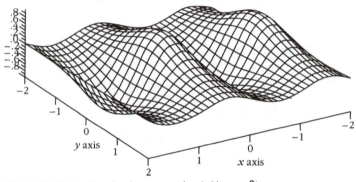

FIGURE 3.3.11. Graph of $z = \sin(\pi x)/(1 + y^2)$.

13. Show that a rectangular box of given volume has minimum surface area when the box is a cube.

14. Show that the rectangular parallelepiped with fixed surface area and maximum volume is a cube.

15. Write the number 120 as a sum of three numbers so that the sum of the products taken two at a time is a maximum.

16. Find the point nearest the origin in the plane $2x - y + 2z = 20$.

17. Suppose that the material for the bottom of the box in Example 6 costs b cents per square centimeter, while that for the sides costs s cents per square centimeter. Find the dimensions that minimize the cost of the material.

3.4
Second
Derivative Test

Our goal in this section is to derive the second derivative test, which determines whether a critical point of a function of two variables is a local minimum, a local maximum or a saddle point, or perhaps none of these. The three fundamental cases are depicted in Figure 3.4.1.

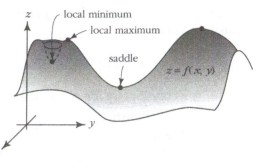

FIGURE 3.4.1. A local minimum, maximum, and a saddle.

Suppose that (x_0, y_0) is a critical point for f. In this case

$$\frac{\partial f}{\partial x}(x_0, y_0) = 0 = \frac{\partial f}{\partial y}(x_0, y_0),$$

so we get the following simpler version of Taylor's formula.

Taylor's Formula Near a Critical Point

If (x_0, y_0) is a critical point for $z = f(x, y)$, then

$$f(x, y) = f(x_0, y_0) + A(x - x_0)^2 + 2B(x - x_0)(y - y_0) + C(y - y_0)^2 + R_2$$

with $R_2/\|\mathbf{h}\|^2 \to 0$ as $\|\mathbf{h}\| \to 0$, where $\mathbf{h} = (x - x_0, y - y_0)$, and where A, B, C are the constants

$$A = \frac{1}{2}\frac{\partial^2 f}{\partial x^2}(x_0, y_0), \ B = \frac{1}{2}\frac{\partial^2 f}{\partial x \partial y}(x_0, y_0), \ C = \frac{1}{2}\frac{\partial^2 f}{\partial y^2}(x_0, y_0).$$

If $(x_0, y_0) = (0, 0)$, Taylor's formula becomes

$$f(x, y) = f(0, 0) + Ax^2 + 2Bxy + Cy^2 + R_2.$$

Since $f(0, 0)$ is a constant and R_2 is very small, the quadratic expression

$$f(0, 0) + Ax^2 + 2Bxy + Cy^2$$

closely approximates $z = f(x,y)$ for (x,y) near $(0,0)$. Thus, the *shape* of the graph of $z = f(x,y)$ should be close to the *shape* of the graph of $z = Ax^2 + 2Bxy + Cy^2$ for (x,y) close to $(0,0)$. This simple idea is the basis for the second derivative test to determine whether critical points are local maxima, local minima, or saddle points.

We now determine when $(0,0)$ is a local maximum, local minimum, or saddle for the quadratic function

$$z = g(x,y) = Ax^2 + 2Bxy + Cy^2.$$

The point $(0,0)$ is a critical point for $g(x,y)$ since

$$\frac{\partial g}{\partial x} = 2Ax + 2By \quad \text{and} \quad \frac{\partial g}{\partial y} = 2Bx + 2Cy$$

and so

$$\frac{\partial g}{\partial x}(0,0) = \frac{\partial g}{\partial y}(0,0) = 0.$$

If $AC - B^2 \neq 0$, then $(0,0)$ is the *only* critical point. To see this, suppose that (x_0, y_0) is a critical point. Then

$$Ax_0 + By_0 = 0 \quad \text{and} \quad Bx_0 + Cy_0 = 0.$$

Multiplying the first equation by C and the second by B and subtracting, we obtain

$$(AC - B^2)x_0 = 0.$$

If $AC - B^2 \neq 0$ then $x_0 = 0$. A similar argument shows that $y_0 = 0$. If $AC - B^2 = 0$, then $(0,0)$ need not be the only critical point. For example, $z = y^2 + 1$, whose graph looks like a sheet of paper folded into a parabolic shape (Figure 3.4.2), has critical points all along the x-axis. Here $B = A = 0$, so $AC - B^2 = 0$.

To avoid degenerate situations like this, we assume that $AC - B^2 \neq 0$. The first case we consider is

$$AC - B^2 > 0.$$

In this case $A \neq 0$ and $C \neq 0$ (if either were zero, we would have $-B^2 > 0$, a contradiction!). We complete the square for $g(x,y) = Ax^2 + 2Bxy + Cy^2$ as

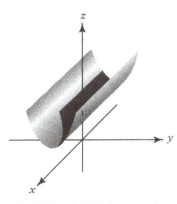

FIGURE 3.4.2. This function has critical points all along the x axis.

follows:

$$
\begin{aligned}
g(x,y) &= A\left(x^2 + \frac{2B}{A}xy + \frac{C}{A}y^2\right) \\
&= A\left(x^2 + \frac{2B}{A}xy + \frac{B^2}{A^2}y^2 - \frac{B^2}{A^2}y^2 + \frac{C}{A}y^2\right) \\
&= A\left(x + \frac{B}{A}y\right)^2 + \frac{AC - B^2}{A}y^2.
\end{aligned}
$$

There are two *subcases* to the case $AC - B^2 > 0$. The first is when $A > 0$. Then C is also positive (why?) and from

$$
g(x,y) = A\left(x + \frac{B}{A}y\right)^2 + \frac{AC - B^2}{A}y^2
$$

we see that

$$
g(x,y) \geq 0
$$

and is equal to zero only when both summands for g are zero; *i.e.*, $(AC - B^2)y^2/A = 0$ (which implies that $y = 0$) and

$$
0 = A\left(x + \frac{B}{A}y\right)^2 = Ax^2,
$$

so $x = 0$ also. Thus, $g(x,y) \geq 0$ and equals zero only when $(x,y) = (0,0)$. Thus, $(0,0)$ is a (strict) minimum for g and the graph of g looks as in Figure 3.4.3(a).

The second subcase is when $A < 0$ (then C is also negative!) and we see that $g(x, y) \leq 0$ and equals 0 only when $(x, y) = (0, 0)$. Thus, $(0, 0)$ is a maximum and the graph of g looks as in Figure 3.4.3(b).

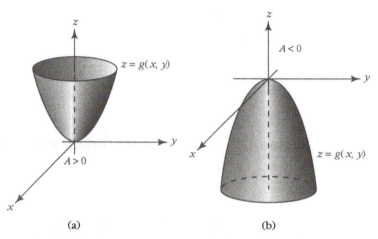

(a)   (b)

FIGURE 3.4.3. Illustrating the subcases (a) $A > 0$ and (b) $A < 0$ in the first case.

Next, consider the case when $AC - B^2 < 0$. If $A \neq 0$, the expression

$$g(x, y) = A \left(x + \frac{B}{A} y \right)^2 + \frac{AC - B^2}{A} y^2$$

is still valid. If $A < 0$, then the first term satisfies $A[x + (B/A)y]^2 \leq 0$, and the second term satisfies

$$\frac{AC - B^2}{A} y^2 \geq 0$$

(since the quotient of two negative numbers is positive). If $x \neq 0$, then

$$g(x, 0) = Ax^2 < 0$$

and on the line $-By/A = x$,

$$g\left(-\frac{B}{A} y, y \right) = \frac{AC - B^2}{A} y^2 > 0.$$

Since $g(x, y)$ can be both positive and negative for (x, y) near $(0, 0)$, the graph of $g(x, y)$ is saddle shaped. Finally, if $A = 0$,

$$g(x, y) = 2Bxy + Cy^2 = y(2Bx + Cy)$$

which again can be both positive (if y and $2Bx + Cy$ have the same sign) and negative (if they have the opposite sign), so the graph again looks like a saddle near $(0,0)$.

We summarize our conclusions in the following box[1]:

The Shape of Graphs of Quadratic Functions

$$g(x,y) = Ax^2 + 2Bxy + Cy^2$$

1. If $AC - B^2 > 0$ and if
   (a) $A > 0$, then $(0,0)$ is a minimum point of g;
   (b) $A < 0$, then $(0,0)$ is a maximum point of g.

2. If $AC - B^2 < 0$, then the graph of g looks like a saddle near $(0,0)$.

Example 1   (a) Analyze the critical point at the origin for $g(x,y) = x^2 + 3xy + y^2$.

(b) Determine whether $(0,0)$ is a maximum point, a minimum point, or neither, of $g(x,y) = 3x^2 - 5xy + 3y^2$.

Solution

(a) Here, $A = 1, B = \frac{3}{2}$, and $C = 1$. Since $AC - B^2 = 1 - \frac{9}{4}$ is negative, g has a saddle point at $(0,0)$.

(b) Here, $A = 3, B = -\frac{5}{2}$, and $C = 3$, so $A = 3 > 0$ and $AC - B^2 = 9 - \frac{25}{4} > 0$. Thus $(0,0)$ is a minimum point.   ◆

If we consider the more general expression

$$g(x,y) = K + A(x - x_0)^2 + 2B(x - x_0)(y - y_0) + C(y - y_0)^2$$

where K, A, B, and C are constants, then we can draw similar conclusions:

---

[1]Students who have studied linear algebra should recognize here the standard test for positive or negative definiteness of a quadratic form.

The Shape of Graphs of General Quadratic Functions

$$g(x, y) = K + A(x - x_0)^2 + 2B(x - x_0)(y - y_0) + C(y - y_0)^2$$

1. If $AC - B^2 > 0$ and if
   (a) $A > 0$, then (x_0, y_0) is a minimum for g;
   (b) $A < 0$, then (x_0, y_0) is a maximum for g.
2. If $AC - B^2 < 0$, then the graph of g looks like a saddle near (x_0, y_0).

This result can now be used to give a second derivative test. Let $z = f(x, y)$ have continuous partial derivatives up to second order and let (x_0, y_0) be a critical point. By Taylor's formula, we can write

$$f(x, y) = f(x_0, y_0) + A(x - x_0)^2 + 2B(x - x_0)(y - y_0) + C(y - y_0)^2 + R_2$$

where R_2 is "very small" for (x, y) near (x_0, y_0) and where

$$A = \frac{1}{2}\frac{\partial^2 f}{\partial x^2}(x_0, y_0), \quad B = \frac{1}{2}\frac{\partial^2 f}{\partial x \partial y}(x_0, y_0), \quad \text{and} \quad C = \frac{1}{2}\frac{\partial^2 f}{\partial y^2}(x_0, y_0).$$

Since R_2 is small, the shape of the graph of f near (x_0, y_0) is "roughly the same" as that of

$$g(x, y) = f(x_0, y_0) + A(x - x_0)^2 + 2B(x - x_0)(y - y_0) + C(y - y_0)^2.$$

Using the values for A, B, C given by Taylor's theorem, we arrive at the following second derivative test.

Second Derivative Test

Let $z = f(x, y)$ have continuous partial derivatives up to second order and suppose that (x_0, y_0) is a critical point of f. Consider the expression

$$D = [f_{xx}(x_0, y_0)][f_{yy}(x_0, y_0)] - [f_{xy}(x_0, y_0)]^2$$

called the **discriminant**.

1. If $D > 0$ and $f_{xx}(x_0, y_0) > 0$, then (x_0, y_0) is a local minimum for f.
2. If $D > 0$ and $f_{xx}(x_0, y_0) < 0$, then (x_0, y_0) is a local maximum for f.
3. If $D < 0$, then (x_0, y_0) is a saddle point for f.
If $D = 0$ the test fails and further examination of the function is necessary.

You will notice that in going from the previous box to this one, some factors of two were dropped. In fact, $D = (2A)(2C) - (2B)^2 = 4(AC - B^2)$ so D is positive exactly when $AC - B^2$ is positive. We have stated the final test *without* the A, B, C notation to avoid carrying these extra factors around and to simplify doing examples.

Example 2   Find the critical points of $f(x, y) = x \sin y$ and determine whether they are local maxima, local minima, or saddle points.

Solution

$$\frac{\partial f}{\partial x} = \sin y, \qquad \frac{\partial f}{\partial y} = x \cos y.$$

If $\partial f / \partial x = 0$, then $y = \pi n$, where n is an integer. At these points $\cos y \neq 0$, so if $\partial f / \partial y = 0$, then $x = 0$. Consequently, all the critical points are of the form $(0, \pi n)$. The second derivatives are $f_{xx} = 0$, $f_{yy} = -x \sin y$ and $f_{xy} = \cos y$. The discriminant is thus

$$D = f_{xx} f_{yy} - (f_{xy})^2 = 0(-x \sin y) - \cos^2 y = -\cos^2 y.$$

At the critical points $\cos y \neq 0$. Therefore $D = -\cos^2 y < 0$, so all critical points are saddle points.   ◆

Example 3   Find the maxima, minima, and saddle points of $z = (x^2 - y^2)e^{(-x^2 - y^2)/2}$.

Solution   First we locate the critical points by setting $\partial z / \partial x = 0$ and $\partial z / \partial y = 0$. Here

$$\frac{\partial z}{\partial x} = \left[2x - x(x^2 - y^2) \right] e^{(-x^2 - y^2)/2}$$

and

$$\frac{\partial z}{\partial y} = \left[-2y - y(x^2 - y^2) \right] e^{(-x^2 - y^2)/2},$$

so the critical points are the solutions of the simultaneous equations

$$x \left[2 - (x^2 - y^2) \right] = 0,$$

$$y \left[-2 - (x^2 - y^2) \right] = 0.$$

These have solutions $(0,0), (\pm\sqrt{2},0)$, and $(0,\pm\sqrt{2})$. The second derivatives are

$$\frac{\partial^2 z}{\partial x^2} = \left[2 - 5x^2 + x^2(x^2 - y^2) + y^2\right] e^{(-x^2-y^2)/2},$$

$$\frac{\partial^2 z}{\partial x \partial y} = xy(x^2 - y^2)e^{(-x^2-y^2)/2},$$

$$\frac{\partial^2 z}{\partial y^2} = \left[5y^2 - 2 + y^2(x^2 - y^2) - x^2\right] e^{(-x^2-y^2)/2}.$$

Using the second derivative test results in the following data:

Point	f_{xx}	f_{xy}	f_{yy}	D	Type
$(0,0)$	2	0	-2	-4	Saddle
$(\sqrt{2},0)$	$-4/e$	0	$-4/e$	$16/e^2$	Maximum
$(-\sqrt{2},0)$	$-4/e$	0	$-4/e$	$16/e^2$	Maximum
$(0,\sqrt{2})$	$4/e$	0	$4/e$	$16/e^2$	Minimum
$(0,-\sqrt{2})$	$4/e$	0	$4/e$	$16/e^2$	Minimum

The results of this example are confirmed by the graph in Figure 3.4.4. ◆

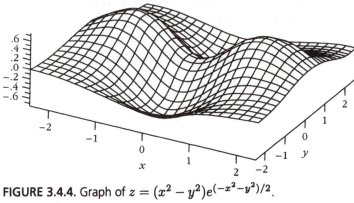

FIGURE 3.4.4. Graph of $z = (x^2 - y^2)e^{(-x^2-y^2)/2}$.

Example 4   Let $z = (x^2 + y^2)\cos(x + 2y)$. Show that $(0,0)$ is a critical point. Is it a local maximum or minimum?

Solution   We compute

$$\frac{\partial z}{\partial x} = 2x\cos(x + 2y) - (x^2 + y^2)\sin(x + 2y),$$

$$\frac{\partial z}{\partial y} = 2y\cos(x + 2y) - 2(x^2 + y^2)\sin(x + 2y).$$

These vanish at $(0,0)$, so $(0,0)$ is a critical point. The second derivatives are:

$$\frac{\partial^2 z}{\partial x^2} = 2\cos(x+2y) - 4x\sin(x+2y) - (x^2+y^2)\cos(x+2y),$$

$$\frac{\partial^2 z}{\partial x \partial y} = -4x\sin(x+2y) - 2y\sin(x+2y) - 2(x^2+y^2)\cos(x+2y),$$

$$\frac{\partial^2 z}{\partial y^2} = 2\cos(x+2y) - 8y\sin(x+2y) - 4(x^2+y^2)\cos(x+2y).$$

Evaluating at $(0,0)$, we get

$$\frac{\partial^2 z}{\partial x^2} = 2, \quad \frac{\partial^2 z}{\partial x \partial y} = 0, \quad \text{and} \quad \frac{\partial^2 z}{\partial y^2} = 2,$$

so $D = 4 - 0 = 4 > 0$ and $\partial^2 z / \partial x^2 > 0$, and thus $(0,0)$ is a local minimum.
♦

Exercises for §3.4

Find the critical points of each function in Exercises 1–4. Decide by inspection whether each of the critical points is a local maximum, minimum, or neither.

1. $f(x,y) = x^2 + 2y^2$

2. $f(x,y) = x^2 - 2y^2$

3. $f(x,y) = \exp(-x^2 - 7y^2 + 3)$

4. $f(x,y) = \exp(x^2 + 2y^2)$

In Exercises 5–10, use the maximum–minimum test for quadratic functions to decide whether $(0,0)$ is a maximum, minimum, or saddle point.

5. $f(x,y) = x^2 + xy + y^2$

6. $f(x,y) = x^2 - xy + y^2 + 1$

7. $f(x,y) = y^2 - x^2 + 3xy$

8. $f(x,y) = x^2 + y^2 - xy$

9. $f(x,y) = y^2$

10. $f(x,y) = 3 + 2x^2 - xy + y^2$

Find the critical points of the functions in Exercises 11–20, and use the second derivative test to classify them as local maxima, local minima, or saddles, or state that the second derivative test fails.

11. $f(x, y) = x^2 + y^2 + 6x - 4y + 13$

12. $f(x, y) = x^2 + y^2 + 3x - 2y + 1$

13. $f(x, y) = x^2 - y^2 + xy - 7$

14. $f(x, y) = x^2 + y^2 + 3xy + 10$

15. $f(x, y) = x^2 - 3xy + 5x - 2y + 6y^2 + 8$

16. $f(x, y) = x^2 + xy^2 + y^4$

17. $f(x, y) = e^{1+x^2-y^2}$

18. $f(x, y) = (x^2 + y^2)e^{x^2-y^2}$

19. $f(x, y) = \log(2 + \sin xy)$ [Consider only the critical point $(0, 0)$.]

20. $f(x, y) = \sin(x^2 + y^2)$ [Consider only the critical point $(0, 0)$.]

21. Let $f(x, y) = x^2 + y^2 + kxy$. If you imagine the graph changing as k increases, at what values of k does the shape of the graph change qualitatively?

22. Find the local maxima and minima for $z = (x^2 + 3y^2)e^{1-x^2-y^2}$. (See Figure 2.1.23.)

23. An examination of the function $f(x, y) = (y - 3x^2)(y - x^2)$ will give an idea of the difficulty of finding conditions that guarantee that a critical point is a local extremum when the second derivative test fails. Show that

    (a) the origin is a critical point of f;

    (b) f has a local minimum at $(0, 0)$ on every straight line through the point $(0, 0)$; that is, if $g(t) = (at, bt)$, then $f \circ g$. R → R has a local minimum at 0, for every choice of a and b;

    (c) the origin is not a relative minimum of f.

24. Let n be an integer greater than 2 and set $f(x, y) = ax^n + cy^n$, where $ac \neq 0$. Determine the nature of the critical points of f.

3.5
Constrained
Extrema and
Lagrange
Multipliers

In some problems we need to maximize a function $f(x, y)$ subject to certain constraints or side conditions. For example, we might want to maximize $f(x, y)$ subject to the condition that $x^2 + y^2 = 1$; *i.e.*, that (x, y) lie on the unit circle. More generally we might need to maximize or minimize $f(x, y)$ subject to the side condition that (x, y) also satisfies an equation $g(x, y) = c$ where g is some function and c equals a constant (in the example above, $g(x, y) = x^2 + y^2$, and $c = 1$). The set of such (x, y) is a level curve for g.

The purpose of this section is to develop some methods for handling this sort of problem. In Figure 3.5.1 we picture a graph of a function $f(x, y)$. In this picture, the maximum of f occurs at $(0, 0)$. However, suppose we are not interested in this maximum but only the maximum of $f(x, y)$ when (x, y) belongs to the unit circle; *i.e.*, when $x^2 + y^2 = 1$. The cylinder over $x^2 + y^2 = 1$ intersects the graph of $z = f(x, y)$ in a curve that lies on this graph. The problem of maximizing or minimizing $f(x, y)$ subject to the constraint $x^2 + y^2 = 1$ amounts to finding the point on this curve where z is the greatest or the least.

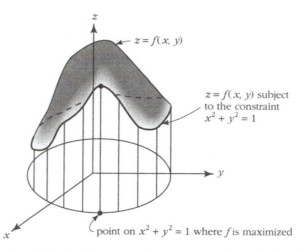

FIGURE 3.5.1. The geometric meaning of maximizing f subject to the constraint $x^2 + y^2 = 1$.

When f is restricted to a level curve $g(x, y) = c$, we again have the notion of local maxima or local minima of f (local extrema), and an absolute maximum (largest value) or absolute minimum (smallest value) must be a local extremum. The following test provides a necessary condition for a constrained extremum:

Critical Point Test for Constrained Extrema

Let f and g be functions of two variables with continuous partial derivatives. Suppose that the function f, when restricted to the level curve C defined by $g(x, y) = c$, has an extremum at (x_0, y_0) and that $\nabla g(x_0, y_0) \neq \mathbf{0}$. Then there is a number λ such that

$$\nabla f(x_0, y_0) = \lambda \nabla g(x_0, y_0).$$

If $\lambda \neq 0$, this formula says that the level curves of f and g through (x_0, y_0) have the same tangent line at (x_0, y_0).

Thus, to find a maximum or a minimum of f on a level curve C we look at those points (x_0, y_0) on C for which there is a constant λ, called a ***Lagrange multiplier***, such that $\nabla f(x_0, y_0) = \lambda \nabla g(x_0, y_0)$. This means we wish to solve the three simultaneous equations

$$\begin{aligned}
f_x(x, y) &= \lambda g_x(x, y), \\
f_y(x, y) &= \lambda g_y(x, y), \\
g(x, y) &= c
\end{aligned}$$

for the three unknowns x, y, and λ. After we find all pairs of points (x, y) satisfying these three equations, say $(x_1, y_1), (x_2, y_2), \ldots, (x_n, y_n)$, we evaluate f at these points obtaining a list of numbers $f(x_1, y_1), f(x_2, y_2), \ldots, f(x_n, y_n)$. If C is a closed curve, the absolute maximum will be the largest of these and the absolute minimum the smallest.

Justification of Critical Point Test Recall that $\nabla g(x_0, y_0)$ is orthogonal to the level curve $g(x, y) = c$ at (x_0, y_0). Parametrize this curve by $\mathbf{c}(t)$, so $\mathbf{c}(0) = (x_0, y_0)$. Since $f(\mathbf{c}(t))$ has an extremum as a function of t at $t = 0$, its derivative vanishes:

$$0 = \frac{d}{dt} f(\mathbf{c}(t)) = \nabla f(x_0, y_0) \cdot \mathbf{c}'(0).$$

The second equality was obtained by the chain rule. Since $\mathbf{c}'(0)$ is orthogonal to $\nabla g(x_0, y_0) \neq 0$ and $\nabla f(x_0, y_0)$ is orthogonal to $\mathbf{c}'(0)$, the two vectors $\nabla f(x_0, y_0)$ and $\nabla g(x_0, y_0)$ are parallel. Therefore, $\nabla f(x_0, y_0) = \lambda \nabla g(x_0, y_0)$ for some constant λ. See Figure 3.5.2. ∎

Here is another way of stating our result:

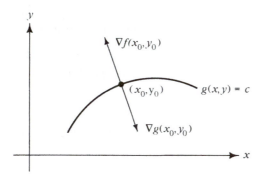

FIGURE 3.5.2. Critical point of a constrained function.

Method of Lagrange Multipliers

To find the extreme points of $f(x, y)$ subject to the constraint $g(x, y) = c$, seek points (x, y) and numbers λ such that $g_x(x, y)$ and $g_y(x, y)$ are not both zero and the **Lagrange multiplier equations** hold:

$$f_x(x, y) = \lambda g_x(x, y),$$
$$f_y(x, y) = \lambda g_y(x, y),$$
$$g(x, y) = c.$$

If the level curve $g(x, y) = c$ is a closed curve, then after finding all such points pick the one where f is largest (this will be the absolute maximum) and the one where f is the smallest (this will be the absolute minimum).

Example 1   Find the extreme values of $f(x, y) = x^2 - y^2$ along the circle C of radius 1 centered at the origin in the xy plane.

Solution   The circle C is the level curve $g(x, y) = x^2 + y^2 = 1$, so we want x, y, and λ such that

$$f_x(x, y) = \lambda g_x(x, y),$$
$$f_y(x, y) = \lambda g_y(x, y),$$
$$g(x, y) = 1.$$

That is

$$2x = \lambda 2x,$$
$$-2y = \lambda 2y,$$
$$x^2 + y^2 = 1.$$

From the first equation, either $x = 0$ or $\lambda = 1$. If $x = 0$, then from the third equation, $y = \pm 1$, and then from the second, $\lambda = -1$. If $\lambda = 1$, then $y = 0$

and $x = \pm 1$; so the eligible points are $(x, y) = (0, \pm 1)$ with $\lambda = -1$ and $(x, y) = (\pm 1, 0)$ with $\lambda = 1$. To determine for extrema, we evaluate f:

$$f(0, 1) = f(0, -1) = -1,$$
$$f(1, 0) = f(-1, 0) = 1,$$

so the maximum and minimum values are 1 and -1. ◆

The next example shows that if C is not closed, the function f might not have any maximum or minimum on C.

Example 2   Let C be the line $y = x + 1$ and let $f(x, y) = x^2 + y^2$. (a) Show that f has no maximum on C. (b) Find the minimum of f on C.

Solution   (a) Substituting $y = x + 1$ in $f(x, y)$ we see that $f(x, y) = x^2 + (x + 1)^2 = 2x^2 + 2x + 1$ when $y = x + 1$. Taking x large, we see that $f(x, y)$ can take arbitrarily large values on C, so there is no maximum.

(b) As x gets large positive (or large negative) $f(x, y)$ gets larger and larger. As in the discussion in §**3.3**, this implies that f must have a minimum on C. Since $\nabla f(x, y) = (2x, 2y)$ and C is the level curve $g(x, y) = 1$, where $g(x, y) = y - x$, the Lagrange multiplier equations give

$$2x = -\lambda,$$
$$2y = \lambda,$$
$$y - x = 1.$$

Solving these gives $x = -\frac{1}{2}, y = \frac{1}{2}$ and $\lambda = 1$. Since $(-\frac{1}{2}, \frac{1}{2})$ is the only possible extreme point for f, it must be the global absolute minimum.

In this case we can find an even simpler solution. As we already observed, the function f on C is given by $f(x, y) = 2x^2 + 2x + 1$, $-\infty < x < \infty$. Minimizing f is the same as finding the minimum of the function of *one variable* $h(x) = 2x^2 + 2x + 1$. Since $h(x) \to \infty$ as $x \to \pm\infty$, this minimum must occur at a critical point of h. But $h'(x) = 4x + 2 = 0$ only when $x = -\frac{1}{2}$. ◆

In §**3.3** we discussed the problem of finding absolute maxima and minima of functions of one variable on an interval. Lagrange multipliers will help us solve analogous problems in higher dimensions. Let D be a region in the plane $\mathbb{R}^2$ whose boundary is a smooth closed curve C (see Figure 3.5.3 for several examples).

Let f be a differentiable function defined on D and on C. How can we find the absolute maximum and minimum of f on D; *i.e.,* those points (x_0, y_0) such that $f(x_0, y_0) \geq f(x, y)$ for all (x, y) in D and those points (x_1, y_1) such that $f(x_1, y_1) \leq f(x, y)$ for all (x, y) in D?

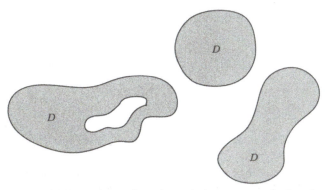

FIGURE 3.5.3. Regions whose boundaries are smooth closed curves.

Finding the Absolute Maximum and Minimum of $f(x, y)$ on a Region D

Let $f(x, y)$ be a differentiable function of two variables defined on a region D in $\mathbb{R}^2$, which is bounded by a smooth closed curve C. To find the absolute maximum and minimum of f on D:

1. Locate all the critical points of f in D.

2. Using the method of Lagrange multipliers or a parametrization of C, find the candidates for the absolute maximum and absolute minimum of f on C.

3. Among all the points found in steps 1 and 2, choose those where f is the largest and those where f is the smallest; the values of f at these points will be the absolute maximum and minimum of f, respectively.

If D is a region bounded by a collection of smooth curves (such as a square), then one follows a similar procedure, but including in step 3 the points where the curves meet (such as the corners of the square).

Example 3 Find the absolute maximum of $f(x, y) = xy$ on the unit disc D, where D is the set of points (x, y) with $x^2 + y^2 \leq 1$.

Solution First we find all the critical points of f in D. Since

$$\frac{\partial f}{\partial x} = y \quad \text{and} \quad \frac{\partial f}{\partial y} = x,$$

$(0, 0)$ is the only critical point of f inside D. Now consider f on the unit circle C, the level curve $g(x, y) = 1$, where $g(x, y) = x^2 + y^2$. To locate the maximum and minimum of f on C, we write down the Lagrange multiplier equations

$\nabla f(x,y) = (y,x) = \lambda \nabla g(x,y) = \lambda(2x, 2y)$ and $x^2 + y^2 = 1$. Rewriting these in component form, we get

$$\begin{aligned} y &= 2\lambda x, \\ x &= 2\lambda y, \\ x^2 + y^2 &= 1. \end{aligned}$$

Thus,

$$y = 4\lambda^2 y$$

or $\lambda = \pm 1/2$ and $y = \pm x$ which means that $x^2 + x^2 = 2x^2 = 1$ or $x = \pm 1/\sqrt{2}, y = \pm 1/\sqrt{2}$. On C we compute four candidates for the absolute maximum and minimum, namely,

$$\left(-\frac{1}{\sqrt{2}}, -\frac{1}{\sqrt{2}}\right), \left(-\frac{1}{\sqrt{2}}, \frac{1}{\sqrt{2}}\right), \left(\frac{1}{\sqrt{2}}, \frac{1}{\sqrt{2}}\right), \left(\frac{1}{\sqrt{2}}, -\frac{1}{\sqrt{2}}\right).$$

The value of f at both $\left(-\frac{1}{\sqrt{2}}, -\frac{1}{\sqrt{2}}\right)$ and $\left(\frac{1}{\sqrt{2}}, \frac{1}{\sqrt{2}}\right)$ is $\frac{1}{2}$. The value of f at $\left(-\frac{1}{\sqrt{2}}, \frac{1}{\sqrt{2}}\right)$ and $\left(\frac{1}{\sqrt{2}}, -\frac{1}{\sqrt{2}}\right)$ is $-1/2$, and the value of f at $(0,0)$ is 0. Therefore the absolute maximum of f is $1/2$ and the absolute minimum is $-1/2$, both occuring on C. At $(0,0), \partial^2 f/\partial x^2 = 0, \partial^2 f/\partial y^2 = 0$ and $\partial^2 f/\partial x \partial y = 1$, so the discriminant is -1 and thus $(0,0)$ is a saddle point.   ◆

Example 4   Find the absolute maximum and minimum of $f(x,y) = \frac{1}{2}x^2 + \frac{1}{2}y^2$ in the elliptical region D defined by $\frac{1}{2}x^2 + y^2 \leq 1$.

Solution   We first locate the critical points of f in D. Since

$$\frac{\partial f}{\partial x} = x, \qquad \frac{\partial f}{\partial y} = y,$$

the only critical point is the origin $(0,0)$.

We now find the maximum and minimum of f on the boundary of D which is the level curve $g(x,y) = 1$, where $g(x,y) = \frac{1}{2}x^2 + y^2$. The Lagrange multiplier equations are

$$\nabla f(x,y) = (x,y) = \lambda \nabla g(x,y) = \lambda(x, 2y)$$

and

$$\frac{x^2}{2} + y^2 = 1.$$

In other words,

$$x = \lambda x$$
$$y = 2\lambda y$$
$$\frac{x^2}{2} + y^2 = 1.$$

If $x = 0$, then $y = \pm 1$ and $\lambda = \frac{1}{2}$. If $y = 0$, then $x = \pm\sqrt{2}$ and $\lambda = 1$. If $x \neq 0$ and $y \neq 0$ we get both $\lambda = 1$ and $\frac{1}{2}$, which is impossible. Thus the candidates for the maxima and minima of f on C are $(0, \pm 1), (\pm\sqrt{2}, 0)$ and for f inside D, the candidate is $(0, 0)$. The value of f at $(0, \pm 1)$ is $\frac{1}{2}$, at $(\pm\sqrt{2}, 0)$ it is 1 and at $(0, 0)$ it is 0. Thus the absolute minimum of f occurs at $(0, 0)$ and is 0. The absolute maximum of f on D is thus 1 and occurs at the points $(\pm\sqrt{2}, 0)$. ◆

A similar procedure applies for functions of three (or more) variables.

Lagrange Multiplier Method in Space

If f has a local maximum or minimum at (x_0, y_0, z_0) when subject to the constraint $g(x, y, z) = c$, and if $\nabla g(x_0, y_0, z_0) \neq \mathbf{0}$, then there is a number λ such that

$$\nabla f(x_0, y_0, z_0) = \lambda \nabla g(x_0, y_0, z_0)$$

or, equivalently, $\nabla f(x_0, y_0, z_0)$ is perpendicular to the level set $g(x, y, z) = c$. See Figure 3.5.4.

If $g(x, y, z) = c$ is a closed surface, then the absolute maximum (or minimum) of f is obtained by finding all points where ∇f is a multiple of ∇g and choosing those where f is largest (or smallest). (A closed surface is one like a sphere—the "shell"—or an ellipsoid that has no edge and is confined to a finite region of space.)

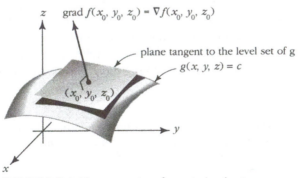

FIGURE 3.5.4. The geometry of constrained extrema.

Example 5   Maximize the function $f(x, y, z) = x + z$ subject to the constraint $x^2 + y^2 + z^2 = 1$.

Solution   Using the Lagrange multiplier method, we seek λ and (x, y, z) such that

$$
\begin{aligned}
1 &= 2x\lambda, \\
0 &= 2y\lambda, \\
1 &= 2z\lambda, \\
x^2 + y^2 + z^2 &= 1.
\end{aligned}
$$

From the first or the third equation, we see that $\lambda \neq 0$. Thus, from the second equation, we get $y = 0$. From the first and third equations, $x = z$, and so from the fourth, $x = \pm 1/\sqrt{2} = z$. Hence our points are $(1/\sqrt{2}, 0, 1/\sqrt{2})$ and $(-1/\sqrt{2}, 0, -1/\sqrt{2})$. Comparing the values of f at these points, we see that the first point is the maximum of f (restricted by the constraint) and the second is the minimum.   ◆

Example 6   Assume that among all rectangular boxes with fixed surface area of 10 square meters there is a box of largest possible volume. Find its dimensions.

Solution   If x, y, and z are the lengths of the sides, $x \geq 0, y \geq 0, z \geq 0$, respectively, and the volume is $f(x, y, z) = xyz$. The constraint is $2(xy + xz + yz) = 10$; that is, $xy + xz + yz = 5$. Thus, the Lagrange multiplier conditions are

$$
\begin{aligned}
yz &= \lambda(y + z) \\
xz &= \lambda(x + z) \\
xy &= \lambda(y + x) \\
xy + xz + yz &= 5.
\end{aligned}
$$

First of all, $x \neq 0$, since $x = 0$ implies $yz = 5$ and $0 = \lambda z$, so that $\lambda = 0$ and we get the contradictory equation $yz = 0$. Similarly, $y \neq 0, z \neq 0, x + y \neq 0$, and so on. Elimination of λ from the first two equations gives $yz/(y + z) = xz/(x + z)$, which gives $x = y$; similarly, $y = z$. Substituting these values into the last equation, we obtain $3x^2 = 5$, or $x = \sqrt{5/3}$. Thus we get the solution $x = y = z = \sqrt{5/3}$, and $xyz = (5/3)^{3/2}$. This (cubical) shape must therefore maximize the volume, assuming there is a box of maximum volume.   ◆

Notes

1. The solution to Example 6 does *not* demonstrate that the cube is the rectangular box of largest volume with a given fixed surface area; it proves that the cube is the only possible candidate for a maximum. We shall sketch a proof that it is the maximum in Note 2 below. The distinction between showing that there is *only one possible solution* to a problem and that, in fact, *a solution exists* is a subtle one that many (even great) mathematicians have overlooked.

For example, Queen Dido (ca. 900 B.C.) realized that among all planar regions with fixed circumference the disc is the region of maximum area. It is not terribly difficult to prove this fact under the assumption that there *is* a region of maximum area; however, proving that such a region of maximum area exists is quite another (difficult) matter. A complete proof was not given until the second half of the nineteenth century by the German mathematician Weierstrass.

Let us consider a nonmathematical parallel to this situation. Put yourself in the place of Lord Peter Wimsey, Dorothy Sayers' famous detective:

> "Undoubtedly," said Wimsey, "but if you think that this identification is going to make life one grand, sweet song for you, you are mistaken. . . . Since we have devoted a great deal of time and thought to the case on the assumption that it was murder, it's a convenience to know that the assumption is correct."

Wimsey has found the body of a dead man, and after some time has located ten suspects. He is sure that no one else other than one of the suspects could be the murderer. By collecting all the evidence and checking alibis, he then reduces the number of suspects one by one, until, finally, only the butler remains; hence he is the murderer! But wait, Peter is a very cautious man. By checking everything once again, he discovers that the man died by suicide; so there is no murder. You see the point: It does not suffice to find a clear and uniquely determined suspect in a criminal case where murder is suspected; you must prove that a murder actually took place.

The same goes for our cube; the fact that it is the only possible candidate for a maximum does not prove that it is a maximum. (For more information see S. Hildebrandt and A. Tromba, *Mathematics and Optimal Form*, Scientific American Books, N.Y. 1984.)

2. The problem in showing that $f(x, y, z) = xyz$ has a maximum lies in the fact that f is a continuous function which is defined on the unbounded surface $S : xy + xz + yz = 5$, and not on a bounded set which includes its boundary. We have already seen problems of this sort for functions of one and two variables (see §**3.3**), but not yet for problems involving maximizing or minimizing functions subject to a constraint.

The way to show that $f(x, y, z) = xyz \geq 0$ does indeed have a maximum on $xy + yz + xz = 5$ is to show that if either x, y, or z tend to ∞, then $f(x, y, z) \to 0$. We may then conclude that the maximum of f on S must exist, as in §**3.3**. So, suppose (x, y, z) lies in S and $x \to \infty$, then $y \to 0$ and $z \to 0$ (why?). Multiplying the equation defining S by z we obtain the equation $xyz + xz^2 + yz^2 = 5z \to 0$ as $x \to \infty$. Since $x, y, z \geq 0, xyz = f(x, y, z) \to 0$. Similarly $f(x, y, z) \to 0$ if either y or z tend to ∞. Thus a box of maximum volume must exist.

3. There is a similar test for local maxima or minima of functions subject to more than one constraint. For instance, if we want to extremize $f(x, y, z)$ subject to $g_1(x, y, z) = c_1$ and $g_2(x, y, z) = c_2$, then $\nabla f = \lambda \nabla g$ is replaced by $\nabla f = \lambda_1 \nabla g_1 + \lambda_2 \nabla g_2$.

4. There are also second derivative tests available for constrained extrema. The reader may consult J. Marsden and A. Tromba, *Vector Calculus*, W.H. Freeman, N.Y., 3rd Ed. (1988).

Exercises for §3.5

In Exercises 1–8, find the extrema of f subject to the stated constraints.

1. $f(x,y) = x,\ x^2 + 2y^2 = 3$

2. $f(x,y) = x - y,\ x^2 - y^2 = 2$

3. $f(x,y) = 3x + 2y,\ 2x^2 + 3y^2 = 3$

4. $f(x,y) = x - y,\ x^2 + y^2 = 2$

5. $f(x,y) = xy,\ x + y = 1$

6. $f(x,y) = \cos^2 x + \cos^2 y,\ x + y = \pi/4$

7. $f(x,y,z) = x - y + z,\ x^2 + y^2 + z^2 = 2$

8. $f(x,y,z) = x + y + z,\ x^2 - y^2 = 1,\ 2x + z = 1$

In Exercises 9–12, find the absolute maxima and minima of the given function on the disk $x^2 + y^2 \leq 1$.

9. $f(x,y) = 2x^2 + 3y^2$

10. $f(x,y) = xy + 5y$

11. $f(x,y) = 5x^2 - 2y^2 + 10$

12. $f(x,y) = 3xy - y + 5$

Find the extrema of f subject to the stated constraints in Exercises 13–16.

13. $f(x,y) = 3x + 2y;\ 2x^2 + 3y^2 \leq 3$

14. $f(x,y) = x - 3y;\ x^2 + y^2 = 1$

15. $f(x,y) = xy;\ 2x + 3y \leq 10, 0 \leq x, 0 \leq y$

16. $f(x,y) = x^2 + y^2;\ x^4 + y^4 = 2$

17. Design a cylindrical can (with a lid) to contain 1 liter of water, using the minimum amount of metal.

18. A rectangular box *with no top* is to have a surface area of 16 square meters. Find the dimensions that maximize its volume.

19. A parcel delivery service requires that the dimensions of a rectangular box be such that the length plus twice the width plus twice the height be no more than 108 inches $(l + 2w + 2h \leq 108)$. What is the volume of the largest-volume box the company will deliver?

20. A rectangular mirror with area A square feet is to have trim along the edges. If the trim along the horizontal edges costs p cents per foot and that for the vertical edges costs q cents per foot, find the dimensions that will minimize the total cost of the trim.

21. Cascade Container Company produces a cardboard shipping crate at three different plants in amounts x, y, z, respectively, producing an annual revenue of $R(x, y, z) = 8xyz^2 - 200,000(x + y + z)$. The company is to produce $100,000$ units annually. How should production be distributed to maximize the revenue?

22. The Baraboo, Wisconsin, plant of International Widget Co. uses aluminum, iron, and magnesium to produce high-quality widgets. The quantity of widgets which may be produced using x tons of aluminum, y tons of iron, and z tons of magnesium is $Q(x, y, z) = xyz$. The cost of raw materials is aluminum, \$6 per ton; iron, \$4 per ton; and magnesium, \$8 per ton. How many tons each of aluminum, iron, and magnesium should be used to manufacture 1000 widgets at the lowest possible cost?

23. A firm uses wool and cotton fiber to produce cloth. The amount of cloth produced is given by $Q(x, y) = xy - x - y + 1$, where x is the number of pounds of wool, y is the number of pounds of cotton, and $x > 1$ and $y > 1$. If wool costs p dollars per pound, cotton costs q dollars per pound, and the firm can spend B dollars on material, what should the mix of cotton and wool be to produce the most cloth?

24. Let P be a point on a surface S in $\mathbf{R}^3$ defined by the equation $f(x, y, z) = 1$, where f is of class C^1 and $\nabla f \neq 0$ at P. Suppose that P is a point where the distance from the origin to S is maximized. Show that the vector emanating from the origin and ending at P is perpendicular to S.

25. The state of Megalomania occupies the region $x^4 + 2y^4 \leq 30,000$. The altitude at point (x, y) is $\frac{1}{8}xy + 200x$ meters above sea level. Where are the highest and lowest points in the state?

Review Exercises for Chapter 3

In Exercises 1–2, calculate all first order partial derivatives of the given function.

1. $u = g(x, y) = \dfrac{\sin(\pi x)}{1 + y^2}$

2. $u = f(x, y, z) = \dfrac{x}{1 + \cos(yz)}$

3. Find $f_x(1, 0)$ if $f(x, y) = \cos(x + e^{yx})$.

4. Find $f_s(-1, 2)$ if $f(r, s) = r + s^2/(1 - r^2 - s^2)$.

In Exercises 5–8, compute all first and second order partial derivatives and verify the equality of mixed partials.

5. $z = 3x^2 + 2y^2$

6. $z = x^2 y^2 c^{2xy}$, $c = $ constant

7. $f(x, y) = \cos \sqrt{x^2 + y^2}$

8. $f(x, y) = x \arctan (x/y)$

9. Verify that

$$\frac{\partial^3 f}{\partial x \partial y \partial z} = \frac{\partial^3 f}{\partial y \partial z \partial x}$$

for $f(x, y, z) = ze^{xy} + yz^3 x^2$.

10. Show that

$$\frac{\partial^4 f}{\partial x^2 \partial y^2} = \frac{\partial^4 f}{\partial x \partial y \partial x \partial y}.$$

11. Prove that the function $f(x, y) = \log (x^2 + y^2)$ satisfies the Laplace equation $f_{xx} + f_{yy} = 0$.

12. Prove that the function

$$h(x, y, z, w) = \frac{1}{x^2 + y^2 + z^2 + w^2}$$

satisfies the Laplace equation $h_{xx} + h_{yy} + h_{zz} + h_{ww} = 0$.

13. Find the first terms in the Taylor expansion of $f(x, y) = e^{xy} \cos x$ about $x = 0, y = 0$, up through the second order.

14. Find the second order Taylor expansion of $f(x, y) = \sin [(x^2 + 1)y]$ about $x = 0, y = \pi/2$.

Find and classify (as maxima, minima, or saddles) the critical points of the functions in Exercises 15–18.

15. $f(x, y) = x^2 - 6xy - y^2$

16. $f(x, y) = 2x^2 - y^2 + 5xy$

17. $f(x, y) = \exp (x^2 - y^2)$

18. $f(x, y) = \sin (x^2 + y^2)$ (Consider only $(0, 0)$).

19. Prove that

$$z = \frac{3x^4 - 4x^3 - 12x^2 + 18}{12(1 + 4y^2)}$$

has one local maximum, one local minimum, and one saddle point. (The graph is shown in Figure 3.R.1.)

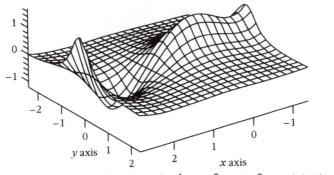

FIGURE 3.R.1. Graph of $z = (3x^4 - 4x^3 - 12x^2 + 18)/12(1 + 4y^2)$.

20. Find the maxima, minima, and saddles of the function $z = (2 + \cos \pi x)(\sin \pi y)$, which is graphed in Figure 3.R.2.

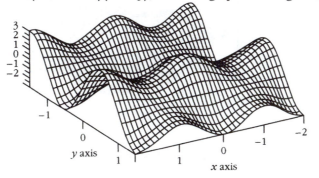

FIGURE 3.R.2. Graph of $z = (2 + \cos \pi x)(\sin \pi y)$.

21. Find and describe the critical points of $f(x,y) = y\sin(\pi x)$ (See Figure 3.R.3.)

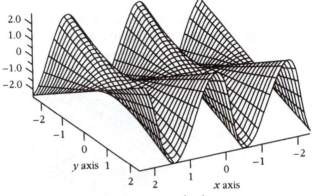

FIGURE 3.R.3. Graph of $z = y\sin(\pi x)$.

22. A graph of the function $z = \sin(\pi x)/(1 + y^2)$ is shown in Figure 3.R.4. Verify that this function has alternating maxima and minima on the x axis, with no other critical points.

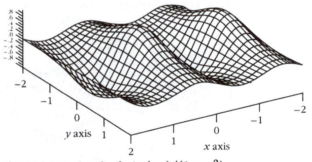

Figure 3.R.4. Graph of $\sin(\pi x)/(1 + y^2)$.

23. Find the minimum distance from the origin in $\mathbb{R}^3$ to the surface $z = \sqrt{x^2 - 1}$.

24. Find the points on the surface $z^2 - xy = 1$ nearest to the origin.

25. Find the extreme points of $z = xy$, subject to the condition $x + y = 1$.

26. Find the extreme points of $z = \cos^2 x + \cos^2 y$ subject to the condition $x + y = \pi/4$.

In Exercises 27–30, find the extrema of the given functions subject to the given constraints.

27. $f(x,y) = x^2 - 2xy + y^2;\ x^2 + y^2 = 1$

28. $f(x, y) = xy - y^2; x^2 + y^2 = 1$

29. $f(x, y) = \cos(x^2 - y^2); x^2 + y^2 = 1$

30. $f(x, y) = \dfrac{x^2 - y^2}{x^2 + y^2}; x + y = 1$

31. Find the shortest distance from the point $(0, b)$ to the parabola $x^2 - 4y = 0$. Solve this problem using the Lagrange multiplier and also without using Lagrange's method.

32. Find the maximum and minimum values of $f(x, y) = x^2 + y$ for (x, y) on a circle of radius 1 centered at the origin in two ways:

    (a) By parametrizing the circle.

    (b) By Lagrange multipliers.

33. Find the maximum of $f(x, y) = xy$ on the curve $(x + 1)^2 + y^2 = 1$.

34. Find the maximum and minimum of $f(x, y) = xy - y + x - 1$ on the set $x^2 + y^2 \le 2$.

35. Suppose that $z = f(x, y)$ is defined, has continuous second partial derivatives, and satisfies Laplace's equation:

$$\frac{\partial^2 z}{\partial x^2} + \frac{\partial^2 z}{\partial y^2} = 0.$$

    Assume that $(\partial^2 z / \partial x^2)(x_0, y_0) \ne 0$. Prove that f cannot have a local maximum or minimum at (x_0, y_0).

Solve the geometric problems in Exercises 36–38 by Lagrange's method.

36. Find the shortest distance from the point (a_1, a_2, a_3) in $\mathbb{R}^3$ to the plane whose equation is given by $b_1 x_1 + b_2 x_2 + b_3 x_3 + b_0 = 0$, where $(b_1, b_2, b_3) \ne (0, 0, 0)$.

37. Find the point on the line of intersection of the two planes $a_1 x_1 + a_2 x_2 + a_3 x_3 = 0$ and $b_1 x_1 + b_2 x_2 + b_3 x_3 + b_0 = 0$ that is nearest to the origin.

38. Show that the volume of the largest rectangular parallelepiped that can be inscribed in the ellipsoid

$$\frac{x^2}{a^2} + \frac{y^2}{b^2} + \frac{z^2}{c^2} = 1$$

    is $8abc/3\sqrt{3}$.

39. Study the nature of the function $f(x, y) = x^3 - 3xy^2$ near $(0, 0)$. Show that the point $(0, 0)$ is a degenerate critical point, that is, $D = 0$. This surface is called a "monkey saddle."

40. Drug reactions can be measured by functions of the form $R(u, t) = u^2(c - u)t^2 e^{-t}, 0 \leq u \leq c, t \geq 0$. The symbols u and t are drug units and time in hours, respectively. Find the dosage u and time t at which R is a maximum.

41. Find the absolute maximum and minimum values for $f(x, y) = \sin x + \cos y$ on the rectangle $R = [0, 2\pi] \times [0, 2\pi]$.

42. Find the absolute maximum and minimum values for the function $f(x, y) = xy$ on the rectangle $R = [-1, 1] \times [-1, 1]$.

4

Vector-Valued Functions

. . .who by vigor of mind almost divine, the motions and figures of the planets, the paths of comets, and the tides of the seas first demonstrated.

Newton's Epitaph

One of our main concerns in Chapter 3 was the study of real-valued functions. This chapter deals with functions whose values are vectors. We begin in §**4.1** with paths, which are maps from $\mathbb{R}$ to $\mathbb{R}^2$ or $\mathbb{R}^3$. Then we go on to vector fields and introduce the main operations of vector differential calculus other than the gradient, namely, the divergence and the curl. We consider some of the geometry associated with these operations, just as we did for the gradient, but the most significant physical applications will have to wait until we have studied integration theory.

4.1
Acceleration

In §**1.6** we studied paths, their velocity vectors and their tangent lines. In this section we study additional properties of paths, and in particular, acceleration and Newton's law $\mathbf{F} = m\mathbf{a}$.

Since a path in $\mathbb{R}^n$ is a map $\mathbf{c} : \mathbb{R} \to \mathbb{R}^n$, if the path is differentiable, its derivative at each time t is an $n \times 1$ matrix. Specifically, if $x_1(t), \ldots, x_n(t)$ are the component functions of $\mathbf{c}$, the derivative matrix $\mathbf{c}'(t)$ is

$$
\begin{bmatrix}
dx_1/dt \\
dx_2/dt \\
\vdots \\
dx_n/dt
\end{bmatrix},
$$

which can also be written in vector form as $(dx_1/dt, \ldots, dx_n/dt)$ or as $\big(x_1'(t), \ldots, x_n'(t) \big)$.

Recall from §**1.6** that $\mathbf{c}'(t)$ can be interpreted as the tangent or the velocity vector. Let us elaborate:

Tangents, Velocity, and Speed

If $\mathbf{c}$ is a path in $\mathbb{R}^n$, its **tangent** or **velocity vector** is $\mathbf{c}'(t)$.
If $\mathbf{c}$ represents the path of a moving particle, its **velocity vector** is

$$\mathbf{v} = \mathbf{c}'(t)$$

and its **speed** is $s = \|\mathbf{v}\|$.

The differentiation of paths is facilitated by the following rules:

> ## Differentiation Rules
>
> Let $\mathbf{b}(t)$ and $\mathbf{c}(t)$ be paths in $\mathbb{R}^3$ and $p(t)$ and $q(t)$ be scalar functions:
>
> | | |
> |---|---|
> | *Sum Rule:* | $\dfrac{d}{dt}[\mathbf{b}(t) + \mathbf{c}(t)] = \mathbf{b}'(t) + \mathbf{c}'(t)$ |
> | *Scalar Multiplication Rule:* | $\dfrac{d}{dt}[p(t)\mathbf{c}(t)] = p'(t)\mathbf{c}(t) + p(t)\mathbf{c}'(t)$ |
> | *Dot Product Rule:* | $\dfrac{d}{dt}[\mathbf{b}(t) \cdot \mathbf{c}(t)] = \mathbf{b}'(t) \cdot \mathbf{c}(t) + \mathbf{b}(t) \cdot \mathbf{c}'(t)$ |
> | *Cross Product Rule:* | $\dfrac{d}{dt}[\mathbf{b}(t) \times \mathbf{c}(t)] = \mathbf{b}'(t) \times \mathbf{c}(t) + \mathbf{b}(t) \times \mathbf{c}'(t)$ |
> | *Chain Rule:* | $\dfrac{d}{dt}[\mathbf{c}(q(t))] = q'(t)\mathbf{c}'(q(t))$ |

These rules follow by applying the usual differentiation rules to the components.

Example 1   Show that if $\mathbf{c}(t)$ is a vector function such that $\|\mathbf{c}(t)\|$ is constant, then $\mathbf{c}'(t)$ is perpendicular to $\mathbf{c}(t)$ for all t.

Solution   Since $\|\mathbf{c}(t)\|$ is constant, so is its square $\|\mathbf{c}(t)\|^2 = \mathbf{c}(t) \cdot \mathbf{c}(t)$. The derivative of this constant is zero, so by the dot product rule,

$$0 = \frac{d}{dt}[\mathbf{c}(t) \cdot \mathbf{c}(t)] = \mathbf{c}'(t) \cdot \mathbf{c}(t) + \mathbf{c}(t) \cdot \mathbf{c}'(t) = 2\mathbf{c}(t) \cdot \mathbf{c}'(t);$$

so $\mathbf{c}(t) \cdot \mathbf{c}'(t) = 0$; that is, $\mathbf{c}'(t)$ is perpendicular to $\mathbf{c}(t)$.   ◆

For a curve describing uniform rectilinear motion, the velocity vector is constant. In general, the velocity vector is a vector function $\mathbf{v} = \mathbf{c}'(t)$ which depends on t. The derivative $\mathbf{a} = d\mathbf{v}/dt = \mathbf{c}''(t)$ is called the ***acceleration vector*** of the curve. If the curve is $(x(t), y(t), z(t))$, then the acceleration vector is

$$\mathbf{a} = x''(t)\mathbf{i} + y''(t)\mathbf{j} + z''(t)\mathbf{k}.$$

Example 2   A particle moves in such a way that its acceleration is constantly equal to $-\mathbf{k}$. If the position when $t = 0$ is $(0, 0, 1)$ and the velocity at $t = 0$ is $\mathbf{i}+\mathbf{j}$, when and where does the particle fall below the plane $z = 0$? Describe the path travelled by the particle.

Solution   Let $(x(t), y(t), z(t))$ be the parametric curve traced out by the particle, so that the velocity vector is $\mathbf{c}'(t) = x'(t)\mathbf{i}+y'(t)\mathbf{j}+z'(t)\mathbf{k}$. The acceleration $\mathbf{c}''(t)$ is $-\mathbf{k}$, so $x''(t) = 0, y''(t) = 0$, and $z''(t) = -1$. It follows that $x'(t)$ and

$y'(t)$ are constant functions, and $z'(t)$ is a linear function with slope -1. Since $\mathbf{c}'(0) = \mathbf{i}+\mathbf{j}$, we get $\mathbf{c}'(t) = \mathbf{i}+\mathbf{j}-t\mathbf{k}$. Integrating again and using the initial position $(0,0,1)$, we find that $(x(t), y(t), z(t)) = (t, t, 1 - \frac{1}{2}t^2)$. The particle drops below the plane $z = 0$ when $1 - \frac{1}{2}t^2 = 0$; that is, $t = \sqrt{2}$. At that time, the position is $(\sqrt{2}, \sqrt{2}, 0)$. The path travelled by the particle is a parabola in the plane $y = x$ (see Figure 4.1.1) because in this plane the equation is described by $z = 1 - \frac{1}{2}x^2$. ◆

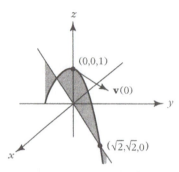

FIGURE 4.1.1. The path of the parabola with initial position $(0, 0, 1)$, initial velocity $\mathbf{i}+\mathbf{j}$, and constant acceleration $-\mathbf{k}$ is a parabola in the plane $y = x$.

If a particle of mass m moves in $\mathbb{R}^3$, the force $\mathbf{F}$ acting on it at the point $\mathbf{c}(t)$ is related to the acceleration by **Newton's second law**:

$$\mathbf{F}(\mathbf{c}(t)) = m\mathbf{a}(t).$$

In particular, if no forces act on a particle, then $\mathbf{a}(t) = \mathbf{0}$, so $\mathbf{c}'(t)$ is constant and the particle follows a straight line.

Acceleration and Newton's Second Law

The **acceleration** of a path $\mathbf{c}(t)$ is

$$\mathbf{a}(t) = \mathbf{c}''(t).$$

If $\mathbf{F}$ is the force acting and m is the mass of the particle, then

$$\mathbf{F} = m\mathbf{a}.$$

In the problem of determining the path $\mathbf{c}(t)$ of a particle, Newton's law becomes a differential equation (*i.e.*, an equation involving derivatives) for $\mathbf{c}(t)$.

For example, a planet moving around the sun (considered to be located at the origin in $\mathbb{R}^3$) along a path $\mathbf{r}(t)$ obeys the law

$$m\mathbf{r}'' = -\frac{GmM}{r^3}\mathbf{r}$$

where M is the mass of the sun, m that of the planet, $r = \|\mathbf{r}\|$, and G is the gravitational constant. The relation used in determining the force, $\mathbf{F} = -GmM\mathbf{r}/r^3$, is called **Newton's law of gravitation** (see Figure 4.1.2). We shall not make a general study of such equations in this book, but content ourselves with the following special case.

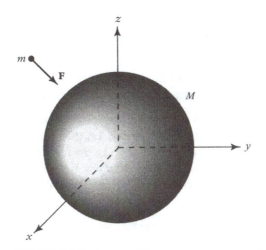

FIGURE 4.1.2. A mass M attracts a mass m with a force $\mathbf{F}$ given by Newton's law of gravitation: $\mathbf{F} = -GmM\mathbf{r}/r^3$.

Consider a particle of mass m moving at constant speed s in a circular path of radius r_0. Supposing that it moves in the xy plane, we can suppress the third component and write

$$\mathbf{r}(t) = \left(r_0\cos\frac{st}{r_0}, r_0\sin\frac{st}{r_0}\right),$$

since this is a circle of radius r_0 and $\|\mathbf{r}'(t)\| = s$. Then

$$\mathbf{a}(t) = \mathbf{r}''(t) = \left(-\frac{s^2}{r_0}\cos\frac{st}{r_0}, -\frac{s^2}{r_0}\sin\frac{st}{r_0}\right) = -\frac{s^2}{r_0^2}\mathbf{r}(t).$$

Thus, the acceleration is in a direction opposite to $\mathbf{r}(t)$; that is, it is directed toward the center of the circle (see Figure 4.1.3). This acceleration multiplied by the mass of the particle is called the **centripetal force**. Even though the speed

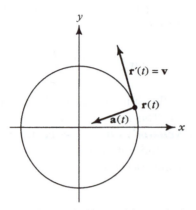

FIGURE 4.1.3. The position, velocity, and acceleration of a particle in circular motion.

is constant, the direction of the velocity is continuously changing—acceleration is a rate of change in either speed or direction or both.

Newton's law helps us discover a relationship between the radius of the orbit of a revolving body and the period, *i.e.*, the time it takes for one complete revolution. Consider a satellite of mass m moving with a speed s around a central body with mass M in a *circular* orbit of radius r_0 (distance from the *center* of the spherical central body). By Newton's second law $\mathbf{F} = m\mathbf{a}$, we get

$$-\frac{s^2 m}{r_0^2}\mathbf{r}(t) = -\frac{GmM}{r_0^3}\mathbf{r}(t).$$

The lengths of the vectors on both sides of this equation must be equal. Hence

$$s^2 = \frac{GM}{r_0}.$$

If T denotes the period, then $s = 2\pi r_0 / T$; substituting this value for s in the above equation and solving for T, we obtain the following:

Kepler's Law

$$T^2 = r_0^3 \frac{(2\pi)^2}{GM}.$$

Thus, the *square of the period is proportional to the cube of the radius.*

We have defined two basic concepts associated with a path; its velocity and its acceleration. Both involve *differential* calculus. The basic concept of the length of a path, which involves *integral* calculus, will be taken up in the next section.

The law of planetary motion in the preceding box is one of the three that Kepler derived before Newton formulated his general laws of motion. It enables one to compute the period of a satellite about the Earth or a planet about the Sun when the radius of its orbit is given, and vice versa. Kepler and Newton discovered and used formulas like this not only for circular orbits, but more generally for elliptical orbits. Newton was able to derive Kepler's three celestial laws from his own law of gravitation. The neat mathematical order of the universe that these laws provided had great impact on eighteenth-century thought.

Newton never wrote down his laws as analytical equations. This was first done by Euler around 1750. Newton made most of his deductions (at least in published form) by geometric methods alone.

Example 3 Suppose that a satellite is to be in circular orbit about the Earth such that it stays fixed in the sky over one point on the equator. What is the radius of such a *geosynchronous* orbit? (The mass of the Earth is 5.98×10^{24} kilograms and $G = 6.67 \times 10^{-11}$ in the meter-kilogram-second system of units.)

Solution The period of the satellite should be 1 day, so $T = 60 \times 60 \times 24 = 86{,}400$ seconds. From the formula $T^2 = r_0^3 (2\pi)^2 / GM$ we get $r_0^3 = T^2 GM / (2\pi)^2$ and so

$$r_0^3 = \frac{T^2 GM}{(2\pi)^2} = \frac{(86{,}400)^2 \times (6.67 \times 10^{-11}) \times (5.98 \times 10^{24})}{(2\pi)^2}$$

$$\approx 7.54 \times 10^{22} \text{ meters}^3.$$

Thus the radius of the orbit is $r_0 \approx 4.23 \times 10^7$ meters $= 42{,}300$ kilometers $\approx 26{,}200$ miles. ◆

Exercises for §4.1

In Exercises 1–4, find the speed and acceleration of the given path.

1. $\mathbf{c}(t) = 6t\mathbf{i} + 3t^2\mathbf{j} + t^3\mathbf{k}$

2. $\mathbf{c}(t) = (\sin 3t)\mathbf{i} + (\cos 3t)\mathbf{j} + 2t^{3/2}\mathbf{k}$

3. $\mathbf{r}(t) = (\cos^2 t, 3t - t^3, t)$

4. $\mathbf{r}(t) = (4e^t, 6t^4, \cos t)$

In Exercises 5–8, let $c_1(t) = e^t i + (\sin t)j + t^3 k$ and $c_2(t) = e^{-t}i + (\cos t)j - 2t^3 k$. Find each of the stated derivatives in two different ways to verify the rules in the box preceding Example 1.

5. $\dfrac{d}{dt}[c_1(t) + c_2(t)]$

6. $\dfrac{d}{dt}[c_1(t) \cdot c_2(t)]$

7. $\dfrac{d}{dt}[c_1(t) \times c_2(t)]$

8. $\dfrac{d}{dt}\{c_1(t) \cdot [2c_2(t) + c_1(t)]\}$

9. If $r(t) = 6ti + 3t^2 j + t^3 k$ (see Exercise 1), what force acts on a particle of mass m moving along r at $t = 0$?

10. A particle of mass of 1 gram follows the path $r(t) = \sin 3t i + \cos 3t j + 2t^{3/2}k$ (see Exercise 2), with units in seconds and centimeters. What force acts on it at $t = 0$? (Give the units in your answer.)

11. A body of mass 2 kilograms moves in a circular path on a circle of radius 3 meters, making one revolution every 5 seconds. Find the centripetal force acting on the body.

12. Find the centripetal force acting on a body of mass 4 kilograms, moving on a circle of radius 10 meters with a frequency of two revolutions per second.

13. Show that if the acceleration of an object is always perpendicular to the velocity, then the speed of the object is constant. (Hint: See Example 1.)

14. Show that, at a local maximum or minimum of $\|r(t)\|$, $r'(t)$ is perpendicular to $r(t)$.

15. A satellite is in a circular orbit 500 miles above the surface of the Earth. What is the period of the orbit? (You may take the radius of the Earth to be 4000 miles.)

16. What is the gravitational acceleration on the satellite in Exercise 15? The centripetal acceleration?

4.2
Arc Length

What is the length of a path $\mathbf{c}(t)$? Since the speed $\|\mathbf{c}'(t)\|$ is the rate of change of distance travelled with respect to time, the distance travelled by a point moving along the curve is the integral of speed with respect to the time over the interval $[t_0, t_1]$ of travel time; *i.e.*, the length of the path, also called its **arc length**, is $\int_{t_0}^{t_1} \|\mathbf{c}'(t)\| dt$.

Example 1 The arc length of the path $\mathbf{c}(t) = (r \cos t, r \sin t)$, for $0 \leq t \leq 2\pi$, is

$$l = \int_0^{2\pi} \sqrt{(-r \sin t)^2 + (r \cos t)^2} dt = 2\pi r,$$

the circumference of a circle of radius r. If we had allowed $0 \leq t \leq 4\pi$, we would have obtained $4\pi r$, because the path traverses the same circle *twice* (Figure 4.2.1). ◆

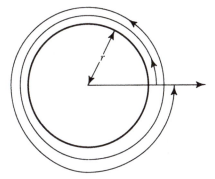

FIGURE 4.2.1. The arc length of a circle traversed twice is $4\pi r$.

Arc Length

The length of the path $\mathbf{c}(t) = (x(t), y(t), z(t))$ for $t_0 \leq t \leq t_1$, is

$$l = \int_{t_0}^{t_1} \sqrt{[x'(t)]^2 + [y'(t)]^2 + [z'(t)]^2} dt.$$

For planar curves one omits the $z'(t)$ term, as in Example 1. Here is an example in $\mathbb{R}^3$.

Example 2 Find the arc length of $(\cos t, \sin t, t^2), 0 \leq t \leq \pi$.

Solution The path $\mathbf{c}(t) = (\cos t, \sin t, t^2)$ has the velocity vector given by $\mathbf{v} = (-\sin t, \cos t, 2t)$. Since

$$\|\mathbf{v}\| = \sqrt{\sin^2 t + \cos^2 t + 4t^2} = \sqrt{1 + 4t^2} = 2\sqrt{t^2 + \left(\frac{1}{2}\right)^2},$$

the arc length is

$$L = \int_0^\pi 2\sqrt{t^2 + \left(\frac{1}{2}\right)^2}\, dt.$$

This integral may be evaluated using the following formula from the table of integrals:

$$\int \sqrt{x^2 + a^2}\, dx = \frac{1}{2}\left[x\sqrt{x^2 + a^2} + a^2 \log\left(x + \sqrt{x^2 + a^2}\right)\right] + C.$$

Thus,

$$
\begin{aligned}
L &= 2 \cdot \frac{1}{2}\left[t\sqrt{t^2 + \left(\frac{1}{2}\right)^2} + \left(\frac{1}{2}\right)^2 \log\left(t + \sqrt{t^2 + \left(\frac{1}{2}\right)^2}\right)\right]\Bigg|_{t=0}^{\pi} \\
&= \pi\sqrt{\pi^2 + \frac{1}{4}} + \frac{1}{4}\log\left(\pi + \sqrt{\pi^2 + \frac{1}{4}}\right) - \frac{1}{4}\log\left(\sqrt{\frac{1}{4}}\right) \\
&= \frac{\pi}{2}\sqrt{1 + 4\pi^2} + \frac{1}{4}\log\left(2\pi + \sqrt{1 + 4\pi^2}\right) \approx 10.63.
\end{aligned}
$$

As a check on our answer, we may note that the path **c** connects the points $(1, 0, 0)$ and $(-1, 0, \pi^2)$. The distance between these points is $\sqrt{4 + \pi^4} \approx 10.06$ which is less than 10.63, so the answer is reasonable. ◆

If a curve is made up of a finite number of pieces each of which is smooth, we compute the arc length by adding the lengths of the component pieces.

Example 3   A billiard ball on a square table follows the path **c**: $[-1, 1] \to \mathbb{R}^3$ defined by $\mathbf{c}(t) = (x(t), y(t), z(t)) = (|t|, |t - \frac{1}{2}|, 0)$. Find the distance travelled by the ball.

Solution   This path is not smooth, because $x(t) = |t|$ is not differentiable at 0, nor is $y(t) = |t - \frac{1}{2}|$ differentiable at $\frac{1}{2}$. However, if we divide the interval $[-1, 1]$ into the pieces $[-1, 0]$, $[0, \frac{1}{2}]$, and $[\frac{1}{2}, 1]$, we see that $x(t)$ and $y(t)$ have continuous derivatives on each of the intervals $[-1, 0]$, $[0, \frac{1}{2}]$, and $[\frac{1}{2}, 1]$. (See Figure 4.2.2.)

On $[-1, 0]$, $x(t) = -t$, $y(t) = -t + \frac{1}{2}$, and $z(t) = 0$, so $\|\mathbf{c}'(t)\| = \sqrt{2}$. Hence, the arc length of **c** between -1 and 0 is $\int_{-1}^0 \sqrt{2}\, dt = \sqrt{2}$. Similarly, on $[0, \frac{1}{2}]$, $x(t) = t$, $y(t) = -t + \frac{1}{2}$, $z(t) = 0$, and again $\|\mathbf{c}'(t)\| = \sqrt{2}$, so that the arc length of **c** between 0 and $\frac{1}{2}$ is $\frac{1}{2}\sqrt{2}$. Finally, on $[\frac{1}{2}, 1]$ we have $x(t) = t$, $y(t) = t - \frac{1}{2}$, $z(t) = 0$, and the arc length of **c** between $\frac{1}{2}$ and 1 is $\frac{1}{2}\sqrt{2}$. Thus the total arc length of **c** is $2\sqrt{2}$. ◆

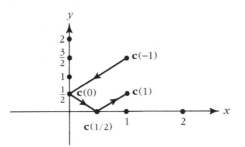

FIGURE 4.2.2. A piecewise smooth path.

Note

We have given a definition of arc length which was motivated by the concept of total distance traversed by a moving particle. However there are situations where a more exacting ("rigorous") treatment is needed. For example, some curves, like a mountain range silhouette or a coastline are quite complex and their "length" perhaps depends on the detail with which they are measured and could even turn out to be infinite! Computers can generate complex curves of this sort, and it is a serious problem to decide what their length might be. See Figure 4.2.3 for an example. You may also refer to B. Mandelbrot, *The Fractal Geometry of Nature*, W.H. Freeman, N.Y. 1982.

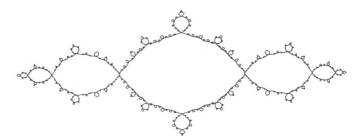

FIGURE 4.2.3. A complicated computer-generated curve. What is its length?

Example 4   Consider the point with position function

$$\mathbf{c}(t) = (t - \sin t, 1 - \cos t)$$

tracing out the cycloid discussed in §**1.6**. Find the velocity, the speed, and the length of one arc.

Solution   The velocity vector is $\mathbf{c}'(t) = (1 - \cos t, \sin t)$, so the speed of the point $\mathbf{c}(t)$ is

$$\|\mathbf{c}'(t)\| = \sqrt{(1 - \cos t)^2 + \sin^2 t} = \sqrt{2 - 2\cos t}.$$

Hence, $\mathbf{c}(t)$ moves at variable speed although the circle rolls at constant speed. Furthermore, the speed of $\mathbf{c}(t)$ is zero when t is an integral multiple of 2π. At these values of t, the y coordinate of the point $\mathbf{c}(t)$ is zero and so the point lies on the x axis. The arc length of one cycle is

$$
\begin{aligned}
l &= \int_0^{2\pi} \sqrt{2 - 2\cos t}\, dt = 2\int_0^{2\pi} \sqrt{\frac{1 - \cos t}{2}}\, dt \\
&= 2\int_0^{2\pi} \sin\frac{t}{2}\, dt \left(\text{since } 1 - \cos t = 2\sin^2\frac{t}{2} \text{ and } \sin\frac{t}{2} \geq 0 \text{ on } [0, 2\pi] \right) \\
&= 4\left(-\cos\frac{t}{2} \right)\Big|_0^{2\pi} = 8. \quad \blacklozenge
\end{aligned}
$$

The arc length formula suggests that one introduce the following notation, which will be useful in Chapter 6 in our discussion of line integrals.

Arc Length Differential

An infinitesimal displacement of a particle following a path $\mathbf{c}(t) = x(t)\mathbf{i} + y(t)\mathbf{j} + z(t)\mathbf{k}$ is

$$
d\mathbf{s} = dx\mathbf{i} + dy\mathbf{j} + dz\mathbf{k} = \left(\frac{dx}{dt}\mathbf{i} + \frac{dy}{dt}\mathbf{j} + \frac{dz}{dt}\mathbf{k} \right) dt
$$

and its length

$$
ds = \sqrt{dx^2 + dy^2 + dz^2} = \sqrt{\left(\frac{dx}{dt}\right)^2 + \left(\frac{dy}{dt}\right)^2 + \left(\frac{dz}{dt}\right)^2}\, dt
$$

is the differential of arc length. See Figure 4.2.4.

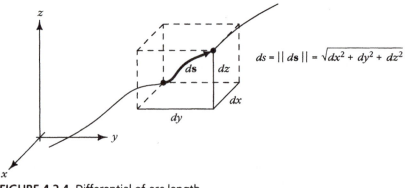

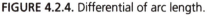

FIGURE 4.2.4. Differential of arc length.

These formulas help us remember the arc length formula as

$$\text{arc length} = \int_{t_0}^{t_1} ds.$$

As we have done before with such geometric concepts as length and angle, we can extend the notion of arc length to paths in n-dimensional space.

Arc Length in $\mathbb{R}^n$

Let $\mathbf{c} : [t_0, t_1] \to \mathbb{R}^n$ be a path. Its **length** is defined to be

$$l = \int_{t_0}^{t_1} \|\mathbf{c}'(t)\| dt.$$

The integrand is the square root of the sum of the squares of the coordinate functions of $\mathbf{c}'(t)$: if

$$\mathbf{c}(t) = (x_1(t), x_2(t), \ldots, x_n(t)),$$

then

$$l = \int_{t_0}^{t_1} \sqrt{(x_1'(t))^2 + (x_2'(t))^2 + \cdots + (x_n'(t))^2} dt.$$

Example 5 Find the length of the path $\mathbf{c}(t) = (\cos t, \sin t, \cos 2t, \sin 2t)$ in $\mathbb{R}^4$, defined on the interval from 0 to π.

Solution We have $\mathbf{c}'(t) = (-\sin t, \cos t, -2\sin 2t, 2\cos 2t)$, so

$$\|\mathbf{c}'(t)\| = \sqrt{\sin^2 t + \cos^2 t + 4\sin^2 2t + 4\cos^2 2t} = \sqrt{1+4} = \sqrt{5},$$

a constant, so the length of the path is

$$\int_0^\pi \sqrt{5} dt = \sqrt{5}\pi. \quad \blacklozenge$$

Exercises for §4.2

Find the arc length of the given curve on the specified interval in Exercises 1–6.

1. $(2\cos t, 2\sin t, t); 0 \le t \le 2\pi$

2. $(1, 3t^2, t^3); 0 \le t \le 1$

3. $(\sin 3t, \cos 3t, 2t^{3/2}); 0 \le t \le 1$

4. $\left(t+1, \frac{2\sqrt{2}}{3}t^{3/2} + 7, \frac{1}{2}t^2\right); 1 \le t \le 2$

5. $(t, t, t^2); 1 \le t \le 2$

6. $(t, t\sin t, t\cos t); 0 \le t \le \pi$

7. Find the length of the path $\mathbf{c}(t)$, defined by $\mathbf{c}(t) = (2\cos t, 2\sin t, t)$, if $0 \le t \le 2\pi$ and $\mathbf{c}(t) = (2, t - 2\pi, t)$, if $2\pi \le t \le 4\pi$.

8. Find the length of the path $\mathbf{c}(t)$, where $\mathbf{c}(t) = (1, 3t^2, t^3)$, if $0 \le t \le 1$ and $\mathbf{c}(t) = (t, 3t, t)$, if $1 \le t \le 2$.

9. The arc length function $s(t)$ for a given path $\mathbf{c}(t)$, defined by $s(t) = \int_a^t \|\mathbf{c}'(\tau)\| d\tau$, represents the distance a particle traversing the trajectory of $\mathbf{c}$ will have traveled by time t if it starts out at time a; that is, it gives the length of $\mathbf{c}$ between $\mathbf{c}(a)$ and $\mathbf{c}(t)$. Find the arc length functions for the curves $\alpha(t) = (\cosh t, \sinh t, t)$ and $\beta(t) = (\cos t, \sin t, t)$, with $a = 0$.

10. Let $\mathbf{c}$ be the path $\mathbf{c}(t) = (2t, t^2, \log t)$, defined for $t > 0$. Find the arc length of $\mathbf{c}$ between the points $(2, 1, 0)$ and $(4, 4, \log 2)$.

11. Find the arc length of the path $\mathbf{c}(t) = (t, t\sin t, t\cos t)$ between $(0, 0, 0)$ and $(\pi, 0, -\pi)$.

12. Let $\mathbf{c}(t)$ be a given path, $a \le t \le b$. Let $s = \alpha(t)$ be a new variable, where α is a strictly increasing C^1 function given on $[a, b]$. For each s in $[\alpha(a), \alpha(b)]$ there is a unique t with $\alpha(t) = s$. Define the function $\mathbf{d} : [\alpha(a), \alpha(b)] \to \mathbb{R}^3$ by $\mathbf{d}(s) = \mathbf{c}(t)$. The path $\mathbf{d}$ is said to be a *reparametrization* of $\mathbf{c}$.

   (a) Argue that the image curves of $\mathbf{c}$ and $\mathbf{d}$ are the same.

   (b) Show that $\mathbf{c}$ and $\mathbf{d}$ have the same arc length.

   (c) Let $s = \alpha(t) = \int_a^t \|\mathbf{c}'(\tau)\| d\tau$. Define $\mathbf{d}$ as above by $\mathbf{d}(s) = \mathbf{c}(t)$. Show that
   $$\left\| \frac{d}{ds}\mathbf{d}(s) \right\| = 1.$$

   $\mathbf{d}$ is called the *arc length reparametrization* of $\mathbf{c}$.

13. Let $\mathbf{c} : [a, b] \to \mathbb{R}^3$ be a smooth path. Assume $\mathbf{c}'(t) \ne \mathbf{0}$ for any t. The vector $\mathbf{c}'(t)/\|\mathbf{c}'(t)\| = \mathbf{T}(t)$ is tangent to $\mathbf{c}$ at $\mathbf{c}(t)$, and, since $\|\mathbf{T}(t)\| = 1$, $\mathbf{T}$ is called the *unit tangent* to $\mathbf{c}$.

   (a) Show that $\mathbf{T}'(t) \cdot \mathbf{T}(t) = 0$. [Hint: Differentiate $\mathbf{T}(t) \cdot \mathbf{T}(t) = 1$.]

   (b) Write down a formula in terms of $\mathbf{c}$ for $\mathbf{T}'(t)$.

14. In special relativity, the **proper time** of a path $\mathbf{c} : [a, b] \to \mathbb{R}^4$ with $\mathbf{c}(\lambda) = (x(\lambda), y(\lambda), z(\lambda), t(\lambda))$ is defined to be the quantity

$$\frac{1}{c} \int_a^b \sqrt{-[x'(\lambda)]^2 - [y'(\lambda)]^2 - [z'(\lambda)]^2 + c^2[t'(\lambda)]^2} d\lambda,$$

where c is the velocity of light, a constant. In Figure 4.2.5, show that, using self-explanatory notation,

$$\text{proper time } (AB) + \text{proper time } (BC) < \text{proper time } (AC).$$

(This inequality is a special case of what is known as the **twin paradox**.)

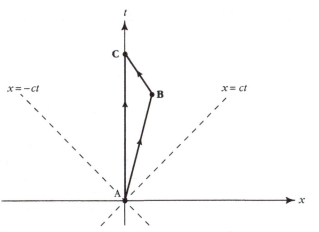

FIGURE 4.2.5. The relativistic triangle inequality.

4.3
Vector Fields

In Chapter 2 we introduced a particular kind of vector field, the gradient. In this section we study *general* vector fields, discussing their geometric and physical significance.

Vector Fields

A **vector field** in $\mathbb{R}^n$ is a map $\mathbf{F} : A \subset \mathbb{R}^n \to \mathbb{R}^n$ that assigns to each point $\mathbf{x}$ in its domain A a vector $\mathbf{F}(\mathbf{x})$. If $n = 2$, $\mathbf{F}$ is called a **vector field in the plane**, and if $n = 3$, $\mathbf{F}$ is a **vector field in space**.

Picture $\mathbf{F}$ as attaching an *arrow* to each point (Figure 4.3.1). By contrast, a map $f : A \subset \mathbb{R}^n \to \mathbb{R}$ that assigns a *number* to each point is a **scalar field**. A vector field $\mathbf{F}(x, y, z)$ on $\mathbb{R}^3$ has three **component scalar fields** F_1, F_2, and F_3, so that $\mathbf{F}(x, y, z) = (F_1(x, y, z), F_2(x, y, z), F_3(x, y, z))$. If each of $F_1, F_2,$

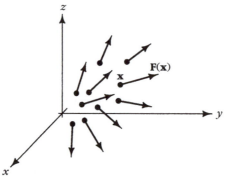

FIGURE 4.3.1. A vector field **F** assigns a vector **F(x)** to each point **x** of its domain.

and F_3 is a C^k function, we say the vector field **F** is of *class C^k*. Vector fields will be assumed to be at least of ***class*** C^1 unless otherwise noted.

In many applications, **F(x)** represents a physical vector quantity (force, velocity, etc.) associated with the position **x**, as in the following examples.

Example 1   This example concerns fluid flow in a pipe. The flow of water through a pipe is said to be ***steady*** if, at each point inside the pipe, the velocity of the fluid passing through that point does not change with time. (Note that this is quite different from saying that the water in the pipe is not moving.) Attaching to each point the fluid velocity at that point, we obtain the ***velocity field*** **V** of the fluid (see Figure 4.3.2). Notice that the length of the arrows (the speed), as well as the direction of flow, changes from point to point.   ◆

FIGURE 4.3.2. A vector field describing the velocity of flow in a pipe.

Example 2   Consider a piece of material that is heated on one side and cooled on another. The temperature at each point within the body is described at a given moment by a scalar field $T(x, y, z)$. The flow of heat may be marked by a field of arrows indicating the direction and magnitude of the flow (Figure 4.3.3). This ***energy*** or ***beat flux vector field*** is given by $\mathbf{J} = -k\nabla T$, where $k > 0$ is a constant called the ***conductivity*** and ∇T is the gradient of the real-valued function T. Note that the heat flows from hot regions toward cold ones, since $-\nabla T$ points in the direction of decreasing T.   ◆

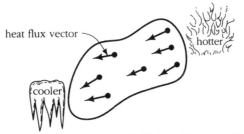

FIGURE 4.3.3. A vector field describing the direction and magnitude of heat flow.

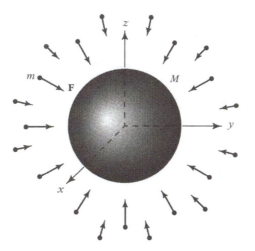

FIGURE 4.3.4. The vector field **F** given by Newton's law of gravitation.

Example 3   The force of attraction of the earth on a mass m can be described by a vector field called the ***gravitational force field***. Place the origin of a coordinate system at the center of the Earth (assumed spherical). According to Newton's law, this field is given by

$$\mathbf{F} = -\frac{mMG}{r^3}\mathbf{r},$$

where $\mathbf{r}(x, y, z) = (x, y, z)$, and $r = \|\mathbf{r}\|$ (see Figure 4.3.4). The domain of this vector field consists of those $\mathbf{r}$ for which $\|\mathbf{r}\|$ is greater than the radius of the Earth. As we saw in §**2.5**, $\mathbf{F}$ is a gradient field, $\mathbf{F} = -\nabla V$, where $V = -(mMG)/r$. Note again that $\mathbf{F}$ points in the direction of decreasing V. Writing $\mathbf{F}$ out in terms of components, we see that

$$\mathbf{F}(x, y, z) = \left(\frac{-mMG}{r^3}x, \frac{-mMG}{r^3}y, \frac{-mMG}{r^3}z\right). \quad \blacklozenge$$

Example 4   Rotary motion (such as the motion of particles on a turntable) is described by the vector field

$$\mathbf{V}(x, y) = -y\mathbf{i} + x\mathbf{j}.$$

See Figure 4.3.5, in which we have shown instead of $\mathbf{V}$ the shorter vector field $\frac{1}{4}\mathbf{V}$ so that the arrows do not overlap. This is a common convention in drawing pictures of vector fields.   ◆

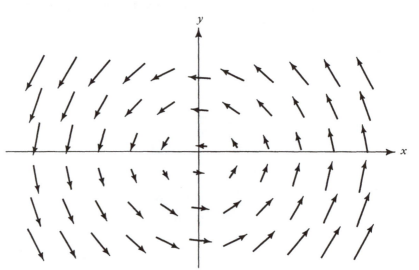

FIGURE 4.3.5. A rotary vector field.

Example 5   In the plane, $\mathbb{R}^2$, the function $\mathbf{V}$ defined by

$$\mathbf{V}(x, y) = \frac{y\mathbf{i}}{x^2 + y^2} - \frac{x\mathbf{j}}{x^2 + y^2} = \left(\frac{y}{x^2 + y^2}, -\frac{x}{x^2 + y^2} \right)$$

is a vector field on $\mathbb{R}^2$, except at the origin, where it is not defined. This vector field is a good approximation to the horizontal part of the velocity of water flowing toward a hole in the bottom of a tub (Figure 4.3.6). Notice that the velocity becomes *larger* as you approach the hole.   ◆

Example 6   According to **Coulomb's law**, the force acting on a charge e at position $\mathbf{r}$ due to a charge Q at the origin is

$$\mathbf{F} = \frac{\varepsilon Q e}{r^3}\mathbf{r} = -\nabla V,$$

where $V = \varepsilon Q e / r$ and ε is a constant that depends on the units used. For $Qe > 0$ (like charges) the force is repulsive [Figure 4.3.7(a)], and for $Qe < 0$

FIGURE 4.3.6. The vector field describing circular flow in a tub.

(unlike charges) the force is attractive [Figure 4.3.7(b)]. Since the potential V is constant on the level surfaces of V, they are called ***equipotential surfaces***. Note that the force field is orthogonal to the equipotential surfaces (the force field is radial and the equipotential surfaces are concentric spheres).   ♦

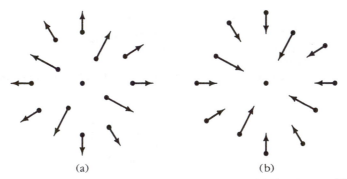

FIGURE 4.3.7. The vector fields associated with (a) like charges ($Qe > 0$) and (b) unlike charges ($Qe < 0$).

The next example shows that not every vector field is a gradient.

Example 7   Show that the vector field **V** on $\mathbb{R}^2$ defined by $\mathbf{V}(x, y) = y\mathbf{i} - x\mathbf{j}$ is not a gradient vector field; *i.e.*, there is no function f such that

$$\mathbf{V}(x, y) = \nabla f(x, y) = \frac{\partial f}{\partial x}\mathbf{i} + \frac{\partial f}{\partial y}\mathbf{j}.$$

Solution   Suppose that such an f exists. Then $\partial f / \partial x = y$ and $\partial f / \partial y = -x$. Since there are C^1 functions, f itself must have continuous first and second

order partial derivatives. But, $\partial^2 f/\partial x \partial y = -1$ and $\partial^2 f/\partial y \partial x = 1$ which violates the equality of mixed partials. Thus **V** cannot be a gradient vector field.
♦

An important concept related to vector fields is that of a flow line.

Flow Lines
If **F** is a vector field, a ***flow line*** for **F** is a path $\mathbf{c}(t)$ such that

$$\mathbf{c}'(t) = \mathbf{F}(\mathbf{c}(t)).$$

That is, **F** yields the velocity field of the path $\mathbf{c}(t)$.

In the context of Example 1, a flow line is the path followed by a small particle suspended in the fluid (Figure 4.3.8). Flow lines are also appropriately called ***streamlines*** or ***integral curves***.

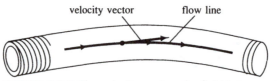

FIGURE 4.3.8. The velocity vector of a fluid is tangent to a flow line.

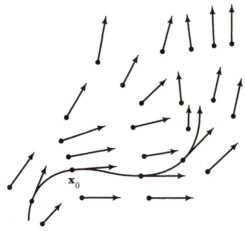

FIGURE 4.3.9. A flow line threading its way through a vector field in the plane.

Geometrically, a flow line for a given vector field **F** is a curve that threads its way through the domain of the vector field in such a way that the tangent vector of the curve coincides with the vector field, as in Figure 4.3.9. Flow lines provide a geometric picture of a vector field, enabling us to understand it more fully.

A flow line may be viewed as a solution of a system of differential equations. Indeed, we can write the definition $\mathbf{c}'(t) = \mathbf{F}(\mathbf{c}(t))$ as

$$
\begin{aligned}
x'(t) &= P(x(t), y(t), z(t)), \\
y'(t) &= Q(x(t), y(t), z(t)), \\
z'(t) &= R(x(t), y(t), z(t)),
\end{aligned}
$$

where

$$\mathbf{F} = P\mathbf{i} + Q\mathbf{j} + R\mathbf{k}.$$

One learns about such systems in courses on differential equations.

Example 8   Show that the path $\mathbf{c}(t) = (\cos t, \sin t)$ is a flow line of $\mathbf{F}(x, y) = -y\mathbf{i} + x\mathbf{j}$. Can you find others?

Solution   We must verify that $\mathbf{c}'(t) = \mathbf{F}(\mathbf{c}(t))$. The left side is $(-\sin t)\mathbf{i} + (\cos t)\mathbf{j}$ while the right side is $\mathbf{F}(\cos t, \sin t) = (-\sin t)\mathbf{i} + (\cos t)\mathbf{j}$, so we have a flow line. As suggested by Figure 4.3.5, the other flow lines are also circles. They have the form

$$\mathbf{c}(t) = (r \cos (t - t_0), r \sin (t - t_0))$$

for constants r and t_0.   ◆

In many cases explicit formulas for flow lines are not available, so one must resort to numerical methods. Figure 4.3.10 shows some output from a program that computes flow lines numerically and plots them on the screen.

Exercises for §4.3

In Exercises 1–8, sketch the given vector field or a small multiple of it.

1. $\mathbf{F}(x, y) = (2, 2)$

2. $\mathbf{F}(x, y) = (4, 0)$

3. $\mathbf{F}(x, y) = (x, y)$

4. $\mathbf{F}(x, y) = (-x, y)$

5. $\mathbf{F}(x, y) = (2y, x)$

6. $\mathbf{F}(x, y) = (y, -2x)$

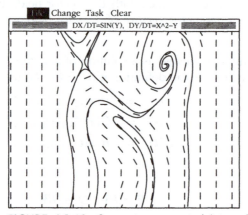

FIGURE 4.3.10. Computer-generated integral curves of the vector field $\mathbf{F}(x, y) = (\sin y)\mathbf{i} + (x^2 - y)\mathbf{j}$ (using "MacMath" by Hubbard and West, Springer-Verlag, N.Y., 1992).

7. $\mathbf{F}(x, y) = \left(\dfrac{x}{\sqrt{x^2 + y^2}}, \dfrac{y}{\sqrt{x^2 + y^2}} \right)$

8. $\mathbf{F}(x, y) = \left(\dfrac{y}{\sqrt{x^2 + y^2}}, \dfrac{x}{\sqrt{x^2 + y^2}} \right)$

In Exercises 9–12, sketch a few flow lines of the given vector field.

9. $\mathbf{F}(x, y) = (y, -x)$

10. $\mathbf{F}(x, y) = (x, -y)$

11. $\mathbf{F}(x, y) = (x, x^2)$

12. $\mathbf{F}(x, y, z) = (y, -x, 0)$

In Exercises 13–16, show that the given curve $\mathbf{c}(t)$ is a flow line of the given velocity vector field $\mathbf{F}(x, y, z)$.

13. $\mathbf{c}(t) = (e^{2t}, \log|t|, 1/t), t \neq 0; \mathbf{F}(x, y, z) = (2x, z, -z^2)$

14. $\mathbf{c}(t) = (t^2, 2t - 1, \sqrt{t}), t > 0; \mathbf{F}(x, y, z) = (y + 1, 2, 1/2z)$

15. $\mathbf{c}(t) = (\sin t, \cos t, e^t); \mathbf{F}(x, y, z) = (y, -x, z)$

16. $\mathbf{c}(t) = \left(\dfrac{1}{t^3}, e^t, \dfrac{1}{t} \right); \mathbf{F}(x, y, z) = (-3z^4, y, -z^2)$

17. Let a particle of mass m move on a path $\mathbf{r}(t)$ according to Newton's law in a force field $\mathbf{F} = -\nabla V$ on $\mathbb{R}^3$, where V is a given potential energy function.

    (a) Prove that the energy $E = \frac{1}{2}m\|\mathbf{r}'(t)\|^2 + V(\mathbf{r}(t))$ is constant in time by calculating dE/dt.

    (b) If the particle moves on an equipotential surface, show that its speed is constant.

18. Let $\mathbf{c}(t)$ be a flow line of a gradient field $\mathbf{F} = -\nabla V$. Prove that $V(\mathbf{c}(t))$ is a decreasing function of t.

19. If $f(\mathbf{x}, t)$ is a real-valued function of $\mathbf{x}$ and t, define the **material deriva-tive** of f relative to a vector field $\mathbf{F}$ as

$$\frac{Df}{Dt} = \frac{\partial f}{\partial t} + \nabla f(\mathbf{x}) \cdot \mathbf{F}$$

    where ∇ denotes the gradient with respect to the variable $\mathbf{x}$. Let $\mathbf{c}(t)$ be a flow line of $\mathbf{F}$ with $\mathbf{c}(t_0) = \mathbf{x}_0$. Prove that

$$\frac{Df}{Dt}(\mathbf{x}_0, t_0) = \left. \frac{d}{dt} f(\mathbf{c}(t), t) \right|_{t=t_0}.$$

4.4
Divergence
and Curl

We now introduce the divergence and curl, two operations on vector fields that will play a basic role. In each of these operations we use the **del operator**

$$\nabla = \mathbf{i}\frac{\partial}{\partial x} + \mathbf{j}\frac{\partial}{\partial y} + \mathbf{k}\frac{\partial}{\partial z}.$$

For functions of one variable, taking a derivative can be thought of as an operation or process; *i.e.*, given a function $y = f(x)$, its derivative is the result of *operating* on y by the derivative *operator* d/dx. Similarly, we can write the gradient as

$$\nabla f = \left(\mathbf{i}\frac{\partial}{\partial x} + \mathbf{j}\frac{\partial}{\partial y} \right) f = \mathbf{i}\frac{\partial f}{\partial x} + \mathbf{j}\frac{\partial f}{\partial y}$$

for functions of two variables, and

$$\nabla f = \left(\mathbf{i}\frac{\partial}{\partial x} + \mathbf{j}\frac{\partial}{\partial y} + \mathbf{k}\frac{\partial}{\partial z} \right) f = \mathbf{i}\frac{\partial f}{\partial x} + \mathbf{j}\frac{\partial f}{\partial y} + \mathbf{k}\frac{\partial f}{\partial z}$$

for three variables. In operational terms, the gradient of f is obtained by taking the ∇ operator and applying it to f.

We define the divergence of a vector field $\mathbf{F}$ by taking the *dot product* of ∇ with $\mathbf{F}$.

Divergence

If $\mathbf{F} = F_1\mathbf{i} + F_2\mathbf{j} + F_3\mathbf{k}$, the **divergence** of $\mathbf{F}$ is the scalar field

$$\operatorname{div} \mathbf{F} = \nabla \cdot \mathbf{F} = \frac{\partial F_1}{\partial x} + \frac{\partial F_2}{\partial y} + \frac{\partial F_3}{\partial z}.$$

Example 1   Compute the divergence of

$$\mathbf{F} = x^2 y\mathbf{i} + z\mathbf{j} + xyz\mathbf{k}.$$

Solution

$$\operatorname{div} \mathbf{F} = \frac{\partial}{\partial x}(x^2 y) + \frac{\partial}{\partial y}(z) + \frac{\partial}{\partial z}(xyz) = 2xy + 0 + xy = 3xy. \quad \blacklozenge$$

The divergence has an important physical interpretation. If we imagine $\mathbf{F}$ as the velocity field of a gas (or a fluid), then div $\mathbf{F}$ *represents the rate of expansion per unit volume under the flow of the gas (or fluid)*. If div $\mathbf{F} < 0$, it *means the gas is compressing*. For a vector field $\mathbf{F}(x, y) = F_1\mathbf{i} + F_2\mathbf{j}$ on the plane, the divergence

$$\nabla \cdot \mathbf{F} = \frac{\partial F_1}{\partial x} + \frac{\partial F_2}{\partial y}$$

measures the rate of expansion of *area*. We will justify this interpretation in Chapter 7, but we will now present some examples.

Example 2   Consider the vector field in the plane given by $\mathbf{V}(x, y) = x\mathbf{i}$. Relate the sign of the divergence of $\mathbf{V}$ with the rate of change of areas under the flow.

Solution   We think of $\mathbf{V}$ as the velocity field of a fluid in the plane. The vector field $\mathbf{V}$ points to the right for $x > 0$ and to the left if $x < 0$ as we see in Figure 4.4.1. The length of $\mathbf{V}$ gets shorter toward the origin. As the fluid moves, it expands (the area of the shaded rectangle increases), so we expect that div $\mathbf{V} > 0$. Indeed, div $\mathbf{V} = 1$. $\quad \blacklozenge$

Example 3   The flow lines of the vector field $\mathbf{F} = x\mathbf{i} + y\mathbf{j}$ are straight lines directed away from the origin (Figure 4.4.2).

If these flow lines are those of a fluid, the fluid is expanding as it moves out from the origin, so div $\mathbf{F}$ should be positive. In fact,

$$\nabla \cdot \mathbf{F} = \frac{\partial}{\partial x}x + \frac{\partial}{\partial y}y = 2 > 0. \quad \blacklozenge$$

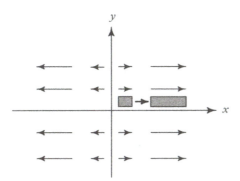

FIGURE 4.4.1. This fluid is expanding.

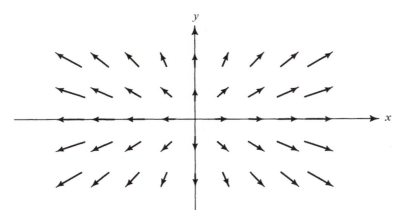

FIGURE 4.4.2. The vector field $\mathbf{F}(x, y) = x\mathbf{i} + y\mathbf{j}$.

Example 4 Consider the vector field $\mathbf{F} = -x\mathbf{i} - y\mathbf{j}$. Here the flow lines point toward the origin instead of away from it (see Figure 4.4.3). Therefore, the fluid is compressing, so we expect $(\text{div } \mathbf{F}) < 0$. Calculating, we see that

$$\nabla \cdot \mathbf{F} = \frac{\partial}{\partial x}(-x) + \frac{\partial}{\partial y}(-y) = -1 - 1 = -2 < 0. \quad \blacklozenge$$

Example 5 As we saw in the last section, the flow lines of $\mathbf{F} = -y\mathbf{i} + x\mathbf{j}$ are concentric circles about the origin, moving counterclockwise (see Figure 4.4.4). It appears that the fluid is neither compressing nor expanding. This is confirmed by calculating

$$\nabla \cdot \mathbf{F} = \frac{\partial}{\partial x}(-y) + \frac{\partial}{\partial y}(x) = 0 + 0 = 0. \quad \blacklozenge$$

Example 6   Some flow lines of $\mathbf{F} = x\mathbf{i} - y\mathbf{j}$ are shown in Figure 4.4.5. Here our intuition about expansion or compression is less clear. However, it is true that the shaded regions shown have the same area and we calculate that

$$\nabla \cdot \mathbf{F} = \frac{\partial}{\partial x}x + \frac{\partial}{\partial y}(-y) = 1 + (-1) = 0. \quad \blacklozenge$$

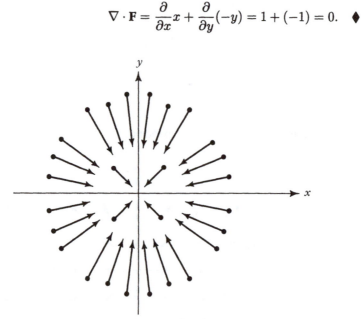

FIGURE 4.4.3. The vector field $\mathbf{F}(x, y) = -x\mathbf{i} - y\mathbf{j}$.

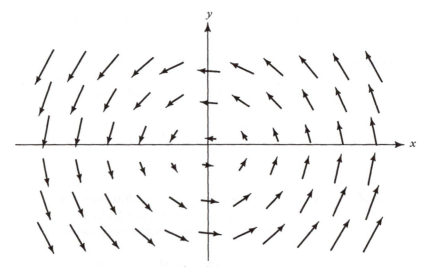

FIGURE 4.4.4. The vector field $\mathbf{F}(x, y) = -y\mathbf{i} + x\mathbf{j}$.

To calculate the curl, the second basic operation performed on vector fields, we take the *cross product* of ∇ with $\mathbf{F}$.

Curl of a Vector Field

If $\mathbf{F} = F_1\mathbf{i} + F_2\mathbf{j} + F_3\mathbf{k}$, the **curl** of $\mathbf{F}$ is the *vector field*

$$\nabla \times \mathbf{F} = \begin{vmatrix} \mathbf{i} & \mathbf{j} & \mathbf{k} \\ \dfrac{\partial}{\partial x} & \dfrac{\partial}{\partial y} & \dfrac{\partial}{\partial z} \\ F_1 & F_2 & F_3 \end{vmatrix}$$

$$= \left(\frac{\partial F_3}{\partial y} - \frac{\partial F_2}{\partial z} \right)\mathbf{i} + \left(\frac{\partial F_1}{\partial z} - \frac{\partial F_3}{\partial x} \right)\mathbf{j} + \left(\frac{\partial F_2}{\partial x} - \frac{\partial F_1}{\partial y} \right)\mathbf{k}.$$

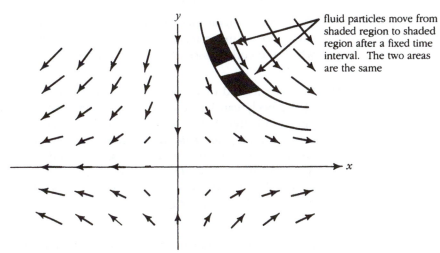

fluid particles move from shaded region to shaded region after a fixed time interval. The two areas are the same

FIGURE 4.4.5. The vector field $\mathbf{F}(x, y) = x\mathbf{i} - y\mathbf{j}$.

We sometimes write curl $\mathbf{F}$ for $\nabla \times \mathbf{F}$.

Example 7   Let $\mathbf{F}(x, y, z) = x\mathbf{i} + xy\mathbf{j} + \mathbf{k}$. Find $\nabla \times \mathbf{F}$.

Solution

$$\nabla \times \mathbf{F} = \begin{vmatrix} \mathbf{i} & \mathbf{j} & \mathbf{k} \\ \dfrac{\partial}{\partial x} & \dfrac{\partial}{\partial y} & \dfrac{\partial}{\partial z} \\ x & xy & 1 \end{vmatrix} = (0 - 0)\mathbf{i} - (0 - 0)\mathbf{j} + (y - 0)\mathbf{k}.$$

Thus $\nabla \times \mathbf{F} = y\mathbf{k}$. ◆

Example 8   Find the curl of $xy\mathbf{i} - \sin z\mathbf{j} + \mathbf{k}$.

Solution   Letting $\mathbf{F} = xy\mathbf{i} - \sin z\mathbf{j} + \mathbf{k}$,

$$\nabla \times \mathbf{F} = \begin{vmatrix} \mathbf{i} & \mathbf{j} & \mathbf{k} \\ \dfrac{\partial}{\partial x} & \dfrac{\partial}{\partial y} & \dfrac{\partial}{\partial z} \\ xy & -\sin z & 1 \end{vmatrix}$$

$$= \begin{vmatrix} \dfrac{\partial}{\partial y} & \dfrac{\partial}{\partial z} \\ -\sin z & 1 \end{vmatrix} \mathbf{i} - \begin{vmatrix} \dfrac{\partial}{\partial x} & \dfrac{\partial}{\partial z} \\ xy & 1 \end{vmatrix} \mathbf{j} + \begin{vmatrix} \dfrac{\partial}{\partial x} & \dfrac{\partial}{\partial y} \\ xy & -\sin z \end{vmatrix} \mathbf{k}$$

$$= \cos z\mathbf{i} - x\mathbf{k}. \quad \blacklozenge$$

Unlike the divergence, which can be defined in $\mathbb{R}^n$ for any n, the curl is defined only in three-dimensional space.

The physical significance of the curl will be discussed in Chapter 7, when we study Stokes' theorem. However, we can now consider a specific situation, in which the curl is associated with rotations.

Example 9   Consider a solid rigid body B rotating about an axis L. The rotational motion of the body can be described by a vector $\mathbf{w}$ along the axis of rotation, the direction being chosen so that the body rotates about $\mathbf{w}$ as in Figure 4.4.6. We call $\mathbf{w}$ the ***angular velocity vector***. The length $\omega = \|\mathbf{w}\|$ is taken to be the angular speed of the body B, that is, the speed of any point in B divided by its distance from the axis L of rotation. The motion of points in the rotating body is described by the vector field $\mathbf{v}$ whose value at each point is the velocity at that point. To find $\mathbf{v}$, let Q be any point in B and let α be the distance from Q to L.

Figure 4.4.6 shows that $\alpha = \|\mathbf{r}\| \sin\theta$, where $\mathbf{r}$ is the vector whose initial point is the origin and whose terminal point is Q and θ is the angle between $\mathbf{r}$ and the axis L of rotation. The tangential velocity $\mathbf{v}$ of Q is directed counterclockwise along the tangent to a circle parallel to the xy plane with radius α and has magnitude

$$\|\mathbf{v}\| = \omega\alpha = \omega\|\mathbf{r}\| \sin\theta = \|\mathbf{w}\| \, \|\mathbf{r}\| \sin\theta.$$

The direction and magnitude of $\mathbf{v}$ imply that $\mathbf{v} = \mathbf{w} \times \mathbf{r}$. Selecting a coordinate system in which L is the z axis, we can write $\mathbf{w} = \omega\mathbf{k}$ and $\mathbf{r} = x\mathbf{i} + y\mathbf{j} + z\mathbf{k}$. Thus,

$$\mathbf{v} = \mathbf{w} \times \mathbf{r} = -\omega y\mathbf{i} + \omega x\mathbf{j}$$

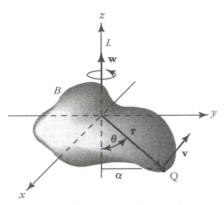

FIGURE 4.4.6. The velocity **v** and angular velocity **w** of a rotating body are related by $\mathbf{v} = \mathbf{w} \times \mathbf{r}$.

and so

$$\text{curl } \mathbf{v} = \begin{vmatrix} \mathbf{i} & \mathbf{j} & \mathbf{k} \\ \dfrac{\partial}{\partial x} & \dfrac{\partial}{\partial y} & \dfrac{\partial}{\partial z} \\ -\omega y & \omega x & 0 \end{vmatrix} = 2\omega\mathbf{k} = 2\mathbf{w}.$$

Hence, for the rotation of a rigid body, the curl of the velocity vector field is a vector field whose value is the same at each point. It is directed along the axis of rotation with magnitude *twice* the angular speed. ◆

If a vector field represents the flow of a *fluid,* then the value of $\nabla \times \mathbf{F}$ at a point is twice the rotation vector of a *rigid* body that rotates as the fluid does near that point. In particular, $\nabla \times \mathbf{F} = \mathbf{0}$ at a point P means that the fluid is free from rigid rotations at P; that is, it has no whirlpools. Another justification of this idea depends on Stokes' theorem from Chapter 7. However, we can say informally that curl $\mathbf{F} = \mathbf{0}$ means that if a *small* rigid paddle wheel is placed in the fluid, it will move with the fluid, but will not rotate around its axis. Such a vector field is called **irrotational**. For example, it has been determined from experiments that fluid draining from a tub is usually irrotational except right at the center, even though the fluid is "rotating" around the drain (see Figure 4.4.7). In Example 10 below, the flow lines of the vector field **V** are circles about the origin, yet the flow is irrotational. Thus the reader should be warned of the possible confusion the word "irrotational" can cause.

Example 10   Verify that the vector field

$$\mathbf{V}(x, y, z) = \frac{y\mathbf{i} - x\mathbf{j}}{x^2 + y^2}$$

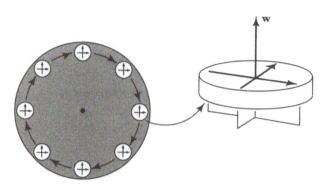

FIGURE 4.4.7. Looking at a paddle wheel from above a moving fluid. The velocity field $\mathbf{V}(x, y, z) = (y\mathbf{i} - x\mathbf{j})/(x^2 + y^2)$ is irrotational; the paddle wheel does not rotate around its axis $\mathbf{w}$.

is irrotational when $(x, y) \neq (0, 0)$, (i.e., except where $\mathbf{V}$ is not defined).

Solution   The curl is

$$\nabla \times \mathbf{V} = \begin{vmatrix} \mathbf{i} & \mathbf{j} & \mathbf{k} \\ \dfrac{\partial}{\partial x} & \dfrac{\partial}{\partial y} & \dfrac{\partial}{\partial z} \\ \dfrac{y}{x^2 + y^2} & \dfrac{-x}{x^2 + y^2} & 0 \end{vmatrix}$$

$$= 0\mathbf{i} + 0\mathbf{j} + \left[\frac{\partial}{\partial x} \left(\frac{-x}{x^2 + y^2} \right) - \frac{\partial}{\partial y} \left(\frac{y}{x^2 + y^2} \right) \right] \mathbf{k}$$

$$= \left[\frac{-(x^2 + y^2) + 2x^2}{(x^2 + y^2)^2} + \frac{-(x^2 + y^2) + 2y^2}{(x^2 + y^2)^2} \right] \mathbf{k} = 0. \quad \blacklozenge$$

The following identity is a basic relation between the gradient and curl, which should be compared with the fact that for any vector $\mathbf{v}$, we have $\mathbf{v} \times \mathbf{v} = \mathbf{0}$.

Curl of a Gradient

For any C^2 function f,
$$\nabla \times (\nabla f) = \mathbf{0}.$$

That is, *the curl of any gradient is the zero vector.*

Proof Since $\nabla f = (\partial f / \partial x, \partial f / \partial y, \partial f / \partial z)$ we have, by definition,

$$\nabla \times \nabla f = \begin{vmatrix} \mathbf{i} & \mathbf{j} & \mathbf{k} \\ \dfrac{\partial}{\partial x} & \dfrac{\partial}{\partial y} & \dfrac{\partial}{\partial z} \\ \dfrac{\partial f}{\partial x} & \dfrac{\partial f}{\partial y} & \dfrac{\partial f}{\partial z} \end{vmatrix}$$

$$= \left(\frac{\partial^2 f}{\partial y \partial z} - \frac{\partial^2 f}{\partial z \partial y} \right) \mathbf{i} + \left(\frac{\partial^2 f}{\partial z \partial x} - \frac{\partial^2 f}{\partial x \partial z} \right) \mathbf{j} + \left(\frac{\partial^2 f}{\partial x \partial y} - \frac{\partial^2 f}{\partial y \partial x} \right) \mathbf{k}.$$

Each component is zero because of the equality of mixed partial derivatives.
∎

Example 11   Let $\mathbf{V}(x, y, z) = y\mathbf{i} - x\mathbf{j}$. Show that $\mathbf{V}$ is not a gradient field.

Solution   If $\mathbf{V}$ were a gradient field, then it would satisfy curl $\mathbf{V} = \mathbf{0}$ by the preceding box. But

$$\text{curl } \mathbf{V} = \begin{vmatrix} \mathbf{i} & \mathbf{j} & \mathbf{k} \\ \dfrac{\partial}{\partial x} & \dfrac{\partial}{\partial y} & \dfrac{\partial}{\partial z} \\ y & -x & 0 \end{vmatrix} = -2\mathbf{k} \neq \mathbf{0}. \; \blacklozenge$$

There is an operation on vector fields in the plane that is closely related to the curl. If $\mathbf{F} = P(x, y)\mathbf{i} + Q(x, y)\mathbf{j}$ is a vector field in the plane, it can also be regarded as a vector field in space for which the $\mathbf{k}$-component is zero and the other two components are independent of z. The curl of $\mathbf{F}$ then reduces to

$$\nabla \times \mathbf{F} = \left(\frac{\partial Q}{\partial x} - \frac{\partial P}{\partial y} \right) \mathbf{k}$$

and always points in the $\mathbf{k}$ direction. The function

$$\frac{\partial Q}{\partial x} - \frac{\partial P}{\partial y}$$

of x and y is called the **scalar curl** of $\mathbf{F}$.

Example 12   Find the scalar curl of $\mathbf{V}(x, y) = -y^2\mathbf{i} + x\mathbf{j}$.

Solution   The curl is

$$\nabla \times \mathbf{V} = \begin{vmatrix} \mathbf{i} & \mathbf{j} & \mathbf{k} \\ \dfrac{\partial}{\partial x} & \dfrac{\partial}{\partial y} & \dfrac{\partial}{\partial z} \\ -y^2 & x & 0 \end{vmatrix} = (1 + 2y)\mathbf{k},$$

so the scalar curl, which is the coefficient of $\mathbf{k}$, is $1 + 2y$.   ◆

A basic relation between the divergence and curl operations is given next.

Divergence of a Curl

For any C^2 vector field $\mathbf{F}$,

$$\text{div curl } \mathbf{F} = \nabla \cdot (\nabla \times \mathbf{F}) = 0.$$

That is, *the divergence of any curl is zero.*

As with the curl of a gradient, the proof rests on the equality of the mixed partial derivatives. The student should write out the details.

Example 13   Show that the vector field $\mathbf{V}(x, y) = x\mathbf{i} + y\mathbf{j} + z\mathbf{k}$ cannot be the curl of some vector field $\mathbf{F}$; *i.e.*, there is no $\mathbf{F}$ with $\mathbf{V} = \text{curl } \mathbf{F}$.

Solution   If this were so, then div $\mathbf{V}$ would be zero by the box above. But

$$\text{div } \mathbf{V} = \frac{\partial x}{\partial x} + \frac{\partial y}{\partial y} + \frac{\partial z}{\partial z} = 3 \neq 0,$$

so $\mathbf{V}$ cannot be curl $\mathbf{F}$ for any $\mathbf{F}$.   ◆

The Laplace Operator

The **Laplace operator** ∇^2, which operates on functions f, is defined as the divergence of the gradient:

$$\nabla^2 f = \nabla \cdot (\nabla f) = \frac{\partial^2 f}{\partial x^2} + \frac{\partial^2 f}{\partial y^2} + \frac{\partial^2 f}{\partial z^2}.$$

This operator plays an important role in many physical laws, as we have mentioned in §3.1.

Example 14   Show that $\nabla^2 f = 0$ for

$$f(x, y, z) = \frac{1}{\sqrt{x^2 + y^2 + z^2}} = \frac{1}{\|\mathbf{r}\|} \quad \text{and} \quad (x, y, z) \neq (0, 0, 0)$$

where $\mathbf{r} = x\mathbf{i} + y\mathbf{j} + z\mathbf{k}$.

Solution   The first derivatives are

$$\frac{\partial f}{\partial x} = \frac{-x}{(x^2 + y^2 + z^2)^{3/2}}, \quad \frac{\partial f}{\partial y} = \frac{-y}{(x^2 + y^2 + z^2)^{3/2}},$$

$$\frac{\partial f}{\partial z} = \frac{-z}{(x^2 + y^2 + z^2)^{3/2}}.$$

Computing the second derivatives, we find that

$$\frac{\partial^2 f}{\partial x^2} = \frac{3x^2}{(x^2 + y^2 + z^2)^{5/2}} - \frac{1}{(x^2 + y^2 + z^2)^{3/2}},$$

$$\frac{\partial^2 f}{\partial y^2} = \frac{3y^2}{(x^2 + y^2 + z^2)^{5/2}} - \frac{1}{(x^2 + y^2 + z^2)^{3/2}},$$

$$\frac{\partial^2 f}{\partial z^2} = \frac{3z^2}{(x^2 + y^2 + z^2)^{5/2}} - \frac{1}{(x^2 + y^2 + z^2)^{3/2}}.$$

Thus,

$$\frac{\partial^2 f}{\partial x^2} + \frac{\partial^2 f}{\partial y^2} + \frac{\partial^2 f}{\partial z^2} = \frac{3(x^2 + y^2 + z^2)}{(x^2 + y^2 + z^2)^{5/2}} - \frac{3}{(x^2 + y^2 + z^2)^{3/2}}$$

$$= \frac{3}{(x^2 + y^2 + z^2)^{3/2}} - \frac{3}{(x^2 + y^2 + z^2)^{3/2}} = 0. \quad \blacklozenge$$

We now have these basic operations on hand: gradient, divergence, curl, and the Laplace operator. The following table lists some basic general formulas that are useful when computing with vector fields.

Some Basic Identities of Vector Analysis

1. $\nabla(f + g) = \nabla f + \nabla g$

2. $\nabla(cf) = c\nabla f$, for a constant c

3. $\nabla(fg) = f\nabla g + g\nabla f$

4. $\nabla(f/g) = (g\nabla f - f\nabla g)/g^2$, at points where $g(\mathbf{x}) \neq 0$

5. $\operatorname{div}(\mathbf{F} + \mathbf{G}) = \operatorname{div}\mathbf{F} + \operatorname{div}\mathbf{G}$

6. $\operatorname{curl}(\mathbf{F} + \mathbf{G}) = \operatorname{curl}\mathbf{F} + \operatorname{curl}\mathbf{G}$

7. $\operatorname{div}(f\mathbf{F}) = f\operatorname{div}\mathbf{F} + \mathbf{F}\cdot\nabla f$

8. $\operatorname{div}(\mathbf{F}\times\mathbf{G}) = \mathbf{G}\cdot\operatorname{curl}\mathbf{F} - \mathbf{F}\cdot\operatorname{curl}\mathbf{G}$

9. $\operatorname{div}\operatorname{curl}\mathbf{F} = 0$

10. $\operatorname{curl}(f\mathbf{F}) = f\operatorname{curl}\mathbf{F} + \nabla f \times \mathbf{F}$

11. $\operatorname{curl}\nabla f = \mathbf{0}$

12. $\nabla^2(fg) = f\nabla^2 g + g\nabla^2 f + 2(\nabla f \cdot \nabla g)$

13. $\operatorname{div}(\nabla f \times \nabla g) = 0$

14. $\operatorname{div}(f\nabla g - g\nabla f) = f\nabla^2 g - g\nabla^2 f$

Example 15   Prove identity 7 in the above table.

Solution   The vector field $f\mathbf{F}$ has components fF_i, for $i = 1, 2, 3$, and so

$$\operatorname{div}(f\mathbf{F}) = \frac{\partial}{\partial x}(fF_1) + \frac{\partial}{\partial y}(fF_2) + \frac{\partial}{\partial z}(fF_3).$$

However, $(\partial/\partial x)(fF_1) = f\partial F_1/\partial x + F_1 \partial f/\partial x$ by the product rule, with similar expressions for the other terms. Therefore

$$\begin{aligned}
\operatorname{div}(f\mathbf{F}) &= f\left(\frac{\partial F_1}{\partial x} + \frac{\partial F_2}{\partial y} + \frac{\partial F_3}{\partial z}\right) + F_1\frac{\partial f}{\partial x} + F_2\frac{\partial f}{\partial y} + F_3\frac{\partial f}{\partial z} \\
&= f(\nabla \cdot \mathbf{F}) + \mathbf{F} \cdot \nabla f. \quad\blacklozenge
\end{aligned}$$

Let us use these identities to redo Example 14.

Example 16   Show that for $\mathbf{r} \neq \mathbf{0}$, $\nabla^2(1/r) = 0$.

Solution   As in the case of the gravitational potential, $\nabla(1/r) = -\mathbf{r}/r^3$. By the identity $\nabla \cdot (f\mathbf{F}) = f\nabla \cdot \mathbf{F} + \nabla f \cdot \mathbf{F}$, we get

$$\nabla \cdot \left(\frac{\mathbf{r}}{r^3}\right) = \frac{1}{r^3}\nabla \cdot \mathbf{r} + \mathbf{r} \cdot \nabla\left(\frac{1}{r^3}\right)$$

$$= \frac{3}{r^3} + \mathbf{r} \cdot \left(\frac{-3\mathbf{r}}{r^5} \right) = 0. \quad \blacklozenge$$

Exercises for §4.4

Find the divergence of the vector fields in Exercises 1–4.

1. $\mathbf{V}(x, y, z) = e^{xy}\mathbf{i} - e^{xy}\mathbf{j} + e^{yz}\mathbf{k}$

2. $\mathbf{V}(x, y, z) = yz\mathbf{i} + xz\mathbf{j} + xy\mathbf{k}$

3. $\mathbf{V}(x, y, z) = x\mathbf{i} + (y + \cos x)\mathbf{j} + (z + e^{xy})\mathbf{k}$

4. $\mathbf{V}(x, y, z) = x^2\mathbf{i} + (x + y)^2\mathbf{j} + (x + y + z)^2\mathbf{k}$

5. Figure 4.4.8 shows some flow lines and moving regions for a fluid moving in the plane with velocity field $\mathbf{V}$. Where is div $\mathbf{V} > 0$ and also where is div $\mathbf{V} < 0$?

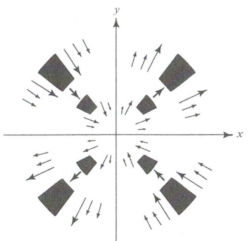

FIGURE 4.4.8. The flow lines of a fluid moving in the plane.

6. Let $V(x, y, z) = x\mathbf{i}$ be the velocity field of a fluid in space. Relate the sign of the divergence with the rate of change of volume under the flow.

7. Sketch a few flow lines for $\mathbf{F}(x, y) = y\mathbf{i}$. Calculate $\nabla \cdot \mathbf{F}$ and explain why your answer is consistent with your sketch.

8. Sketch a few flow lines for $\mathbf{F}(x, y) = -3x\mathbf{i} - y\mathbf{j}$. Calculate $\nabla \cdot \mathbf{F}$ and explain why your answer is consistent with your sketch.

Calculate the divergence of the vector fields in Exercises 9–12.

9. $\mathbf{F}(x, y) = x^3\mathbf{i} - x\sin(xy)\mathbf{j}$

10. $\mathbf{F}(x, y) = y\mathbf{i} - x\mathbf{j}$

11. $\mathbf{F}(x, y) = \sin(xy)\mathbf{i} - \cos(x^2 y)\mathbf{j}$

12. $\mathbf{F}(x, y) = xe^y\mathbf{i} - [y/(x + y)]\mathbf{j}$

Compute the curl, $\nabla \times \mathbf{F}$, of the vector fields in Exercises 13–16.

13. $\mathbf{F}(x, y, z) = x\mathbf{i} + y\mathbf{j} + z\mathbf{k}$

14. $\mathbf{F}(x, y, z) = yz\mathbf{i} + xz\mathbf{j} + xy\mathbf{k}$

15. $\mathbf{F}(x, y, z) = (x^2 + y^2 + z^2)(3\mathbf{i} + 4\mathbf{j} + 5\mathbf{k})$

16. $\mathbf{F}(x, y, z) = \dfrac{yz\mathbf{i} - xz\mathbf{j} + xy\mathbf{k}}{x^2 + y^2 + z^2}$

Calculate the scalar curl of each of the vector fields in Exercises 17–20.

17. $\mathbf{F}(x, y) = \sin x\mathbf{i} + \cos x\mathbf{j}$

18. $\mathbf{F}(x, y) = y\mathbf{i} - x\mathbf{j}$

19. $\mathbf{F}(x, y) = xy\mathbf{i} + (x^2 - y^2)\mathbf{j}$

20. $\mathbf{F}(x, y) = x\mathbf{i} + y\mathbf{j}$

Verify that $\nabla \times (\nabla f) = \mathbf{0}$ for the functions in Exercises 21–24.

21. $f(x, y, z) = \sqrt{x^2 + y^2 + z^2}$

22. $f(x, y, z) = xy + yz + xz$

23. $f(x, y, z) = 1/(x^2 + y^2 + z^2)$

24. $f(x, y, z) = x^2 y^2 + y^2 z^2$

25. Show that $\mathbf{F} = y(\cos x)\mathbf{i} + x(\sin y)\mathbf{j}$ is *not* a gradient vector field.

26. Show that $\mathbf{F} = (x^2 + y^2)\mathbf{i} - 2xy\mathbf{j}$ is *not* a gradient field.

27. Prove identity 10 in the table of vector identities.

28. Suppose that $\nabla \cdot \mathbf{F} = 0$ and $\nabla \cdot \mathbf{G} = 0$. Which of the following necessarily have zero divergence?
    (a) $\mathbf{F} + \mathbf{G}$ (b) $\mathbf{F} \times \mathbf{G}$

29. Let $\mathbf{F} = 2xz^2\mathbf{i} + \mathbf{j} + y^3zx\mathbf{k}$ and $f = x^2y$. Compute the following quantities:
    (a) ∇f (b) $\nabla \times \mathbf{F}$ (c) $\mathbf{F} \times \nabla f$ (d) $\mathbf{F} \cdot (\nabla f)$

30. Let $\mathbf{r}(x, y, z) = (x, y, z)$ and $r = \sqrt{x^2 + y^2 + z^2} = \|\mathbf{r}\|$. Prove the following identities.

    (a) $\nabla(1/r) = -\mathbf{r}/r^3, r \neq 0$; and, in general, $\nabla(r^n) = nr^{n-2}\mathbf{r}$ and $\nabla(\log r) = \mathbf{r}/r^2$.

    (b) $\nabla^2(1/r) = 0, r \neq 0$; and, in general, $\nabla^2 r^n = n(n+1)r^{n-2}$.

    (c) $\nabla \cdot (\mathbf{r}/r^3) = 0$; and, in general, $\nabla \cdot (r^n\mathbf{r}) = (n+3)r^n$.

    (d) $\nabla \times \mathbf{r} = \mathbf{0}$; and, in general, $\nabla \times (r^n\mathbf{r}) = \mathbf{0}$.

31. Does $\nabla \times \mathbf{F}$ have to be perpendicular to $\mathbf{F}$?

32. Let $\mathbf{F}(x, y, z) = 3x^2y\mathbf{i} + (x^3 + y^3)\mathbf{j}$.

    (a) Verify that curl $\mathbf{F} = \mathbf{0}$.

    (b) Find a function f such that $\mathbf{F} = \nabla f$. (Techniques for constructing f in general are given in Chapter 7. The one in this problem should be sought by trial and error.)

Review Exercises for Chapter 4

For Exercises 1–4, at the indicated point, compute the velocity vector, the acceleration vector, the speed, and the equation of the tangent line.

1. $\mathbf{c}(t) = (t^3 + 1, e^{-t}, \cos(\pi t/2)); t = 1$

2. $\mathbf{c}(t) = (t^2 - 1, \cos(t^2), t^4); t = \sqrt{\pi}$

3. $\mathbf{c}(t) = (e^t, \sin t, \cos t); t = 0$

4. $\mathbf{c}(t) = \dfrac{t^2}{1 + t^2}\mathbf{i} + t\mathbf{j} + \mathbf{k}; t = 2$

5. Calculate the tangent and acceleration vectors for the helix $\mathbf{c}(t) = (\cos t, \sin t, t)$ at $t = \pi/4$.

6. Calculate the tangent and acceleration vectors for the cycloid $\mathbf{c}(t) = (t - \sin t, 1 - \cos t)$ at $t = \pi/4$ and sketch.

7. Let a particle of mass m move on the path $\mathbf{c}(t) = (t^2, \sin t, \cos t)$. Compute the force acting on the particle at $t = 0$.

8.(a) Let $\mathbf{c}(t)$ be a path with $\|\mathbf{c}(t)\| = $ constant; $i.e.$, the curve lies on a sphere. Show that $\mathbf{c}'(t)$ is orthogonal to $\mathbf{c}(t)$.

(b) Let $\mathbf{c}$ be a path whose speed is never zero. Show that $\mathbf{c}$ has constant speed if and only if the acceleration vector $\mathbf{c}''$ is always perpendicular to the velocity vector $\mathbf{c}'$.

9. Express as an integral the arc length of the curve $x^2 = y^3 = z^5$ between $x = 1$ and $x = 4$, using a suitable parametrization.

10. Find the arc length of $\mathbf{c}(t) = t\mathbf{i} + (\log t)\mathbf{j} + 2\sqrt{2t}\mathbf{k}$ for $1 \leq t \leq 2$.

11. A particle is constrained to move around the unit circle in the xy plane according to the formula $(x, y, z) = (\cos(t^2),\ \sin(t^2), 0), t \geq 0$.

(a) What are the velocity vector and speed of the particle as functions of t?

(b) At what point on the circle should the particle be released to hit a target at $(2, 0, 0)$? (Be careful about which direction the particle is moving around the circle.)

(c) At what time t should the release take place? (Use the smallest $t > 0$ that will work.)

(d) What are the velocity and speed at the time of release?

(e) At what time is the target hit?

12. A particle of mass m is subject to the force law $\mathbf{F} = -k\mathbf{r}$, where k is a constant.

(a) Write down differential equations for the components of $\mathbf{r}(t)$.

(b) Solve the equations in (a) subject to the initial conditions $\mathbf{r}(0) = \mathbf{0}, \mathbf{r}'(0) = 2\mathbf{j} + \mathbf{k}$.

13. Write in parametric form the curve described by the equations $x - 1 = 2y + 1 = 3z + 2$.

14. Write in parametric form the curve $x = y^3 = z^2 + 1$.

15. Show that $\mathbf{c}(t) = (1/(1 - t), 0, e^t/(1 - t))$ is a flow line of the vector field $\mathbf{F}(x, y, z) = (x^2, 0, z(1 + x))$.

16. Let $\mathbf{F}(x, y) = f(x^2 + y^2)[-y\mathbf{i} + x\mathbf{j}]$ for a function f of one variable. What equation must $g(t)$ satisfy for

$$\mathbf{c}(t) = [\cos g(t)]\mathbf{i} + [\sin g(t)]\mathbf{j}$$

to be a flow line for $\mathbf{F}$?

Compute $\nabla \cdot \mathbf{F}$ and $\nabla \times \mathbf{F}$ for the vector fields in Exercises 17–20.

17. $\mathbf{F} = 2x\mathbf{i} + 3y\mathbf{j} + 4z\mathbf{k}$

18. $\mathbf{F} = x^2\mathbf{i} + y^2\mathbf{j} + z^2\mathbf{k}$

19. $\mathbf{F} = (x + y)\mathbf{i} + (y + z)\mathbf{j} + (z + x)\mathbf{k}$

20. $\mathbf{F} = x\mathbf{i} + 3xy\mathbf{j} + z\mathbf{k}$

Compute the divergence and curl of the vector fields in Exercises 21 and 22 at the points indicated.

21. $\mathbf{F}(x, y, z) = y\mathbf{i} + z\mathbf{j} + x\mathbf{k}; (1, 1, 1)$

22. $\mathbf{F}(x, y, z) = (x + y)^3\mathbf{i} + (\sin xy)\mathbf{j} + (\cos xyz)\mathbf{k}; (2, 0, 1)$

Calculate the gradients of the functions in Exercises 23–26, and verify that $\nabla \times \nabla f = \mathbf{0}$.

23. $f(x, y) = e^{xy} + \cos(xy)$

24. $f(x, y) = \dfrac{x^2 - y^2}{x^2 + y^2}$

25. $f(x, y) = e^{x^2} - \cos(xy^2)$

26. $f(x, y) = \tan^{-1}(x^2 + y^2)$

27.(a) Let $f(x, y, z) = xyz^2$; compute ∇f.

(b) Let $\mathbf{F}(x, y, z) = xy\mathbf{i} + yz\mathbf{j} + zy\mathbf{k}$; compute $\nabla \times \mathbf{F}$.

(c) Compute $\nabla \times (f\mathbf{F})$ using identity 10 of the table of vector identities. Compare with a direct computation.

28.(a) Let $\mathbf{F} = 2xye^z\mathbf{i} + e^z x^2\mathbf{j} + (x^2ye^z + z^2)\mathbf{k}$. Compute $\nabla \cdot \mathbf{F}$ and $\nabla \times \mathbf{F}$.

(b) Find a function $f(x, y, z)$ such that $\mathbf{F} = \nabla f$.

29. Let $\mathbf{F}(x, y) = f(x^2 + y^2)[-y\mathbf{i} + x\mathbf{j}]$, as in Exercise 16. Calculate div $\mathbf{F}$ and curl $\mathbf{F}$ and discuss your answers in view of the results of Exercise 16.

30.(a) Write in parametric form the curve that is the intersection of the surfaces $x^2 + y^2 + z^2 = 3$ and $y = 1$.

(b) Find the equation of the line tangent to this curve at $(1, 1, 1)$.

(c) Write an integral expression for the arc length of this curve. What is the value of this integral?

31. In meteorology, the **negative pressure gradient** G is a vector quantity that points from regions of high pressure to regions of low pressure, normal to the lines of constant pressure (**isobars**).

(a) In an xy coordinate system,

$$\mathbf{G} = -\frac{\partial P}{\partial x}\mathbf{i} - \frac{\partial P}{\partial y}\mathbf{j}.$$

Write a formula for the magnitude of the negative pressure gradient.

(b) If the horizontal pressure gradient provided the only horizontal force acting on the air, the wind would blow directly across the isobars in the direction of $\mathbf{G}$, and for a given air mass, with acceleration proportional to the magnitude of $\mathbf{G}$. Explain, using Newton's second law.

(c) Because of the rotation of the Earth, the wind does not blow in the direction that (b) would suggest. Instead, it obeys **Buys–Ballot's law** which states: "If in the Northern Hemisphere, you stand with your back to the wind, the high pressure is on your right and the low pressure is on your left." Draw a figure and introduce xy coordinates so that $\mathbf{G}$ points in the proper direction.

(d) State and graphically illustrate Buys–Ballot's law for the Southern Hemisphere, in which the orientation of high and low pressure is reversed.

32. A sphere of mass m, radius a, and uniform density has potential u and gravitational force $\mathbf{F}$, at a distance r from the center $(0,0,0)$, given by

$$u = \frac{3m}{2a} - \frac{mr^2}{2a^3}, \quad \mathbf{F} = -\frac{m}{a^3}\mathbf{r} \quad (r \le a);$$
$$u = \frac{m}{r}, \quad \mathbf{F} = -\frac{m}{r^3}\mathbf{r} \quad (r > a).$$

Here, $r = \|\mathbf{r}\|, \mathbf{r} = x\mathbf{i} + y\mathbf{j} + z\mathbf{k}$.

(a) Verify that $\mathbf{F} = \nabla u$ on the inside and outside of the sphere.

(b) Check that u satisfies Poisson's equation: $\partial^2 u/\partial x^2 + \partial^2 u/\partial y^2 + \partial^2 u/\partial z^2 =$ constant inside the sphere.

(c) Show u satisfies Laplace's equation: $\partial^2 u/\partial x^2 + \partial^2 u/\partial y^2 + \partial^2 u/\partial z^2 = 0$ outside the sphere.

33. A circular helix that lies on the cylinder $x^2 + y^2 = R^2$ with pitch ρ may be described parametrically by

$$x = R\cos\theta, \; y = R\sin\theta, \; z = \rho\theta, \; \theta \ge 0.$$

A particle slides under the action of gravity (which acts parallel to the z axis) without friction along the helix. If the particle starts out at the height $z_0 > 0$, then when it reaches the height z along the helix, its speed is given by

$$\frac{ds}{dt} = \sqrt{(z_0 - z)2g},$$

where s is arc length along the helix, g is the constant of gravity, t is time, and $0 \le z \le z_0$.

(a) Find the length of the part of the helix between the planes $z = z_0$ and $z = z_1, 0 \le z_1 < z_0$.

(b) Compute the time T_0 it takes the particle to reach the plane $z = 0$.

Interlude: Where We Are Headed

In Chapter 4 we studied the concepts of divergence and curl of vector fields. In the next section we will learn how to "integrate" vector fields over curves, surfaces, and regions in space. This will set the stage for the main theorems of this book in Chapter 7—the theorems of Green, Stokes, and Gauss.

Green's and Stokes' theorems equate the integral of a vector field $\mathbf{F}$ along the boundary of a surface S with that of the integral of curl $\mathbf{F}$ over the surface. Gauss' theorem equates the integral of a vector field $\mathbf{F}$ over the boundary of a three-dimensional region with the integral of div $\mathbf{F}$ over the region. These theorems are the highlights of any course in multivariable calculus, and are one of the "jewels" of the first two years of college mathematics. They are the generalizations of the "Fundamental Theorem" of one-variable calculus to the vector setting.

To state these theorems we must first develop the concepts of integration of vector fields over curves, surfaces, and regions in space, which we begin to do in the next chapter.

5

Multiple Integrals

It is to Archimedes himself (c.225 B.C.) that we owe the nearest approach to actual integration to be found among the Greeks. His first noteworthy advance in this direction was concerned with his proof that the area of a parabolic segment is four-thirds that of a triangle with the same base and vertex, or two-thirds of the circumscribed parallelogram.

D.E. Smith *(History of Mathematics)*

In this chapter we study the integration of real-valued functions of two and three variables over regions in the plane and in space. In these cases, the analogues of the integral $\int_a^b f(x)dx$ of a function of one variable are called double and triple integrals, respectively. In the next chapter, we study the integral of functions and vector functions over curves and surfaces.

We begin the chapter with a geometric introduction to the double integral in terms of volume and we end with a discussion of other applications. One of the main tools for evaluating double and triple integrals is the method of iterated integrals, introduced informally in §**5.1** and in more detail in §**5.2**, **5.3** and **5.4**. This method enables us to utilize our knowledge of one-variable calculus and all the techniques we learned there to evaluate integrals. (Yes, you will have to bone up on these methods to do multiple integrals!) The second main tool that helps us evaluate integrals is the change of variables formula, given in §**5.5**.

5.1
Volume and Cavalieri's Principle

The geometric interpretation of the integral from one-variable calculus is that the integral of a nonnegative function is the area under its graph. Specifically, if f is a function that takes nonnegative values at all points of an interval $[a, b]$, the integral

$$\int_a^b f(x)dx$$

equals the area of the region under the graph of f on $[a, b]$. This region is the set of all points (x, y) such that $a \le x \le b$ and $0 \le y \le f(x)$, as in Figure 5.1.1. If the function f is both negative and positive, then the integral is interpreted as a signed area, with the areas below the x axis being counted negative.

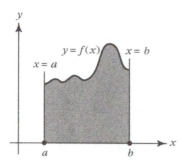

FIGURE 5.1.1. The area under the graph of a nonnegative continuous function f from $x = a$ to $x = b$ is $\int_a^b f(x)dx$.

The student should be familiar with the idea of integration from one-variable calculus. We also recall that the interpretation of the integral as an area leads to the precise *definition* of the integral, usually as a limit of sums, as well as to the most important means of *calculating* integrals, namely, by antidifferentiation, using the fundamental theorem of calculus.

The operation of *double integration* assigns to a function $f(x, y)$ defined on a region D in the plane a number

$$\iint_D f(x, y)dxdy, \text{ also denoted } \iint_D f(x, y)dA,$$

which, if $f(x, y) \geq 0$ for all (x, y) in D, is the *volume of the region under the graph of f*. This region is the set W of points (x, y, z) in $\mathbb{R}^3$ for which (x, y) lies in D and $0 \leq z \leq f(x, y)$, as in Figure 5.1.2. Occasionally, abbreviated notations such as $\int_D f$ or $\int_D f(x, y)dA$ are used for the double integral.

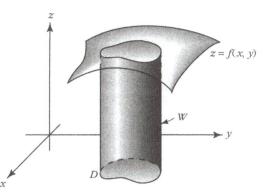

FIGURE 5.1.2. The double integral of f over the two-dimensional region D equals the volume of the three-dimensional region W.

When $D = R$ is a rectangle whose sides are parallel to the x and y axes, the solid region W appears as in Figure 5.1.3.

Notice that the rectangle R is determined by the two closed intervals $[a, b]$ and $[c, d]$, representing the sides of R along the x and y axes, respectively, as in Figure 5.1.3. We say that R is the **Cartesian product** of $[a, b]$ and $[c, d]$ and write $R = [a, b] \times [c, d]$.

<u>*Note*</u>

The symbol "$\times$" in the cartesian product is used because the *area* of $[a, b] \times [c, d]$ is found by multiplying the *lengths* of the intervals $[a, b]$ and $[c, d]$.

Example 1   Find

$$\iint_R f(x, y)dA$$

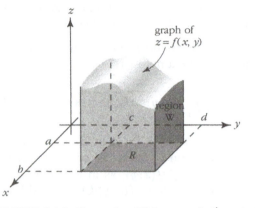

FIGURE 5.1.3. The region W in space is bounded by the graph of f, the rectangle R, and the four vertical sides indicated.

if $f(x, y)$ is a constant function equal to the positive number k and $R = [a, b] \times [c, d]$.

Solution   We have $\iint_R f(x, y)dA = k(b-a)(d-c)$, since the integral is equal to the volume of a rectangular box with base R and height k.   ◆

Example 2   Find the integral

$$\iint_R (1 - x)dx dy$$

where the region R is $[0, 1] \times [0, 1]$.

Solution   The integral equals the volume of the wedge-shaped region shown in Figure 5.1.4. This region is half of a unit cube, so

$$\iint_R (1 - x)dx dy = \frac{1}{2}.   ◆$$

Example 3   Sketch the solid whose volume is

$$\iint_R (x^2 + y^2)dx dy,$$

where the region R is $[-1, 1] \times [0, 1]$.

Solution   The solid is shown in Figure 5.1.5. We will compute its volume in Example 4.   ◆

The problem of how to define precisely the volume of W is a difficult one, and we do not indicate how it is solved until the next section. The problem

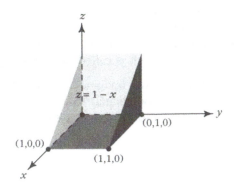

FIGURE 5.1.4. Volume under the graph $z = 1 - x$ and over $R = [0, 1] \times [0, 1]$.

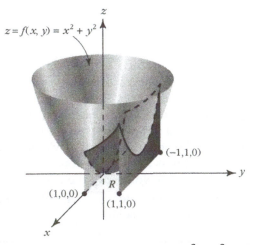

FIGURE 5.1.5. Volume under $z = x^2 + y^2$ and over $R = [-1, 1] \times [0, 1]$.

may seem to be of little importance at this stage, since we are most familiar with solids that clearly have well-defined volumes given by formulas from elementary geometry. However, a number of interesting solids have shapes whose volume formulas are not obvious, although it is clear they should have a well-defined volume. On the other hand, there are solids that are so complicated (say with infinitely many holes in them) that it is not clear even how the volume should be defined. Some have said that these are pathological, and only of interest to the pure mathematician. However, this is not the case: worrying about such things forces mathematicians to formulate very precise definitions, which guide them to a deeper understanding and that eventually leads to practical results, such as better computer programs to find everyday volumes!

As with areas in one-variable calculus, the volume of W is not computed directly from its definition but from a more effective method which, in the case of multiple integration, is called the ***method of iterated integrals***. This method is based on a geometric method found by the Italian mathematician Bonaventura Cavalieri (1598–1647), which is called ***Cavalieri's principle***.

This method proceeds as follows. Imagine a solid that is sliced by a family of parallel planes which are perpendicular to an axis in space. The plane at a distance x from a reference point on the axis is denoted P_x, as in Figure 5.1.6.

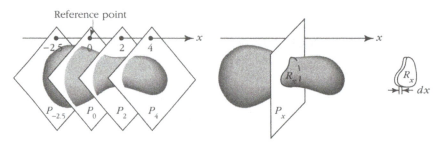

FIGURE 5.1.6. Slicing a solid using planes.

The plane P_x cuts the solid in a plane region R_x; the corresponding "infinitesimal piece" of the solid is a slab whose base is a region R_x and whose thickness is dx (Figure 5.1.6). The volume of such a cylinder is equal to the area of the base R_x times the thickness dx. If we denote the area of R_x by $A(x)$, then this volume is $A(x)dx$. Thus the volume of the entire solid, obtained by summing the infinitesimal pieces, is the integral $\int_a^b A(x)dx$, where the limits a and b are determined by the ends of the solid.

The Slice Method—Cavalieri's Principle

Let S be a solid and P_x be a family of parallel planes such that:

1. S lies between P_a and P_b;

2. the area of the slice of S cut by P_x is $A(x)$.

Then the volume of S is equal to

$$\int_a^b A(x)dx.$$

Historical Note

Bonaventura Cavalieri (1598–1647) was a pupil of Galileo and a professor in Bologna. His investigations into area and volume were important building blocks of the foundation of calculus. Although his methods were criticized by his contemporaries, similar ideas had been used by Archimedes in antiquity and were later taken up by the "founders" of calculus, Newton and Leibniz.

We now use Cavalieri's principle to evaluate double integrals over rectangles. Consider the solid region under a graph $z = f(x, y)$ defined on the region $[a, b] \times [c, d]$, where f is continuous and greater than zero. There are two cross-sectional area functions: one obtained by using cutting planes perpendicular to the x axis, and the other obtained by using cutting planes perpendicular to the y axis. The cross section determined by a cutting plane $x = x_0$ is the plane region under the graph of $z = f(x_0, y)$ from $y = c$ to $y = d$ (Figure 5.1.7). When we fix $x = x_0$, the cross-sectional area $A(x_0)$ is the integral $\int_c^d f(x_0, y)dy$. Thus the cross-sectional area function is $A(x) = \int_c^d f(x, y)dy$. By Cavalieri's principle, the volume V of the region under $z = f(x, y)$ equals

$$V = \int_a^b A(x)dx = \int_a^b \left[\int_c^d f(x, y)dy \right] dx.$$

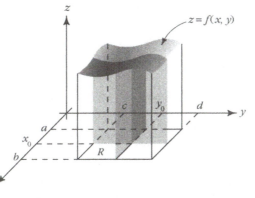

FIGURE 5.1.7. Two different cross sections sweeping out the volume under $z = f(x, y)$.

Note

Here is a "down to earth" explanation of what we have done: Imagine a stack of infinitely thin cards. Using one-dimensional calculus, we can figure out the area of each card. To discover the volume, we have to add up all the cards, a process you can think of as "vertical" integration. We can write a formula for this process that looks like

$$\int \left[\int f(x, y)dy \right] dx.$$

The "inner" integral represents the area of the cards. The "outer" integral represents adding up all these cards.

The integral $\int_a^b \left[\int_c^d f(x,y) dy \right] dx$ is known as an **iterated integral**, because it is obtained by carrying out the one-variable integration process twice. First, one ignores the outer integral and evaluates $\int_c^d f(x,y) dy$ to get a function of x alone. Second, we integrate this function of x from a to b to get a number. Since $\iint_R f(x,y) dA$ is equal to the volume V, we get the result in the following box; it turns out to be valid even when f is allowed to take on negative values.

Double and Iterated Integrals

If f is a continuous function on a rectangle R,

$$\iint_R f(x,y) dA = \int_a^b \left[\int_c^d f(x,y) dy \right] dx.$$

If we reverse the roles of x and y, we obtain

$$\iint_R f(x,y) dA = \int_c^d \left[\int_a^b f(x,y) dx \right] dy.$$

As the following examples illustrate, iterated integrals provide a powerful method for computing the double integral of a function of two variables.

Example 4   Let $z = f(x,y) = x^2 + y^2$ and let $R = [-1, 1] \times [0, 1]$. Evaluate the integral $\iint_R (x^2 + y^2) dx dy$.

Solution   By the second equation in the box above,

$$\iint_R (x^2 + y^2) dx dy = \int_0^1 \left[\int_{-1}^1 (x^2 + y^2) dx \right] dy.$$

To find $\int_{-1}^1 (x^2 + y^2) dx$, we treat y as a constant and integrate with respect to x. Using methods of one-variable calculus, we obtain

$$\int_{-1}^1 (x^2 + y^2) dx = \left[\frac{x^3}{3} + y^2 x \right]_{x=-1}^1 = \frac{2}{3} + 2y^2.$$

Next we integrate $\frac{2}{3} + 2y^2$ with respect to y from 0 to 1, to obtain

$$\int_0^1 \left(\frac{2}{3} + 2y^2 \right) dy = \left[\frac{2}{3} y + \frac{2}{3} y^3 \right]_{y=0}^1 = \frac{4}{3}.$$

Hence the volume of the solid in Figure 5.1.5 is 4/3.

For completeness, let us evaluate $\iint_R (x^2+y^2)dx\,dy$ by integrating with respect to y and then with respect to x. The first equation in the box above gives

$$\iint_R (x^2 + y^2)dx\,dy = \int_{-1}^{1} \left[\int_0^1 (x^2 + y^2)dy \right] dx.$$

Treating x as a constant in the y integration, we obtain

$$\int_0^1 (x^2 + y^2)dy = \left[x^2 y + \frac{y^3}{3} \right]_{y=0}^{1} = x^2 + \frac{1}{3}.$$

Next we evaluate $\int_{-1}^{1}(x^2 + \frac{1}{3})dx$ to obtain

$$\int_{-1}^{1} \left(x^2 + \frac{1}{3} \right) dx = \left[\frac{x^3}{3} + \frac{x}{3} \right]_{x=-1}^{1} = \frac{4}{3},$$

which agrees with our previous answer.   ◆

Example 5   Compute $\displaystyle\iint_S \cos x \sin y\, dx\, dy$, where S is the square $[0, \pi/2] \times [0, \pi/2]$ (see Figure 5.1.8).

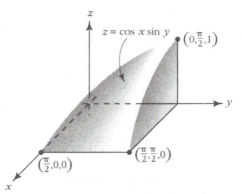

FIGURE 5.1.8. Volume under the surface $z = \cos x \sin y$ and over the rectangle $[0, \pi/2] \times [0, \pi/2]$.

Solution   By the second equation in the preceding box,

$$\iint_S \cos x \sin y\, dx\, dy = \int_0^{\pi/2} \left[\int_0^{\pi/2} \cos x \sin y\, dx \right] dy$$

$$= \int_0^{\pi/2} \sin y \left[\int_0^{\pi/2} \cos x\, dx \right] dy$$

$$= \int_0^{\pi/2} \sin y \, dy = 1. \quad \blacklozenge$$

In the next section, we shall define the double integral for a wide class of functions of two variables without recourse to the notion of volume. Although we shall drop the requirement that $f(x, y) \geq 0$, the equations in the preceding box will remain valid. Therefore, the iterated integral will again provide the key to computing the double integral. In §5.3 we treat double integrals over regions more general than rectangles.

Finally, we remark that it is common to delete the brackets in iterated integrals and write

$$\int_a^b \int_c^d f(x, y) dy dx \quad \text{instead of} \quad \int_a^b \left[\int_c^d f(x, y) dy \right] dx$$

and

$$\int_c^d \int_a^b f(x, y) dx dy \quad \text{instead of} \quad \int_c^d \left[\int_a^b f(x, y) dx \right] dy.$$

Note

Often the notation $\int_R f(x, y) dx dy$ is used for a double integral. That notation, with a *single* integral sign is motivated by the fact that $\int_R f dA$ is really a "single operation," which can be *evaluated* as a double-iterated integral in several ways.

Exercises for §5.1

Find $\iint_R f(x, y) dA$ for the given functions and regions in Exercises 1 and 2.

1. $f(x, y) = 7$ and $R = [3, 5] \times [-1, 9]$

2. $f(x, y) = x$ and $R = [0, 1] \times [0, 1]$

In Exercises 3–6, evaluate the given integral.

3. $\int_{-1}^1 \int_0^1 (x^4 y + y^2) dy dx$

4. $\int_0^{\pi/2} \int_0^1 (y \cos x + 2) dy dx$

5. $\displaystyle\int_0^1\int_0^1 (xye^{x+y})\,dy\,dx$

6. $\displaystyle\int_0^1\int_1^2 (-x\log y)\,dy\,dx$

Evaluate the double integrals in Exercises 7 and 8, where R is the rectangle $[0,2] \times [-1,0]$.

7. $\displaystyle\iint_R (x^2y^2 + x)\,dy\,dx$

8. $\displaystyle\iint_R \left(|y|\cos\frac{1}{4}\pi x\right)\,dy\,dx$

9. Find the integral of $xe^x \sin\left(\frac{1}{2}\pi y\right)$ over the rectangle $[0,2] \times [-1,0]$.

10. Find the volume under the graph of $f(x,y) = 1 + 2x + 3y$, over the rectangle $[1,2] \times [0,1]$.

11. Repeat Exercise 10 for the surface $f(x,y) = x^4 + y^2$ and the rectangle $[-1,1] \times [-3,-2]$.

12. Use Cavalieri's principle to show that the volume of two cylinders with the same base and height are equal (see Figure 5.1.9).

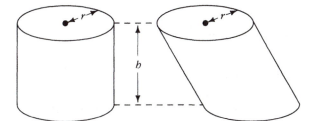

FIGURE 5.1.9. Two cylinders with same base and height have the same volume.

13. Using Cavalieri's principle, compute the volume of the structure shown in Figure 5.1.10; each cross section is a rectangle of length 5 and width 3.

14. A lumberjack cuts out a wedge-shaped piece of a cylindrical tree of radius r obtained by making two saw cuts to the tree's center, one horizontally and one at an angle θ. Compute the volume of the wedge W using Cavalieri's principle (see Figure 5.1.11).

15.(a) Use Cavalieri's principle to show that the volume of the solid of revolution shown in Figure 5.1.12(a) is

$$\pi \int_a^b [f(x)]^2\,dx.$$

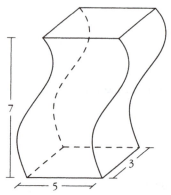

FIGURE 5.1.10. Compute this volume.

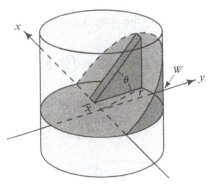

FIGURE 5.1.11. Find the volume of W.

(b) Show that the volume of the region obtained by rotating the region under the graph of the parabola $y = -x^2 + 2x + 3, -1 \leq x \leq 3$, about the x axis is $512\pi/15$ [see Figure 5.1.12(b)].

5.2
The Double Integral Over a Rectangle

In the previous section, our treatment of double integrals was based on the geometric notion of volume. Since multiple integration has many applications besides this geometric one, and since we need to integrate functions whose values may be negative as well as positive, we need a more general definition. The result will not only be more general, but also addresses other fundamental issues, such as "how is the volume of a complicated region *defined*?"

In one-variable calculus, we do *not* use the definition to calculate integrals. Rather, we use antidifferentiation and the fundamental theorem of calculus. Similarly, for double integrals, the techniques for calculating are those illustrated in the last section—reduction to iterated integrals, combined with one-variable techniques.

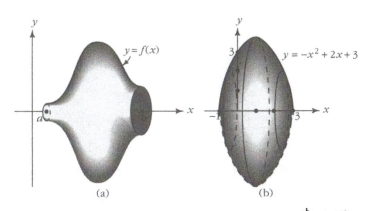

FIGURE 5.1.12. The solid of revolution (a) has volume $\pi \int_a^b [f(x)]^2 dx$. Part (b) shows the region between the graph of $y = -x^2 + 2x + 3$ and the x axis rotated about the x axis.

Note

You may ask why the "technical" definition given below is necessary at all, when all calculations are done by a simpler method. One answer to this question is that the definitions help us to *understand* the fundamentals of the integration process. This understanding is important; for example, there are many integrals that are difficult or even impossible to evaluate by the methods of antidifferentiation. Such integrals can be estimated, however, by numerical methods, many of which are based on these definitions.

The definition of the integral in one variable involves subdividing a given interval into smaller and smaller subintervals. In two variables, the sets that play the role of intervals are the **rectangles** whose sides are parallel to the axes. The cartesian product $R = [a, b] \times [c, d]$ of two closed intervals shown in Figure 5.2.1 is such a rectangle.

By the **regular partition** of R of order n we mean the two ordered collections of $n + 1$ *equally spaced points* $\{x_j\}_{j=0}^n$ and $\{y_k\}_{k=0}^n$; satisfying

$$a = x_0 < x_1 < \cdots < x_n = b, \quad c = y_0 < y_1 < \cdots < y_n = d$$

and

$$x_{j+1} - x_j = \frac{b - a}{n}, \quad y_{k+1} - y_k = \frac{d - c}{n}$$

(see Figure 5.2.2). One could modify the condition that the points be equally spaced, but it would complicate the exposition a little.

For given j and k, let R_{jk} be the rectangle $[x_j, x_{j+1}] \times [y_k, y_{k+1}]$, and let $\mathbf{c}_{jk}$ be any point in R_{jk}. Suppose f is a bounded real-valued function defined

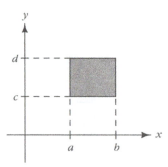

FIGURE 5.2.1. The rectangle $[a, b] \times [c, d]$ consists of (x, y)'s such that $a \leq x \leq b$ and $c \leq y \leq d$.

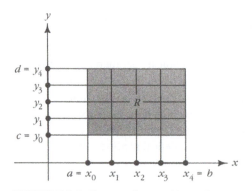

FIGURE 5.2.2. The regular partition of a rectangle R, with $n = 4$.

on R. (A function is called **bounded** if there is a number $M > 0$ such that $-M \leq f(x, y) \leq M$ for all (x, y) in the domain of f.) Form the sum

$$S_n = \sum_{j,k=0}^{n-1} f(\mathbf{c}_{jk}) \Delta x \Delta y = \sum_{j,k=0}^{n-1} f(\mathbf{c}_{jk}) \Delta A,$$

where

$$\Delta x = x_{j+1} - x_j = \frac{b-a}{n}, \quad \Delta y = y_{k+1} - y_k = \frac{d-c}{n},$$

and

$$\Delta A = \Delta x \Delta y.$$

This sum is taken over all j's and k's from 0 to $n-1$, and so there are n^2 terms. A sum of this type is called a **Riemann sum** for f.

Definition of the Double Integral

If the sequence $\{S_n\}$ converges to a limit S as $n \to \infty$ and the limit S is the same for any choice of points $\mathbf{c}_{jk}$ in the rectangles R_{jk}, then we say that f is **integrable** over R and we write

$$\iint_R f, \quad \iint_R f(x,y)dA, \quad \text{or} \quad \iint_R f(x,y)dxdy$$

for the limit S.

Thus we can describe integrability in the following way:

$$\lim_{n \to \infty} \sum_{j,k=0}^{n-1} f(\mathbf{c}_{jk})\Delta x \Delta y = \iint_R f$$

for any choice of $\mathbf{c}_{jk} \in R_{jk}$.

If $f(x,y) \geq 0$, the existence of $\lim_{n \to \infty} S_n$ has a straightforward geometric meaning. Consider the graph of $z = f(x,y)$ as the top of a solid whose base is the rectangle R. If we take each $\mathbf{c}_{jk}$ to be a point where $f(x,y)$ has its minimum value on R_{jk}, then $f(\mathbf{c}_{jk})\Delta x \Delta y$ represents the volume of a rectangular box with base R_{jk}. The sum $\sum_{j,k=0}^{n-1} f(\mathbf{c}_{jk})\Delta x \Delta y$ equals the volume of an *inscribed* solid, part of which is shown in Figure 5.2.3. Similarly, if $\mathbf{c}_{jk}$ is a point where $f(x,y)$ has its maximum on R_{jk}, then the sum $\sum_{j,k=0}^{n-1} f(\mathbf{c}_{jk})\Delta x \Delta y$ is equal to the volume of a *circumscribed* solid. Therefore, if $\lim_{n \to \infty} S_n$ exists and is independent of $\mathbf{c}_{jk} \in R_{jk}$, the volumes of the inscribed and circumscribed solids approach the same limit as $n \to \infty$.

Geometry of the Double Integral

If $f(x,y) \geq 0$, then

$$\iint_R f(x,y)dxdy$$

is the volume of the region under the graph of f and above the rectangle R in the xy plane.

As we shall see shortly, there are other interpretations possible. For example, if a rectangular plate R has mass density $\rho(x,y)$ grams per square centimeter, then $\iint_R \rho(x,y)dxdy$ is the total mass of the plate.

It is usually difficult to prove from the definition that a function is integrable. There is a theorem proved in more advanced courses guaranteeing the existence of the integral of continuous and certain discontinuous functions. We shall be specifically interested in functions whose discontinuities comprise

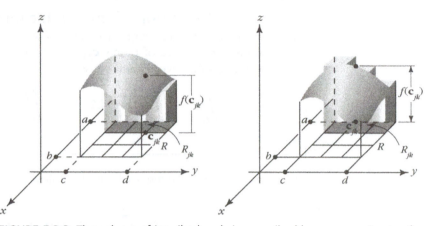

FIGURE 5.2.3. The volume of inscribed and circumscribed boxes approximates the volume under $z = f(x, y)$.

curves in the xy plane. Figure 5.2.4 shows two functions defined on a rectangle R whose discontinuities lie along curves. In other words, f is continuous at each point that is in R but not on the curve. Useful curves are graphs of functions $y = \phi(x), a \le x \le b$, or $x = \psi(y), c \le y \le d$, or finite unions of such graphs.

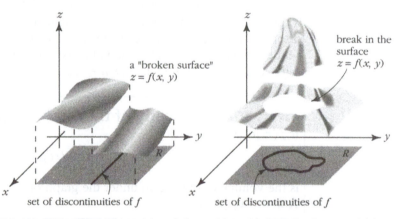

FIGURE 5.2.4. What the graphs of discontinuous functions of two variables might look like.

Existence of Integrals

If a function f is bounded on R and is continuous except possibly for discontinuities along a finite number of graphs of continuous functions, then f is integrable on R.

A continuous function on a closed rectangle is always bounded, but, for example, $f(x,y) = y + 1/x$ on $(0,1] \times [0,1]$ is not bounded, because $1/x$ becomes arbitrarily large for x near 0.

Now that we have defined the double integral and have a large supply of functions to integrate, we turn to its properties, stated without proof.

Properties of the Double Integral

1. Every continuous function is integrable. (See the previous box for a more general statement.)

2. If a rectangle R is divided by a line segment into two rectangles R_1 and R_2 (Figure 5.2.5), and if $f(x,y)$ is integrable on R_1 and R_2, then f is integrable on R and

$$\iint_R f(x,y)\,dx\,dy = \iint_{R_1} f(x,y)\,dx\,dy + \iint_{R_2} f(x,y)\,dx\,dy.$$

3. If f_1 and f_2 are integrable on R and if $f_1(x,y) \leq f_2(x,y)$ for all (x,y) in R, then

$$\iint_R f_1(x,y)\,dx\,dy \leq \iint_R f_2(x,y)\,dx\,dy.$$

4. If $f(x,y) = k$ for all (x,y) in R,

$$\iint_R f(x,y)\,dx\,dy = k\,(\text{area of } R).$$

5.

$$\iint_R [f_1(x,y) + f_2(x,y)]\,dx\,dy = \iint_R f_1(x,y)\,dx\,dy + \iint_R f_2(x,y)\,dx\,dy.$$

6.

$$\iint_R cf(x,y)\,dx\,dy = c \iint_R f(x,y)\,dx\,dy.$$

7.

$$\left| \iint_R f(x,y)\,dx\,dy \right| \leq \iint_R |f(x,y)|\,dx\,dy.$$

As we promised earlier, we will now explain how many double integrals can be computed by the method of iterated integration. We recall from §**5.1** that the iterated integral

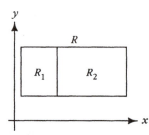

FIGURE 5.2.5. The rectangle R divided into smaller rectangles R_1 and R_2.

$$\int_c^d \left[\int_a^b f(x,y)dx \right] dy$$

is evaluated, like most parenthesized expressions, from the inside out. One first holds y fixed and evaluates the integral $\int_a^b f(x,y)dx$ with respect to x; the resulting function of y is then integrated from c to d. The expression

$$\int_a^b \left[\int_c^d f(x,y)dy \right] dx$$

is defined similarly; this time the integral with respect to y is evaluated first.

In the previous section we used Cavalieri's principle (slicing by planes perpendicular to the x or y axis) to show that the volume of the region under a graph can be evaluated as an iterated integral. A similar result can be proved, without recourse to geometry, for the integrals of functions which may take both positive or negative values.

Note

The notation $\iint_R f(x,y)dxdy$ is used to denote the double integral, whereas

$$\int_c^d \left[\int_a^b f(x,y)dx \right] dy \quad \text{or} \quad \int_a^b \left[\int_c^d f(x,y)dy \right] dx$$

are used for *iterated* integrals. We emphasize that each iterated integral is the result of a calculation that involves two one-variable integrations, whereas the double integral is a quantity that "involves both variables at once." Thus, the reduction to iterated integrals is a significant result which equates essentially different concepts. Without a result like this together with the fundamental theorem of calculus, how would we be able to *evaluate* double integrals?

It is customary to remove the brackets in the iterated integrals. Thus, one writes

$$\int_a^b \int_c^d f(x,y)dydx \quad \text{and} \quad \int_c^d \int_a^b f(x,y)dxdy.$$

Reduction to Iterated Integrals

If f is integrable on the rectangle $R = [a, b] \times [c, d]$, then either of the iterated integrals

$$\int_c^d \left[\int_a^b f(x, y) dx \right] dy \quad \text{or} \quad \int_a^b \left[\int_c^d f(x, y) dy \right] dx,$$

if it exists, equals the double integral

$$\iint_R f(x, y) dx dy.$$

As a consequence, we see that (when the iterated integrals exist)

$$\int_c^d \left[\int_a^b f(x, y) dx \right] dy = \int_a^b \left[\int_c^d f(x, y) dy \right] dx.$$

Example 1 Compute $\iint_R (x^2 + y) dA$, where R is the square $[0, 1] \times [0, 1]$.

Solution By reduction to iterated integrals,

$$\iint_R (x^2 + y) dA = \int_0^1 \int_0^1 (x^2 + y) dx dy = \int_0^1 \left[\int_0^1 (x^2 + y) dx \right] dy.$$

The x integration is performed first:

$$\int_0^1 (x^2 + y) dx = \left[\frac{x^3}{3} + yx \right]_{x=0}^1 = \frac{1}{3} + y.$$

Thus

$$\iint_R (x^2 + y) dA = \int_0^1 \left[\frac{1}{3} + y \right] dy = \left[\frac{1}{3} y + \frac{y^2}{2} \right]_0^1 = \frac{5}{6}.$$

What we have done is hold y fixed, integrate with respect to x, and then evaluate the result between the given limits for the x variable. Next we integrated the remaining function (of y alone) with respect to y to obtain the final answer.

♦

Example 2   A consequence of the reduction to iterated integrals is that interchanging the order of integration in the iterated integrals does not change the answer. Verify this for Example 1.

Solution   We carry out the integration in the other order:

$$
\int_0^1 \int_0^1 (x^2 + y)\, dy\, dx \;=\; \int_0^1 \left[x^2 y + \frac{y^2}{2} \right]_{y=0}^1 dx = \int_0^1 \left[x^2 + \frac{1}{2} \right] dx
$$

$$
=\; \left[\frac{x^3}{3} + \frac{x}{2} \right]_0^1 = \frac{5}{6}. \quad \blacklozenge
$$

Example 3   Verify that $\displaystyle \int_0^2 \int_1^3 x^2 y \, dy\, dx = \int_1^3 \int_0^2 x^2 y \, dx\, dy.$

Solution

$$
\int_0^2 \int_1^3 x^2 y \, dy\, dx \;=\; \int_0^2 \left(\int_1^3 x^2 y \, dy \right) dx = \int_0^2 \left(\frac{x^2 y^2}{2} \Big|_{y=1}^3 \right) dx
$$

$$
=\; \int_0^2 x^2 \left(\frac{9}{2} - \frac{1}{2} \right) dx = 4 \int_0^2 x^2 \, dx = 4 \frac{x^3}{3} \Big|_0^2 = \frac{32}{3}.
$$

In the other order,

$$
\int_1^3 \int_0^2 x^2 y \, dx\, dy \;=\; \int_1^3 \left(\int_0^2 x^2 y \, dx \right) dy = \int_1^3 \left(\frac{x^3 y}{3} \Big|_{x=0}^2 \right) dy
$$

$$
=\; \int_1^3 \frac{8}{3} y \, dy = \frac{4}{3} y^2 \Big|_1^3 = \frac{4}{3}(3^2 - 1^2) = \frac{32}{3}. \quad \blacklozenge
$$

Example 4   Compute $\iint_R \sin(x + y)\, dx\, dy$, where $R = [0, \pi] \times [0, 2\pi]$.

Solution

$$
\iint_R \sin(x + y)\, dx\, dy \;=\; \int_0^{2\pi} \left[\int_0^\pi \sin(x + y)\, dx \right] dy
$$

$$
=\; \int_0^{2\pi} \left[-\cos(x + y)\big|_{x=0}^\pi \right] dy
$$

$$
=\; \int_0^{2\pi} \left[\cos y - \cos(y + \pi) \right] dy
$$

$$
=\; \left[\sin y - \sin(y + \pi) \right] \big|_{y=0}^{2\pi} = 0. \quad \blacklozenge
$$

Example 5   Find the volume under the graph of $f(x, y) = x^2 + y^2$ between the planes $x = 0, x = 3, y = -1$, and $y = 1$.

Solution   The volume is

$$
\int_{-1}^{1} \int_{0}^{3} (x^2 + y^2) dx dy = \int_{-1}^{1} \left(\frac{x^3}{3} + y^2 x \Big|_{x=0}^{3} \right) dy = \int_{-1}^{1} (9 + 3y^2) dy
$$

$$
= (9y + y^3)|_{-1}^{1} = 20. \quad \blacklozenge
$$

The definition of the double integral leads to interpretations other than the volume of the region under the graph. For example, if $\rho(x, y)$ is the mass per unit area of a rectangle R, then

$$
\iint_{R} \rho(x, y) dx dy
$$

represents the *total mass* of the rectangle R.

Example 6   If D is a plate defined by $1 \le x \le 2, 0 \le y \le 1$ (measured in centimeters), and the mass density is $\rho(x, y) = y e^{xy}$ grams per square centimeter, integrate ρ over D to find the mass of the plate.

Solution   The total mass is

$$
\iint_{D} \rho(x, y) dx dy = \int_{0}^{1} \int_{1}^{2} y e^{xy} \, dx dy = \int_{0}^{1} (e^{xy}|_{x=1}^{2}) dy
$$

$$
= \int_{0}^{1} (e^{2y} - e^{y}) dy = \left(\frac{e^{2y}}{2} - e^{y} \right) \Big|_{y=0}^{1}
$$

$$
= \frac{e^2}{2} - e + \frac{1}{2} \approx 1.4762 \, \text{grams.} \quad \blacklozenge
$$

As a conclusion to this section we illustrate the process of summation that is represented by the double integral by discussing an example connected with solar energy. The intensity of solar radiation is a "local" quantity, which may be measured at any point on the Earth's surface. Since the solar intensity is really a rate of power input, we can measure it in units of watts per square meter.

If the solar intensity is uniform over a region, the total power received is equal to the intensity times the area of the region. In practice, the intensity is a function of position (in particular, it is a function of latitude), so we cannot just multiply a value by the area of the region. Instead, we must *integrate* the intensity over the region. Thus, the method of this section would allow us to find (at least in principle) the total power received by the state of Colorado, which is a rectangle in longitude–latitude coordinates. The problem for Utah is also tractable, since that state is composed of two rectangles, but what happens if we are interested in Michigan or Florida? For this problem, we need to integrate over regions that are not rectangles: The method for doing this is presented in the next section.

Exercises for §5.2

In Exercises 1–4, evaluate the integrals if $R = [0,1] \times [0,1]$.

1. $\iint_R (x^3 + y^2)dA$

2. $\iint_R ye^{xy}dA$

3. $\iint_R (xy)^2 \cos x^3 dA$

4. $\iint_R \log[(x+1)(y+1)]dA$

Evaluate $\iint_R f(x,y)dxdy$ for the indicated functions and rectangles in Exercises 5–8.

5. $f(x,y) = (x + 2y)^2; R = [-1,2] \times [0,2]$

6. $f(x,y) = y^3 \cos^2 x; R = [-\pi/2, \pi] \times [1,2]$

7. $f(x,y) = xy^3 e^{x^2 v^2}; R = [1,3] \times [1,2]$

8. $f(x,y) = xy + x/(y+1); R = [1,4] \times [1,2]$

Evaluate the iterated integrals in Exercises 9–12, and verify that one gets the same answer if the order of integration is changed.

9. $\int_0^2 \int_{-1}^1 (yx)^2 \, dydx$

10. $\int_{-1}^1 \int_0^2 (yx)^3 \, dxdy$

11. $\int_{-1}^1 \int_0^1 ye^x \, dydx$

12. $\int_{-1}^1 \int_0^3 y^5 e^{xy^3} \, dxdy$

Find the volume under the graph of f between the planes $x = a, x = b, y = c$, and $y = d$ in Exercises 13 and 14.

13. $f(x,y) = x^3 + y^2 + 2; a = -1, b = 1, c = 1, d = 3$

14. $f(x,y) = 2x + 3y^2 + 2; a = 0, b = 3, c = -2, d = 1$

15. Compute the volume of the solid bounded by the surface $z = \sin y$, the planes $x = -1, x = 0, y = 0$ and $y = \pi/2$, and the xy plane.

16. Compute the volume of the solid below the graph $z = x^2 + y$ and lying above the rectangle $R = [0, 1] \times [1, 2]$.

17. The density at each point of a 1 centimeter square (*i.e.,* each side has length 1 centimeter) microchip is $4 + r^2$ grams per square centimeter, where r is the distance in centimeters from the point to the center of the chip. What is the mass of the chip?

18. Do as in Exercise 17, but now let r be the distance to the lower left-hand corner of the plate.

19. Let f be continuous on $[a, b]$ and g continuous on $[c, d]$. Show that

$$\iint_R [f(x)g(y)]dxdy = \left[\int_a^b f(x)dx\right]\left[\int_c^d g(y)dy\right],$$

where $R = [a, b] \times [c, d]$.

20. The state of Colorado occupies the region between 37° and 41° latitude and 102° and 109° longitude. A degree of latitude is about 110 kilometers and a degree of longitude at the latitude of Colorado is about 83 kilometers. The intensity of solar radiation at time t on day T at latitude l is:

$$I = \cos l\sqrt{1 - \sin^2\alpha\cos^2\left(\frac{2\pi T}{365}\right)}\cos\left(\frac{2\pi t}{24}\right) + \sin l\sin\alpha\cos\left(\frac{2\pi T}{365}\right).$$

$\alpha = 23.5°$ is the tilt of the Earth's axis.

(a) What is the integrated intensity of solar energy over Colorado at time t on day T?

(b) Suppose that the result of part (a) is integrated with respect to t from t_1 to t_2. What does the integral represent?

5.3
The Double Integral Over Regions

Many applications involve double integrals $\iint_D f(x,y)dxdy$ over regions D that are not rectangles. For instance, the volume of a hemisphere, the mass of an elliptical plate, or the total solar power received by France can be expressed as such integrals.

We begin this section by giving a definition of the double integral over a region that is not necessarily a rectangle. Then we show how these integrals can be evaluated by a modified form of iterated integration.

If f is a function defined on a region D in the plane, we want

$$\iint_D f(x,y)dxdy$$

to agree with our previous definition when D is a rectangle. In addition, when $f(x,y) \geq 0$ on D, this integral should be the volume of the three-dimensional region under the graph of f and above D.

Our first step uses the "mathematician's trick" of reducing a problem to one which has already been solved. We replace the original function f by the function $\bar{f}$ defined on the entire plane $\mathbb{R}^2$ by

$$\bar{f}(x,y) = \begin{cases} f(x,y), & \text{if } (x,y) \text{ is in } D; \\ 0, & \text{if } (x,y) \text{ is not in } D \end{cases}$$

(see Figure 5.3.1).

The second step is to integrate the modified function $\bar{f}$ over any rectangle R that is large enough to contain D. (Since $\bar{f}$ is zero outside D, it doesn't matter which such rectangle we choose, since the part of R outside D doesn't contribute to the integral.) If $f(x,y) \geq 0$ on D, then $\bar{f}(x,y) \geq 0$ on R, and the region W under the graph of f on D differs from the region $\bar{W}$ under the graph of $\bar{f}$ on R only by a flat "plate" consisting of those points (x,y,z) such that (x,y) is in R but not in D, and $z = 0$. This plate has no volume, so $\iint_R \bar{f}(x,y)dxdy$, which we know to be equal to the volume of $\bar{W}$, is also equal to the volume of W; hence it makes good sense to *define* $\iint_D f(x,y)dxdy$ as $\iint_R \bar{f}(x,y)dxdy$.

The Double Integral over a Region D

Extend f to a rectangle R containing D by defining $\bar{f}(x,y)$ to be $f(x,y)$ for (x,y) in D and $\bar{f}(x,y) = 0$ outside D. If $\bar{f}$ is integrable on R, we say that f is **integrable** on D, and we define

$$\iint_D f(x,y)dxdy = \iint_R \bar{f}(x,y)dxdy.$$

This definition is not *directly* useful for evaluating integrals; however, knowing that the integral is a volume, we can evaluate a few simple ones.

Example 1   Let D be the disc $x^2 + y^2 \leq 1$, and let $f(x,y)$ be the constant function whose value is 3 for all (x,y). Find $\iint_D f(x,y)dxdy$.

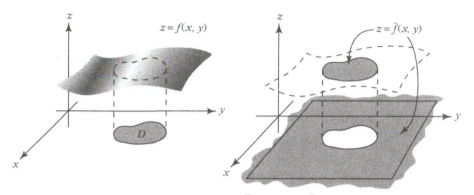

FIGURE 5.3.1. Given f and D, we construct $\bar{f}$ by setting $\bar{f}(x, y)$ equal to zero outside D.

Solution   The integral is the volume of the region under the graph of f on D. This region is a right circular cylinder with base area π and height 3, so its volume is 3π; hence $\iint_D f(x, y)dx dy = 3\pi$.   ◆

Having defined double integrals over nonrectangular regions, we will now develop a fairly general method to calculate them. We do this for special regions, called *elementary regions*. Consider two continuous real-valued functions $\phi_1 : [a, b] \to \mathbb{R}$ and $\phi_2 : [a, b] \to \mathbb{R}$ that satisfy $\phi_1(x) \le \phi_2(x)$ for $a \le x \le b$. Let D be the set of all points (x, y) such that

$$a \le x \le b \quad \text{and} \quad \phi_1(x) \le y \le \phi_2(x).$$

This region D is said to be of *type 1*. Figure 5.3.2 shows various examples of regions of type 1. The curves and straight-line segments that bound the region constitute the *boundary* of D, denoted ∂D.

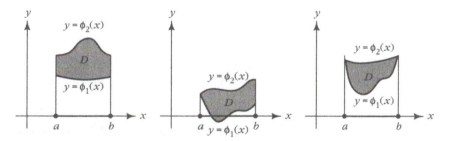

FIGURE 5.3.2. Some regions of type 1.

A region D is of *type 2* if there are continuous functions $\psi_1, \psi_2 : [c, d] \to \mathbb{R}$ such that D is the set of points (x, y) satisfying

$$c \le y \le d \quad \text{and} \quad \psi_1(y) \le x \le \psi_2(y),$$

where $\psi_1(y) \le \psi_2(y)$ for $c \le y \le d$. Again, the curves that bound the region D constitute its boundary ∂D. Figure 5.3.3 shows some examples of type 2 regions.

Finally, a region of **type 3** is one that can be described both as a region of type 1 and as a region of type 2. An example of a type 3 region is the unit disk (Figure 5.3.4). Regions of types 1, 2, or 3 are called **elementary regions**.

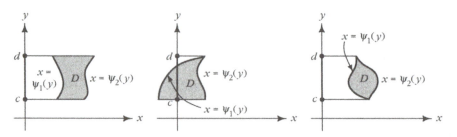

FIGURE 5.3.3. Some regions of type 2.

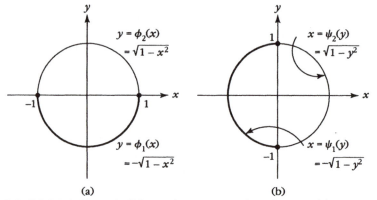

FIGURE 5.3.4. The unit disk, an elementary region of type 3: (a) as a type 1 region, and (b) as a type 2 region.

In the last section we saw that a function that has discontinuities only along curves in a rectangle is integrable. Applying this to $\bar{f}$ gives:

Continuity Implies Integrability

If f is continuous on an elementary region D, then f is integrable on D.

This fact, combined with the reduction of iterated integrals for rectangles, enables us to evaluate integrals over elementary regions by iterated integration.

If $R = [a, b] \times [c, d]$ is a rectangle containing D, then

$$\iint_D f(x, y)dxdy \;=\; \iint_R \bar{f}(x, y)dxdy = \int_a^b \int_c^d \bar{f}(x, y)dydx$$

$$=\; \int_c^d \int_a^b \bar{f}(x, y)dxdy,$$

where $\bar{f}$ equals f in D and is zero outside D. If D is a region of type 1 determined by functions ϕ_1 and ϕ_2 on $[a, b]$, the integral $\int_c^d \bar{f}(x, y)dy$ becomes (see Figure 5.3.5):

$$\int_c^d \bar{f}(x, y)dy = \int_{\phi_1(x)}^{\phi_2(x)} \bar{f}(x, y)dy = \int_{\phi_1(x)}^{\phi_2(x)} f(x, y)dy.$$

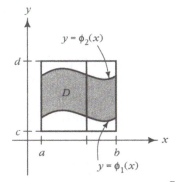

FIGURE 5.3.5. The integral of $\bar{f}$ along the vertical line $x = $ constant equals that of f from $\phi_1(x)$ to $\phi_2(x)$.

These equalities hold since $\bar{f}(x, y) = f(x, y)$ for $\phi_1(x) \leq y \leq \phi_2(x)$ and $\bar{f}(x, y) = 0$ otherwise.

Therefore,

$$\iint_D f(x, y)dxdy = \int_a^b \left[\int_{\phi_1(x)}^{\phi_2(x)} f(x, y)dy \right] dx.$$

Using a similar calculation for type 2 regions, we get the second formula in the following box.

Iterated Integrals for Elementary Regions

If D is a region of type 1,

$$\iint_D f(x,y)\,dx\,dy = \int_a^b \left[\int_{\phi_1(x)}^{\phi_2(x)} f(x,y)\,dy \right] dx.$$

If D is of type 2,

$$\iint_D f(x,y)\,dx\,dy = \int_c^d \left[\int_{\psi_1(y)}^{\psi_2(y)} f(x,y)\,dx \right] dy.$$

If D is of both types, either formula is applicable.

If $f(x,y) \geq 0$ on D, we may interpret the procedure of iterated integration as Cavalieri's principle for finding volumes by slicing. Suppose, for instance, that D is of type 1, determined by $\phi_1(x)$ and $\phi_2(x)$ on $[a,b]$. If we fix x and slice the volume under the graph of $f(x,y)$ on D by the plane which passes through the point $(x,0,0)$ and which is parallel to the yz plane, we obtain the region in the yz plane defined by the inequalities $\phi_1(x) \leq y \leq \phi_2(x)$ and $0 \leq z \leq f(x,y)$. The area $A(x)$ of this region is $\int_{\phi_1(x)}^{\phi_2(x)} f(x,y)\,dy$ (Figure 5.3.6). The double integral $\iint_D f(x,y)\,dx\,dy$, which is the volume of the entire solid, equals

$$\int_a^b A(x)\,dx = \int_a^b \left[\int_{\phi_1(x)}^{\phi_2(x)} f(x,y)\,dy \right] dx,$$

the iterated integral. Thus we get the first formula in the preceding box.

If D is of type 2, then slicing by planes parallel to the xz plane produces the second formula in the preceding box.

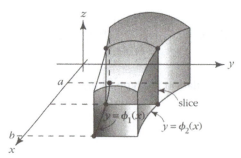

FIGURE 5.3.6. The area of the slice with fixed x is $\int_{\phi_1(x)}^{\phi_2(x)} f(x,y)\,dy$.

Example 2 Find $\iint_D (x+y)dxdy$, where D is the shaded region in Figure 5.3.7.

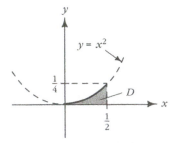

FIGURE 5.3.7. Find $\iint_D(x+y)dxdy$.

Solution D is a region of type 1, with $[a,b] = [0, \frac{1}{2}], \phi_1(x) = 0$, and $\phi_2(x) = x^2$. By the first formula in the preceding box,

$$\iint_D (x+y)dxdy \;=\; \int_0^{1/2} \int_0^{x^2} (x+y)dydx$$

$$=\; \int_0^{1/2} \left[\left(xy + \frac{y^2}{2}\right)\Big|_{y=0}^{x^2} \right] dx$$

$$=\; \int_0^{1/2} \left(x^3 + \frac{x^4}{2}\right) dx = \left(\frac{x^4}{4} + \frac{x^5}{10}\right)\Big|_0^{1/2}$$

$$=\; \frac{1}{64} + \frac{1}{320} = \frac{3}{160}.$$

D is also a region of type 2, with $[c,d] = [0, \frac{1}{4}]$, $\psi_1(y) = \sqrt{y}$, and $\psi_2(y) = \frac{1}{2}$. We leave it to you to verify that the double integral calculated by the second formula in the preceding box is also $3/160$. ◆

Example 3 Evaluate $\displaystyle\int_0^1 \int_{x^3}^{x^2} xy\, dydx$. Sketch the region D for which this iterated integral gives the double integral $\iint_D xy\, dxdy$.

Solution Here y ranges from x^3 to x^2, while x goes from 0 to 1. Hence the region is as shown in Figure 5.3.8. The integral is

$$\int_0^1 \left(\frac{xy^2}{2}\Big|_{y=x^3}^{x^2} \right) dx \;=\; \int_0^1 \left(\frac{x^5}{2} - \frac{x^7}{2}\right) dx$$

$$=\; \left(\frac{x^6}{12} - \frac{x^8}{16}\right)\Big|_{x=0}^1 = \frac{1}{12} - \frac{1}{16} = \frac{1}{48}. \;◆$$

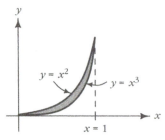

FIGURE 5.3.8. The region of integration for $\int_0^1 \int_{x^3}^{x^2} xy\,dy\,dx$.

If a region D is of type 3, we have

$$\iint_D f(x,y)dx\,dy = \int_a^b \int_{\phi_1(x)}^{\phi_2(x)} f(x,y)dy\,dx = \int_c^d \int_{\psi_1(y)}^{\psi_2(y)} f(x,y)dx\,dy,$$

so the two iterated integrals are equal. If we want to find $\iint_D f(x,y)dx\,dy$ but cannot integrate one of the iterated integrals above, we may try to evaluate the other integral, a technique called ***changing the order of integration***.

Example 4   Write $\displaystyle\int_0^1 \int_0^{\sqrt{1-x^2}} \sqrt{1-y^2}\,dy\,dx$ as an integral over a region. Sketch the region and show that it is of types 1 and 2. Reverse the order of integration and evaluate.

Solution   The region D is shown in Figure 5.3.9. D is a type 1 region with $\phi_1(x) = 0, \phi_2(x) = \sqrt{1-x^2}$ and a type 2 region with $\psi_1(y) = 0, \psi_2(y) = \sqrt{1-y^2}$. Thus

$$
\begin{aligned}
\int_0^1 \int_0^{\sqrt{1-x^2}} \sqrt{1-y^2}\,dy\,dx &= \int_0^1 \int_0^{\sqrt{1-y^2}} \sqrt{1-y^2}\,dx\,dy \\
&= \int_0^1 \left[\sqrt{1-y^2}\,x \Big|_{x=0}^{\sqrt{1-y^2}} \right] dy \\
&= \int_0^1 (1-y^2)dy = \left(y - \frac{y^3}{3} \right) \Big|_0^1 = \frac{2}{3}.
\end{aligned}
$$

Evaluating the integral in the original order is much harder!   ◆

Example 5   Calculate the integral of $f(x,y) = (x+y)^2$ over the region shown in Figure 5.3.10.

Solution   In this example, the region is of type 1 but not of type 2. Therefore there is a preferred order of integration. We could use the other order of integration, *i.e.*, x first, but this requires us to break up the region into two regions

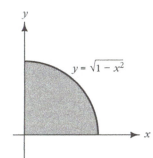

FIGURE 5.3.9. The region of integration for Example 4.

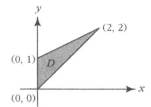

FIGURE 5.3.10. The region of integration for Example 5.

of type 2 by drawing the line $y = 1$; we then would integrate over each region and add the results. If we use the other order, we can perform the integration in one step:

$$\iint_D f(x,y)\,dx\,dy = \int_0^2 \int_x^{\frac{1}{2}x+1} f(x,y)\,dy\,dx.$$

(The lines bounding D on the bottom and top are $y = x$ and $y = \frac{1}{2}x + 1$.) The integral is thus

$$
\begin{aligned}
\int_0^2 \int_x^{\frac{1}{2}x+1} (x+y)^2 \, dy\,dx &= \int_0^2 \left[\frac{1}{3}(x+y)^3 \Big|_{y=x}^{\frac{1}{2}x+1} \right] dx \\
&= \frac{1}{3} \int_0^2 \left[\left(\frac{3}{2}x + 1 \right)^3 - (2x)^3 \right] dx \\
&= \frac{1}{3} \left[\frac{1}{6} \left(\frac{3}{2}x + 1 \right)^4 \Big|_0^2 - 2x^4 \Big|_0^2 \right] \\
&= \frac{1}{3} \left[\frac{1}{6}(4^4 - 1) - 2 \cdot 16 \right] = \frac{1}{3} \left[\frac{21}{2} \right] = \frac{7}{2}. \quad \blacklozenge
\end{aligned}
$$

General regions can often be broken into elementary regions, and double integrals over these regions can be computed one piece at a time.

Example 6   Find $\iint_D x^2 dxdy$, where D is the shaded region in Figure 5.3.11.

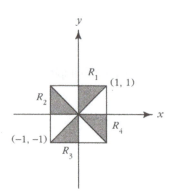

FIGURE 5.3.11. Integrate x^2 over the pinwheel.

Solution   Each of the regions R_1, R_2, R_3, R_4 is of types 1 and 2, so we may integrate over each one separately and sum the results:

$$
\iint_{R_1} x^2\, dxdy = \int_0^1 \int_x^1 x^2\, dydx = \int_0^1 (x^2 y)\Big|_{y=x}^{1}\, dx
$$
$$
= \int_0^1 (x^2 - x^3)dx = \left(\frac{x^3}{3} - \frac{x^4}{4}\right)\Big|_0^1 = \frac{1}{3} - \frac{1}{4} = \frac{1}{12}.
$$

Similarly, we find

$$
\iint_{R_2} x^2\, dxdy = \int_{-1}^0 \int_0^{-x} x^2\, dydx = \frac{1}{4}.
$$

Since the integrand is unchanged under reflection through the origin, the integrals over R_1 and R_3 are equal, as are those over R_2 and R_4. Therefore,

$$
\iint_{R_3} x^2\, dxdy = \frac{1}{12} \quad \text{and} \quad \iint_{R_4} x^2\, dxdy = \frac{1}{4},
$$

so adding these:

$$
\iint_D x^2\, dxdy = \frac{1}{12} + \frac{1}{4} + \frac{1}{12} + \frac{1}{4} = \frac{2}{3}. \quad \blacklozenge
$$

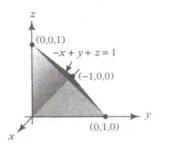

FIGURE 5.3.12. A tetrahedron bounded by the planes $y = 0, z = 0, x = 0$, and $-x + y + z = 1$.

Example 7 Find the volume of the tetrahedron bounded by the planes $y = 0, z = 0, x = 0$, and $-x + y + z = 1$ (Figure 5.3.12).

Solution The tetrahedron has a triangular base D, whose points (x, y) satisfy $-1 \leq x \leq 0$ and $0 \leq y \leq 1 + x$; hence D is a region of type 1 (Figure 5.3.13).

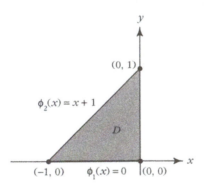

FIGURE 5.3.13. The base of the tetrahedron in Figure 5.3.12 represented as a region of type 1.

For any point (x, y) in D, the height of the surface z above (x, y) is $1 - y + x$. Thus, the volume we seek is the double integral

$$\iint_D (1 - y + x)\,dx\,dy.$$

Integrating with respect to y first, we get

$$
\iint_D (1 - y + x)\,dx\,dy = \int_{-1}^0 \int_0^{1+x} (1 - y + x)\,dy\,dx
$$

$$
= \int_{-1}^0 \left[(1 + x)y - \frac{y^2}{2} \right]_{y=0}^{1+x} dx
$$

$$= \int_{-1}^{0}\left[\frac{(1+x)^2}{2}\right]dx = \left[\frac{(1+x)^3}{6}\right]_{-1}^{0} = \frac{1}{6}. \quad \blacklozenge$$

We conclude this section with the mean value theorem for integrals. For one-variable calculus, we recall that the result states:

Mean Value Theorem for Single Integrals

Let f be continuous on $[a,b]$. Then there is a point x_0 in (a,b) such that

$$f(x_0) = \frac{1}{b-a}\int_a^b f(x)dx.$$

The right-hand side of this equation is called the **average value** of f on $[a,b]$.

Geometrically, the average value is the height of the rectangle with base $[a,b]$ that has the same area as the region under the graph of f (see Figure 5.3.14). If the graph of f is a picture of the surface of wavy water in a narrow channel, then the average value of f is the height of the water when it settles.

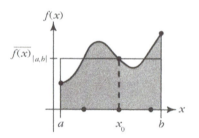

FIGURE 5.3.14. The average value is defined so that the area of the rectangle equals the area under the graph. The dots on the x axis indicate places where the average value is attained.

There is a similar result for double integrals as follows:

Mean Value Theorem for Double Integrals

Suppose that $f : D \to \mathbb{R}$ is continuous and D is an elementary region. Then for some point (x_0, y_0) in D we have

$$f(x_0, y_0) = \frac{1}{A(D)}\iint_D f(x,y)dA$$

where $A(D)$ denotes the area of D.

Main Ideas of the Proof. Since f is continuous on D, it has a maximum value M and a minimum value m as in §3.3. Thus, $m \leq f(x, y) \leq M$ for all (x, y) in D. Furthermore, $f(x_1, y_1) = m$ and $f(x_2, y_2) = M$ for some pairs (x_1, y_1) and (x_2, y_2) in D. Thus,

$$mA(D) = \iint_D m\, dA \leq \iint_D f(x, y)\, dA \leq \iint_D M\, dA = MA(D).$$

Dividing through by $A(D)$, we get

$$m \leq \frac{1}{A(D)} \iint_D f(x, y)\, dA \leq M.$$

Join (x_1, y_1) and (x_2, y_2) by a continuous curve $\mathbf{c}(t), a \leq t \leq b$, in D. By the intermediate value theorem from one-variable calculus, every value between the two assumed values m and M is achieved by the function $g(t) = f(\mathbf{c}(t))$. In particular, $(1/A(D)) \iint_D f(x, y)\, dA$ is achieved at some intermediate point t_0, so $(x_0, y_0) = \mathbf{c}(t_0)$ gives the desired point in D. ∎

Example 8 Show that

$$0 \leq \iint_R \frac{\sin x}{1 + (xy)^4}\, dx\, dy \leq \pi$$

where $R = [0, \pi] \times [0, 1]$.

Solution Since $\sin x \geq 0$ for $0 \leq x \leq \pi$ and $1 + (xy)^4 > 0$, the integrand is nonnegative, so the integral is nonnegative. Since

$$\frac{\sin x}{1 + (xy)^4} \leq \sin x \leq 1,$$

the integral is $\leq 1 \times$ (area of $[0, \pi] \times [0, 1]) = \pi$. ◆

Exercises for §5.3

In Exercises 1–4, sketch each region and tell whether it is of type 1, type 2, both, or neither.

1. (x, y) such that $0 \leq y \leq 3x, 0 \leq x \leq 1$

2. (x, y) such that $y^2 \leq x \leq y, 0 \leq y \leq 1$

3. (x, y) such that $x^4 + y^4 \leq 1$

4. (x, y) such that $\frac{1}{2} \leq x^4 + y^4 \leq 1$

5. Find $\iint_D (x + y)^2\, dx\, dy$, where D is the shaded region in Figure 5.3.15.

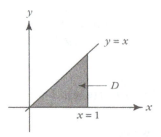

FIGURE 5.3.15. The region of integration for Exercises 5 and 6.

6. Find $\iint_D (1 - \sin \pi x) y \, dx dy$, where D is the region in Figure 5.3.15.

7. Find $\iint_D (x - y)^2 dx dy$, where D is the region in Figure 5.3.16.

8. Find $\iint_D y(1 - \cos(\pi x/4)) dx dy$, where D is the region in Figure 5.3.16.

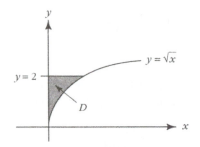

FIGURE 5.3.16. The region of integration for Exercises 7 and 8.

Evaluate the integrals in Exercises 9–16. Sketch and identify the type of the region (corresponding to the way the integral is written).

9. $\displaystyle\int_0^\pi \int_{\sin x}^{3\sin x} x(1 + y) dy dx$

10. $\displaystyle\int_0^1 \int_{x-1}^{x \cos 2\pi x} (x^2 + xy + 1) dy dx$

11. $\displaystyle\int_{-1}^1 \int_{y^{2/3}}^{(2-y)^2} \left(\frac{3}{2}\sqrt{x} - 2y\right) dx dy$

12. $\displaystyle\int_0^2 \int_{-3(\sqrt{4-x^2})/2}^{3(\sqrt{4-x^2})/2} \left(\frac{5}{\sqrt{2+x}} + y^3\right) dy dx$

13. $\displaystyle\int_0^1 \int_0^{x^2} (x^2 + xy - y^2)\,dy\,dx$

14. $\displaystyle\int_2^4 \int_{y^2-1}^{y^3} 3\,dx\,dy$

15. $\displaystyle\int_0^1 \int_{x^2}^x (x + y)^2\,dy\,dx$

16. $\displaystyle\int_0^1 \int_0^{3y} e^{x+y}\,dx\,dy$

In Exercises 17–20, sketch the region of integration, interchange the order, and evaluate.

17. $\displaystyle\int_0^1 \int_x^1 xy\,dy\,dx$

18. $\displaystyle\int_0^{\pi/2} \int_0^{\cos x} \cos x\,dy\,dx$

19. $\displaystyle\int_0^1 \int_{1-y}^1 (x + y^2)\,dx\,dy$

20. $\displaystyle\int_1^4 \int_1^{\sqrt{x}} (x^2 + y^2)\,dy\,dx$

In Exercises 21–24, integrate the function f over the region D.

21. $f(x,y) = x - y$; D is the triangle with vertices $(0,0), (1,0)$, and $(2,1)$.

22. $f(x,y) = x^3 y + \cos x$; D is the triangle defined by $0 \le x \le \pi/2, 0 \le y \le x$.

23. $f(x,y) = (x^2 + 2xy^2 + 2)$; D is the region bounded by the graph of $y = -x^2 + x$, the x axis, and the lines $x = 0$ and $x = 2$.

24. $f(x,y) = \sin x \cos y$; D is the pinwheel in Figure 5.3.11.

25. If $D = [-1, 1] \times [-1, 2]$, show that $1 \le \displaystyle\iint_D \frac{dx\,dy}{x^2 + y^2 + 1} \le 6$.

26. If $f(x,y) = e^{\sin(x+y)}$ and $D = [-\pi, \pi] \times [-\pi, \pi]$, show that

$$\frac{1}{e} \le \frac{1}{4\pi^2} \iint_D f(x,y)\,dA \le e.$$

27. Show that evaluting $\iint_D dx\,dy$, where D is a region of type 1, reproduces the formula for the area between curves from one-variable calculus.

28. Let D be the region defined by $x^2 + y^2 \leq 1$.

    (a) Estimate $\iint_D dx\,dy$ (the area of D) within 0.1 by taking a rectangle grid in the plane and counting the number of rectangles: (i) contained entirely in D; (ii) intersecting D.

    (b) Compute $\iint_D dx\,dy$ exactly by using an iterated integral.

29. Prove: $\displaystyle\int_0^x \left[\int_0^t F(u)\,du \right] dt = \int_0^x (x - u) F(u)\,du.$ (Hint: Sketch the region of integration.)

30. Show that

$$\frac{d}{dx} \int_a^x \int_c^d f(x, y, z)\,dz\,dy = \int_c^d f(x, x, z)\,dz + \int_a^x \int_c^d f_x(x, y, z)\,dz\,dy.$$

5.4
Triple
Integrals

We now extend the integral from regions in the plane to regions in space. As with double integrals, triple integrals are often evaluated using iterated integrals, with three integrations instead of two. A second important technique is to use the change of variables formula; this is discussed in the next section.

Triple integrals are needed for many physical problems. For example, if the temperature inside an oven is not uniform, determining the average temperature involves first "summing" the values of the temperature function at all points in the solid region enclosed by the oven walls and then dividing the answer by the total volume of the oven. Such a sum is expressed mathematically as a triple integral.

Our object now is to define the triple integral of a function $f(x, y, z)$ over a box (rectangular parallelepiped) $B = [a, b] \times [c, d] \times [p, q]$. Proceeding as in double integrals, we partition the three sides of B into n equal parts and form the sum

$$S_n = \sum_{i=0}^{n-1} \sum_{j=0}^{n-1} \sum_{k=0}^{n-1} f(\mathbf{c}_{ijk}) \Delta V,$$

where $\mathbf{c}_{ijk}$ is a point in B_{ijk}, the ijkth rectangular parallelepiped (or box) in the partition of B, and ΔV is the volume of B_{ijk} (see Figure 5.4.1).

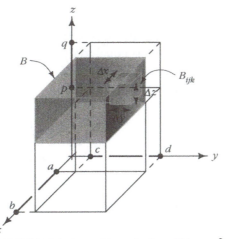

FIGURE 5.4.1. A partition of a box B into n^3 subboxes B_{ijk}.

The Triple Integral over a Box

Let f be a bounded function of three variables defined on B. If $S = \lim_{n\to\infty} S_n$ exists (for any choices of $\mathbf{c}_{ijk}$), we say that f is ***integrable*** and call S the ***triple integral*** (or simply the integral) of f over B and denote it by

$$\iiint_B f\, dV, \ \iiint_B f(x,y,z)dV, \text{or} \iiint_B f(x,y,z)dxdydz.$$

As before, one can prove that continuous functions defined on B are integrable. Moreover, bounded functions whose discontinuities are confined to graphs of continuous functions [such as $x = \alpha(y,z), y = \beta(x,z)$, or $z = \gamma(x,y)$] are integrable. The other basic properties (such as the fact that the integral of a sum is the sum of the integrals) for double integrals also hold for triple integrals. Especially important is the reduction to iterated integrals:

Reduction to Iterated Integrals

Let $f(x, y, z)$ be integrable on the box $B = [a, b] \times [c, d] \times [p, q]$. Then any iterated integral that exists is equal to the triple integral; that is,

$$\iiint_B f(x, y, z)\,dxdydz \;=\; \int_p^q \int_c^d \int_a^b f(x, y, z)\,dxdydz \quad (5.4.1)$$

$$=\; \int_p^q \int_a^b \int_c^d f(x, y, z)\,dydxdz \quad (5.4.2)$$

$$=\; \int_a^b \int_p^q \int_c^d f(x, y, z)\,dydzdx, \quad (5.4.3)$$

and so on. (There are six possible orders altogether.)

Example 1   (a) Let B be the box $[0, 1] \times [-\frac{1}{2}, 0] \times [0, \frac{1}{3}]$. Evaluate

$$\iiint_B (x + 2y + 3z)^2\,dxdydz.$$

(b) Verify that we get the same answer if the integration is done in the order y first, then z, and then x.

Solution

(a) According to the principle of reduction to iterated integrals, this integral may be evaluated as

$$\int_0^{1/3} \int_{-1/2}^0 \int_0^1 (x + 2y + 3z)^2\,dxdydz$$

$$=\; \int_0^{1/3} \int_{-1/2}^0 \left[\left. \frac{(x + 2y + 3z)^3}{3} \right|_{x=0}^1 \right] dydz$$

$$=\; \int_0^{1/3} \int_{-1/2}^0 \frac{1}{3} \left[(1 + 2y + 3z)^3 - (2y + 3z)^3 \right] dydz$$

$$=\; \int_0^{1/3} \frac{1}{24} \left[(1 + 2y + 3z)^4 - (2y + 3z)^4 \right] \Big|_{y=-1/2}^0 dz$$

$$=\; \int_0^{1/3} \frac{1}{24} \left[(3z + 1)^4 - 2(3z)^4 + (3z - 1)^4 \right] dz$$

$$=\; \frac{1}{24 \cdot 15} \left[(3z + 1)^5 - 2(3z)^5 + (3z - 1)^5 \right] \Big|_{z=0}^{1/3}$$

$$=\; \frac{1}{24 \cdot 15} (2^5 - 2) = \frac{1}{12}.$$

(b)

$$\iiint_B (x + 2y + 3z)^2 dydzdx$$

$$= \int_0^1 \int_0^{1/3} \int_{-1/2}^0 (x + 2y + 3z)^2 dydzdx$$

$$= \int_0^1 \int_0^{1/3} \left[\frac{(x + 2y + 3z)^3}{6} \Big|_{y=-1/2}^0 \right] dzdx$$

$$= \int_0^1 \int_0^{1/3} \frac{1}{6} \left[(x + 3z)^3 - (x + 3z - 1)^3 \right] dzdx$$

$$= \int_0^1 \frac{1}{6} \left\{ \left[\frac{(x + 3z)^4}{12} - \frac{(x + 3z - 1)^4}{12} \right] \Big|_{z=0}^{1/3} \right\} dx$$

$$= \int_0^1 \frac{1}{72} \left[(x + 1)^4 + (x - 1)^4 - 2x^4 \right] dx$$

$$= \frac{1}{72} \frac{1}{5} \left[(x + 1)^5 + (x - 1)^5 - 2x^5 \right]_{x=0}^1 = \frac{1}{12}. \quad \blacklozenge$$

Example 2   Integrate e^{x+y+z} over the box $[0,1] \times [0,1] \times [0,1]$.

Solution

$$\int_0^1 \int_0^1 \int_0^1 e^{x+y+z} dxdydz = \int_0^1 \int_0^1 (e^{x+y+z}|_{x=0}^1) dydz$$

$$= \int_0^1 \int_0^1 (e^{1+y+z} - e^{y+z}) dydz = \int_0^1 \left[e^{1+y+z} - e^{y+z} \right]_{y=0}^1 dz$$

$$= \int_0^1 \left[e^{2+z} - 2e^{1+z} + e^z \right] dz = \left[e^{2+z} - 2e^{1+z} + e^z \right]_0^1$$

$$= e^3 - 3e^2 + 3e - 1 = (e - 1)^3. \quad \blacklozenge$$

As in the two-variable case, we define the integral of a function f over a bounded region W by defining a new function $\bar{f}$, equal to f on W and zero outside W, and then setting

$$\iiint_W f(x, y, z)dxdydz = \iiint_B \bar{f}(x, y, z)dxdydz,$$

where B is any box containing the region W.

As before, we restrict our attention to particularly simple regions. An **elementary region** in three-dimensional space is one defined by restricting one of the variables to be between two functions of the remaining variables, the domains of these functions being an elementary region in the plane. For example, if D

is an elementary region in the xy plane and if $\gamma_1(x,y)$ and $\gamma_2(x,y)$ are two functions with $\gamma_2(x,y) \geq \gamma_1(x,y)$, an elementary region consists of all (x,y,z) such that (x,y) lies in D and $\gamma_1(x,y) \leq z \leq \gamma_2(x,y)$. Figure 5.4.2 shows two elementary regions.

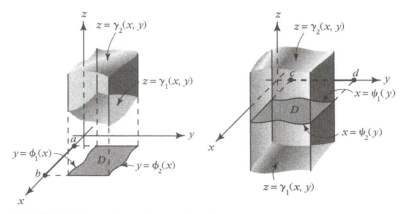

FIGURE 5.4.2. Two elementary regions in space.

Example 3   Describe the unit ball $x^2 + y^2 + z^2 \leq 1$ as an elementary region.

Solution   This can be done in several ways. One is:

$$-1 \leq x \leq 1,$$

$$-\sqrt{1 - x^2} \leq y \leq \sqrt{1 - x^2},$$

$$-\sqrt{1 - x^2 - y^2} \leq z \leq \sqrt{1 - x^2 - y^2}.$$

In doing this, we first write the top and bottom hemispheres as $z = \sqrt{1 - x^2 - y^2}$ and $z = -\sqrt{1 - x^2 - y^2}$, respectively, where x and y vary over the unit disk (that is, $-\sqrt{1 - x^2} \leq y \leq \sqrt{1 - x^2}$ and x varies between -1 and 1). (See Figure 5.4.3.) We can describe the region in other ways by interchanging the roles of x, y, and z in the defining inequalities.   ◆

As with integrals in the plane, any function of three variables that is continuous over an elementary region is integrable on that region. An argument like that for double integrals shows that a triple integral over an elementary region can be rewritten as an iterated integral in which the limits of integration are functions. The formulas for such iterated integrals are given in the following box.

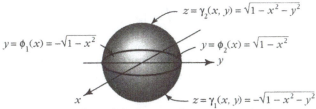

FIGURE 5.4.3. The unit ball as an elementary region in space.

Triple Integrals by Iterated Integration

Suppose that W is an elementary region described by bounding z between two functions of x and y. Then either

$$\iiint_W f(x,y,z)\,dxdydz = \int_a^b \int_{\phi_1(x)}^{\phi_2(x)} \int_{\gamma_1(x,y)}^{\gamma_2(x,y)} f(x,y,z)\,dzdydx$$

[see Figure 5.4.2 (left)] or

$$\iiint_W f(x,y,z)\,dxdydz = \int_c^d \int_{\psi_1(y)}^{\psi_2(y)} \int_{\gamma_1(x,y)}^{\gamma_2(x,y)} f(x,y,z)\,dzdxdy$$

[see Figure 5.4.2 (right)].

If $f = 1$, $\iiint_W dxdydz$ has a geometric interpretation as the *volume* of the region W.

Example 4 Verify the volume formula for the ball of radius 1:

$$\iiint_W dxdydz = \frac{4}{3}\pi,$$

where W is the set of (x,y,z) with $x^2 + y^2 + z^2 \leq 1$.

Solution We use the description of the unit ball from Example 3. From the first formula in the preceding box, the integral is

$$\int_{-1}^1 \int_{-\sqrt{1-x^2}}^{\sqrt{1-x^2}} \int_{-\sqrt{1-x^2-y^2}}^{\sqrt{1-x^2-y^2}} dzdydx.$$

Holding y and x fixed and integrating with respect to z yields

$$\int_{-1}^1 \int_{-\sqrt{1-x^2}}^{\sqrt{1-x^2}} \left[z \Big|_{-\sqrt{1-x^2-y^2}}^{\sqrt{1-x^2-y^2}} \right] dydx$$

$$= 2 \int_{-1}^{1} \left[\int_{-\sqrt{1-x^2}}^{\sqrt{1-x^2}} (1 - x^2 - y^2)^{1/2} dy \right] dx.$$

Since x is fixed in the y-integral, it can be expressed as $\int_{-a}^{a} (a^2 - y^2)^{1/2} dy$, where $a = (1 - x^2)^{1/2}$. This integral is the area of a semicircular region of radius a, so that

$$\int_{-a}^{a} (a^2 - y^2)^{1/2} dy = \frac{a^2}{2} \pi.$$

(We could also have used a trigonometric substitution or a table of integrals.) Thus

$$\int_{-\sqrt{1-x^2}}^{\sqrt{1-x^2}} (1 - x^2 - y^2)^{1/2} dy = \frac{1 - x^2}{2} \pi,$$

and so

$$2 \int_{-1}^{1} \int_{-\sqrt{1-x^2}}^{\sqrt{1-x^2}} (1 - x^2 - y^2)^{1/2} dy dx = 2 \int_{-1}^{1} \pi \frac{1 - x^2}{2} dx$$

$$= \pi \int_{-1}^{1} (1 - x^2) dx = \pi \left(x - \frac{x^3}{3} \right) \Big|_{x=-1}^{1} = \frac{4}{3} \pi. \quad \blacklozenge$$

Other types of elementary regions are shown in Figure 5.4.4. For instance, in the second region, (y, z) lies in an elementary region in the yz plane and x lies between two graphs:

$$\rho_1(y, z) \leq x \leq \rho_2(y, z).$$

As shown in Figure 5.4.5, some elementary regions can be simultaneously described in all three ways.

Corresponding to each description of a region as an elementary region is an integration formula. For instance, if W is expressed as the set of all (x, y, z) such that

$$c \leq y \leq d, \quad \psi_1(y) \leq z \leq \psi_2(y), \quad \rho_1(y, z) \leq x \leq \rho_2(y, z),$$

then

$$\iiint_W f(x, y, z) dx dy dz = \int_c^d \int_{\psi_1(y)}^{\psi_2(y)} \int_{\rho_1(y,z)}^{\rho_2(y,z)} f(x, y, z) dx dz dy.$$

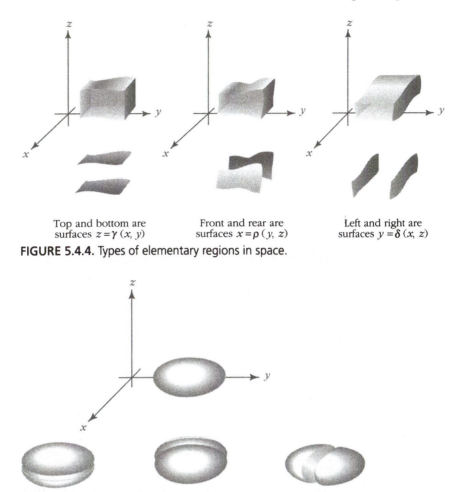

FIGURE 5.4.4. Types of elementary regions in space.

Top and bottom are
surfaces $z = \gamma\,(x,\,y)$

Front and rear are
surfaces $x = \rho\,(\,y,\,z)$

Left and right are
surfaces $y = \delta\,(x,\,z)$

FIGURE 5.4.5. An elementary region that can be described in three overall ways.

Example 5   Let W be the region bounded by the planes $x = 0, y = 0$, and $z = 2$, and the surface $z = x^2 + y^2$ and lying in the quadrant $x \geq 0, y \geq 0$. Compute $\iiint_W x\,dx\,dy\,dz$ and sketch the region.

Solution   *Method 1.* The region W is sketched in Figure 5.4.6. As indicated in the figure, we may describe this region by the inequalities

$$0 \leq x \leq \sqrt{2}, \quad 0 \leq y \leq \sqrt{2 - x^2}, \quad x^2 + y^2 \leq z \leq 2.$$

Therefore,

$$\iiint_W x\,dx\,dy\,dz \;=\; \int_0^{\sqrt{2}} \left[\int_0^{\sqrt{2-x^2}} \left(\int_{x^2+y^2}^2 x\,dz \right) dy \right] dx$$

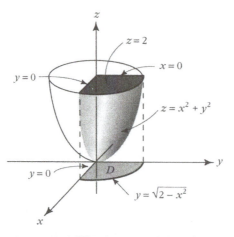

FIGURE 5.4.6. W is the region below the plane $z = 2$, above the paraboloid $z = x^2 + y^2$, and on the positive sides of the planes $x = 0, y = 0$.

$$= \int_0^{\sqrt{2}} \int_0^{\sqrt{2-x^2}} x(2 - x^2 - y^2) dy dx$$

$$= \int_0^{\sqrt{2}} x\left[(2 - x^2)^{3/2} - \frac{(2 - x^2)^{3/2}}{3}\right] dx$$

$$= \int_0^{\sqrt{2}} \frac{2x}{3}(2 - x^2)^{3/2} dx = \left.\frac{-2(2 - x^2)^{5/2}}{15}\right|_0^{\sqrt{2}}$$

$$= 2 \cdot \frac{2^{5/2}}{15} = \frac{8\sqrt{2}}{15}.$$

Method 2. We can also limit x first to describe W by $0 \le x \le (z - y^2)^{1/2}$ and (y, z) in D, where D is the subset of the yz plane with $0 \le z \le 2$ and $0 \le y \le z^{1/2}$ (see Figure 5.4.7).

Therefore,

$$\iiint_W x\, dx dy dz = \iint_D \left(\int_0^{(z-y^2)^{1/2}} x dx\right) dy dz$$

$$= \int_0^2 \left[\int_0^{z^{1/2}} \left(\int_0^{(z-y^2)^{1/2}} x dx\right) dy\right] dz$$

$$= \int_0^2 \int_0^{z^{1/2}} \left(\frac{z - y^2}{2}\right) dy dz$$

$$= \frac{1}{2}\int_0^2 \left(z^{3/2} - \frac{z^{3/2}}{3}\right) dz = \frac{1}{2}\int_0^2 \frac{2}{3}z^{3/2} dz$$

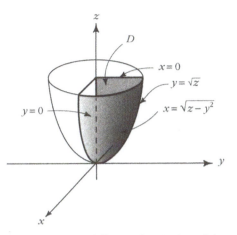

FIGURE 5.4.7. A different description of the region in Example 5.

$$= \left[\frac{2}{15}z^{5/2}\right]_0^2 = \frac{2}{15}2^{5/2} = \frac{8\sqrt{2}}{15},$$

which agrees with our other answer.   ◆

Example 6   Evaluate

$$\int_0^1 \int_0^x \int_{x^2+y^2}^2 dz\,dy\,dx.$$

Sketch the region W of integration and interpret.

Solution

$$\int_0^1 \int_0^x \int_{x^2+y^2}^2 dz\,dy\,dx = \int_0^1 \int_0^x (2 - x^2 - y^2)dy\,dx$$

$$= \int_0^1 \left(2x - x^3 - \frac{x^3}{3}\right) dx = 1 - \frac{1}{4} - \frac{1}{12} = \frac{2}{3}.$$

This integral is the volume of the region sketched in Figure 5.4.8.   ◆

Exercises for §5.4

In Exercises 1–4, perform the indicated integration over the given box.

1. $\displaystyle\iiint_B x^2 dx\,dy\,dz, B = [0,1] \times [0,1] \times [0,1]$

2. $\displaystyle\iiint_B e^{-xy} y\,dx\,dy\,dz, B = [0,1] \times [0,1] \times [0,1]$

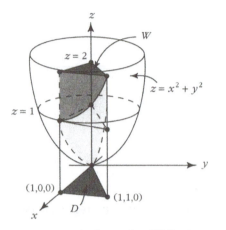

FIGURE 5.4.8. The region W lies between the paraboloid $z = x^2 + y^2$ and the plane $z = 2$ and above the region D.

3. $\displaystyle\iiint_B (2x + 3y + z)\,dxdydz,\ B = [0, 2] \times [-1, 1] \times [0, 1]$

4. $\displaystyle\iiint_B ze^{x+y}\,dxdydz,\ B = [0, 1] \times [0, 1] \times [0, 1]$

In Exercises 5–8, describe the given region as an elementary region.

5. The region between the cone $z = \sqrt{x^2 + y^2}$ and the paraboloid $z = x^2 + y^2$

6. The region cut out of the ball $x^2 + y^2 + z^2 \le 4$ by the elliptic cylinder $2x^2 + z^2 = 1$, *i.e.*, the region inside the cylinder and the ball

7. The region inside the ellipsoid $x^2 + y^2 + z^2 = 1$ and above the plane $z = 0$

8. The region bounded by the planes $x = 0, y = 0, z = 0, x + y = 4$, and $x = z - y - 1$

Find the volumes of the region in Exercises 9–12.

9. The region bounded by $z = x^2 + y^2$ and $z = 10 - x^2 - 2y^2$

10. The solid bounded by $x^2 + 2y^2 = 2, z = 0$, and $x + y + 2z = 2$

11. The solid bounded by $x = y, z = 0, y = 0, x = 1$, and $x + y + z = 0$

12. The region common to the intersecting cylinders $x^2 + y^2 \le a^2$ and $x^2 + z^2 \le a^2$

Evaluate the integrals in Exercises 13–21.

13. $\displaystyle\int_0^1 \int_1^2 \int_2^3 \cos\left[\pi(x+y+z)\right]dxdydz$

14. $\displaystyle\int_0^1 \int_0^x \int_0^y (y+xz)dzdydx$

15. $\displaystyle\iiint_W (x^2+y^2+z^2)dxdydz$; W is the region bounded by $x+y+z=a$ (where $a>0$), $x=0, y=0$, and $z=0$.

16. $\displaystyle\iiint_W z\,dxdydz$; W is the region bounded by the planes $x=0, y=0, z=0, z=1$, and the cylinder $x^2+y^2=1$, with $x\geq 0, y\geq 0$.

17. $\displaystyle\iiint_W x^2\cos z\,dxdydz$; W is the region bounded by $z=0, z=\pi, y=0, y=1, x=0$, and $x+y=1$.

18. $\displaystyle\int_0^2 \int_0^x \int_0^{x+y} dzdydx$

19. $\displaystyle\iiint_W (1-z^2)dxdydz$; W is the pyramid with top vertex at $(0,0,1)$ and base vertices at $(0,0,0), (1,0,0), (0,1,0)$, and $(1,1,0)$.

20. $\displaystyle\iiint_W (x^2+y^2)dxdydz$; W is the same pyramid as in Exercise 19.

21. $\displaystyle\int_0^1 \int_0^{2x} \int_{x^2+y^2}^{x+y} dzdydx.$

22.(a) Sketch the region for the integral

$$\int_0^1 \int_0^x \int_0^y f(x,y,z)dzdydx.$$

(b) Write the integral with the integration order $dxdydz$.

For the regions in Exercises 23 and 24, find the appropriate limits $\phi_1(x), \phi_2(x)$, $\gamma_1(x,y)$, and $\gamma_2(x,y)$, and write the triple integral over the region W as an iterated integral in the form

$$\iiint_W f\,dV = \int_a^b \left\{ \int_{\phi_1(x)}^{\phi_2(x)} \left[\int_{\gamma_1(x,y)}^{\gamma_2(x,y)} f(x,y,z)dz \right] dy \right\} dx.$$

23. $W = \{(x,y,z) \mid \sqrt{x^2+y^2} \leq z \leq 1\}$

24. $W = \{(x, y, z) \mid \frac{1}{2} \le z \le 1 \text{ and } x^2 + y^2 + z^2 \le 1\}$

25. Show that the formula using triple integrals for the volume under the graph of a positive function $f(x, y)$, on an elementary region D in the plane, reduces to the double integral of f over D.

26. Let W be the region bounded by the planes $x = 0, y = 0, z = 0$, $x + y = 1$, and $z = x + y$.

   (a) Find the volume of W.

   (b) Evaluate $\iiint_W x\,dx\,dy\,dz$.

   (c) Evaluate $\iiint_W y\,dx\,dy\,dz$.

27. Let f be continuous and let B_ϵ be the ball of radius ϵ centered at the point (x_0, y_0, z_0). Let vol (B_ϵ) be the volume of B_ϵ. Prove that

$$\lim_{\epsilon \to 0} \frac{1}{\text{vol}(B_\epsilon)} \iiint_{B_\epsilon} f(x, y, z)dV = f(x_0, y_0, z_0).$$

5.5
Change of Variables, Cylindrical and Spherical Coordinates

Integration in one variable is often made easier by a change of variables. For instance, $\int_2^3 \sin(x^2)x\,dx$ is evaluated using the new variable $u = x^2$ instead of x to produce $\int_4^9 \sin u(\frac{1}{2}du)$, which is readily evaluated. In multivariable calculus, changes of variable are also useful. Polar coordinates in the plane (new variables r and θ) and cylindrical and spherical coordinates in space (new variables r, θ, z and ρ, θ, ϕ) are especially helpful.

The main purpose of this section is to explain how multiple integral problems may be transformed into polar, cylindrical, or spherical coordinates. As an application, we will see how to evaluate the **Gaussian integral** $\int_{-\infty}^{\infty} e^{-x^2}dx$, which is important in probability, statistics, and quantum mechanics. Although this is a one-variable integral, its evaluation requires the use of polar coordinates in the plane! At the end of the section, we explain how *general* changes of variable can be applied to multiple integrals.

Polar coordinates (r, θ) in the plane are related to rectangular coordinates (x, y) by the formulas

$$x = r\cos\theta \quad \text{and} \quad y = r\sin\theta,$$

where $r \ge 0$ and $0 \le \theta < 2\pi$. (See Figure 5.5.1.)

To evaluate the double integral $\iint_D f(x, y)dx\,dy$ of an integrable function f over a region D in the xy plane using polar coordinates, one needs to do the following:

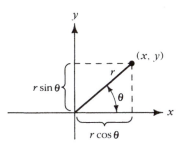

FIGURE 5.5.1. The polar coordinates of (x, y) are (r, θ).

1. Express $f(x, y)$ in terms of r and θ.

2. Replace $dx\,dy$ by $r\,dr\,d\theta$. (The extra factor of r is analogous to the factor dx/du when we replace dx by $(dx/du)du$ in one-variable integrals.)

3. Replace the region of integration D in the xy plane by a new region D^* in the $r\theta$ plane (just as one changes the limits of integration in one variable).

Step 1 is straightforward. For example, the function xy becomes

$$(r\cos\theta)(r\sin\theta) = r^2\cos\theta\sin\theta = \frac{1}{2}r^2\sin 2\theta.$$

The function $\sqrt{1 + x^2 + y^2}$ becomes $\sqrt{1 + r^2}$, not involving θ at all.

We will explain shortly where the "r" comes from in step 2. For step 3, we take D^* to be the region in the $r\theta$ plane *corresponding* to D; that is, the region consisting of those (r, θ) for which $(r\cos\theta, r\sin\theta)$ lies in the given region D.

Double Integrals in Polar Coordinates

If D is a region in the xy plane,

$$\iint_D f(x, y)dx\,dy = \iint_{D^*} f(r\cos\theta, r\sin\theta)r\,dr\,d\theta$$

where D^* is the corresponding region in the $r\theta$ plane.

Example 1 Find and sketch the region D^* in the (r, θ) plane corresponding to each of the following regions in the (x, y) plane. (a) The disk D_a of radius a, defined by $x^2 + y^2 \leq a^2$. (b) The triangle with vertices $(0, 0), (1, 0)$, and $(1, 1)$.

Solution (a) From the geometric meaning of polar coordinates, we find that the disk D_a is described in the $r\theta$ plane by the region D_a^* given by $0 \leq r \leq$

$a, 0 \le \theta \le 2\pi$, which is a rectangle in the $r\theta$ plane. Thus a disc around the origin in the xy plane corresponds to a rectangle in the $r\theta$ plane under the change to polar coordinates. We illustrate D_a and D_a^* in Figure 5.5.2.

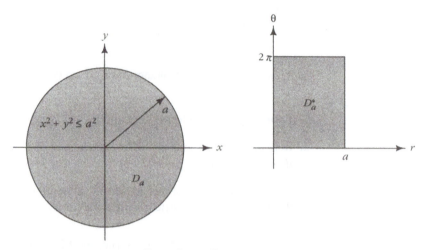

FIGURE 5.5.2. The disk $x^2 + y^2 \le a^2$ in the xy plane corresponds to the rectangle $[0, a] \times [0, 2\pi]$ in the $r\theta$ plane.

(b) From the left-hand figure in Figure 5.5.3, we see that points in D all have polar coordinates with θ between 0 and $\frac{\pi}{4}$. For a given θ, the smallest value of r is 0, whereas the largest is that for which $x = r \cos \theta = 1$, so $r = 1/\cos \theta$. Thus, the region D^* is defined by $0 \le r \le 1/\cos \theta, 0 \le \theta \le \pi/4$. ◆

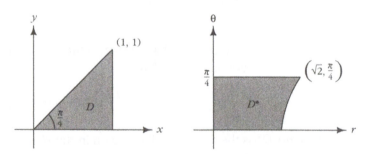

FIGURE 5.5.3. The triangular region D in the xy plane corresponds to the region D^* in the $r\theta$ plane.

The reason that $dxdy$ is replaced by $r\,drd\theta$ is explained in Figure 5.5.4 and the next paragraph.

Why this formula for the double integral in polar coordinates? An intuitive explanation: A double integral $\iint_D f(x, y)dxdy$ may be thought of as a "sum"

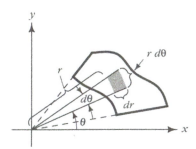

FIGURE 5.5.4. The area of the infinitesimal shaded region is $r\,dr\,d\theta$.

of the values of f over infinitesimal rectangles with area $(dx) \cdot (dy)$. However, we can also describe a region using polar coordinates and can use infinitesimal regions appropriate to those coordinates. The area of such a region is $r\,dr\,d\theta$, as is evident from Figure 5.5.4. If we "sum" over these regions with area $r\,dr\,d\theta$, it is plausible that the same answer is obtained as summing over the rectangular regions with area $dx\,dy$. Thus one arrives at $\iint_D f(x,y)\,dx\,dy = \iint_{D^*} f(r\cos\theta, r\sin\theta)\,r\,dr\,d\theta$. ∎

Example 2 Evaluate $\displaystyle\iint_D \log(x^2 + y^2)\,dx\,dy$, where D is the region in the first quadrant lying between the circles $x^2 + y^2 = 1$ and $x^2 + y^2 = 4$.

Solution The region D^* in the (r, θ) plane that corresponds to D is the rectangle $1 \le r \le 2, 0 \le \theta \le \pi/2$ (Figure 5.5.5). Hence

$$\iint_D \log(x^2 + y^2)\,dx\,dy = \int_0^{\pi/2} \int_1^2 \log(r^2)\,r\,dr\,d\theta$$

$$= \int_0^{\pi/2} \int_1^2 2(\log r) \cdot r\,dr\,d\theta$$

$$= \int_0^{\pi/2} \left(\frac{r^2}{2}(2\log r - 1)\Big|_{r=1}^2\right) d\theta \quad \text{(integration by parts)}$$

$$= \int_0^{\pi/2} \left(4\log 2 - \frac{3}{2}\right) d\theta = \frac{\pi}{2}\left(4\log 2 - \frac{3}{2}\right). \quad \blacklozenge$$

Example 3 Evaluate $\displaystyle\iint_{D_a} e^{-(x^2+y^2)}\,dx\,dy$, where D_a is the disk $x^2 + y^2 \le a^2$.

Solution Using Example 1(a), $r^2 = x^2 + y^2$, and $dx\,dy = r\,dr\,d\theta$,

$$\iint_{D_a} e^{-(x^2+y^2)}\,dx\,dy = \int_0^{2\pi} \int_0^a e^{-r^2} r\,dr\,d\theta = \int_0^{2\pi} \left(-\frac{1}{2}e^{-r^2}\right)\Big|_0^a d\theta$$

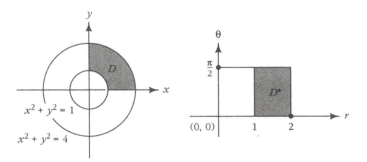

FIGURE 5.5.5. The region D and the corresponding region D^* for Example 2.

$$= -\frac{1}{2}\int_0^{2\pi}(e^{-a^2}-1)d\theta = \pi(1-e^{-a^2}). \quad \blacklozenge$$

If we let $a \to \infty$ in this example, we get

$$\iint_{\mathbb{R}^2} e^{-(x^2+y^2)}\,dx\,dy = \pi.$$

Assuming (and it can be shown!) that we can also evaluate this improper integral as the limit of the integrals over the rectangles $R_a = [-a, a] \times [-a, a]$ as $a \to \infty$, we get

$$\lim_{a\to\infty}\iint_{R_a} e^{-(x^2+y^2)}\,dx\,dy = \pi.$$

By reduction to iterated integrals, we can write this as

$$\lim_{a\to\infty}\left[\int_{-a}^a e^{-x^2}\,dx\int_{-a}^a e^{-y^2}\,dy\right] = \left[\lim_{a\to\infty}\int_{-a}^a e^{-x^2}\,dx\right]^2 = \pi,$$

i.e.,

$$\left[\int_{-\infty}^\infty e^{-x^2}\,dx\right]^2 = \pi,$$

so taking square roots we get (see Figure 5.5.6):

The Gaussian Integral

$$\int_{-\infty}^\infty e^{-x^2}\,dx = \sqrt{\pi}.$$

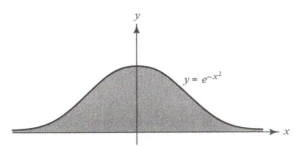

FIGURE 5.5.6. The shaded area is $\pi^{1/2}$.

Example 4   Find $\displaystyle\int_{-\infty}^{\infty} e^{-2x^2}\, dx$.

Solution   Use the change of variables $y = \sqrt{2}x$ to reduce the problem to the Gaussian integral just computed:

$$\int_{-\infty}^{\infty} e^{-2x^2}\, dx \;=\; \lim_{a \to \infty} \int_{-a}^{a} e^{-2x^2}\, dx = \lim_{a \to \infty} \int_{-\sqrt{2}a}^{\sqrt{2}a} e^{-y^2}\, \frac{dy}{\sqrt{2}}$$

$$=\; \frac{1}{\sqrt{2}}\int_{-\infty}^{\infty} e^{-y^2}\, dy = \frac{1}{\sqrt{2}}\sqrt{\pi} = \sqrt{\frac{\pi}{2}}. \;\blacklozenge$$

Next we turn our attention to cylindrical coordinates, which combine polar coordinates (r, θ) with the third Cartesian coordinate z.

Cylindrical Coordinates

The ***cylindrical coordinates*** (r, θ, z) of a point (x, y, z) are defined by

$$x = r\cos\theta, \quad y = r\sin\theta, \quad z = z,$$

where $r \geq 0$ and $0 \leq \theta < 2\pi$ (see Figure 5.5.7).

We use the term "cylindrical coordinates," because if a is some positive constant, then the surface defined by $r = a$ is a cylinder of radius a (see Figure 5.5.8).

Example 5   (a) Find and plot the cylindrical coordinates of $(6, 6, 8)$. (b) If a point has cylindrical coordinates $(8, 2\pi/3, -3)$, what are its Cartesian coordinates? Plot.

Solution   For part (a), we have $r = \sqrt{6^2 + 6^2} = 6\sqrt{2}$ and $\theta = \tan^{-1}(6/6)$ $= \tan^{-1}(1) = \pi/4$. Thus the cylindrical coordinates are $(6\sqrt{2}, \pi/4, 8)$, which is the point P in Figure 5.5.9. For part (b), we have

$$x = r\cos\theta = 8\cos\frac{2\pi}{3} = -\frac{8}{2} = -4$$

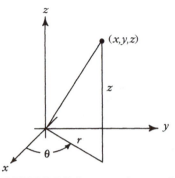

FIGURE 5.5.7. Representing a point (x, y, z) in terms of its cylindrical coordinates r, θ, and z.

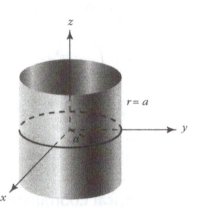

FIGURE 5.5.8. The points whose cylindrical coordinates satisfy $r = a$ form a cylinder.

and

$$y = r \sin \theta = 8 \sin \frac{2\pi}{3} = 8 \frac{\sqrt{3}}{2} = 4\sqrt{3}.$$

Thus the Cartesian coordinates are $(-4, 4\sqrt{3}, -3)$, which is the point Q in the figure. ◆

The infinitesimal "volume element" in cylindrical coordinates has the volume $r\, dr d\theta dz$, as in Figure 5.5.10.

As in the case of polar coordinates, this leads to a formula for multiple integrals, presented in the next box.

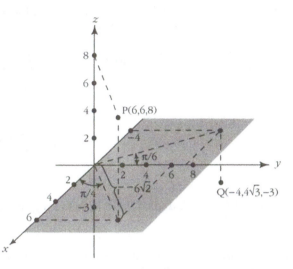

FIGURE 5.5.9. Some examples of the conversion between Cartesian and cylindrical coordinates.

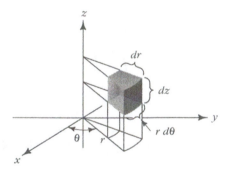

FIGURE 5.5.10. The infinitesimal shaded region has volume $r\,dr\,d\theta\,dz$.

Triple Integrals in Cylindrical Coordinates

If W is a region in space,

$$\iiint_W f(x,y,z)\,dx\,dy\,dz = \iiint_{W^*} f(r\cos\theta, r\sin\theta, z)\,r\,dr\,d\theta\,dz$$

where W^* is the corresponding region in (r,θ,z) space.

Example 6   Evaluate $\displaystyle\iiint_W (z^2 x^2 + z^2 y^2)\,dx\,dy\,dz$, where W is the cylindrical region determined by $x^2 + y^2 \leq 1, -1 \leq z \leq 1$.

Solution   The region W is described in cylindrical coordinates as $0 \le r \le 1, 0 \le \theta \le 2\pi, -1 \le z \le 1$, so

$$\iiint_W (z^2 x^2 + z^2 y^2)\,dx\,dy\,dz = \int_{-1}^1 \int_0^{2\pi} \int_0^1 (z^2 r^2)r\,dr\,d\theta\,dz$$

$$= \int_{-1}^1 \int_0^{2\pi} z^2 \frac{r^4}{4}\Big|_{r=0}^1 \,d\theta\,dz$$

$$= \int_{-1}^1 \frac{2\pi}{4} z^2 \,dz = \frac{\pi}{3}. \quad \blacklozenge$$

Example 7   Evaluate $\iiint_W xyz\,dx\,dy\,dz$, where W is the region defined by $x^2 + y^2 \le 1, 0 \le z \le x^2 + y^2$.

Solution   In the $r\theta$ plane, the region that corresponds to the unit disk $x^2 + y^2 \le 1$ is the rectangle $0 \le r \le 1, 0 \le \theta \le 2\pi$, so in cylindrical coordinates, the region W^* is described by $0 \le z \le x^2 + y^2 = r^2, 0 \le r \le 1$, $0 \le \theta \le 2\pi$. Therefore,

$$\iiint_W xyz\,dx\,dy\,dz = \int_0^{2\pi} \int_0^1 \int_0^{r^2} (r\cos\theta)(r\sin\theta)zr\,dz\,dr\,d\theta$$

$$= \int_0^{2\pi} \int_0^1 r^3 \cos\theta\sin\theta \cdot \frac{r^4}{2}\,dr\,d\theta$$

$$= \int_0^{2\pi} \frac{\cos\theta\sin\theta}{16}\,d\theta = \int_0^{2\pi} \frac{1}{32}\sin 2\theta\,d\theta$$

$$= -\frac{1}{64}\cos 2\theta\Big|_0^{2\pi} = 0. \quad \blacklozenge$$

Could you have predicted the answer 0 in Example 7 using symmetry?

We now turn to the spherical coordinate system, which is useful for problems possessing *spherical* symmetry. Given a point (x, y, z) in space, let $\rho = \sqrt{x^2 + y^2 + z^2}$ and represent x and y by polar coordinates in the xy plane: $x = r\cos\theta, y = r\sin\theta$. The coordinate z is given by $z = \rho\cos\phi$, where ϕ is the angle (between 0 and π, inclusive) that the position vector $x\mathbf{i} + y\mathbf{j} + z\mathbf{k}$ makes with the z axis in the plane containing this vector and the z axis (see Figure 5.5.11).

We take as our coordinates the quantities (ρ, θ, ϕ). Since $r = \rho\sin\phi$, we find (x, y, z) in terms of the spherical coordinates (ρ, θ, ϕ) as follows:

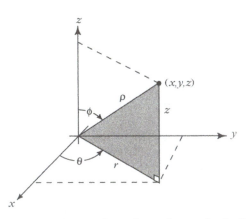

FIGURE 5.5.11. Spherical coordinates (ρ, θ, ϕ); the set of points satisfying $\rho = a$ is a sphere.

Spherical Coordinates

The ***spherical coordinates*** of (x, y, z) are defined as follows:

$$x = \rho \sin \phi \cos \theta, \quad y = \rho \sin \phi \sin \theta, \quad z = \rho \cos \phi$$

where

$$\rho \geq 0, \quad 0 \leq \theta < 2\pi, \quad 0 \leq \phi \leq \pi.$$

The spherical coordinates θ and ϕ are similar to the geographic coordinates of longitude and latitude if we take the Earth's axis to be the z axis. There are differences, though. The geographical longitude is $|\theta|$ and takes values between 0 and π. It is called east or west longitude according to whether θ is positive or negative; the geographical latitude is $|\pi/2 - \phi|$ and is called north or south latitude according to whether $\pi/2 - \phi$ is positive or negative. Of course, the geographic coordinates are normally expressed in degrees rather than radians.

Example 8 (a) Find the spherical coordinates of $(1, -1, 1)$ and plot.

(b) Find the Cartesian coordinates of $(3, \pi/6, \pi/4)$ and plot.

(c) Let a point have Cartesian coordinates $(2, -3, 6)$. Find its spherical coordinates and plot.

(d) Let a point have spherical coordinates $\left(1, \frac{3\pi}{2}, \frac{\pi}{4}\right)$. Find its Cartesian coordinates and plot.

Solution

(a)

$$\rho = \sqrt{x^2 + y^2 + z^2} = \sqrt{1^2 + (-1)^2 + 1^2} = \sqrt{3},$$

$$\theta = \tan^{-1}\left(\frac{y}{x}\right) = \tan^{-1}\left(\frac{-1}{1}\right) = \frac{7\pi}{4},$$

$$\phi = \cos^{-1}\left(\frac{z}{\rho}\right) = \cos^{-1}\left(\frac{1}{\sqrt{3}}\right) \approx 0.955 \approx 54.74°.$$

See Figure 5.5.12.

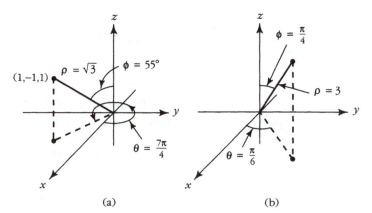

(a)   (b)

FIGURE 5.5.12. Finding (a) the spherical coordinates of the point $(1, -1, 1)$ and (b) the Cartesian coordinates of $(3, \pi/6, \pi/4)$.

(b)

$$x = \rho \sin\phi \cos\theta = 3\sin\left(\frac{\pi}{4}\right)\cos\left(\frac{\pi}{6}\right) = 3\left(\frac{1}{\sqrt{2}}\right)\frac{\sqrt{3}}{2} = \frac{3\sqrt{3}}{2\sqrt{2}},$$

$$y = \rho \sin\phi \sin\theta = 3\sin\left(\frac{\pi}{4}\right)\sin\left(\frac{\pi}{6}\right) = 3\left(\frac{1}{\sqrt{2}}\right)\left(\frac{1}{2}\right) = \frac{3}{2\sqrt{2}},$$

$$z = \rho \cos\phi = 3\cos\left(\frac{\pi}{4}\right) = \frac{3}{\sqrt{2}} = \frac{3\sqrt{2}}{2}.$$

See Figure 5.5.12.

(c)

$$\rho = \sqrt{x^2 + y^2 + z^2} = \sqrt{2^2 + (-3)^2 + 6^2} = \sqrt{49} = 7,$$

$$\theta = \tan^{-1}\left(\frac{y}{x}\right) = \tan^{-1}\left(\frac{-3}{2}\right) \approx 5.3 \approx 304°,$$

$$\phi = \cos^{-1}\left(\frac{z}{\rho}\right) = \cos^{-1}\left(\frac{6}{7}\right) \approx 0.541 \approx 31.0°.$$

See Figure 5.5.13.

(d)

$$x = \rho \sin \phi \cos \theta = 1 \sin \left(\frac{\pi}{4} \right) \cos \left(\frac{3\pi}{2} \right) = \left(\frac{\sqrt{2}}{2} \right) \cdot 0 = 0,$$

$$y = \rho \sin \phi \sin \theta = 1 \sin \left(\frac{\pi}{4} \right) \sin \left(\frac{3\pi}{2} \right) = \left(\frac{\sqrt{2}}{2} \right) (-1) = -\frac{\sqrt{2}}{2},$$

$$z = \rho \cos \phi = 1 \cos \left(\frac{\pi}{4} \right) = \frac{\sqrt{2}}{2}.$$

See Figure 5.5.13.

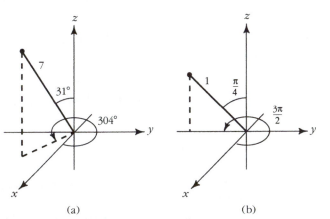

(a)   (b)

FIGURE 5.5.13. Finding (a) the spherical coordinates of $(2, -3, 6)$ and (b) the Cartesian coordinates of $(1, -\pi/2, \pi/4)$.

Example 9 Express (a) the surface $xz = 1$ and (b) the surface $x^2 + y^2 - z^2 = 1$ in spherical coordinates.

Solution Since $x = \rho \sin \phi \cos \theta$ and $z = \rho \cos \phi$, the surface (a) consists of all (ρ, θ, ϕ) such that

$$\rho^2 \sin \phi \cos \theta \cos \phi = 1, \quad \text{i.e.,} \quad \rho^2 \sin 2\phi \cos \theta = 2.$$

For part (b) we can write

$$x^2 + y^2 - z^2 = x^2 + y^2 + z^2 - 2z^2 = \rho^2 - 2\rho^2 \cos^2 \phi,$$

so that the surface is $\rho^2 (1 - 2 \cos^2 \phi) = 1$, or $-\rho^2 \cos (2\phi) = 1$. ◆

The volume in space corresponding to infinitesimal changes $d\rho, d\theta,$ and $d\phi$ is shown in Figure 5.5.14. The sides of this "box" have lengths given by $d\rho, r d\theta$ $(= \rho \sin \phi d\theta)$, and $\rho d\phi$ as shown. Therefore its volume is $\rho^2 \sin \phi d\rho d\theta d\phi$. As

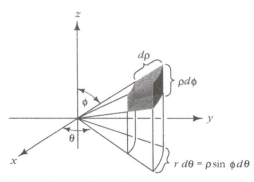

FIGURE 5.5.14. The infinitesimal shaded region has volume $\rho^2 \sin\phi\, d\rho\, d\theta\, d\phi$.

in the case of polar and cylindrical coordinates, we obtain the change of variables formula in the next box.

Triple Integrals in Spherical Coordinates

$$\iiint_W f(x, y, z)\,dx\,dy\,dz$$

$$= \iiint_{W^*} f(\rho\sin\phi\cos\theta, \rho\sin\phi\sin\theta, \rho\cos\phi)\rho^2 \sin\phi\, d\rho\, d\theta\, d\phi,$$

where W^* is the region in ρ, θ, ϕ space corresponding to W.

Example 10   Find the volume of the ball $x^2 + y^2 + z^2 \le R^2$ using spherical coordinates.

Solution   The ball is described in spherical coordinates by $0 \le \theta \le 2\pi, 0 \le \phi \le \pi$, and $0 \le \rho \le R$. Therefore,

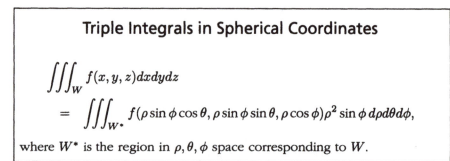

which is the familiar formula for the volume of a ball. Compare the effort involved here with that in Example 4, §**5.4**.   ◆

Example 11   Evaluate $\iiint_W \exp[(x^2 + y^2 + z^2)^{3/2}]dx\,dy\,dz$, where W is the unit ball; *i.e.*, the set of (x, y, z) satisfying $x^2 + y^2 + z^2 \le 1$.

Solution   In spherical coordinates, W is described by the three inequalities $0 \le \rho \le 1, 0 \le \phi \le \pi, 0 \le \theta \le 2\pi$. Hence

$$\iiint_W \exp\left[(x^2 + y^2 + z^2)^{3/2}\right] dx\,dy\,dz$$

$$= \int_0^{2\pi} \int_0^{\pi} \int_0^1 \exp(\rho^3) \cdot \rho^2 \sin\phi \, d\rho\,d\phi\,d\theta$$

$$= \frac{1}{3} \int_0^{2\pi} \int_0^{\pi} \left[\exp(\rho^3)\big|_0^1 \right] \sin\phi \, d\phi\,d\theta$$

$$= \frac{1}{3} \int_0^{2\pi} \int_0^{\pi} (e - 1) \sin\phi \, d\phi\,d\theta = \frac{1}{3}(e - 1) \int_0^{2\pi} \left[(-\cos\phi)\big|_{\phi=0}^{\pi} \right] d\theta$$

$$= \frac{1}{3}(e - 1) \int_0^{2\pi} 2\,d\theta = \frac{2}{3}(e - 1)(2\pi - 0) = \frac{4\pi}{3}(e - 1). \quad \blacklozenge$$

It is important to spend a few moments with each triple integral problem to decide whether cylindrical, spherical, or rectangular coordinates are most useful. Usually the symmetry of the problem provides the needed clue.

<u>Note</u>

When doing integrals in spherical or cylindrical coordinates it does not matter if we use, for example, $0 \le \theta < 2\pi$ or $-\pi < \theta \le \pi$.

Example 12   Find the volumes of the following regions:

(a) The solid bounded by the circular cylinder $r = 2a\cos\theta$, the cone $z = r$, and the plane $z = 0$.

(b) The solid bounded by the cone $z = \sqrt{x^2 + y^2}$ and the paraboloid of revolution $z = x^2 + y^2$.

(c) The region bounded by $y = x^2, y = x + 2, 4z = x^2 + y^2$, and $z = x + 3$.

Solution

(a) Since we can write $r = 2a\cos\theta$ as $r^2 = 2ax$ or $(x - a)^2 + y^2 = a^2$, we see that the base of the solid is a disc in the xy plane centered at $(a, 0)$ with radius a (see Figure 5.5.15). The xz plane is a plane of symmetry, so the total volume is twice the volume of the part lying above the shaded region. In cylindrical coordinates, the total volume is

$$2 \int_0^{\pi/2} \int_0^{2a\cos\theta} \int_0^r r \, dz\,dr\,d\theta$$

$$= 2 \int_0^{\pi/2} \int_0^{2a\cos\theta} (rz\big|_{z=0}^r)\,dr\,d\theta$$

$$= 2 \int_0^{\pi/2} \int_0^{2a\cos\theta} r^2 \, dr\,d\theta = 2 \int_0^{\pi/2} \left(\frac{r^3}{3}\bigg|_{r=0}^{2a\cos\theta} \right) d\theta$$

$$= 2 \int_0^{\pi/2} \frac{8a^3 \cos^3\theta}{3} d\theta = \left(\frac{16a^3}{3}\right) \int_0^{\pi/2} (1 - \sin^2\theta) \cos\theta \, d\theta.$$

Letting $u = \sin\theta$, we get

$$\frac{16a^3}{3} \int_0^1 (1 - u^2) du = \frac{16a^3}{3} \left(u - \frac{u^3}{3}\right)\Big|_0^1 = \frac{32a^3}{9}.$$

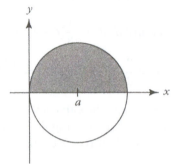

FIGURE 5.5.15. The base of W for Example 12(a).

(b) In cylindrical coordinates, the solid is bounded by $z = r$ and $z = r^2$ (Figure 5.5.16). The solid is obtained by rotating the shaded area around the z axis. Thus, the volume is

$$\int_0^{2\pi} \int_0^1 \int_{r^2}^r r \, dz dr d\theta$$

$$= \int_0^{2\pi} \int_0^1 (rz|_{z=r^2}^r) dr d\theta = \int_0^{2\pi} \int_0^1 (r^2 - r^3) dr d\theta$$

$$= \int_0^{2\pi} \left(\frac{r^3}{3} - \frac{r^4}{4}\right)\Big|_{r=0}^1 d\theta = \frac{1}{12} \int_0^{2\pi} d\theta = \frac{\pi}{6}.$$

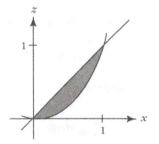

FIGURE 5.5.16. A cross section of W for Example 12(b).

(c) This part does *not* require cylindrical or spherical coordinates; $y = x^2 = x + 2$ has the solutions $x = -1$ and $x = 2$, so the volume is

$$\int_{-1}^{2} \int_{x^2}^{x+2} \int_{(x^2+y^2)/4}^{x+3} dz\,dy\,dx$$

$$= \int_{-1}^{2} \int_{x^2}^{x+2} \left[(x+3) - \frac{x^2+y^2}{4} \right] dy\,dx$$

$$= \int_{-1}^{2} \left\{ \left[\left(x + 3 - \frac{x^2}{4} \right) y - \frac{y^3}{12} \right] \Big|_{y=x^2}^{x+2} \right\} dx$$

$$= \int_{-1}^{2} \left(\frac{16}{3} + 4x - 3x^2 - \frac{4x^3}{3} + \frac{x^4}{4} + \frac{x^6}{12} \right) dx$$

$$= \left(\frac{16x}{3} + 2x^2 - x^3 - \frac{x^4}{3} + \frac{x^5}{20} + \frac{x^7}{84} \right) \Big|_{-1}^{2} = \frac{783}{70}. \quad \blacklozenge$$

Next we state the *general* change of variables formula for two and three variables. Suppose we have an integral $\iint_D f(x, y)\,dx\,dy$ of an integrable function f of two variables and we express x and y in new variables, say u and v; for example, $x = u + v$, $y = u - v$. Suppose that the domain D in the xy plane corresponds to a domain D^* in the uv plane. How can we express $\iint_D f(x, y)\,dx\,dy$ as an integral with respect to u and v? As with polar coordinates, the central question is: what replaces $dx\,dy$? If the transformation is linear, it follows from our work in §1.4, that its determinant is the factor by which it changes areas. For general transformations a similar fact can be deduced using the linear approximation. (We will elaborate on some of these ideas in our discussion of the area of parametrized surfaces in §6.3.) This leads one to introduce the determinant of the derivative matrix.

The Jacobian Determinant

If x and y are functions of u and v, their *Jacobian determinant* is the determinant of the derivative matrix:

$$\frac{\partial(x, y)}{\partial(u, v)} = \begin{vmatrix} \dfrac{\partial x}{\partial u} & \dfrac{\partial x}{\partial v} \\[2mm] \dfrac{\partial y}{\partial u} & \dfrac{\partial y}{\partial v} \end{vmatrix}.$$

Example 13   Compute the Jacobian determinant of $x = u + v, y = u - v$.

Solution

$$\begin{vmatrix} \dfrac{\partial x}{\partial u} & \dfrac{\partial x}{\partial v} \\[3mm] \dfrac{\partial y}{\partial u} & \dfrac{\partial y}{\partial v} \end{vmatrix} = \begin{vmatrix} 1 & 1 \\ 1 & -1 \end{vmatrix} = -2. \quad \blacklozenge$$

The next example replaces u and v by r and θ.

Example 14   Compute the Jacobian determinant of $x = r \cos \theta, y = r \sin \theta$.

Solution

$$\begin{vmatrix} \dfrac{\partial x}{\partial r} & \dfrac{\partial x}{\partial \theta} \\[3mm] \dfrac{\partial y}{\partial r} & \dfrac{\partial y}{\partial \theta} \end{vmatrix} = \begin{vmatrix} \cos \theta & -r \sin \theta \\ \sin \theta & r \cos \theta \end{vmatrix} = r \cos^2 \theta + r \sin^2 \theta = r. \quad \blacklozenge$$

Change of Variables Formula

Let D be a region in the xy plane that corresponds to a region D^* under a transformation in which x and y are functions of u and v. Then

$$\iint_D f(x, y)\, dx\, dy = \iint_{D^*} f(x(u, v), y(u, v)) \left| \frac{\partial(x, y)}{\partial(u, v)} \right| du\, dv.$$

Here the relation between D and D^* is similar to what we saw with polar coordinates (see Figure 5.5.17). Except possibly for points on the boundary, each point of D^* is assumed to correspond to *just one* point of D, and vice versa.

For linear transformations

$$x = au + bv, \qquad y = cu + dv$$

the situation is simplified by the fact that straight lines correspond to straight lines and parallelograms to parallelograms. Note that the Jacobian determinant is $ad - bc$. The technique is illustrated by the following example.

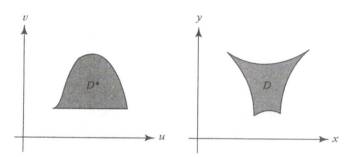

FIGURE 5.5.17. Corresponding regions in the uv and xy planes.

Example 15 Evaluate $\displaystyle\iint_D (x^2 - y^2)\,dx\,dy$ where D is the square with vertices $(0,0)$, $(1,-1)$, $(1,1)$ and $(2,0)$ using the change of variables $x = u + v, y = u - v$ (Figure 5.5.18).

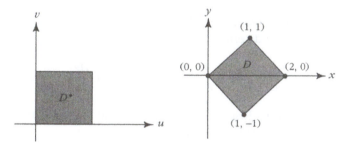

FIGURE 5.5.18. The transformation $x = u + v, y = u - v$ takes D^* to D.

Solution   This transformation is linear and takes $(0,0)$ to $(0,0)$, $(1,0)$ to $(1,1)$, $(1,1)$ to $(2,0)$, and $(0,1)$ to $(1,-1)$, so it takes the parallelogram D^* to D, as in the figure. The integrand is

$$x^2 - y^2 = (u + v)^2 - (u - v)^2 = 4uv$$

and the Jacobian determinant is

$$\frac{\partial(x, y)}{\partial(u, v)} = \begin{vmatrix} 1 & 1 \\ 1 & -1 \end{vmatrix} = -2,$$

so the change of variables formula gives

$$\iint_D (x^2 - y^2)\,dx\,dy = \iint_{D^*} 4uv \cdot |-2|\,du\,dv = 8 \int_0^1 \int_0^1 uv\,du\,dv = 2. \quad \blacklozenge$$

There is also a change of variables formula for triple integrals, which we state below. First we must define the Jacobian of a transformation from $\mathbb{R}^3$ to $\mathbb{R}^3$—it is an extension of the two-variable case.

Jacobians

Let $T : W \subset \mathbb{R}^3 \rightarrow \mathbb{R}^3$ be a C^1 function defined by the equations $x = x(u, v, w), y = y(u, v, w), z = z(u, v, w)$. Then the **Jacobian determinant** of T, which is denoted $\partial(x, y, z)/\partial(u, v, w)$, is the determinant of the derivative matrix:

$$\frac{\partial(x, y, z)}{\partial(u, v, w)} = \begin{vmatrix} \dfrac{\partial x}{\partial u} & \dfrac{\partial x}{\partial v} & \dfrac{\partial x}{\partial w} \\[2mm] \dfrac{\partial y}{\partial u} & \dfrac{\partial y}{\partial v} & \dfrac{\partial y}{\partial w} \\[2mm] \dfrac{\partial z}{\partial u} & \dfrac{\partial z}{\partial v} & \dfrac{\partial z}{\partial w} \end{vmatrix}.$$

The absolute value of this determinant is equal to the volume of the parallelepiped determined by the vectors

$$\frac{\partial x}{\partial u}\mathbf{i} + \frac{\partial y}{\partial u}\mathbf{j} + \frac{\partial z}{\partial u}\mathbf{k}, \quad \frac{\partial x}{\partial v}\mathbf{i} + \frac{\partial y}{\partial v}\mathbf{j} + \frac{\partial z}{\partial v}\mathbf{k}, \quad \text{and} \quad \frac{\partial x}{\partial w}\mathbf{i} + \frac{\partial y}{\partial w}\mathbf{j} + \frac{\partial z}{\partial w}\mathbf{k}$$

(see §**1.4**). Just as in the two-variable case, the Jacobian gives the factor by which T changes infinitesimal volumes.

Change of Variables in Triple Integrals

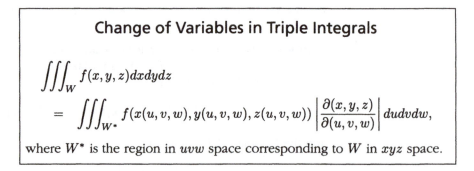

$$\iiint_W f(x, y, z)\,dx\,dy\,dz$$

$$= \iiint_{W^*} f(x(u, v, w), y(u, v, w), z(u, v, w)) \left| \frac{\partial(x, y, z)}{\partial(u, v, w)} \right| du\,dv\,dw,$$

where W^* is the region in uvw space corresponding to W in xyz space.

Our change of variables formulas for cylindrical and spherical coordinates are instances of this general formula.

Exercises for §5.5

In Exercises 1 and 2, find the region D^* in the (r, θ) plane corresponding to the given region described in the xy plane.

1. $-1 \le x \le 1, x^2 + y^2 \le 1, y \ge 0$

2. $0 \le x \le 1, y \le x$

In Exercises 3 and 4, evaluate the given integral by changing to polar coordinates.

3. $\iint_D (x^2 + y^2)^{3/2} dx dy$, where D is the disk $x^2 + y^2 \le 4$.

4. $\iint_D e^{x^2 + y^2} dx dy$, where D is the semicircle in the first two quadrants bounded by $x^2 + y^2 = 9$.

5. Evaluate $\int_{-\infty}^{\infty} e^{-10x^2} dx$.

6. Evaluate $\int_{-\infty}^{\infty} 3e^{-8x^2} dx$.

7. Find the *normalizing constant c*, depending on σ, such that
$$\int_{-\infty}^{\infty} ce^{-x^2/\sigma} dx = 1.$$

8. Evaluate $\int_{-\infty}^{\infty} xe^{-x^4} dx$. (Think first!)

In Exercises 9–12, change the given point from rectangular to cylindrical coordinates.

9. $(2, 1, -2)$

10. $(0, 3, 4)$

11. $(\sqrt{2}, 1, 1)$

12. $(-2\sqrt{3}, -2, 3)$

In Exercises 13–16, convert from cylindrical to Cartesian coordinates.

13. $(1, \pi/2, 0)$

14. $(3, 45°, 82)$

15. $(1, 7\pi/6, 4)$

16. $(2, 0, 1)$

17. Integrate $ze^{x^2+y^2}$ over the cylinder $x^2 + y^2 \leq 4, 2 \leq z \leq 3$.

18. Integrate $x^2 + y^2 + z^2$ over the cylinder given by $x^2 + z^2 \leq 2, -2 \leq y \leq 3$.

In Exercises 19–22, convert from Cartesian to spherical coordinates.

19. $(0, 1, 1)$

20. $(1, 0, 1)$

21. $(-2, 1, -3)$

22. $(1, 2, 3)$

In Exercises 23–26, convert from spherical to Cartesian coordinates.

23. $(3, \pi/3, \pi)$

24. $(2, -\pi/6, \pi/3)$

25. $(3, 2\pi, 0)$

26. $(1, \pi/6, \pi/3)$

27. Evaluate
$$\iiint_W \frac{dx\,dy\,dz}{\sqrt{1 + x^2 + y^2 + z^2}},$$
where W is the ball $x^2 + y^2 + z^2 \leq 1$.

28. Evaluate $\iiint_W (x^2 + y^2 + z^2)^{5/2} dx\,dy\,dz$; W is the ball $x^2 + y^2 + z^2 \leq 1$.

29. Evaluate
$$\iiint_S \frac{dx\,dy\,dz}{(x^2 + y^2 + z^2)^{3/2}},$$
where S is the solid bounded by the spheres $x^2 + y^2 + z^2 = a^2$ and $x^2 + y^2 + z^2 = b^2$, where $a > b > 0$.

30. Integrate $\sqrt{x^2 + y^2 + z^2}\, e^{-(x^2+y^2+z^2)}$ over the region in Exercise 29.

31. Find the volume of the region inside the sphere $x^2 + y^2 + z^2 = 1$ and outside the cylinder $x^2 + y^2 = \frac{1}{4}$.

32. Find the volume of the region enclosed by the cones $z = \sqrt{x^2 + y^2}$ and $z = 1 - 2\sqrt{x^2 + y^2}$.

33. Find the volume inside the ellipsoid $x^2 + y^2 + 4z^2 = 6$.

34. Find the volume of the intersection of the ellipsoid $x^2 + 2(y^2 + z^2) \le 10$ and the cylinder $y^2 + z^2 \le 1$.

35. Find $\int_{-1}^{1} \int_{-\sqrt{1-x^2}}^{\sqrt{1-x^2}} \sin(x^2 + y^2) dy dx$ by converting to polar coordinates.

36. Integrate $(x^2 + y^2)z^2$ over the part of the cylinder $x^2 + y^2 \le 1$ inside the sphere $x^2 + y^2 + z^2 = 4$.

37. Let $T(u, v) = (x(u, v), y(u, v))$ be the mapping defined by $(x, y) = (4u, 2u + 3v)$. Let D^* be the rectangle $[0, 1] \times [1, 2]$. Find the parallelogram D in the (x, y) plane that corresponds to D^*. Evaluate

    (a) $\iint_D xy \, dx dy$ and (b) $\iint_D (x - y) dx dy$

    by making a change of variables to evaluate them as integrals over D^*.

38. Evaluate

$$\iint_D \frac{dx dy}{\sqrt{1 + x + 2y}}$$

where $D = [0, 1] \times [0, 1]$, by setting $(x(u, v), y(u, v)) = (u, v/2)$ and evaluating an integral over the parallelogram D^*, which corresponds to D.

39. By using the change of variables $u = x + y, y = uv$, show that

$$\int_0^1 \int_0^{1-x} e^{y/(x+y)} dy dx = \frac{e - 1}{2}.$$

    Note This is really an improper integral and these have not been discussed. However, proceed assuming that the steps can be justified.

40. Let D be the region bounded by $x + y = 1, x = 0, y = 0$. Show that

$$\iint_D \cos\left(\frac{x - y}{x + y}\right) dx dy = \frac{\sin 1}{2},$$

    using $u = x - y$ and $v = x + y$.

5.6
Applications of Multiple Integrals

The applications treated in this section involve how one *interprets* the integral in a variety of contexts. We begin with the notion of an *average*.

The average value of a function on an interval involves an integral, just as the average, or mean, of a list $a_1, \ldots, a_n$ of n numbers is given in terms of a sum as $(1/n) \sum_{i=1}^{n} a_i$.

If a grain dealer buys wheat from n farmers, buying b_i bushels from the ith farmer at the price of p_i dollars per bushel, the average price is determined not by taking the average of the p_i's, but rather by the "weighted average":

$$p_{\text{average}} = \frac{\sum_{i=1}^{n} p_i b_i}{\sum_{i=1}^{n} b_i} = \frac{\text{total dollars}}{\text{total bushels}}.$$

Note that if the b_i's are all equal, then the average value is the usual average of the p_i's.

Motivated by examples like this, one defines in one variable calculus the average value of a function by

$$[f]_{\text{av}} = \frac{\int_a^b f(x)\,dx}{b-a}.$$

We make a similar definition for a function defined on a plane region D.

Average Value

If f is an integrable function on D, the ratio of the integral to the area of D,

$$[f]_{\text{av}} = \frac{\iint_D f(x,y)\,dx\,dy}{\iint_D dx\,dy},$$

is called the **average value** of f on D.

Example 1   Find the average value of $f(x,y) = x\sin^2(xy)$ on $D = [0,\pi] \times [0,\pi]$.

Solution   First we compute the numerator:

$$
\begin{aligned}
\iint_D f(x,y)\,dx\,dy &= \int_0^\pi \int_0^\pi x\sin^2(xy)\,dx\,dy \\
&= \int_0^\pi \left[\int_0^\pi \frac{1-\cos(2xy)}{2} x\,dy \right] dx \\
&= \int_0^\pi \left[\frac{y}{2} - \frac{\sin(2xy)}{4x} \right] x \Big|_{y=0}^{\pi} dx \\
&= \int_0^\pi \left[\frac{\pi x}{2} - \frac{\sin(2\pi x)}{4} \right] dx \\
&= \left[\frac{\pi x^2}{4} + \frac{\cos(2\pi x)}{8\pi} \right]\Big|_0^\pi \\
&= \frac{\pi^3}{4} + \frac{\cos(2\pi^2)-1}{8\pi}.
\end{aligned}
$$

Thus the average value of f is

$$\frac{\pi^3/4 + [\cos(2\pi^2) - 1]/8\pi}{\pi^2} = \frac{\pi}{4} + \frac{\cos(2\pi^2) - 1}{8\pi^3} \approx 0.7839. \quad \blacklozenge$$

The notion of a weighted average also has a generalization from sums to integrals with an important physical application. If masses $m_1, \ldots, m_n$ are placed at points $x_1, \ldots, x_n$ on the x axis, their **center of mass** is defined to be

$$\bar{x} = \frac{m_1 x_1 + \cdots + m_n x_n}{m_1 + \cdots + m_n}.$$

This definition arises from the following observation: If mass points are distributed on a lever (Figure 5.6.1), the balance point occurs along at the weighted (according to mass) average of the positions of the individual points.

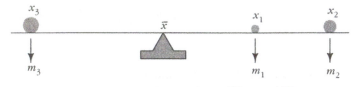

FIGURE 5.6.1. The lever is balanced if $\bar{x} = \sum m_i x_i / \sum m_i$.

For a continuous mass density $\rho(x)$ along the lever, the analog of this formula for the center of mass is the weighted average of the function x:

$$\bar{x} = \frac{\displaystyle\int x\rho(x)dx}{\displaystyle\int \rho(x)dx}.$$

For two-dimensional plates to find the center of mass we must take the weighted average of *each* coordinate (see Figure 5.6.2):

Coordinates of the Center of Mass

$$\bar{x} = \frac{\displaystyle\iint_D x\rho(x,y)dxdy}{\displaystyle\iint_D \rho(x,y)dxdy} \quad \text{and} \quad \bar{y} = \frac{\displaystyle\iint_D y\rho(x,y)dxdy}{\displaystyle\iint_D \rho(x,y)dxdy}.$$

Example 2   Find the center of mass of the rectangle $[0,1] \times [0,1]$ if the mass density is e^{x+y}.

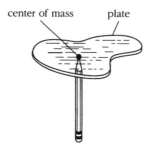

center of mass   plate

FIGURE 5.6.2. The plate balances when supported at its center of mass.

Solution   First we compute the total mass:

$$\iint_D e^{x+y} dx\,dy = \int_0^1 \int_0^1 e^{x+y} dx\,dy$$

$$= \int_0^1 (e^{x+y}|_{x=0}^1)dy = \int_0^1 (e^{1+y} - e^y)dy$$

$$= (e^{1+y} - e^y)|_{y=0}^1 = e^2 - e - (e-1) = e^2 - 2e + 1 = (e-1)^2.$$

The numerator in the formula for $\bar{x}$ is

$$\int_0^1 \int_0^1 xe^{x+y} dx\,dy = \int_0^1 (xe^{x+y} - e^{x+y})|_{x=0}^1 dy$$

$$= \int_0^1 [e^{1+y} - e^{1+y} - (0e^y - e^y)]dy$$

$$= \int_0^1 e^y dy = e^y|_{y=0}^1 = e - 1,$$

so that

$$\bar{x} = \frac{e-1}{(e-1)^2} = \frac{1}{e-1} \approx 0.582.$$

Since the roles of x and y may be interchanged in all these calculations, $\bar{y} = 1/(e-1) \approx 0.582$ as well.   ◆

The formulas above carry over from double to triple integrals. We can compute the volume, mass, and center of mass of a region W with variable density $\rho(x, y, z)$ by the formulas in the following box.

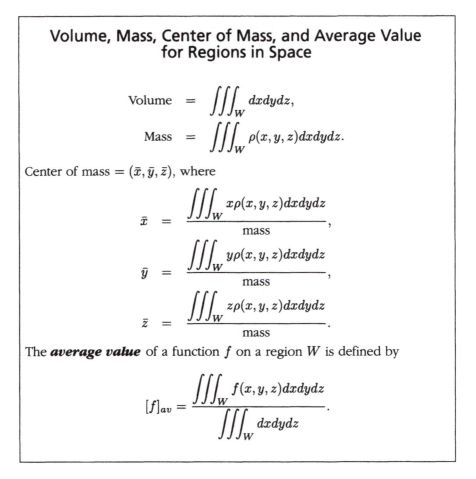

Volume, Mass, Center of Mass, and Average Value for Regions in Space

$$\text{Volume} = \iiint_W dx\,dy\,dz,$$

$$\text{Mass} = \iiint_W \rho(x, y, z)\,dx\,dy\,dz.$$

Center of mass $= (\bar{x}, \bar{y}, \bar{z})$, where

$$\bar{x} = \frac{\iiint_W x\rho(x, y, z)\,dx\,dy\,dz}{\text{mass}},$$

$$\bar{y} = \frac{\iiint_W y\rho(x, y, z)\,dx\,dy\,dz}{\text{mass}},$$

$$\bar{z} = \frac{\iiint_W z\rho(x, y, z)\,dx\,dy\,dz}{\text{mass}}.$$

The **average value** of a function f on a region W is defined by

$$[f]_{av} = \frac{\iiint_W f(x, y, z)\,dx\,dy\,dz}{\iiint_W dx\,dy\,dz}.$$

Example 3 The cube $[1, 2] \times [1, 2] \times [1, 2]$ has mass density $\rho(x, y, z) = (1 + x)e^z y$. Find the mass of the box.

Solution The mass of the box is the integral of the mass density:

$$\int_1^2 \int_1^2 \int_1^2 (1 + x)e^z y\,dx\,dy\,dz$$

$$= \int_1^2 \int_1^2 \left[\left(x + \frac{x^2}{2}\right)e^z y\right]_{x=1}^{x=2} dy\,dz = \int_1^2 \int_1^2 \frac{5}{2}e^z y\,dy\,dz$$

$$= \int_1^2 \frac{15}{4}e^z\,dz = \left[\frac{15}{4}e^z\right]_{z=1}^{z=2} = \frac{15}{4}(e^2 - e). \quad \blacklozenge$$

The following principle often simplifies finding centers of mass:

If a region and its mass density are reflection-symmetric across a plane, then the center of mass lies on that plane. For example, in the formula for $\bar{x}$, if the region and mass density are symmetric in the yz plane, then the integrand is an odd function of x, and so $\bar{x} = 0$. This use of symmetry is illustrated in the next example.

Example 4   Find the center of mass of the hemispherical region W defined by the inequalities $x^2 + y^2 + z^2 \leq 1, z \geq 0$. (Assume that the density is constant.)

Solution   By symmetry, the center of mass lies on the z axis, and so $\bar{x} = \bar{y} = 0$. To find $\bar{z}$, we first compute $I = \iiint_W z\, dxdydz$. The hemisphere is an elementary region defined by $-\sqrt{1 - y^2 - z^2} \leq x \leq \sqrt{1 - y^2 - z^2}$ for (y, z) in the unit disc, so we get the iterated integral

$$I = \int_0^1 \int_{-\sqrt{1-z^2}}^{\sqrt{1-z^2}} \int_{-\sqrt{1-y^2-z^2}}^{\sqrt{1-y^2-z^2}} z\, dxdydz.$$

Since z is a constant in the x and y integrations, we can bring it out from under the integral signs to obtain

$$I = \int_0^1 z \left(\int_{-\sqrt{1-z^2}}^{\sqrt{1-z^2}} \int_{-\sqrt{1-y^2-z^2}}^{\sqrt{1-y^2-z^2}} dxdy \right) dz.$$

Instead of calculating the innter two integrals explicitly, we observe that they equal the double integral $\iint_D dxdy$ over the disk $x^2 + y^2 \leq 1 - z^2$. The area of this disk is $\pi(1 - z^2)$, and so

$$I = \pi \int_0^1 z(1 - z^2)dz = \pi \int_0^1 (z - z^3)dz = \pi \left[\frac{z^2}{2} - \frac{z^4}{4} \right]_0^1 = \frac{\pi}{4}.$$

The volume of the hemisphere is half that of the ball, *i.e.*, $\frac{2}{3}\pi$, and so $\bar{z} = (\pi/4)/(\frac{2}{3}\pi) = \frac{3}{8}$. ◆

Example 5   The temperature at points in the cube $W = [-1, 1] \times [-1, 1] \times [-1, 1]$ is proportional to the square of the distance from the origin.
(a) What is the average temperature? (b) At which points of the cube is the temperature equal to the average temperature?

Solution   (a) Let c be the constant of proportionality, so $T = c(x^2 + y^2 + z^2)$. Since the volume of the cube is 8, the average temperature is $[T]_{av} = \frac{1}{8} \iiint_W T dxdydz$. Thus

$$[T]_{av} = \frac{c}{8} \int_{-1}^1 \int_{-1}^1 \int_{-1}^1 (x^2 + y^2 + z^2)dxdydz.$$

The triple integral is the sum of the integrals of x^2, y^2, and z^2. Since x, y, and z enter symmetrically into the description of the cube, the three integrals will be equal, so that

$$[T]_{av} = \frac{3c}{8} \int_{-1}^{1} \int_{-1}^{1} \int_{-1}^{1} z^2 dx dy dz = \frac{3c}{8} \int_{-1}^{1} z^2 \left(\int_{-1}^{1} \int_{-1}^{1} dx dy \right) dz.$$

The inner integral is equal to the area of the square $[-1, 1] \times [-1, 1]$. The area of that square is 4, and so

$$[T]_{av} = \frac{3c}{8} \int_{-1}^{1} 4z^2 dz = \frac{3c}{2} \left(\frac{z^3}{3} \right) \Big|_{-1}^{1} = c.$$

(b) The temperature is equal to the average temperature when $c(x^2+y^2+z^2) = c$, that is, on the sphere $x^2 + y^2 + z^2 = 1$. ◆

The next concept coming from mechanics is used in studying the dynamics of a rotating rigid body.

Moments of Inertia

If the solid body W has uniform density ρ, the **moment of inertia** about the x axis is defined by

$$I_x = \iiint_W \rho(y^2 + z^2) dx dy dz.$$

Similarly,

$$I_y = \iiint_W \rho(x^2 + z^2) dx dy dz \quad \text{and} \quad I_z = \iiint_W \rho(x^2 + y^2) dx dy dz.$$

The moment of inertia measures a body's response to efforts to spin it; it is analogous to the mass of a body, which measures its response to efforts to translate it. In contrast to translational motion, however, a body's resistance to spin depends on the shape and not just the total mass. (It is harder to spin a large spherical mass than a compact ball of the same total mass.)

Example 6   Compute the moment of inertia I_z for the solid above the xy plane bounded by the paraboloid $z = x^2 + y^2$ and the cylinder $x^2 + y^2 = b^2$, assuming b and the mass density to be constants.

Solution   The paraboloid and cylinder intersect along a circle in the plane $z = b^2$ (see Figure 5.6.3). Using cylindrical coordinates, we find

$$I_z = \int_0^b \int_0^{2\pi} \int_0^{r^2} \rho r^2 \cdot r\,dz\,d\theta\,dr = \rho \int_0^b \int_0^{2\pi} \int_0^{r^2} r^3 dz\,d\theta\,dr = \frac{\pi \rho b^6}{3}. \quad \blacklozenge$$

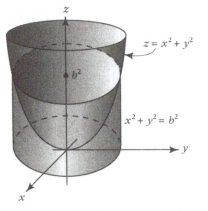

FIGURE 5.6.3. Finding the moment of inertia of the region between the paraboloid and the cylinder.

An interesting physical application of triple integration is the determination of the gravitational fields of solid objects. In Chapter 2 we showed that the gravitational force field $\mathbf{F}(x, y, z)$ of a particle is the negative of the gradient of a function $V(x, y, z)$ called the **gravitational potential**. If there is a point mass M at (x, y, z), then the gravitational potential acting on a mass m at (x_1, y_1, z_1) due to this mass is $GmM[(x - x_1)^2 + (y - y_1)^2 + (z - z_1)^2]^{-1/2}$, where G is the universal gravitational constant.

If our attracting object is an extended domain W with mass density $\rho(x, y, z)$, we may think of it as made of infinitesimal box-shaped regions with masses $dM = \rho(x, y, z)dx\,dy\,dz$ located at points (x, y, z). The total gravitational potential for W is then obtained by "summing" the potentials from the infinitesimal masses, that is, as a triple integral (see Figure 5.6.4):

$$V(x_1, y_1, z_1) = Gm \iiint_W \frac{\rho(x, y, z)dx\,dy\,dz}{\sqrt{(x - x_1)^2 + (y - y_1)^2 + (z - z_1)^2}}.$$

Example 7   Let W be a region of constant density and total mass M. Show that the gravitational potential at the point (x_1, y_1, z_1) is given by

$$V(x_1, y_1, z_1) = \left[\frac{1}{r}\right]_{\text{av}} GMm,$$

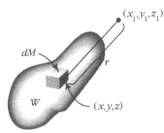

FIGURE 5.6.4. The gravitational potential acting on a mass m at (x_1, y_1, z_1) arising from the mass $dM = \rho(x, y, z)dxdydz$ at (x, y, z) is $[Gm\rho(x, y, z)dxdydz]/r$.

where $[1/r]_{\text{av}}$ is the average over W of

$$f(x, y, z) = \frac{1}{\sqrt{(x - x_1)^2 + (y - y_1)^2 + (z - z_1)^2}}.$$

Solution   According to the formula preceding Example 7,

$$
\begin{aligned}
& V(x_1, y_1, z_1) \\
&= Gm \iiint_W \frac{\rho\, dxdydz}{\sqrt{(x - x_1)^2 + (y - y_1)^2 + (z - z_1)^2}} \\
&= Gm\rho \iiint_W \frac{dxdydz}{\sqrt{(x - x_1)^2 + (y - y_1)^2 + (z - z_1)^2}} \\
&= Gm[\rho\ \text{volume}\ (W)] \frac{\displaystyle\iiint_W \frac{dxdydz}{\sqrt{(x - x_1)^2 + (y - y_1)^2 + (z - z_1)^2}}}{\text{volume}\ (W)} \\
&= GmM \left[\frac{1}{r}\right]_{\text{av}}.
\end{aligned}
$$

In the last step, we used the definition of the average of f.   ◆

Historical Note

The theory of gravitational force fields and gravitational potentials was developed by Isaac Newton (1642−1727). Newton withheld publication of his gravitational theories for quite some time. His finding that a spherical planet has the same gravitational field that it would have if its mass were all concentrated at the planet's center first appeared in his famous *Philosophia Naturalis, Principia Mathematica*, the first edition of which appeared in 1687. One can use multiple integrals and spherical coordinates to solve Newton's problem; remarkably, Newton's published solution used only Euclidean geometry. A very readable and lively account of Newton's work on this and related problems may be found in V.I. Arnold's little book, *Huygens and Barrow, Newton and Hooke*, Birkhäuser, Boston, 1990.

An important problem for Newton was that of finding the gravitational potential for a spherically symmetric planet whose density need not be uniform. One finds that outside such a planet the potential is still $V = GMm/R$, so *the body attracts just as if all of its mass were concentrated at the center.* A second surprising result of Newton is that inside a *hollow* planet, V is constant, so the gravitational force is zero. Both of these results can be proved by a slight elaboration of the techniques above.

Example 8   Find the gravitational potential of a spherical star with a mass $M = 3.02 \times 10^{30}$ kg acting on a unit mass at a distance 2.25×10^{11} m from its center ($G = 6.67 \times 10^{-11}$ N m^2/kg^2).

Solution   Since m = 1, the potential is

$$V = \frac{GM}{R} = \frac{6.67 \times 10^{-11} \times 3.02 \times 10^{30}}{2.25 \times 10^{11}} = 8.95 \times 10^8 \text{ m}^2/\text{s}^2. \quad \blacklozenge$$

Exercises for §5.6

In Exercises 1–4, find the average value of f over D (or W).

1. $f(x, y) = y \sin xy$ and $D = [0, \pi] \times [0, \pi]$

2. $f(x, y) = e^{x+y}$ and D is the triangle with vertices $(0, 0), (0, 1)$, and $(1, 0)$

3. $f(x, y, z) = \sin^2(\pi z) \cos^2(\pi x)$ and $W = [0, 2] \times [0, 4] \times [0, 6]$

4. $f(x, y, z) = e^{-z}$ and W is the ball $x^2 + y^2 + z^2 \leq 1$

5. Find the mass of the box $[0, 1/2] \times [0, 1] \times [0, 2]$ assuming a density $\rho(x, y, z) = x^2 + 3y^2 + z + 1$.

6. A sculptured gold plate D is defined by $0 \leq x \leq 2\pi$ and $0 \leq y \leq \pi$ (centimeters) and has mass density $\rho(x, y) = y^2 \sin^2 4x + 2$ (grams per cm^2). If gold sells for \$7 per gram, how much is the gold in the plate worth?

In Exercises 7–10, find the center of mass of the region described.

7. The region bounded by $x + y + z = 2, x = 0, y = 0$, and $z = 0$, constant density

8. The region between $y = 0, y = x^2$, where $0 \leq x \leq 1/2$, constant density

9. The region between $y = x^2$ and $y = x$ if the density is $x + y$

10. The disk determined by $(x - 1)^2 + y^2 \leq 1$ if the density is x^2

11. A solid with constant density is bounded above by the plane $z = a$ and below by the cone described in spherical coordinates by $\phi = k$, where k is a constant $0 < k < \pi/2$. Set up an integral for its moment of inertia about the z axis. Do not try to evaluate the integral.

12. Find the moment of inertia around the y axis for the ball $x^2 + y^2 + z^2 \leq R^2$ if the mass density is constant, and the total mass is M.

13. Find the gravitational potential due to a spherical planet with mass $M = 3 \times 10^{26}$ kg, on a mass m at a distance of 2×10^8 m from its center.

14. Find the gravitational force exerted on a 70-kg object at the position in Exercise 13.

15. A body W in xyz coordinates is **symmetric** *with respect to a plane* if for every particle one one side of the plane there is a particle of equal mass located at its mirror image through the plane.

    (a) Discuss the planes of symmetry for the body of an automobile.

    (b) Let a plane of symmetry be the xy plane, and denote by W^+ and W^- the portions of W above and below the plane, respectively. By our assumption, the mass density $\rho(x, y, z)$ satisfies $\rho(x, y, -z) = \rho(x, y, z)$. Justify these steps:

$$\bar{z} \iiint_W \rho(x, y, z)\,dxdydz$$

$$= \iiint_W z\rho(x, y, z)\,dxdydz$$

$$= \iiint_{W+} z\rho(x, y, z)\,dxdydz + \iiint_{W-} z\rho(x, y, z)\,dxdydz$$

$$= \iiint_{W+} z\rho(x, y, z)\,dxdydz + \iiint_{W+} -w\rho(u, v, -w)\,dudvdw$$

$$= 0.$$

    (c) Explain why part (b) proves that if a body is symmetrical with respect to a plane, then its center of mass lies in that plane.

    (d) Derive this law of mechanics: *If a body is symmetrical in two planes, then its center of mass lies on their line of intersection.*

16. As is well known, the density of a typical planet is not constant throughout the planet. Assume that the planet MAX (including its atmosphere) has a

radius of 5×10^8 cm and a mass density (in grams per cubic centimeter)

$$\rho(x,y,z) = \begin{cases} \dfrac{3 \times 10^4}{r}, & r \geq 10^4 \text{ cm}, \\ 3, & r \leq 10^4 \text{ cm}, \end{cases}$$

where $r = \sqrt{x^2 + y^2 + z^2}$. Find a formula for the gravitational potential outside MAX.

Review Exercises for Chapter 5

For Exercises 1 and 2, evaluate $\iint_R f(x,y)dA$ for the given function and region.

1. $f(x,y) = x + y, R = [-1,2] \times [1,2]$

2. $f(x,y) = ye^{xy}, R = [-1,1] \times [-1,0]$

In Exercises 3–6, evaluate the double integral for the given region and function.

3. $\iint_R (x^2y^2 + x)dydx; R = [0,2] \times [-1,0]$

4. $\iint_R (x^3 + y^2)dA; R = [0,1] \times [0,1]$

5. $\iint_R \sin(x+y)dxdy; R = [0,1] \times [0,1]$

6. $\iint_R (ax + by + c)dxdy; R = [0,1] \times [0,4]$

In Exercises 7–10, evaluate the given iterated integral.

7. $\int_0^3 \int_0^2 x^3y\,dxdy$

8. $\int_0^2 \int_{-1}^1 (yx)^2dydx$

9. $\int_{-1}^1 \int_0^1 ye^x\,dydx$

10. $\int_0^1 \int_{-1}^1 ye^x\,dydx$

In Exercises 11–14, identify the type of the region of integration (corresponding to the way the integral is written), and evaluate the integral.

11. $\displaystyle\int_0^\pi \int_{\sin x}^{3\sin x} x(1+y)\,dy\,dx$

12. $\displaystyle\int_0^1 \int_0^{x^2} (x^2 + xy - y^2)\,dy\,dx$

13. $\displaystyle\int_2^4 \int_{y^2-1}^{y^3} 3\,dx\,dy$

14. $\displaystyle\int_0^1 \int_0^{3y} e^{x+y}\,dx\,dy$

In Exercises 15 and 16, change the order of integration and compute the integral.

15. $\displaystyle\int_0^2 \int_{y/2}^1 (x + y)^2\,dx\,dy$

16. $\displaystyle\int_0^2 \int_{y/2}^2 e^{x^3} y\,dx\,dy$

Evaluate the integrals in Exercises 17–20.

17. $\displaystyle\int_1^2 \int_2^3 \int_3^4 (x + y + z)\,dx\,dy\,dz$

18. $\displaystyle\iiint_W [e^x + (y + z)^5]\,dx\,dy\,dz$ where W is the cube $[0,1] \times [0,1] \times [0,1]$

19. $\displaystyle\int_0^1 \int_0^x \int_0^y xyz\,dz\,dy\,dx$

20. $\displaystyle\iiint_W (x^2 + y^2 + z^2)\,dx\,dy\,dz$ where W is the solid hemisphere defined by $x^2 + y^2 + z^2 \le 1$, and $z \ge 0$.

In Exercises 21–24, set up the integral that you would use to find the volume of the solid described. Evaluate the integral.

21. The region bounded by the five planes $x = 0, x = 1, y = 0, y = 2, z = 0$, and the paraboloid $z = x^2 + y^2$

22. The ellipsoid $\dfrac{x^2}{a^2} + \dfrac{y^2}{b^2} + \dfrac{z^2}{c^2} \le 1$

23. The region below the plane $z + y = 1$ and inside the cylinder $x^2 + y^2 \le 1, 0 \le z \le 1$

24. The solid bounded by $x^2 + y^2 + z^2 = 1$ and $z^2 \geq x^2 + y^2$

25. Evaluate $\displaystyle\iiint_W \frac{dx\,dy\,dz}{\sec[(x^2 + y^2 + z^2)^{3/2}]}$, where W is the solid bounded by the spheres $x^2 + y^2 + z^2 = a^2$ and $x^2 + y^2 + z^2 = b^2$, where $a > b > 0$.

26. Write the iterated integral $\displaystyle\int_0^1 \int_{1-x}^1 \int_x^1 f(x, y, z)\,dz\,dy\,dx$ as an integral over a region in $\mathbb{R}^3$ and then rewrite it using the five other possible orders of integration.

In Exercises 27 and 28, convert the given point from Cartesian to cylindrical and spherical coordinates and plot.

27. $(-1, 0, 1)$

28. $(-\sqrt{2}, 1, 0)$

In Exercises 29 and 30, convert the given point from cylindrical to Cartesian and spherical coordinates and plot.

29. $(1, \pi/4, 1)$

30. $(3, \pi/6, -4)$

In Exercises 31 and 32, convert the given point from spherical to Cartesian and cylindrical coordinates and plot.

31. $(1, \pi/2, \pi)$

32. $(0, \pi/8, \pi/35)$

In Exercises 33–36, make the indicated change of variables. (Do not evaluate the integral.)

33. $\displaystyle\int_0^1 \int_{-1}^1 \int_{-\sqrt{1-y^2}}^{\sqrt{1-y^2}} (x^2 + y^2)^{1/2}\,dx\,dy\,dz$, cylindrical coordinates

34. $\displaystyle\int_{-1}^1 \int_{-\sqrt{1-y^2}}^{\sqrt{1-y^2}} \int_{-\sqrt{4-x^2-y^2}}^{\sqrt{4-x^2-y^2}} xyz\,dz\,dx\,dy$, cylindrical coordinates

35. $\displaystyle\int_{-\sqrt{2}}^{\sqrt{2}} \int_{-\sqrt{2-y^2}}^{\sqrt{2-y^2}} \int_{-\sqrt{4-x^2-y^2}}^{\sqrt{4-x^2-y^2}} z^2\,dz\,dx\,dy$, spherical coordinates

36. $\int_0^1 \int_0^{\pi/4} \int_0^{2\pi} \rho^3 \sin 2\phi \, d\theta \, d\phi \, d\rho$, rectangular coordinates

Evaluate the integrals in Exercises 37–40.

37. $\iint_D \sec(x^2 + y^2) dx dy$, where D is the region defined by $x^2 + y^2 \leq 1$.

38. $\iint_D (x^3 + y^2 x) dx dy$, where D is the region under the graph of $y = x^2$ from $x = 0$ to $x = 2$.

39. $\int_0^1 \int_0^{\pi/2} \int_0^{\cos\theta} rz \, dr d\theta dz$

40. $\int_0^{\pi} \int_0^{\pi/2} \int_0^{\sin\phi} \rho^2 \sin\phi \, d\rho d\phi d\theta$

41. Find $\int_{-\infty}^{\infty} e^{-5x^2} dx$.

42. Find $\int_{-\infty}^{\infty} (5 + x^2) e^{-2x^2} dx$.

43. Find the average value of $f(x, y) = x^2 + \sin(2y) + 1$ on $D = [-3, 1] \times [0, \pi]$.

44. Find the average value of $f(x, y) = 10 - x^2 - y^2$ on $D = [-2, 2] \times [-1, 1]$.

45. Suppose the density of a solid of radius R is given by $(1 + d^3)^{-1}$ where d is the distance to the center of the sphere. Find the total mass of the sphere.

46. The density of the material of a spherical shell whose inner radius is 1 m and whose outer radius is 2 m is $0.4d^2$ g/cm^3, where d is the distance to the center of the sphere in meters.

   (a) Find the total mass of the shell.
   (b) Would the shell float in pure water? (Assume that the density of water is exactly 1g/cm^3.)

47. Use cylindrical coordinates to find the center of mass of the region defined by
$$y^2 + z^2 \leq 1/4, \ (x - 1)^2 + y^2 + z^2 \leq 1, \ x \leq 1.$$

48. Find the center of mass of the solid hemisphere
$$V = \{(x, y, z) | x^2 + y^2 + z^2 \leq a^2 \quad \text{and} \quad z \geq 0\}$$
if the density is constant.

49. The *flexural rigidity* EI of a uniform beam is the product of its Young's modulus of elasticity E and the moment of inertia I of the cross section of the beam at x with respect to a horizontal line l passing through the center of gravity of this cross section. Here

$$I = \iint_R [d(x, y)]^2 \, dx \, dy,$$

where $d(x, y) =$ the distance from (x, y) to l and $R =$ the cross section of the beam being considered.

(a) Assume the cross section R is the rectangle $-1 \le x \le 1, -1 \le y \le 2$ and that l is the x axis. Find I.

(b) Assume that the cross section R is a circle of radius 4 and that l is the x axis. Find I, using polar coordinates.

50. Evaluate $\displaystyle\iint_B \exp\left(\frac{y - x}{y + x}\right) dx \, dy$ where B is the inside of the triangle with vertices at $(0, 0), (0, 1)$, and $(1, 0)$.

6

Integrals Over Curves and Surfaces

The surface integral N=∫∫(la + mb + nc)dS, where a,b,c are the components of the magnetic induction, represents the quantity of magnetic induction through the shell, or in the language of Faraday, the number of lines of magnetic induction.

James Clerk Maxwell *(A Treatise on Electricity and Magnetism, 1891)*

In Chapter 5, we learned how to integrate *real*-valued functions over regions in two- and three-dimensional space. In this chapter we learn how to integrate *vector*-valued functions over curves and surfaces.

6.1
Line Integrals

Given a vector field $\mathbf{F}$ and a path $\mathbf{c}(t)$ in the plane or space, we define the line integral of $\mathbf{F}$ along this path as follows:

Line Integrals

Let $\mathbf{F}$ be a vector field in the plane or space and $\mathbf{c}$ be a continuously differentiable path defined on the interval $[a, b]$. The **line integral** of $\mathbf{F}$ over $\mathbf{c}$ is defined by

$$\int_{\mathbf{c}} \mathbf{F} \cdot d\mathbf{s} = \int_a^b \mathbf{F}(\mathbf{c}(t)) \cdot \mathbf{c}'(t) dt.$$

The physical and geometrical meaning of the line integral will be discussed shortly.

Example 1   Let $\mathbf{c}(t) = (\sin t, \cos t, t)$, with $0 \leq t \leq 2\pi$, and let $\mathbf{F}(x, y, z) = x\mathbf{i} + y\mathbf{j} + z\mathbf{k}$. Compute $\displaystyle\int_{\mathbf{c}} \mathbf{F} \cdot d\mathbf{s}$.

Solution   Here, $\mathbf{F}(\mathbf{c}(t)) = \mathbf{F}(\sin t,\ \cos t, t) = (\sin t)\mathbf{i} + (\cos t)\mathbf{j} + t\mathbf{k}$, and $\mathbf{c}'(t) = (\cos t)\mathbf{i} - (\sin t)\mathbf{j} + \mathbf{k}$. Therefore,

$$\mathbf{F}(\mathbf{c}(t)) \cdot \mathbf{c}'(t) = \sin t \cos t - \cos t \sin t + t = t,$$

and so

$$\int_{\mathbf{c}} \mathbf{F} \cdot d\mathbf{s} = \int_0^{2\pi} t\, dt = 2\pi^2. \quad \blacklozenge$$

We now do some "formal" manipulations to get another commonly used notation for the line integral. We work in 3-space, but similar calculations apply in the plane (by deleting the z component) and in fact in $\mathbb{R}^n$ for any n.

The expression for the derivative

$$\mathbf{c}'(t) = \frac{dx}{dt}\mathbf{i} + \frac{dy}{dt}\mathbf{j} + \frac{dz}{dt}\mathbf{k}$$

can be "formally" written, by multiplying by "dt," as

$$\mathbf{c}'(t)dt = dx\mathbf{i} + dy\mathbf{j} + dz\mathbf{k}.$$

By "formal" we mean that the manipulations are based on the assumption that the rules of algebra and calculus are valid, even though we are applying them to infinitesimal objects for which this has not been justified. Since we are using the manipulations only to explain why a certain notation is used, and not to prove anything, their formality causes no difficulties. In short, expressions like dx, dy, dz, and dt have not been carefully defined, but you needn't worry about it.

The expression $d\mathbf{s} = \mathbf{c}'(t)dt$ is thought of as an *infinitesimal vector displacement along the curve*. (See §4.2 and Figure 4.2.4 for the reasons for this notation.) Writing the vector field $\mathbf{F}$ as $\mathbf{F} = F_1\mathbf{i} + F_2\mathbf{j} + F_3\mathbf{k}$ and taking the dot product with $d\mathbf{s}$ gives

$$\mathbf{F} \cdot d\mathbf{s} = F_1 dx + F_2 dy + F_3 dz,$$

so we get

$$\int_{\mathbf{c}} \mathbf{F} \cdot d\mathbf{s} = \int_{\mathbf{c}} F_1 dx + F_2 dy + F_3 dz.$$

We call the expression $F_1 dx + F_2 dy + F_3 dz$ a **differential form**.

Line Integrals—Differential Form Notation

The line integral of a vector field $\mathbf{F} = F_1\mathbf{i} + F_2\mathbf{j} + F_3\mathbf{k}$ along a path $\mathbf{c}(t) = x(t)\mathbf{i} + y(t)\mathbf{j} + z(t)\mathbf{k}, a \le t \le b$ is written

$$\int_{\mathbf{c}} \mathbf{F} \cdot d\mathbf{s} = \int_{\mathbf{c}} F_1 dx + F_2 dy + F_3 dz$$

$$= \int_a^b \left(F_1 \frac{dx}{dt} + F_2 \frac{dy}{dt} + F_3 \frac{dz}{dt} \right) dt.$$

Example 2   Evaluate the line integral $\int_{\mathbf{c}} x^2 dx + xy\, dy + dz$, where $\mathbf{c} : [0, 1] \to \mathbb{R}^3$ is given by $\mathbf{c}(t) = (x(t), y(t), z(t)) = (t, t^2, 1)$.

Solution   We compute $dx/dt = 1, dy/dt = 2t, dz/dt = 0$; therefore,

$$\int_{\mathbf{c}} x^2 dx + xy\, dy + dz = \int_0^1 \left([x(t)]^2 \frac{dx}{dt} + [x(t)]y(t)] \frac{dy}{dt} \right) dt$$

$$= \int_0^1 (t^2 + 2t^4)dt = \left[\frac{1}{3}t^3 + \frac{2}{5}t^5 \right]_0^1 = \frac{11}{15}. \quad \blacklozenge$$

Example 3   Evaluate the integral $\int_{\mathbf{c}} \cos z \, dx + e^x dy + e^y dz$, where $\mathbf{c}(t) = (1, t, e^t)$ and $0 \leq t \leq 2$.

Solution   We compute $dx/dt = 0, dy/dt = 1, dz/dt = e^t$, and so

$$\int_{\mathbf{c}} \cos z \, dx + e^x dy + e^y dz = \int_0^2 (0 + e + e^{2t}) dt = \left[et + \frac{1}{2} e^{2t} \right]_0^2 = 2e + \frac{1}{2} e^4 - \frac{1}{2}. \blacklozenge$$

If one happens to recognize a vector field as a gradient, then its line integral can be readily evaluated:

Line Integrals of Gradient Fields

If $f(x, y, z)$ is a given real-valued function and $\mathbf{c}(t)$ is a path joining (x_0, y_0, z_0) to (x_1, y_1, z_1) then

$$\int_{\mathbf{c}} \nabla f \cdot d\mathbf{s} = f(x_1, y_1, z_1) - f(x_0, y_0, z_0).$$

Proof   In one-variable calculus, a function $g(t)$ may be recovered at a point b from its derivative and its value at another point a by the *fundamental theorem of calculus*:

$$g(b) = g(a) + \int_a^b \left[\frac{d}{dt} g(t) \right] dt.$$

Given the path $\mathbf{c}(t)$ in the plane or space with $\mathbf{c}(a) = \mathbf{x}_0$ and $\mathbf{c}(b) = \mathbf{x}_1$, and a given real-valued function $f(\mathbf{x})$, consider the composite function $g(t) = f(\mathbf{c}(t))$. By the preceding formula and the chain rule, we get

$$g(b) = g(a) + \int_a^b \nabla f(\mathbf{c}(t)) \cdot \mathbf{c}'(t) dt$$

i.e.,

$$f(\mathbf{x}_1) = f(\mathbf{x}_0) + \int_a^b \nabla f(\mathbf{c}(t)) \cdot \mathbf{c}'(t) dt. \quad \blacksquare$$

Example 4   Evaluate $\int_{\mathbf{c}} y \, dx + x \, dy$ if $\mathbf{c}$ is the path $(t^9, \sin^9(\pi t/2)), 0 \leq t \leq 1$.

Solution   We recognize the vector field $\mathbf{F} = y\mathbf{i} + x\mathbf{j}$ (by guesswork) as the gradient of $f(x, y) = xy$, since $f_x = y$ and $f_y = x$. The path of integration goes from $(0,0)$ at $t = 0$ to $(1,1)$ at $t = 1$, so the formula in the preceding box gives

$$\int_{\mathbf{c}} \mathbf{F} \cdot d\mathbf{s} = f(1,1) - f(0,0) = 1. \quad \blacklozenge$$

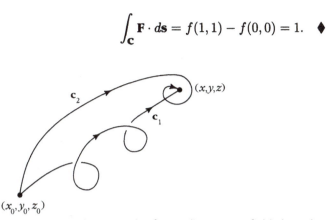

FIGURE 6.1.1. The integrals of a gradient vector field along these two paths are equal.

An important observation about line integrals of gradient fields, which follows from the rule

$$\int_{\mathbf{c}} \nabla f \cdot d\mathbf{s} = f\,(\text{final point}) - f\,(\text{initial point})$$

is that $\int_{\mathbf{c}} \nabla f \cdot d\mathbf{s}$ depends only on the endpoints of $\mathbf{c}$ and *not* on the details of the trajectory of the path $\mathbf{c}$ between its endpoints. In Figure 6.1.1, $\int_{\mathbf{c}_1} \nabla f \cdot d\mathbf{s} = \int_{\mathbf{c}_2} \nabla f \cdot d\mathbf{s}$. The next example shows that for a *nongradient* vector field the line integral *can* depend on the path chosen between two points in space.

Example 5 Find the line integral of the vector field $\mathbf{F} = y\mathbf{i} - x\mathbf{j} + \mathbf{k}$ along each of the following paths joining $(1,0,0)$ to $(1,0,1)$.

$$\mathbf{c}_1(t) = (\cos t)\mathbf{i} + (\sin t)\mathbf{j} + \frac{t}{2\pi}\mathbf{k}, \qquad 0 \le t \le 2\pi$$

$$\mathbf{c}_2(t) = (\cos t^3)\mathbf{i} + (\sin t^3)\mathbf{j} + \frac{t^3}{2\pi}\mathbf{k}, \qquad 0 \le t \le \sqrt[3]{2\pi}$$

$$\mathbf{c}_3(t) = (\cos t)\mathbf{i} - (\sin t)\mathbf{j} + \frac{t}{2\pi}\mathbf{k}, \qquad 0 \le t \le 2\pi.$$

Solution The (helical) paths are sketched in Figure 6.1.2.

Differentiating $\mathbf{c}_1$, we get $\mathbf{c}_1' = -(\sin t)\mathbf{i} + (\cos t)\mathbf{j} + \frac{1}{2\pi}\mathbf{k}$ and so

$$\int_{\mathbf{c}_1} \mathbf{F} \cdot d\mathbf{s} = \int_0^{2\pi} (\sin t\,\mathbf{i} - \cos t\,\mathbf{j} + \mathbf{k}) \cdot \left(-\sin t\,\mathbf{i} + \cos t\,\mathbf{j} + \frac{1}{2\pi}\mathbf{k}\right) dt$$

$$= \int_0^{2\pi} \left(-\sin^2 t - \cos^2 t + \frac{1}{2\pi}\right) dt$$

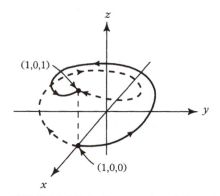

FIGURE 6.1.2. Paths $\mathbf{c}_1$ and $\mathbf{c}_2$ follow the solid line; path $\mathbf{c}_3$ follows the dashed line.

$$= 2\pi\left(-1+\frac{1}{2\pi}\right) = -2\pi + 1 \approx -5.28.$$

Next,

$$\mathbf{c}_2'(t) = -(\sin t^3)(3t^2)\mathbf{i} + (\cos t^3)(3t^2)\mathbf{j} + \frac{3t^2}{2\pi}\mathbf{k}$$

$$= 3t^2\left(-\sin t^3\,\mathbf{i} + \cos t^3\,\mathbf{j} + \frac{1}{2\pi}\mathbf{k}\right),$$

so

$$\int_{\mathbf{c}_2} \mathbf{F}\cdot d\mathbf{s}$$

$$= \int_0^{\sqrt[3]{2\pi}} (\sin t^3\mathbf{i} - \cos t^3\mathbf{j} + \mathbf{k})\cdot(3t^2)\left(-\sin t^3\mathbf{i} + \cos t^3\mathbf{j} + \frac{1}{2\pi}\mathbf{k}\right)dt$$

$$= \int_0^{\sqrt[3]{2\pi}} \left(-\sin^2 t^3 - \cos^2 t^3 + \frac{1}{2\pi}\right) 3t^2\,dt$$

$$= \int_0^{\sqrt[3]{2\pi}} \left(-1+\frac{1}{2\pi}\right) 3t^2\,dt = \left(-1+\frac{1}{2\pi}\right) t^3\Big|_{t=0}^{\sqrt[3]{2\pi}}$$

$$= \left(-1+\frac{1}{2\pi}\right)(2\pi) = -2\pi + 1,$$

which equals the integral along $\mathbf{c}_1$. Finally,

$$\mathbf{c}_3'(t) = -\sin t\mathbf{i} - \cos t\mathbf{j} + \frac{1}{2\pi}\mathbf{k},$$

so

$$\int_{\mathbf{c}_3} \mathbf{F}\cdot d\mathbf{s} = \int_0^{2\pi}(-\sin t\mathbf{i} - \cos t\mathbf{j} + \mathbf{k})\cdot\left(-\sin t\mathbf{i} - \cos t\mathbf{j} + \frac{1}{2\pi}\mathbf{k}\right)dt$$

$$= \int_0^{2\pi} \left(\sin^2 t + \cos^2 t + \frac{1}{2\pi} \right) dt$$

$$= 2\pi \left(1 + \frac{1}{2\pi} \right) = 2\pi + 1 \approx 7.28. \quad \blacklozenge$$

We now make some observations about this example.

Observation 1. The equality

$$\int_{\mathbf{c}_1} \mathbf{F} \cdot d\mathbf{s} = \int_{\mathbf{c}_2} \mathbf{F} \cdot d\mathbf{s}$$

is not a lucky accident. It reflects a *general* fact called *independence of parametrization*.

Observation 2. The fact that

$$\int_{\mathbf{c}_1} \mathbf{F} \cdot d\mathbf{s} \neq \int_{\mathbf{c}_3} \mathbf{F} \cdot d\mathbf{s}$$

shows that $\mathbf{F}$ *is not a gradient vector field.* If $\mathbf{F}$ were equal to ∇f for some function f, we would have had $\int_{\mathbf{c}} \mathbf{F} \cdot d\mathbf{s} = f(1,0,1) - f(1,0,0)$ for any path $\mathbf{c}$ joining $(1,0,0)$ to $(1,0,1)$, and the results for $\mathbf{c}_1$ and $\mathbf{c}_3$ would be the same.

Observation 3. The fact that the integral of $\mathbf{F}$ along $\mathbf{c}_1$ and $\mathbf{c}_2$ is negative, while that along $\mathbf{c}_3$ is positive, has a geometric explanation. For the paths $\mathbf{c}_1$ and $\mathbf{c}_2$, the dot product $\mathbf{F} \cdot d\mathbf{s}$ is everywhere negative; the projection of $\mathbf{F}$ along the tangent vector to $\mathbf{c}_1$ or $\mathbf{c}_2$ is opposite to the direction of motion. In the case of $\mathbf{c}_3$, the projection is *in* the direction of motion.

We now consider the general explanation of independence of parametrization. If $\mathbf{c}(t)$ is a parametrized curve in $\mathbb{R}^3$ defined for t in $[t_1, t_2]$ and if $t = h(u)$ is a (differentiable) function defined on the interval $[u_1, u_2]$ such that $h(u_1) = t_1$, and $h(u_2) = t_2$, then the new parametric curve $\mathbf{b}(u) = \mathbf{c}(h(u))$ defined for u in $[u_1, u_2]$ is called a ***reparametrization*** of $\mathbf{c}$. Using the chain rule, which tells us that $\mathbf{b}'(u) = \mathbf{c}'(h(u))h'(u)$, we calculate the line integral of a vector field $\mathbf{F}$ along $\mathbf{b}$ by changing variables:

$$\int_{\mathbf{b}} \mathbf{F} \cdot d\mathbf{s} = \int_{u_1}^{u_2} \mathbf{F}(\mathbf{b}(u))\mathbf{b}'(u) du$$

$$= \int_{u_1}^{u_2} \mathbf{F}(\mathbf{c}(h(u)))\mathbf{c}'(h(u))h'(u) du.$$

Changing variables from u to t:

$$\int_{\mathbf{b}} \mathbf{F} \cdot d\mathbf{s} = \int_{h(u_1)}^{h(u_2)} \mathbf{F}(\mathbf{c}(t)) \cdot \mathbf{c}'(t) dt$$

$$= \int_{t_1}^{t_2} \mathbf{F}(\mathbf{c}(t)) \cdot \mathbf{c}'(t) dt = \int_{\mathbf{c}} \mathbf{F} \cdot d\mathbf{s}.$$

We conclude that the integral of a vector field (gradient or not) is unchanged when the curve is reparametrized.

Independence of Parametrization

If the path $\mathbf{b}$ is a reparametrization of the path $\mathbf{c}$, then for any vector field $\mathbf{F}$

$$\int_{\mathbf{c}} \mathbf{F} \cdot d\mathbf{s} = \int_{\mathbf{b}} \mathbf{F} \cdot d\mathbf{s}.$$

To exploit this result, we introduce some concepts that are independent of the parametrization. A ***geometric curve*** C is a set of points in the plane or in space that can be traversed by a parametrized path. The *direction* of travel along C is specified but C is *not* specifically parametrized. See Figure 6.1.3.

Although a geometric curve admits many different parametrizations, it may be subdivided into simple pieces such that, on each piece, any two parametrizations are reparametrizations of each other (at least if the parametrizations have nonzero velocity). The result in the preceding box then guarantees that the line integral depends only on the geometric curve and not on the particular parametrization chosen.

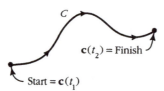

FIGURE 6.1.3. A geometric curve must be parametrized in a specific direction.

Line Integrals Along Geometric Curves

To evaluate $\int_C \mathbf{F} \cdot d\mathbf{s}$ for a geometric curve C, choose a convenient parametrization $\mathbf{c}$ of C traversing C in the given direction and calculate

$$\int_C \mathbf{F} \cdot d\mathbf{s} = \int_{\mathbf{c}} \mathbf{F} \cdot d\mathbf{s} = \int_{t_1}^{t_2} \mathbf{F}(\mathbf{c}(t)) \cdot \mathbf{c}'(t)\,dt,$$

being careful that

1. the parametrization does indeed go in the correct direction, and

2. if the curve is closed, the parametrization traverses it exactly once.

Example 6 Let C be the line segment from $(0,0,0)$ to $(1,0,0)$ and let $\mathbf{c}_1(t) = (t,0,0)$, where $0 \le t \le 1$. Find the line integral of $\mathbf{F}(x,y,z) = \mathbf{i}$ along the curve. Also, find the line integral if C is parametrized by $\mathbf{c}_2(t) = (1-t,0,0)$, where $0 \le t \le 1$.

Solution Here $\mathbf{c}_1'(t) = \mathbf{i}$, $t_1 = 0, t_2 = 1$, and $\mathbf{F}(\mathbf{c}_1(t)) = \mathbf{i}$, so

$$\int_{t_1}^{t_2} \mathbf{F}(\mathbf{c}_1(t)) \cdot \mathbf{c}_1'(t)\,dt = \int_0^1 \mathbf{i} \cdot \mathbf{i}\,dt = \int_0^1 dt = 1.$$

For $\mathbf{c}_2$, we similarly have $t_1 = 0, t_2 = 1, \mathbf{c}_2'(t) = -\mathbf{i}$, and $\mathbf{F}(\mathbf{c}_2(t)) = \mathbf{i}$, so

$$\int_{t_1}^{t_2} \mathbf{F}(\mathbf{c}_2(t)) \cdot \mathbf{c}_2'(t)\,dt = \int_0^1 -\mathbf{i} \cdot \mathbf{i}\,dt = -\int_0^1 dt = -1.$$

Here the curve C is the "same," but the two parametrizations, $\mathbf{c}_1$ and $\mathbf{c}_2$, traverse C in *opposite* directions. See Figure 6.1.4. ◆

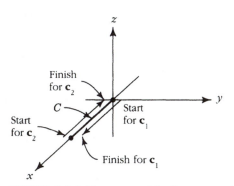

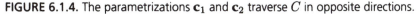

FIGURE 6.1.4. The parametrizations $\mathbf{c}_1$ and $\mathbf{c}_2$ traverse C in opposite directions.

A parametric curve $\mathbf{c}(t)$ defined on $[t_1, t_2]$ is called **closed** if its endpoints coincide, that is, if $\mathbf{c}(t_1) = \mathbf{c}(t_2)$. A geometric curve is **closed** if it has a parametrization that is closed. When C is a closed curve, any point of C may be taken as the initial point for the parametrization, but we must be sure to go around C just once (see Figure 6.1.5).

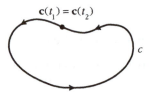

$\mathbf{c}(t_1) = \mathbf{c}(t_2)$

C

FIGURE 6.1.5. Any point may be taken as the starting point for integration around a closed curve.

Example 7   Let C be the circle given by $x^2 + y^2 = 1, z = 0$. Let $\mathbf{c}_1(t) = (\cos t,\ \sin t, 0)$, where $0 \le t \le 2\pi$. Find the line integral of $\mathbf{F}(x, y, z) = -y\mathbf{i} + x\mathbf{j}$ along this curve. Also, find the line integral if C is parametrized by $\mathbf{c}_2(t) = (\cos t,\ \sin t, 0), 0 \le t \le 4\pi$.

Solution   The line integral for $\mathbf{c}_1$ is obtained by using $t_1 = 0, t_2 = 2\pi, \mathbf{c}_1(t) = (\cos t,\ \sin t, 0), \mathbf{F}(x, y, z) = -y\mathbf{i} + x\mathbf{j}, \mathbf{F}(\mathbf{c}_1(t)) = -\sin t\mathbf{i} + \cos t\mathbf{j}, \mathbf{c}_1(t) = \cos t\mathbf{i} + \sin t\mathbf{j}$, and $\mathbf{c}_1'(t) = -\sin t\mathbf{i} + \cos t\mathbf{j}$, as follows:

$$\int_{t_1}^{t_2} \mathbf{F}(\mathbf{c}_1(t)) \cdot \mathbf{c}_1'(t)dt = \int_0^{2\pi} (\sin^2 t + \cos^2 t)dt = 2\pi.$$

We get the same result if we parametrize C by the equation $\mathbf{c}(t) = (\cos(t + \theta),\ \sin(t + \theta), 0), 0 \le t \le 2\pi$, where θ is a constant; this will start and finish at $(\cos \theta,\ \sin \theta, 0)$. If we go backwards along C, using the parametrization $\mathbf{c}(t) = (\cos(-t),\ \sin(-t), 0)$, we will get the *negative* of our earlier answer.

If we use $\mathbf{c}_2(t)$, the only change is that t_2 is changed to 4π, so we get

$$\int_{t_1}^{t_2} \mathbf{F}(\mathbf{c}_2(t)) \cdot \mathbf{c}_2'(t)dt = 4\pi.$$

This is double our first answer, since $\mathbf{c}_2$ *traverses* C *twice*. See Figure 6.1.6.   ◆

Example 8   Let C be the straight-line segment joining $(2, 1, 3)$ to $(-4, 6, 8)$. Find

$$\int_C \mathbf{F} \cdot d\mathbf{s}$$

where $\mathbf{F}(x, y, z) = x\mathbf{i} - y\mathbf{j} + xy\mathbf{k}$.

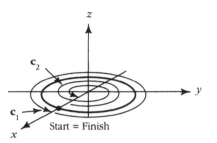

FIGURE 6.1.6. Parametrization $\mathbf{c}_1$ goes around C once, whereas $\mathbf{c}_2$ goes around twice.

Solution   We may choose *any* parametrization of C; one of the simplest is

$$
\begin{aligned}
\mathbf{c}(t) &= (1-t)(2,1,3) + t(-4,6,8) \\
&= (2-6t, 1+5t, 3+5t),\ 0 \le t \le 1.
\end{aligned}
$$

As t varies from 0 to 1, $\mathbf{c}(t)$ moves along C from $(2,1,3)$ to $(-4,6,8)$. By the preceding box, we get

$$
\begin{aligned}
\int_C \mathbf{F} \cdot d\mathbf{s} &= \int_0^1 [(2-6t)\mathbf{i} - (1+5t)\mathbf{j} + (2-6t)(1+5t)\mathbf{k}] \cdot (-6\mathbf{i} + 5\mathbf{j} + 5\mathbf{k})dt \\
&= \int_0^1 (-7 + 31t - 150t^2)dt = -\frac{83}{2}. \quad \blacklozenge
\end{aligned}
$$

The formulas

$$
\int_a^b f(x)dx + \int_b^c f(x)dx = \int_a^c f(x)dx \quad \text{and} \quad \int_a^b f(x)dx = -\int_b^a f(x)dx
$$

for ordinary integrals have their counterparts in line integration. If we choose a point on a curve C, it divides C into two curves, C_1 and C_2 [see Figure 6.1.7(a)]. We write $C = C_1 + C_2$. Then the definition of line integral and the additivity of one-variable integrals gives

$$
\int_{C_1+C_2} \mathbf{F} \cdot d\mathbf{s} = \int_{C_1} \mathbf{F} \cdot d\mathbf{s} + \int_{C_2} \mathbf{F} \cdot d\mathbf{s}.
$$

Let $-C$ be the curve C traversed in the reverse direction, as in Figure 6.1.7 (b). Then we get

$$
\int_{-C} \mathbf{F} \cdot d\mathbf{s} = -\int_C \mathbf{F} \cdot d\mathbf{s}.
$$

The formula for the integral over $C_1 + C_2$ suggests a way to define line integrals over curves with "corners," that is, continuous curves on which the tangent vector is undefined at certain points. If C is such a curve, we write $C =$

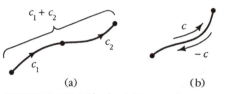

FIGURE 6.1.7. "Algebraic" operations on curves.

$C_1 + C_2 + C_3 + \cdots + C_n$ by dividing it at the corner points, and we define (see Figure 6.1.8)

$$\int_C \mathbf{F} \cdot d\mathbf{s} = \int_{C_1} \mathbf{F} \cdot d\mathbf{s} + \cdots + \int_{C_n} \mathbf{F} \cdot d\mathbf{s}.$$

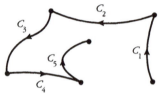

FIGURE 6.1.8. If $C = C_1 + \cdots + C_n$, then $\int_C \mathbf{F} \cdot d\mathbf{s} = \sum_{i=1}^n \int_{C_i} \mathbf{F} \cdot d\mathbf{s}$.

Example 9   Let C be the perimeter of the unit square $[0,1] \times [0,1]$ in the plane, traversed in the counterclockwise direction (see Figure 6.1.9). Evaluate the line integral $\int_C \mathbf{F} \cdot d\mathbf{s}$, where $\mathbf{F}(x,y) = x^2 \mathbf{i} + xy\mathbf{j}$.

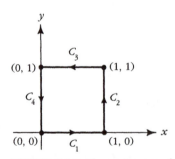

FIGURE 6.1.9. The perimeter of the unit square broken into four pieces.

Solution   We do this problem by integrating along each of the sides C_1, C_2, C_3, C_4 separately and adding the results. The parametrizations are

$$C_1: \quad (t, 0), 0 \leq t \leq 1; \mathbf{c}_1'(t) = \mathbf{i}; \mathbf{F}(\mathbf{c}_1(t)) = t^2\mathbf{i}.$$

$$C_2: \quad (1, t), 0 \leq t \leq 1; \mathbf{c}_2'(t) = \mathbf{j}; \mathbf{F}(\mathbf{c}_2(t)) = \mathbf{i} + t\mathbf{j}.$$

$$C_3: \quad (1 - t, 1), 0 \leq t \leq 1; \mathbf{c}_3'(t) = -\mathbf{i}; \mathbf{F}(\mathbf{c}_3(t)) = (1 - t)^2\mathbf{i} + (1 - t)\mathbf{j}.$$

$$C_4: \quad (0, 1 - t), 0 \leq t \leq 1; \mathbf{c}_4'(t) = -\mathbf{j}; \mathbf{F}(\mathbf{c}_4(t)) = 0.$$

Thus

$$\int_{C_1} \mathbf{F} \cdot d\mathbf{s} = \int_0^1 t^2 dt = \frac{1}{3},$$

$$\int_{C_2} \mathbf{F} \cdot d\mathbf{s} = \int_0^1 t \, dt = \frac{1}{2},$$

$$\int_{C_3} \mathbf{F} \cdot d\mathbf{s} = \int_0^1 (1 - t)^2(-1)dt = -\frac{1}{3},$$

$$\int_{C_4} \mathbf{F} \cdot d\mathbf{s} = \int_0^1 0 \, dt = 0.$$

Adding, we get

$$\int_C \mathbf{F} \cdot d\mathbf{s} = \frac{1}{3} + \frac{1}{2} - \frac{1}{3} + 0 = \frac{1}{2}. \quad \blacklozenge$$

We now rewrite the line integral in a way that sheds additional light on its geometric meaning. If $\mathbf{c}$ is a parametrization of a geometric curve C, and if $\mathbf{c}'(t) \neq \mathbf{0}$ for all t, we may form the **unit tangent vector**

$$\mathbf{T}(t) = \frac{\mathbf{c}'(t)}{\|\mathbf{c}'(t)\|}.$$

The element of arc length is

$$ds = \|\mathbf{c}'(t)\|dt = \sqrt{(dx)^2 + (dy)^2 + (dz)^2},$$

as we saw in §4.2. Thus, $\mathbf{c}'(t)dt = \mathbf{T}(t)ds$, so the line integral of a vector field may be written

$$\int_C \mathbf{F} \cdot d\mathbf{s} = \int_a^b \mathbf{F}(\mathbf{c}(t)) \cdot \mathbf{T}(t)ds.$$

Since $\mathbf{F} \cdot \mathbf{T}$ is the projection of $\mathbf{F}$ in the direction of the tangent vector to the curve, we obtain: *The line integral of $\mathbf{F}$ is the integral, with respect to arc length, of the tangential component $\mathbf{F} \cdot \mathbf{T}$ of $\mathbf{F}$.*

The line integral may be written as a limit of sums, which is important for the additional geometric insight and physical interpretation it provides.

Choose a partition $a = t_0 < t_1, < \cdots < t_k = b$ of the interval on which the curve $\mathbf{c}$ is defined, together with a point u_i in each interval $[t_{i-1}, t_i]$. Instead of looking at $\mathbf{F}(\mathbf{c}(t))$ for all values of t, we take the "sample" values $\mathbf{F}(\mathbf{c}(u_1)), \ldots, \mathbf{F}(\mathbf{c}(u_k))$, multiply them by the displacement vectors $\Delta \mathbf{s}_i = \mathbf{c}(t_i) - \mathbf{c}(t_{i-1})$, and form the ***Riemann sum***

$$\sum_{i=1}^{k} \mathbf{F}(\mathbf{c}(u_i)) \cdot (\mathbf{c}(t_i) - \mathbf{c}(t_{i-1})) = \sum_{i=1}^{k} \mathbf{F}(\mathbf{c}(u_i)) \cdot \Delta \mathbf{s}_i.$$

As the length of all the intervals $[t_{i-1}, t_i]$ goes to zero (as k goes to infinity), the displacement $\Delta \mathbf{s}_i$ is better and better approximated by $\mathbf{c}'(u_i)\Delta t_i$, and the Riemann sum approaches the line integral

$$\int_a^b \mathbf{F}(\mathbf{c}(t)) \cdot \mathbf{c}'(t)dt = \int_{\mathbf{C}} \mathbf{F} \cdot d\mathbf{s}.$$

(A proof of this fact is beyond the scope of this book.)

The physical notion of *work* provides an important application of the line integral. If $\mathbf{F}$ is a constant force acting on a particle, the work done by $\mathbf{F}$ if the particle undergoes a displacement $\mathbf{d}$ is $\mathbf{F} \cdot \mathbf{d}$. If $\mathbf{F}$ is not constant but the displacement $\Delta \mathbf{s}$ is small, the work is approximately $\mathbf{F} \cdot \Delta \mathbf{s}$. The work done during successive displacements $\Delta \mathbf{s}_1$ and $\Delta \mathbf{s}_2$ is the sum $\mathbf{F} \cdot \Delta \mathbf{s}_1 + \mathbf{F} \cdot \Delta \mathbf{s}_2$. The above discussion of Riemann sums then leads us to make the following definition:

Work

The ***work done*** by a force field $\mathbf{F}$ on a particle traversing a path C is the line integral

$$\text{Work} = \int_C \mathbf{F} \cdot d\mathbf{s}.$$

Example 10   Suppose $\mathbf{F}$ is the force field $\mathbf{F}(x, y, z) = x^3 \mathbf{i} + y\mathbf{j} + z\mathbf{k}$. Find the work done by $\mathbf{F}$ along the circle of radius a in the yz plane, as in Figure 6.1.10.

Solution   Parametrize the circle of radius a in the yz plane by letting $\mathbf{c}(\theta)$ have components

$$x = 0, \quad y = a\cos\theta, \quad z = a\sin\theta, \quad 0 \le \theta \le 2\pi.$$

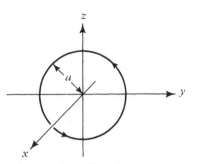

FIGURE 6.1.10. Find the work done by $\mathbf{F}$ along this circle in the yz plane.

Since $\mathbf{F}(\mathbf{c}(\theta)) \cdot \mathbf{c}'(\theta) = 0$, the force $\mathbf{F}$ is normal to the circle at every point on the circle, so $\mathbf{F}$ will not do any work on a particle moving along the circle. The work done by $\mathbf{F}$ must therefore be 0. We can verify this by a direct computation:

$$\begin{aligned} W &= \int_{\mathbf{c}} \mathbf{F} \cdot d\mathbf{s} = \int_{\mathbf{c}} x^3 \, dx + y \, dy + z \, dz \\ &= \int_0^{2\pi} (0 - a^2 \cos\theta \sin\theta + a^2 \cos\theta \sin\theta) d\theta = 0. \quad \blacklozenge \end{aligned}$$

Example 11   Consider the gravitational force field defined [for $(x, y, z) \neq (0, 0, 0)$] by

$$\mathbf{F}(x, y, z) = \frac{-GMm}{(x^2 + y^2 + z^2)^{3/2}}(x\mathbf{i} + y\mathbf{j} + z\mathbf{k}).$$

Show that the work done by the gravitational force as a particle moves from (x_1, y_1, z_1) to (x_2, y_2, z_2) depends only on the radii $r_1 = \sqrt{x_1^2 + y_1^2 + z_1^2}$ and $r_2 = \sqrt{x_2^2 + y_2^2 + z_2^2}$.

Solution   We first observe, or remember from Chapter 2, that $\mathbf{F} = \nabla f$ where $f = GMm/r$ and $r = \sqrt{x^2 + y^2 + z^2}$. (The **gravitational potential** is $V = -GMm/r$.) Thus, by the formula for the line integral of a gradient,

$$\begin{aligned} W &= \int_{\mathbf{c}} \mathbf{F} \cdot d\mathbf{s} = \int_{\mathbf{c}} \nabla f \cdot d\mathbf{s} = f(x_2, y_2, z_2) - f(x_1, y_1, z_1) \\ &= GMm \left(\frac{1}{r_2} - \frac{1}{r_1} \right) \end{aligned}$$

which depends only on r_1 and r_2. $\blacklozenge$

The line integral finds application to many other physical problems. One of these occurs in the formulation of Ampere's law, which relates electric currents to their magnetic effects. Suppose that $\mathbf{H}$ denotes a magnetic field in $\mathbb{R}^3$, and let C be a closed oriented curve in $\mathbb{R}^3$. In appropriate physical units, ***Ampere's law*** states that

$$\int_C \mathbf{H} \cdot d\mathbf{s} = I,$$

where I is the net current that passes through any surface bounded by C (see Figure 6.1.11).

Another physical theory which uses line integrals is *fluid mechanics*. If $\mathbf{V}$ is the velocity field of a fluid, the line integral $\int_C \mathbf{V} \cdot d\mathbf{s}$ around a *closed* curve C is called the ***circulation*** of $\mathbf{V}$ around C. We shall return to this context in §**7.2** when we discuss the physical meaning of the curl.

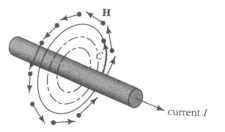

FIGURE 6.1.11. The magnetic field $\mathbf{H}$ surrounding a wire carrying a current I satisfies Ampere's law: $\int_C \mathbf{H} \cdot d\mathbf{s} = I$.

The discussion so far has focused on integrals of *vector* functions along paths. However, we can also integrate *scalar* functions. Since $ds = \|\mathbf{c}'(t)\| dt$ as we saw above, we are led to the following:

Integrals of Scalar Functions Along Paths

Let $f(x, y, z)$ be a (continuous) scalar function and $\mathbf{c} : [a, b] \to \mathbb{R}^3$ a (C^1) path. The ***path integral*** of f along $\mathbf{c}$ is

$$\int_{\mathbf{c}} f\, ds = \int_a^b f(x(t), y(t), z(t)) \|\mathbf{c}'(t)\| dt$$

where $\mathbf{c}(t) = (x(t), y(t), z(t))$.

Sometimes this integral is written

$$\int_{\mathbf{c}} f(x, y, z)\, ds \quad \text{or} \quad \int_a^b f(\mathbf{c}(t)) \|\mathbf{c}'(t)\| dt.$$

If $\mathbf{c}(t)$ is only piecewise C^1 or $f(\mathbf{c}(t))$ is piecewise continuous, we define $\int_{\mathbf{c}} f\,ds$ by breaking $[a, b]$ into pieces over which $f(\mathbf{c}(t))\|\mathbf{c}'(t)\|$ is continuous, and summing the integrals over the pieces.

When $f = 1$, we get the formula for the arc length of $\mathbf{c}$ (see §4.2).

Example 12 Let $\mathbf{c}$ be the helix $\mathbf{c} : [0, 2\pi] \to \mathbb{R}^3, \mathbf{c}(t) = (\cos t, \sin t, t)$ and let $f(x, y, z) = x^2 + y^2 + z^2$. Evaluate the integral $\int_{\mathbf{c}} f(x, y, z)\,ds$.

Solution

$$\|\mathbf{c}'(t)\| = \sqrt{\left[\frac{d(\cos t)}{dt}\right]^2 + \left[\frac{d(\sin t)}{dt}\right]^2 + \left[\frac{dt}{dt}\right]^2}$$

$$= \sqrt{\sin^2 t + \cos^2 t + 1} = \sqrt{2}.$$

We substitute for x, y, and z to obtain

$$f(x, y, z) = x^2 + y^2 + z^2 = \cos^2 t + \sin^2 t + t^2 = 1 + t^2$$

along $\mathbf{c}$. This yields

$$\int_{\mathbf{c}} f(x, y, z)\,ds = \int_0^{2\pi} (1 + t^2)\sqrt{2}\,dt = \sqrt{2}\left[t + \frac{t^3}{3}\right]_0^{2\pi} = \frac{2\sqrt{2}\pi}{3}(3 + 4\pi^2). \; \blacklozenge$$

Exercises for §6.1

In Exercises 1–4, evaluate the integral of the given vector field $\mathbf{F}$ along the given path.

1. $\mathbf{c}(t) = (\sin t, \cos t, t), 0 \le t \le 2\pi, \mathbf{F}(x, y, z) = x\mathbf{i} + y\mathbf{j} + z\mathbf{k}$

2. $\mathbf{c}(t) = (t, t, t), 0 \le t \le 1, \mathbf{F}(x, y, z) = x\mathbf{i} - y\mathbf{j} + z\mathbf{k}$

3. $\mathbf{c}(t) = (\cos t, \sin t, 0), 0 \le t \le \pi/2, \mathbf{F}(x, y, z) = x\mathbf{i} - y\mathbf{j} + z\mathbf{k}$

4. $\mathbf{c}(t) = (\cos t, \sin t, 0), 0 \le t \le \pi/2, \mathbf{F}(x, y, z) = x\mathbf{i} - y\mathbf{j} + 2\mathbf{k}$

5. Let C be parametrized by $x = \cos^3 \theta, y = \sin^3 \theta, z = \theta, 0 \le \theta \le 7\pi/2$. Evaluate the line integral $\int_C \sin z\,dx + \cos z\,dy - (xy)^{1/3}dz$.

6. Evaluate the line integral $\int_C x^2 dx + xy\,dy + dz$, where C is parametrized by $\mathbf{c}(t) = (t, t^2, 1), 0 \le t \le 1$.

7. Evaluate $\int_C \mathbf{F}(\mathbf{r}) \cdot d\mathbf{r}$, where $\mathbf{F}(x, y, z) = \sin z\mathbf{i} + \cos \sqrt{y}\mathbf{j} + x^3\mathbf{k}$ and C is the line segment from $(1, 0, 0)$ to $(0, 0, 3)$.

8. Evaluate $\int_C e^{x+y-z}(\mathbf{i} + \mathbf{j} - \mathbf{k}) \cdot d\mathbf{r}$, where C is the path parametrized by $(\log t, t, t)$ for $2 \leq t \leq 4$.

Let $\mathbf{F}(x, y, z) = x^2\mathbf{i} - xy\mathbf{j} + \mathbf{k}$. Evaluate the line integral of $\mathbf{F}$ along each of the curves in Exercises 9–12.

9. The straight line joining $(0, 0, 0)$ to $(1, 1, 1)$

10. The circle of radius 1, with center at the origin and lying in the yz plane, traversed counterclockwise as viewed from the positive x axis

11. The parabola $z = x^2$, $y = 0$, between $(-1, 0, 1)$ and $(1, 0, 1)$

12. The straight line between $(-1, 0, 1)$ and $(1, 0, 1)$

13. Calculate the work done by the force field $\mathbf{F}(x, y, z) = x\mathbf{i} + y\mathbf{j}$ when a particle is moved along the path $(3t^2, t, 1), 0 \leq t \leq 1$.

14. Find the work done by the force field in Exercise 13 when a particle is moved along the straight-line segment from $(0, 0, 1)$ to $(3, 1, 1)$.

15. Find the work which is done by the force field $\mathbf{F}(x, y) = (x^2 + y^2)(\mathbf{i} + \mathbf{j})$ around the loop $(x, y) = (\cos t, \sin t), 0 \leq t \leq 2\pi$.

16. Find the work done by the force field in Exercise 15 around the loop $(x, y) = (1 + \cos t, 1 + \sin t), 0 \leq t \leq 2\pi$.

17. Evaluate $\int_C 2xyz\,dx + x^2z\,dy + x^2y\,dz$, where C is an oriented curve connecting $(1, 1, 1)$ to $(1, 2, 4)$.

18. Suppose $\nabla f(x, y, z) = 2xyze^{x^2}\mathbf{i} + ze^{x^2}\mathbf{j} + ye^{x^2}\mathbf{k}$. If $f(0, 0, 0) = 5$, find $f(1, 1, 2)$.

19. The image of the curve $\mathbf{c}(t) = (\cos^3 t, \sin^3 t), 0 \leq t \leq 2\pi$, in the plane is shown in Figure 6.1.12. Evaluate the integral of the vector field $\mathbf{F}(x, y) = x\mathbf{i} + y\mathbf{j}$ around the curve.

20. Let $\mathbf{F} = (z^3 + 2xy)\mathbf{i} + x^2\mathbf{j} + 3xz^2\mathbf{k}$. Show that the integral of $\mathbf{F}$ around the square with vertices $(\pm 1, \pm 1, 0)$ is zero.

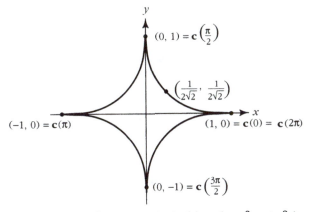

FIGURE 6.1.12. The hypocycloid $\mathbf{c}(t) = (\cos^3 t, \sin^3 t)$.

21. Let L be a very long wire, a planar section of which (with the plane perpendicular to the wire) is shown in Figure 6.1.13. Suppose that this plane is the xy plane. Experiments show that the magnetic field $\mathbf{H}$ produced by a current in L is tangent to every circle in the xy plane whose center is on the axis of L, and that the magnitude of $\mathbf{H}$ is constant on every such circle C. Thus $\mathbf{H} = H\mathbf{T}$, where $\mathbf{T}$ is a unit tangent vector to C and H is some scalar. Using this information and Ampere's law, show that $H = I/2\pi r$, where r is the radius of the circle C and I is the current flowing in the wire.

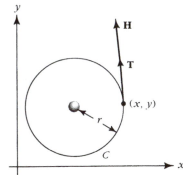

FIGURE 6.1.13. A planar section of a long wire and a curve C about the wire.

22. A cyclist rides up a mountain along the path shown in Figure 6.1.14. He makes one complete revolution around the mountain in reaching the top. The path followed is $\mathbf{c}(t) = (\sqrt{(2\pi - t)} \cos t, \sqrt{(2\pi - t)} \sin t, t)$ for $0 \le t \le 2\pi$. During the trip, he exerts a constant net force of 50 units parallel to his direction of travel. What is the work done by the cyclist in traveling from A to B? You may express your answer as an integral.

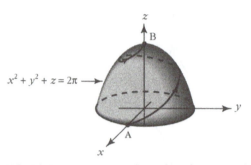

FIGURE 6.1.14. How much work is done in cycling up this mountain?

In Exercises 23–26, evaluate the path integrals $\int_{\mathbf{c}} f(x, y, z)\,ds$, where

23. $f(x, y, z) = x + y + z$ and $\mathbf{c}(t) = (\sin t,\ \cos t, t), 0 \le t \le 2\pi$

24. $f(x, y, z) = \cos z, \mathbf{c}(t) = (\sin t,\ \cos t, t), 0 \le t \le 2\pi$

25. $f(x, y, z) = x \cos z, \mathbf{c}(t) = t\mathbf{i} + t^2\mathbf{j}, 0 \le t \le 1$

26. $f(x, y, z) = e^{\sqrt{z}}$, and $\mathbf{c}(t) = (1, 2, t^2), 0 \le t \le 1$

27.(a) Show that the path integral of $f(x, y)$ along a path given in polar co-ordinates by $r = r(\theta)$, $\theta_1 \le \theta \le \theta_2$, is

$$\int_{\theta_1}^{\theta_2} f(r \cos \theta, r \sin \theta)\sqrt{r^2 + \left(\frac{dr}{d\theta}\right)^2}\,d\theta.$$

(b) Compute the arc length of $r = 1 + \cos \theta, 0 \le \theta \le 2\pi$.

28. Suppose that $\mathbf{c}$ has length l, and $\|\mathbf{F}\| \le M$. Prove that

$$\left|\int_{\mathbf{c}} \mathbf{F} \cdot d\mathbf{s}\right| \le Ml.$$

6.2
Parametrized
Surfaces

In §**6.1** we studied line integrals. To prepare for the analogous concept of surface integrals in §**6.4**, we develop in this section the geometry of surfaces. Recall from Chapter 2 that surfaces may be viewed either as graphs of functions or as level surfaces. Not every surface is a graph. For example, if the surface S is the set of points (x, y, z) where $x - z + z^3 = 0$, it is a sheet that (relative to the xy plane) doubles back on itself (see Figure 6.2.1); it is *not* the graph of a real-valued function $z = f(x, y)$.

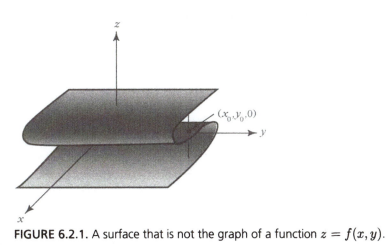

FIGURE 6.2.1. A surface that is not the graph of a function $z = f(x, y)$.

Another example is the torus, or surface of a doughnut, which is depicted in Figure 6.2.2. A torus should certainly qualify as a surface, yet, by the same reasoning as above, a torus cannot be the graph of a differentiable function of two variables. In our study of arc length, we found it useful to describe a curve in $\mathbb{R}^n$, not as a graph or as a level curve, but as the image of a *parametrization* $\mathbf{c} : \mathbb{R} \to \mathbb{R}^n$. For surfaces, we replace the single parameter t in $\mathbb{R}$ by two parameters.

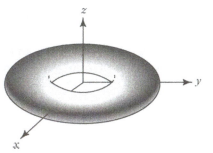

FIGURE 6.2.2. The torus is not the graph of a function $z = f(x, y)$.

Parametrized Surface

A ***parametrized surface*** is a vector-valued function $\Phi : D \subset \mathbb{R}^2 \to \mathbb{R}^3$, where D is some domain in $\mathbb{R}^2$. The ***geometric surface*** S corresponding to the function Φ is its image: $S = \Phi(D)$. We write

$$\Phi(u, v) = (x(u, v), y(u, v), z(u, v)).$$

We can think of Φ as twisting or bending the region D in the plane to yield the surface S (see Figure 6.2.3). Thus each point (u, v) in D becomes a label for a point $(x(u, v), y(u, v), z(u, v))$ on S.

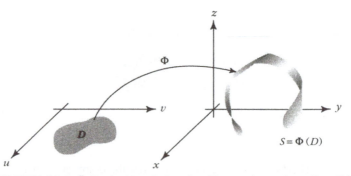

FIGURE 6.2.3. Φ "twists" and "bends" D onto the surface $S = \Phi(D)$.

With surfaces, just as with curves, we distinguish a map (a parametrization) from its image (a geometric object).

Example 1   Show that the graph of a real-valued function $f(x, y)$ is the image of a parametrized surface.

Solution   If $z = f(x, y)$ is the given surface and f is defined on a domain $D \subset \mathbb{R}^2$, we let $\Phi : D \subset \mathbb{R}^2 \to \mathbb{R}^3$ be the vector-valued function defined by

$$\Phi(u, v) = (u, v, f(u, v)).$$

In other words, we let $x(u, v) = u, y(u, v) = v$ and $z(u, v) = f(u, v)$. Thus, each point $\Phi(u, v)$ is on the graph of f so we have realized the graph as the image of a parametrized surface. See Figure 6.2.4.   ◆

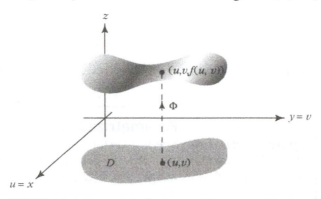

FIGURE 6.2.4. A graph is the image of a parametrized surface; it has domain D, and $\Phi(D)$ is the graph of f.

Fixing u at u_0, we consider the parametrized curve $\mathbf{c}(v) = \Phi(u_0, v)$, whose image is a curve on the surface, as in Figure 6.2.5. The tangent vector to this curve is the derivative with respect to v:

$$\Phi_v = \frac{\partial \Phi}{\partial v} = \frac{\partial x}{\partial v}(u_0, v_0)\mathbf{i} + \frac{\partial y}{\partial v}(u_0, v_0)\mathbf{j} + \frac{\partial z}{\partial v}(u_0, v_0)\mathbf{k}.$$

Similarly, if we fix v and consider the curve $\mathbf{b}(u) = \Phi(u, v_0)$, we obtain the tangent vector to this curve at $\Phi(u_0, v_0)$, given by

$$\Phi_u = \frac{\partial \Phi}{\partial u} = \frac{\partial x}{\partial u}(u_0, v_0)\mathbf{i} + \frac{\partial y}{\partial u}(u_0, v_0)\mathbf{j} + \frac{\partial z}{\partial u}(u_0, v_0)\mathbf{k}.$$

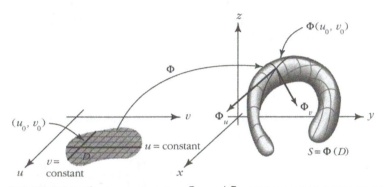

FIGURE 6.2.5. The tangent vectors Φ_u and Φ_v are tangent to curves on a surface S, and hence tangent to S.

Since Φ_u and Φ_v are tangent to the surface, $\Phi_u \times \Phi_v$ is normal to it, which enables us to find the tangent plane.

Tangent Plane

Let Φ be a parametrized surface. To compute the tangent plane at the point

$$(x_0, y_0, z_0) = \Phi(u_0, v_0),$$

form the vector

$$\mathbf{n} = \Phi_u \times \Phi_v = \frac{\partial \Phi}{\partial u} \times \frac{\partial \Phi}{\partial v}$$

evaluated at (u_0, v_0). Assume $\mathbf{n} \neq \mathbf{0}$. The equation of the tangent plane is

$$\mathbf{n} \cdot (x - x_0, y - y_0, z - z_0) = 0$$

i.e.,

$$A(x - x_0) + B(y - y_0) + C(z - z_0) = 0,$$

where $\mathbf{n} = A\mathbf{i} + B\mathbf{j} + C\mathbf{k}$.

Example 2   Suppose that S is the graph of $f(x, y)$. Calculate the tangent plane at the point $(x_0, y_0, f(x_0, y_0))$.

Solution   As in Example 1, we set $x = u$, $y = v$, and $\Phi(u, v) = (u, v, f(u, v))$. Thus,

$$\Phi_u = \mathbf{i} + \frac{\partial f}{\partial u}(u_0, v_0)\mathbf{k} \quad \text{and} \quad \Phi_v = \mathbf{j} + \frac{\partial f}{\partial v}(u_0, v_0)\mathbf{k},$$

and therefore, for $(u_0, v_0) \in \mathbb{R}^2$,

$$\mathbf{n} = \Phi_u \times \Phi_v = -\frac{\partial f}{\partial u}(u_0, v_0)\mathbf{i} - \frac{\partial f}{\partial v}(u_0, v_0)\mathbf{j} + \mathbf{k}.$$

Thus, the tangent plane at $(x_0, y_0, z_0) = (u_0, v_0, f(u_0, v_0))$ is given by the preceding box as

$$(x - x_0, y - y_0, z - z_0) \cdot \left(-\frac{\partial f}{\partial u}, -\frac{\partial f}{\partial v}, 1\right) = 0,$$

where the partial derivatives are evaluated at (u_0, v_0). Remembering that $x = u$ and $y = v$, we can write this as

$$z - z_0 = \left(\frac{\partial f}{\partial x}\right)(x - x_0) + \left(\frac{\partial f}{\partial y}\right)(y - y_0),$$

where $\partial f/\partial x$ and $\partial f/\partial y$ are evaluated at (x_0, y_0).   ◆

This example also shows the definition of the tangent plane for parametrized surfaces agrees with the one for graphs which we obtained in Chapter 2.

Example 3   Let $\Phi : \mathbb{R}^2 \to \mathbb{R}^3$ be given by

$$x = u \cos v, \quad y = u \sin v, \quad z = u, \quad \text{where} \quad u \geq 0.$$

Find the tangent plane at $\Phi(1, 0)$.

Solution   These equations describe the surface $z = \sqrt{x^2 + y^2}$ (square the equations for x, y, and z to check this) and note that $z \geq 0$, which is shown in Figure 6.2.6. This surface is a cone with a vertex at $(0, 0, 0)$. (All points of the form $(0, v)$ are mapped by Φ to the vertex.)

We compute the partial derivatives:

$$\begin{aligned} \Phi_u &= (\cos v)\mathbf{i} + (\sin v)\mathbf{j} + \mathbf{k}, \\ \Phi_v &= -u(\sin v)\mathbf{i} + u(\cos v)\mathbf{j} \end{aligned}$$

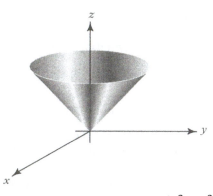

FIGURE 6.2.6. The surface $z = (x^2 + y^2)^{1/2}$ is a cone.

and thus the tangent plane at $\Phi(u, v)$ is the set of vectors through $\Phi(u, v)$ perpendicular to

$$\mathbf{n} = \Phi_u \times \Phi_v = (-u \cos v, -u \sin v, u),$$

if this vector is nonzero. Since $\Phi_u \times \Phi_v$ is equal to $\mathbf{0}$ when $u = 0$, there is no tangent plane at $\Phi(0, v) = (0, 0, 0)$. However, we can find an equation of the tangent plane at all the points where $\Phi_u \times \Phi_v \neq \mathbf{0}$. At the point $\Phi(1, 0) = (1, 0, 1)$,

$$\mathbf{n} = \Phi_u \times \Phi_v = (-1, 0, 1) = -\mathbf{i} + \mathbf{k}.$$

Since we have the vector $\mathbf{n}$ normal to the surface and a point $(1, 0, 1)$ on the surface, we use the preceding box to obtain an equation of the tangent plane:

$$-(x - 1) + (z - 1) = 0;$$

that is,

$$z = x. \quad \blacklozenge$$

Note

Example 3 illustrates out a caution that is needed for parametrized surfaces. For there to be a "good" tangent plane at a point, the vector $\Phi_u \times \Phi_v$ should be nonzero at that point. (The justification of this statement rests on the implicit function theorem; see §**2.6**.) A parametrization with this property at every point is said to be ***regular***.

The next example will use hyperbolic functions, so you may wish to review them before proceeding.

Example 4   Find a parametrization for the hyperboloid of one sheet:

$$x^2 + y^2 - z^2 = 1,$$

and find the tangent plane at the point $(1, 3, 3)$.

Solution   Since x and y appear in the combination $x^2 + y^2$, the surface is invariant under rotation about the z axis, and so we introduce polar coordinates

$$x = r\cos\theta, \qquad y = r\sin\theta.$$

Then $x^2 + y^2 - z^2 = 1$ becomes $r^2 - z^2 = 1$, which we parametrize by hyperbolic functions $r = \cosh u$, $z = \sinh u$. Thus a parametrization is $\Phi(u, \theta) = (x, y, z)$, where

$$x = \cosh u \cos\theta,$$
$$y = \cosh u \sin\theta,$$
$$z = \sinh u,$$

where $0 \le \theta \le 2\pi$, $-\infty < u < \infty$. To find the tangent plane, we compute

$$\Phi_u = (\sinh u \cos\theta, \sinh u \sin\theta, \cosh u),$$
$$\Phi_\theta = (-\cosh u \sin\theta, \cosh u \cos\theta, 0).$$

The point $(x, y, z) = (1, 3, 3)$ corresponds to the parameter values given by $u = \cosh^{-1} r = \cosh^{-1}(\sqrt{x^2 + y^2}) = \cosh^{-1}\sqrt{10}$ and $\theta = \tan^{-1}(y/x) = \tan^{-1}(3)$. At this point, $\sinh u = \sqrt{\cosh^2 u - 1} = \sqrt{9} = 3$, $\sin\theta = 3/\sqrt{10}$, and $\cos\theta = 1/\sqrt{10}$, so

$$\Phi_u = \left(\frac{3}{\sqrt{10}}, \frac{9}{\sqrt{10}}, \sqrt{10}\right), \qquad \Phi_\theta = (-3, 1, 0),$$

and

$$\mathbf{n} = \Phi_u \times \Phi_\theta = -\sqrt{10}\mathbf{i} - 3\sqrt{10}\mathbf{j} + \frac{30}{\sqrt{10}}\mathbf{k},$$

which is proportional to $\mathbf{i} + 3\mathbf{j} - 3\mathbf{k}$, so the equation of the tangent plane through $(1, 3, 3)$ is

$$(x - 1) + 3(y - 3) - 3(z - 3) = 0, \ i.e., \ x + 3y - 3z - 1 = 0.$$

Notice that the hypersurface is a level surface of the function $h(x, y, z) = x^2 + y^2 - z^2$, so that a normal to the tangent plane is given by the gradient $\nabla h = 2x\mathbf{i} + 2y\mathbf{j} - 2z\mathbf{k}$ which at $(1, 3, 3)$ equals $2\mathbf{i} + 6\mathbf{j} - 6\mathbf{k}$, again proportional to $\mathbf{i}+3\mathbf{j}-3\mathbf{k}$. Thus we get the same equation for the tangent plane using *either* a parametrization or the formula for the tangent plane to a level surface. ◆

Example 5   Write the equation of the plane through (x_0, y_0, z_0) and containing the vectors $\mathbf{v}$ and $\mathbf{w}$ as a parametrized surface.

Solution   From Example 12. §1.1, we can choose $\Phi : \mathbb{R}^2 \to \mathbb{R}^3$ given by $\Phi(s,t) = (x_0, y_0, z_0) + s\mathbf{v} + t\mathbf{w}$. ◆

Exercises for §6.2

In Exercises 1–4, find an equation for the plane tangent to the given parametrized surface at the specified point.

1. $x = 2u$, $y = u^2 + v$, $z = v^2$, at $(0, 1, 1)$

2. $x = u^2 - v^2$, $y = u + v$, $z = u^2 + 4v$, at $\left(-\frac{1}{4}, \frac{1}{2}, 2\right)$

3. $x = u^2$, $y = u \sin e^v$, $z = \frac{1}{3}u \cos e^v$, at $(13, -2, 1)$

4. $x = u^2$, $y = v^2$, $z = u^2 + v^2$, $u = 1, v = 1$

5. Find an expression for a unit vector normal to the surface

$$x = \cos v \sin u, \quad y = \sin v \sin u, \quad z = \cos u$$

for u in $[0, \pi]$ and v in $[0, 2\pi]$. Identify this surface.

6. Repeat Exercise 5 for the surface

$$x = 3 \cos \theta \sin \phi, \quad y = 2 \sin \theta \sin \phi, \quad z = \cos \phi$$

for θ in $[0, 2\pi]$ and ϕ in $[0, \pi]$.

7. Given a sphere of radius 2 centered at the origin, find the equation for the plane that is tangent to it at the point $(1, 1, \sqrt{2})$ by considering the sphere as:

(a) a surface parametrized by

$$\Phi(\theta, \phi) = (2 \cos \theta \sin \phi, 2 \sin \theta \sin \phi, 2 \cos \phi);$$

(b) a level surface of $f(x, y, z) = x^2 + y^2 + z^2$;

(c) the graph of $g(x, y) = \sqrt{4 - x^2 - y^2}$.

8.(a) Find a parametrization for the hyperboloid $x^2 + y^2 - z^2 = 25$.

(b) Find an expression for a unit normal to this surface.

(c) Find an equation for the plane tangent to the surface at $(x_0, y_0, 0)$, where $x_0^2 + y_0^2 = 25$.

(d) Show that the lines $(x_0, y_0, 0) + t(-y_0, x_0, 5)$ and $(x_0, y_0, 0) + t(y_0, -x_0, 5)$ lie in the surface *and* in the tangent plane found in part (c).

9. Write a surface $x = h(y, z)$ as a parametrized surface and find a formula for its tangent plane at a typical point.

10. Find a parametrization of a torus (Figure 6.2.2) and a formula for its tangent plane.

11. A parametrized surface is described by a differentiable function $\Phi : \mathbb{R}^2 \to \mathbb{R}^3$. According to Chapter 2, the derivative should give a linear approximation that supplies a representation of the tangent plane. This exercise demonstrates that this is indeed the case.

   (a) Show that the columns of the matrix $\mathbf{D}\Phi(u_0, v_0)$ are Φ_u and Φ_v. [Φ_u and Φ_v are evaluated at (u_0, v_0).]

   (b) Show that tangent plane as defined in this section is the same as the "parametrized plane"

$$(u, v) \mapsto \Phi(u_0, v_0) + \mathbf{D}\Phi(u_0, v_0) \begin{bmatrix} u - u_0 \\ v - v_0 \end{bmatrix}.$$

6.3
Area of a
Surface

In one-variable calculus, we learn that the area of the surface generated by revolving the graph of a function $y = f(x)$ about the x axis is given by

$$A_x = 2\pi \int_a^b |f(x)| \sqrt{1 + [f'(x)]^2} \, dx.$$

If the graph is revolved about the y axis, the corresponding formula is

$$A_y = 2\pi \int_a^b |x| \sqrt{1 + [f'(x)]^2} \, dx.$$

In this section, we study the area of more general surfaces—the formulas above will be special cases of the ones that we derive next.

We begin with surfaces that are graphs. To find the area of the graph $z = f(x, y)$ of a function f over a plane region D, we divide D into "infinitesimal rectangles" which are of the form $[x, x + dx] \times [y, y + dy]$. The image of this infinitesimal rectangle on the graph of f is approximately an "infinitesimal parallelogram" with vertices at

$$
\begin{aligned}
P_1 &= (x, y, f(x, y)), \\
P_2 &= (x + dx, y, f(x + dx, y)) \approx (x + dx, y, f(x, y) + f_x(x, y)dx), \\
P_3 &= (x, y + dy, f(x, y + dy)) \approx (x, y + dy, f(x, y) + f_y(x, y)dy), \\
P_4 &= (x + dx, y + dy, f(x + dx, y + dy)), \\
&\approx (x + dx, y + dy, f(x, y) + f_x(x, y)dx + f_y(x, y)dy),
\end{aligned}
$$

where we have used the linear approximations $f(x + dx, y) \approx f(x, y) + f_x(x, y)dx$, etc. (see Figure 6.3.1).

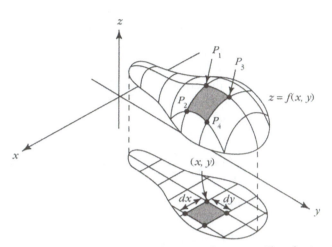

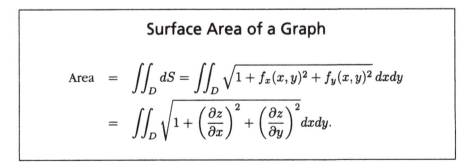

FIGURE 6.3.1. The "image" on the surface $z = f(x, y)$ of an infinitesimal rectangle in the plane is the infinitesimal parallelogram $P_1 P_2 P_4 P_3$.

The area dS of the approximate parallelogram $P_1 P_2 P_4 P_3$ is the length of the cross product of the vectors from P_1 to P_2 and from P_1 to P_3 (see §**1.4**). The vectors in question are $dx\mathbf{i} + f_x(x, y)dx\mathbf{k}$ and $dy\mathbf{j} + f_y(x, y)dy\mathbf{k}$; their cross product is

$$
\begin{vmatrix}
\mathbf{i} & \mathbf{j} & \mathbf{k} \\
dx & 0 & f_x(x, y)dx \\
0 & dy & f_y(x, y)dy
\end{vmatrix} = -f_x(x, y)dxdy\mathbf{i} - f_y(x, y)dxdy\mathbf{j} + dxdy\mathbf{k},
$$

and the length of this vector is $dS = \sqrt{1 + f_x(x, y)^2 + f_y(x, y)^2}dxdy$. To get the area of the surface, we "sum" the areas of the infinitesimal parallelograms by integrating over D. In this way, we have established the following:

Surface Area of a Graph

$$
\begin{aligned}
\text{Area} &= \iint_D dS = \iint_D \sqrt{1 + f_x(x, y)^2 + f_y(x, y)^2}\, dxdy \\
&= \iint_D \sqrt{1 + \left(\frac{\partial z}{\partial x}\right)^2 + \left(\frac{\partial z}{\partial y}\right)^2}\, dxdy.
\end{aligned}
$$

As with arc length, the square root makes the explicit evaluation of surface area integrals difficult or even impossible except for some simple cases.

Example 1   Find the surface area of the part of the sphere $x^2 + y^2 + z^2 = 1$ lying above the elliptical region $x^2 + (y^2/a^2) \leq 1$; (a is a constant satisfying $0 < a \leq 1$).

Solution   The region described by $x^2 + (y^2/a^2) \leq 1$ is type 1 with $\phi_1(x) = -a\sqrt{1-x^2}$ and $\phi_2(x) = a\sqrt{1-x^2}$; $-1 \leq x \leq 1$. The upper hemisphere may be described by the equation $z = f(x,y) = \sqrt{1-x^2-y^2}$. The partial derivatives of f are $\partial z/\partial x = -x/\sqrt{1-x^2-y^2}$ and $\partial z/\partial y = -y/\sqrt{1-x^2-y^2}$, so the area integrand is

$$\sqrt{1 + \frac{x^2}{1-x^2-y^2} + \frac{y^2}{1-x^2-y^2}} = \frac{1}{\sqrt{1-x^2-y^2}}$$

and thus the area is

$$A = \iint_D \frac{dx\,dy}{\sqrt{1-x^2-y^2}} = \int_{-1}^{1} \left(\int_{-a\sqrt{1-x^2}}^{a\sqrt{1-x^2}} \frac{dy}{\sqrt{1-x^2-y^2}} \right) dx.$$

After the substitution $u^2 = 1 - x^2$, this becomes

$$A = \int_{-1}^{1} \left(\int_{-au}^{au} \frac{dy}{\sqrt{u^2-y^2}} \right) dx = \int_{-1}^{1} \left(\sin^{-1} \frac{y}{u} \Big|_{-au}^{au} \right) dx$$

$$= \int_{-1}^{1} \left(\sin^{-1} \frac{y}{\sqrt{1-x^2}} \Big|_{-a\sqrt{1-x^2}}^{a\sqrt{1-x^2}} \right) dx = 2 \int_{-1}^{1} \sin^{-1}a\,dx = 4\sin^{-1}a.$$

(see Figure 6.3.2). As a check on our answer, note that if $a = 1$, we get $4\sin^{-1}1 = 4 \cdot \pi/2 = 2\pi$, the correct formula for the area of a hemisphere of radius 1 (the surface area of a full sphere of radius r is $4\pi r^2$). ◆

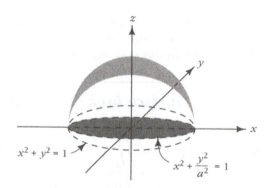

FIGURE 6.3.2. The area of the hemisphere above the ellipse $x^2 + y^2/a^2 \leq 1$ is $4\sin^{-1}a$.

Here is a convenient formula for the surface area element.

Area Element on a Graph

Letting φ be the angle between the unit normal and $\mathbf{k}$ as in Figure 6.3.3, we have

$$dS = \frac{dx\,dy}{\cos\varphi}.$$

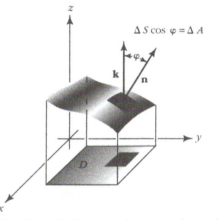

FIGURE 6.3.3. The area of a patch of area ΔS over a patch ΔA is $\Delta S = \Delta A/\cos\varphi$, where φ is the angle the normal $\mathbf{n}$ makes with $\mathbf{k}$.

Proof Let $g(x,y,z) = z - f(x,y)$ so that our surface is $g = 0$. A normal vector is ∇g, *i.e.*,

$$\mathbf{n} = -\frac{\partial f}{\partial x}\mathbf{i} - \frac{\partial f}{\partial y}\mathbf{j} + \mathbf{k}.$$

Thus

$$\cos\varphi = \frac{\mathbf{n}\cdot\mathbf{k}}{\|\mathbf{n}\|} = \frac{1}{\sqrt{(\partial f/\partial x)^2 + (\partial f/\partial y)^2 + 1}}.$$

Substitution of this formula into $dS = (1 + f_x^2 + f_y^2)^{1/2}dx\,dy$ gives the result. ∎

The formula for the area element on a graph can also be understood geometrically: If a small rectangle in the xy plane has area ΔA, then the area of the portion of the surface above it is $\Delta S = \Delta A/\cos\varphi$.

Example 2   Find the area of the triangle with vertices $(1,0,0), (0,1,0)$ and $(0,0,1)$.

Solution   The triangle is contained in a surface, namely, the plane described by the equation $x + y + z = 1$. Since the surface is a plane, the angle φ is constant and a unit normal vector is $\mathbf{n} = (1/\sqrt{3}, 1/\sqrt{3}, 1/\sqrt{3})$. Thus, $\cos\varphi = \mathbf{n} \cdot \mathbf{k} = 1/\sqrt{3}$, and so the area is

$$A = \iint_D dS = \iint_D \frac{dx\,dy}{1/\sqrt{3}} = \sqrt{3}\,(\text{area of } D) = \frac{\sqrt{3}}{2},$$

where D is as shown in Figure 6.3.4. ◆

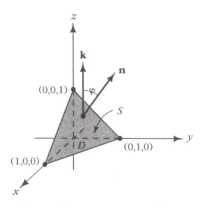

FIGURE 6.3.4. In computing the area of a surface, one finds a formula for the normal $\mathbf{n}$ and computes the angle φ.

We now turn from graphs to parametrized surfaces. Let $\Phi : D \to \mathbb{R}^3$ be such a surface, where D is a region in $\mathbb{R}^2$. We will consider only those parametrizations Φ that do not map any two points of D to the same point (except possibly on the boundary of D); *i.e.,* Φ does not *overlap* itself.

We recall that Φ maps the rectangular grid in the (u, v) plane to a curvilinear grid on the surface $S = \Phi(D)$ (see Figure 6.3.5). If the grid lines are taken "infinitesimally close," then the rectangle with vertices at $(u, v), (u + du, v), (u, v + dv)$, and $(u + du, v + dv)$ is mapped to the "infinitesimal parallelogram" through the point $\Phi(u, v)$ and spanned by the vectors $\Phi_u(u, v)\,du$ and $\Phi_v(u, v)\,dv$, as in Figure 6.3.5.

As in the case of graphs, the area of the surface S is found by "summing" the areas of its infinitesimal pieces to obtain the double integral

$$A(S) = \iint_D \|\Phi_u \times \Phi_v\|\,du\,dv.$$

By writing out the components of the cross product, we see that the length $\|\Phi_u \times \Phi_v\|$ equals

$$\sqrt{\left[\frac{\partial(y, z)}{\partial(u, v)}\right]^2 + \left[\frac{\partial(x, z)}{\partial(u, v)}\right]^2 + \left[\frac{\partial(x, y)}{\partial(u, v)}\right]^2},$$

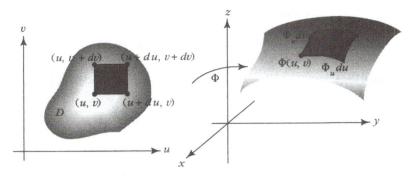

FIGURE 6.3.5. The "infinitesimal parallelogram" through the four points $\Phi(u,v)$, $\Phi(u+du,v)$, $\Phi(u,v+dv)$, and $\Phi(u+du,v+dv)$ has area $\|\Phi_u \times \Phi_v\|dudv$.

the square root of the sum of the squares of the determinants of three derivative matrices.

This result, along with a convenient differential notation for the surface area element is contained in the next box.

Surface Area

For a geometric surface S that is the image of the parametrization Φ, we write $\Phi_u = \partial\Phi/\partial u, \Phi_v = \partial\Phi/\partial v$. Then,

$$d\mathbf{S} = (\Phi_u \times \Phi_v)\,dudv,$$

and

$$dS = \|d\mathbf{S}\| = \|\Phi_u \times \Phi_v\|dudv$$

so that

$$\mathbf{n} = d\mathbf{S}/dS \quad \text{and} \quad d\mathbf{S} = \mathbf{n}\,dS$$

where $\mathbf{n}$ is a *unit* normal vector to the surface. The **surface area** is

$$A = \iint_D dS = \iint_D \sqrt{\left[\frac{\partial(y,z)}{\partial(u,v)}\right]^2 + \left[\frac{\partial(x,z)}{\partial(u,v)}\right]^2 + \left[\frac{\partial(x,y)}{\partial(u,v)}\right]^2}\,dudv$$

where

$$\frac{\partial(x,y)}{\partial(u,v)} = \begin{vmatrix} \dfrac{\partial x}{\partial u} & \dfrac{\partial x}{\partial v} \\[2mm] \dfrac{\partial y}{\partial u} & \dfrac{\partial y}{\partial v} \end{vmatrix}, \text{ and so on.}$$

Since a geometric surface can be parametrized in many different ways, there are many ways to compute its area. A theorem, whose proof is based on the change of variables formula for double integrals, says that all parametrizations (that do not overlap themselves) give the same area.

Example 3   Verify the area formula $2\pi rh$ for a cylinder of height h and radius r.

Solution   We parametrize the cylinder by $x = r\cos\theta, y = r\sin\theta, z = z$ where $0 \le \theta \le 2\pi, 0 \le z \le h$ and r, h are *fixed*. By the preceding box, with $u = \theta$ and $v = z$,

$$A = \int_0^h \int_0^{2\pi} \sqrt{\begin{vmatrix} r\cos\theta & 0 \\ 0 & 1 \end{vmatrix}^2 + \begin{vmatrix} -r\sin\theta & 0 \\ 0 & 1 \end{vmatrix}^2 + \begin{vmatrix} -r\sin\theta & 0 \\ r\cos\theta & 0 \end{vmatrix}^2}\, d\theta\, dz$$

$$= \int_0^h \int_0^{2\pi} \sqrt{r^2 \cos^2\theta + r^2 \sin^2\theta}\, d\theta\, dz$$

$$= \int_0^h \int_0^{2\pi} r\, d\theta\, dz = 2\pi rh. \quad \blacklozenge$$

Example 4   Find the area of the sphere of radius R by using the spherical coordinate parametrization

$$\Phi(\phi, \theta) = (R\sin\phi\cos\theta, R\sin\phi\sin\theta, R\cos\phi).$$

Solution   First, we note that the appropriate region in the (ϕ, θ) plane is the rectangle D defined by $0 \le \phi \le \pi, 0 \le \theta \le 2\pi$. See Figure 6.3.6.

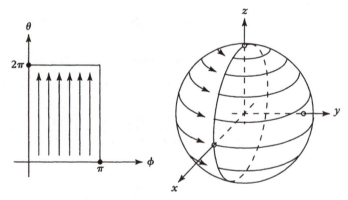

FIGURE 6.3.6. The rectangle D is mapped to the sphere of radius R by the functions $x = R\sin\phi\cos\theta, y = R\sin\phi\sin\theta, z = R\cos\phi$.

Next, we compute $dS = \|\Phi_\phi \times \Phi_\theta\| d\phi d\theta$. First,

$$\Phi_\phi \times \Phi_\theta = (R\cos\phi\cos\theta\mathbf{i} + R\cos\phi\sin\theta\mathbf{j} - R\sin\phi\mathbf{k})$$
$$\times(-R\sin\phi\sin\theta\mathbf{i} + R\sin\phi\cos\theta\mathbf{j})$$

$$\begin{aligned} = \ & R^2 \sin^2\phi \cos\theta \mathbf{i} + R^2 \sin^2\phi \sin\theta \mathbf{j} \\ & + (R^2 \sin\phi \cos\phi \cos^2\theta + R^2 \sin\phi \cos\phi \sin^2\theta)\mathbf{k} \\ = \ & R^2 \sin^2\phi(\cos\theta \mathbf{i} + \sin\theta \mathbf{j}) + (R^2 \sin\phi \cos\phi)\mathbf{k}. \end{aligned}$$

Thus,

$$\| \Phi_\phi \times \Phi_\theta \|^2 = R^4 \sin^4\phi + R^4 \sin^2\phi \cos^2\phi = R^4 \sin^2\phi,$$

so

$$dS = R^2 \sin\phi \, d\phi d\theta,$$

and the area of the sphere is therefore

$$\int_0^{2\pi} \int_0^\pi R^2 \sin\phi \, d\phi d\theta = R^2 \int_0^{2\pi} (\cos 0 - \cos\pi)d\theta = 4\pi R^2. \quad \blacklozenge$$

The expression $dS = R^2 \sin\phi \, d\phi d\theta$ for the "element of area" on the sphere of radius R is useful for many integral calculations in spherical coordinates. Recall that the volume element in spherical coordinates is $dV = \rho^2 \sin\phi \, d\rho d\theta d\phi$, so (with $\rho = R$),

$$dV = dSd\rho$$

which is reasonable in view of Figure 5.5.14.

Another derivation of the formula for dS on a sphere is as follows: Start with

$$dS = \frac{dxdy}{\cos\phi} = \frac{r \, drd\theta}{\cos\phi}.$$

Noticing that $r = R\sin\phi$, we get $dr = R\cos\phi \, d\phi$ and so

$$dS = R\sin\phi \cdot R \, d\phi d\theta = R^2 \sin\phi \, d\phi d\theta.$$

Example 5   Let D be the region determined by $0 \le \theta \le 2\pi, 0 \le r \le 1$ and let the function $\Phi : D \to \mathbb{R}^3$ defined by

$$x = r\cos\theta, \quad y = r\sin\theta, \quad z = r$$

be a parametrization of a cone S (see Figure 6.2.6). Find its surface area.

Solution   We use the preceding box:

$$\frac{\partial(x,y)}{\partial(r,\theta)} = \begin{vmatrix} \cos\theta & -r\sin\theta \\ \sin\theta & r\cos\theta \end{vmatrix} = r, \qquad \frac{\partial(y,z)}{\partial(r,\theta)} = \begin{vmatrix} \sin\theta & r\cos\theta \\ 1 & 0 \end{vmatrix} = -r\cos\theta,$$

and

$$\frac{\partial(x, z)}{\partial(r, \theta)} = \begin{vmatrix} \cos\theta & -r\sin\theta \\ 1 & 0 \end{vmatrix} = r\sin\theta,$$

and so the area element is

$$dS = \|\Phi_r \times \Phi_\theta\| dr d\theta = \sqrt{r^2 + r^2\cos^2\theta + r^2\sin^2\theta}\, dr d\theta = r\sqrt{2}\, dr d\theta.$$

Clearly, $\|\Phi_r \times \Phi_\theta\|$ vanishes for $r = 0$, but $\Phi(0, \theta) = (0, 0, 0)$ for any θ. Thus $(0, 0, 0)$ is the only point where the tangent plane is not defined. We have thus

$$\iint_D dS = \int_0^{2\pi} \int_0^1 \sqrt{2}r\, dr d\theta = \int_0^{2\pi} \frac{1}{2}\sqrt{2}\, d\theta = \sqrt{2}\pi. \quad \blacklozenge$$

Example 6   Derive the formula for the area of a surface of revolution about the x axis using a suitable parametrization.

Solution   Define the parametrization by

$$x = u, \quad y = f(u)\cos v, \quad z = f(u)\sin v$$

over the region D given by

$$a \leq u \leq b, \qquad 0 \leq v \leq 2\pi.$$

This is a parametrization of a surface of revolution, because for fixed u,

$$(u, f(u)\cos v, f(u)\sin v)$$

traces out a circle of radius $|f(u)|$ with the center $(u, 0, 0)$ (Figure 6.3.7). We calculate

$$\frac{\partial(y, z)}{\partial(u, v)} = f(u)f'(u), \quad \frac{\partial(x, z)}{\partial(u, v)} = f(u)\cos v, \quad \frac{\partial(x, y)}{\partial(u, v)} = -f(u)\sin v$$

and so

$$A(S)$$

$$= \iint_D \sqrt{\frac{\partial(y, z)}{\partial(u, v)}^2 + \frac{\partial(x, z)}{\partial(u, v)}^2 + \frac{\partial(x, y)}{\partial(u, v)}^2}\, du dv$$

$$= \iint_D \sqrt{[f(u)]^2[f'(u)]^2 + [f(u)]^2\cos^2 v + [f(u)]^2\sin^2 v}\, du dv$$

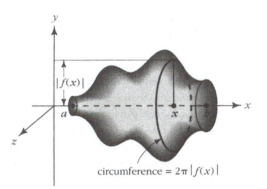

FIGURE 6.3.7. The curve $y = f(x)$ rotated about the x axis.

$$= \iint_D |f(u)|\sqrt{1 + [f'(u)]^2}\, du\,dv = \int_a^b \int_0^{2\pi} |f(u)|\sqrt{1 + [f'(u)]^2}\, dv\,du$$

$$= 2\pi \int_a^b |f(u)|\sqrt{1 + [f'(u)]^2}\, du,$$

which is the correct formula for the area of a surface of revolution.   ◆

Historical Note—Surfaces of Minimal Area

Calculus was invented (or discovered?) by Isaac Newton (1642–1727) about 1669 and by Gottfried Wilhelm Leibniz (1646–1716) about 1684. In the beginning of the eighteenth century, mathematicians were interested in the problem of finding paths of shortest length on a surface by using the methods of calculus. At this time, surfaces were regarded as boundaries of solids that were defined by inequalities (the ball $x^2 + y^2 + z^2 \le 1$ is bounded by the sphere $x^2 + y^2 + z^2 = 1$).

Christian Huygens (1629–1695) was the first person since Archimedes to give results on the areas of special surfaces beyond the sphere, and he obtained the areas of portions of surfaces of revolution, such as the paraboloid and hyperboloid.

The brilliant and prolific mathematician Leonhard Euler (1707–1783) presented the first fundamental work on the curvature of surfaces around 1760; it may have been in this work that a surface was first defined as a graph $z = f(x, y)$. Euler was interested in studying the curvature of surfaces, and in 1771 he introduced the notion of the parametrized surfaces that are described in this section.

After the rapid development of calculus in the early eighteenth century, formulas for the lengths of curves and areas of surfaces were developed. Although we do not know when the area formulas presented in this section first appeared, they were certainly common by the end of the eighteenth century. The underlying concepts of the length of a curve and the area of a surface were understood intuitively before this time, and the use of formulas from calculus to compute areas was considered a great achievement.

Augustin Louis Cauchy (1789–1857) was the first to take the step of defining the quantities of length and surface areas by integrals as we have done in this book. Later, mathematicians took up the question of defining surface area independent of integrals, but this posed many difficult problems that were not properly resolved until the twentieth century.

The Belgian physicist Joseph Plateau (1801–1883) carried out many experiments from 1830 to 1869 on surface tension and capillary phenomena. If a wire is dipped into a soap or glycerine solution and then removed, one usually finds a soap film attached to the wire. The underlying physical principle is that nature tends to minimize area; that is, the surface that forms should be the one of least area among all possible surfaces having the given curve as boundary. Figure 6.3.8 shows some examples, although readers might like to perform the experiment for themselves. Plateau raised the mathematical question: For a given boundary (wire), how does one prove the existence of such a surface (soap film); how many surfaces can there be?

For more information on this subject, the reader may consult S. Hildebrandt and A. Tromba, *Mathematics and Optimal Form*, Scientific American Books, New York, 1985.

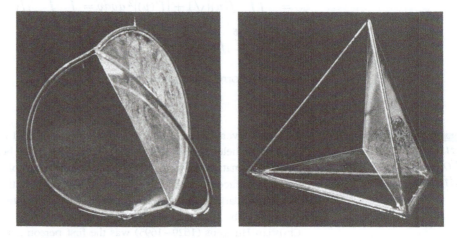

FIGURE 6.3.8. Two soap films spanning wires.

Finally in this section we consider the integral of a *scalar* function $f(x, y, z)$ over a surface. The special case $f(x, y, z) = 1$ will give the *area*. The next section will treat the more important case of the integral of a *vector field* over a surface.

Integral of a Scalar Function over a Surface

Let S be a surface parametrized by a mapping $\Phi : D \to S \subset \mathbb{R}^3, \Phi(u,v) = (x(u,v), y(u,v), z(u,v))$. If $f(x,y,z)$ is a real-valued continuous function defined on S, we define the ***integral of*** f ***over*** S to be

$$\iint_S f(x,y,z)dS = \iint_S f \, dS = \iint_D f(\Phi(u,v)) \|\Phi_u \times \Phi_v\| du \, dv$$

i.e.,

$$\iint_S f \, dS = \iint_D f(x(u,v), y(u,v), z(u,v))$$
$$\times \sqrt{\left[\frac{\partial(x,y)}{\partial(u,v)}\right]^2 + \left[\frac{\partial(y,z)}{\partial(u,v)}\right]^2 + \left[\frac{\partial(x,z)}{\partial(u,v)}\right]^2} \, du \, dv.$$

Thus if f is identically 1, we recover the area formula for a surface. Like surface area, the surface integral is independent of the particular parametrization used.

If S is a union of parametrized surfaces $S_i, i = 1, \ldots, N$, that do not intersect except possibly along curves defining their boundaries, then the integral of f over S is defined by

$$\iint_S f \, dS = \sum_{i=1}^{N} \iint_{S_i} f \, dS,$$

as we should expect. For example, the integral over the surface of a cube may be expressed as the sum of the integrals over the six sides.

Example 7   Consider the helicoid $x = r\cos\theta, y = r\sin\theta, z = \theta$, where $0 \le \theta \le 2\pi$ and $0 \le r \le 1$. Let f be given by $f(x,y,z) = \sqrt{x^2 + y^2 + 1}$. Find $\iint_S f \, dS$.

Solution

$$\frac{\partial(x,y)}{\partial(r,\theta)} = r, \quad \frac{\partial(y,z)}{\partial(r,\theta)} = \sin\theta, \quad \frac{\partial(x,z)}{\partial(r,\theta)} = \cos\theta.$$

Also, $f(r\cos\theta, r\sin\theta, \theta) = \sqrt{r^2 + 1}$. Therefore

$$\iint_S f(x,y,z)dS = \iint_D f(\Phi(r,\theta)) \|\Phi_r \times \Phi_\theta\| \, dr \, d\theta$$
$$= \int_0^{2\pi} \int_0^1 \sqrt{r^2 + 1} \sqrt{r^2 + 1} \, dr \, d\theta = \int_0^{2\pi} \frac{4}{3} d\theta = \frac{8}{3}\pi. \quad \blacklozenge$$

When S is a graph $z = g(x, y)$, then from the relation $dS = dxdy/\cos\varphi$ (see Figure 6.3.3), one gets

$$\iint_S f(x, y, z)dS = \iint_D \frac{f(x, y, g(x, y))}{\cos\varphi}dxdy,$$

where φ is the angle between the unit vector $\mathbf{k}$ and the normal to the surface at the point $(x, y, g(x, y))$.

Example 8   Compute $\iint_S xdS$, where S is the triangle with vertices $(1, 0, 0), (0, 1, 0), (0, 0, 1)$, shown in Figure 6.3.9.

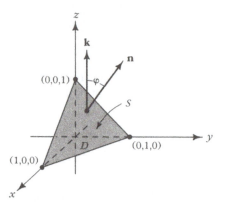

FIGURE 6.3.9. Compute the integral of x over this surface.

Solution   This surface is the plane described by the equation $x + y + z = 1$. Since the surface is a plane, the angle φ is constant and a unit normal vector is $\mathbf{n} = (1/\sqrt{3}, 1/\sqrt{3}, 1/\sqrt{3})$. Thus, $\cos\varphi = \mathbf{n} \cdot \mathbf{k} = 1/\sqrt{3}$, and so

$$\iint_S x\, dS = \sqrt{3} \iint_D x\, dxdy,$$

where D is the domain in the xy plane. But

$$\sqrt{3} \iint_D x\, dxdy = \sqrt{3} \int_0^1 \int_0^{1-x} x\, dydx = \sqrt{3} \int_0^1 x(1-x)dx = \frac{\sqrt{3}}{6}. \blacklozenge$$

Exercises for §6.3

In Exercises 1 and 2, express the surface area of the graph over the indicated region D as a double integral. Do not evaluate.

1. $z = xy^3 e^{x^2 y^2}$; D = unit circle centered at the origin

2. $z = y^3 \cos{}^2 x$; D = triangle with vertices $(-1, 1), (0, 2),$ and $(1, 1)$

3. Find the area of the graph of the function $f(x, y) = \frac{2}{3}(x^{3/2} + y^{3/2})$ over the domain $D = [0, 1] \times [0, 1]$.

4. Find the area cut out of the cylinder $x^2 + z^2 = 1$ by the cylinder $x^2 + y^2 = 1$.

5. Calculate the area of the part of the cone $z^2 = x^2 + y^2$ lying in the region of space defined by $x \geq 0, y \geq 0, 0 \leq z \leq 1$.

6. Find the area of the portion of the cylinder $x^2 + z^2 = 4$ that lies above the rectangle defined by $-1 \leq x \leq 1, 0 \leq y \leq 2$.

7. Let $\Phi(u, v) = (u - v, u + v, u)$, and let D be the unit disk in the uv plane. Find the area of $\Phi(D)$.

8. Find the area of the portion of the unit sphere that is in the cone $z \geq \sqrt{x^2 + y^2}$.

9. Represent the ellipsoid E

$$\frac{x^2}{a^2} + \frac{y^2}{b^2} + \frac{z^2}{c^2} = 1$$

parametrically and write out the integral for its surface area $A(E)$. (Do not evaluate the integral.)

10. Use the methods of this section to find the area of the surface obtained by rotating the curve $y = x^2$, $0 \leq x \leq 1$, about the y axis.

11. Find the area of the surface defined by $x + y + z = 1, x^2 + 2y^2 \leq 1$.

12. The torus T can be represented parametrically by the function $\Phi: D \to \mathbb{R}^3$, where Φ is given by the coordinate functions

$$x = (R + \cos \phi) \cos \theta, \quad y = (R + \cos \phi) \sin \theta, \quad z = \sin \phi;$$

D is the rectangle $[0, 2\pi] \times [0, 2\pi]$, that is, $0 \leq \theta \leq 2\pi, 0 \leq \phi \leq 2\pi$; and $R > 1$ is fixed (see Figure 6.3.10). Show that $A(T) = (2\pi)^2 R$.

13. A cylindrical hole of radius 1 is bored through a solid ball of radius 2 to form a ring coupler as shown in Figure 6.3.11. Find the volume and outer surface area of this coupler.

14.(a) Prove that the area on a sphere of radius r cut out by a cone of angle ϕ is $2\pi r^2 (1 - \cos \phi)$ (Figure 6.3.12).

   (b) A sphere of radius 1 sits with its center on the surface of a sphere of radius $r > 1$. Show that the surface area of the part of the second sphere cut out by the first sphere is π. (Does something about this result surprise you?)

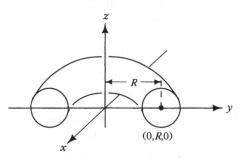

FIGURE 6.3.10. A cross section of a torus.

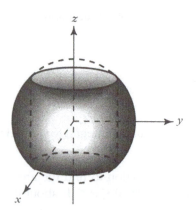

FIGURE 6.3.11. Find the surface area and volume of the shaded region.

15. Sketch and compute the area of the surface given by

$$x = r\cos\theta, \quad y = 2r\cos\theta, \quad z = \theta, \quad 0 \le r \le 1, \quad 0 \le \theta \le 2\pi.$$

16. Prove **Pappus' theorem**: Let $\mathbf{c} : [a, b] \to \mathbb{R}^2$ be a C^1 path whose image lies in the right half-plane and is a simple closed curve. The area of the lateral surface generated by rotating the image of $\mathbf{c}$ about the y axis is equal to $2\pi\bar{x}l(\mathbf{c})$, where $\bar{x}$ is the average value of x coordinates of points on $\mathbf{c}$ and $l(\mathbf{c})$ is the length of $\mathbf{c}$.

Find the surface integrals in Exercises 17–20.

17. Compute $\iint_S xy\, dS$, where S is the surface of the tetrahedron with sides $z = 0, y = 0, x = 0$, and $z = 1 - x - y$.

18. Evaluate $\iint_S z\, dS$, where S is the surface $z = x^2 + y^2, x^2 + y^2 \le 1$.

FIGURE 6.3.12. The area of the cap is $2\pi r^2(1 - \cos\phi)$.

19. Evaluate $\displaystyle\iint_S z\,dS$, where S is the upper hemisphere of radius a; that is, the set of (x, y, z) with $z = \sqrt{a^2 - x^2 - y^2}$.

20. Evaluate $\displaystyle\iint_S (x + y + z)dS$, where S is the boundary of the unit ball B; that is, S is the set of (x, y, z) with $x^2 + y^2 + z^2 = 1$. (Hint: Use the symmetry of the problem.)

21. The cylinder $x^2 + y^2 = r^2$ divides the sphere $x^2 + y^2 + z^2 = R^2$ into two regions S_1 and S_2, where S_1 is inside the cylinder and S_2 is outside. Find the ratio of areas $A(S_2)/A(S_1)$.

22. Consider a surface S that is the graph of a function $z = f(x, y)$, for (x, y) in a region $D \subset \mathbb{R}^2$. Assume that S can also be described as the set of points (x, y, z) in $\mathbb{R}^3$ with $F(x, y, z) = 0$ (a level surface). Derive a formula for $A(S)$ that involves only F. (Hint: By implicit differentiation, derive

$$\frac{\partial z}{\partial x} = -\frac{\partial F}{\partial x}\Big/\frac{\partial F}{\partial z} \quad \text{and} \quad \frac{\partial z}{\partial y} = -\frac{\partial F}{\partial y}\Big/\frac{\partial F}{\partial z},$$

and substitute in the formula for the surface area of a graph.)

23. Calculate the area of the frustum shown in Figure 6.3.13 using (a) geometry alone and (b) a surface area formula.

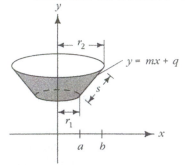

FIGURE 6.3.13. A line segment revolved around the y axis becomes a frustum of a cone.

6.4
Surface
Integrals

The line integral of a vector field $\mathbf{F}$ along a curve involved the dot product of the vector field with a tangent vector to the curve. When we integrate $\mathbf{F}$ over a *surface*, we take the dot product of $\mathbf{F}$ with a *normal* vector to the surface.

Since $d\mathbf{S} = \Phi_u \times \Phi_v \, dudv$ for a parametrized surface, we make the following definition.

Surface Integral

The **surface integral** of a vector field $\mathbf{F}$ on $\mathbb{R}^3$ over a parametrized surface $\Phi : D \to \mathbb{R}^3$ is the *number*

$$\iint_S \mathbf{F} \cdot d\mathbf{S} = \iint_D \mathbf{F}(\Phi(u,v)) \cdot (\Phi_u \times \Phi_v) \, dudv.$$

Since $d\mathbf{S} = \mathbf{n} dS$, where $\mathbf{n} = \Phi_u \times \Phi_v / \|\Phi_u \times \Phi_v\|$ is a unit normal, we can write the surface integral as

$$\iint_S \mathbf{F} \cdot d\mathbf{S} = \iint_S \mathbf{F} \cdot \mathbf{n} \, dS,$$

so *the surface integral is the integral of the normal component of* $\mathbf{F}$ *over the surface.* See Figure 6.4.1.

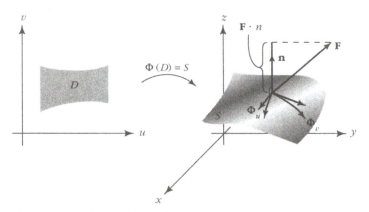

FIGURE 6.4.1. The ingredients for the surface integral.

Our study of surface integrals in this section has two aspects—calculation and interpretation. The first part is, in principle, straightforward; given formulas for $\mathbf{F}$ and Φ, the definition gives a double integral over a plane region D, which is evaluated by the methods of Chapter 5. (In Chapter 7, we will learn some other techniques.)

Example 1   Let D be the rectangle in the $\theta\phi$ plane defined by

$$0 \le \theta \le 2\pi, \qquad 0 \le \phi \le \pi,$$

and let the surface S be defined by the parametrization $\Phi : D \to \mathbb{R}^3$ given by

$$x = \cos\theta\sin\phi, \quad y = \sin\theta\sin\phi, \quad z = \cos\phi.$$

Here θ and ϕ are the angles of spherical coordinates, and S is the unit sphere parametrized by Φ. Let $\mathbf{r}$ be the position vector $\mathbf{r}(x, y, z) = x\mathbf{i} + y\mathbf{j} + z\mathbf{k}$. Compute $\iint_S \mathbf{r} \cdot d\mathbf{S}$.

Solution   First we find

$$\begin{aligned}
\Phi_\theta &= (-\sin\phi\sin\theta)\mathbf{i} + (\sin\phi\cos\theta)\mathbf{j} \\
\Phi_\phi &= (\cos\theta\cos\phi)\mathbf{i} + (\sin\theta\cos\phi)\mathbf{j} - (\sin\phi)\mathbf{k}
\end{aligned}$$

and hence

$$\Phi_\theta \times \Phi_\phi = (-\sin^2\phi\cos\theta)\mathbf{i} - (\sin^2\phi\sin\theta)\mathbf{j} - (\sin\phi\cos\phi)\mathbf{k}.$$

Then we evaluate

$$\begin{aligned}
\mathbf{r} \cdot (\Phi_\theta \times \Phi_\phi) &= (x\mathbf{i} + y\mathbf{j} + z\mathbf{k}) \cdot (\Phi_\theta \times \Phi_\phi) \\
&= [(\cos\theta\sin\phi)\mathbf{i} + (\sin\theta\sin\phi)\mathbf{j} + (\cos\phi)\mathbf{k}] \\
&\quad \cdot(-\sin\phi)[(\sin\phi\cos\theta)\mathbf{i} + (\sin\phi\sin\theta)\mathbf{j} + (\cos\phi)\mathbf{k}] \\
&= (-\sin\phi)(\sin^2\phi\cos^2\theta + \sin^2\phi\sin^2\theta + \cos^2\phi) \\
&= -\sin\phi.
\end{aligned}$$

Thus,

$$\begin{aligned}
\iint_S \mathbf{r} \cdot d\mathbf{S} &= \iint_D \mathbf{r} \cdot (\Phi_\theta \times \Phi_\phi)\,d\theta\,d\phi \\
&= \int_0^{2\pi} \int_0^{\pi} -\sin\phi\,d\phi\,d\theta = \int_0^{2\pi} (-2)\,d\theta = -4\pi. \quad \blacklozenge
\end{aligned}$$

We saw in §**6.1** that line integrals depend on the orientation of the curve but are otherwise independent of the parametrization. There is a similar result for surface integrals. To state it, we first consider the notion of orientation.

Orientation

An **oriented surface** is a two-sided surface with one side specified as the **outside** or **positive side**; we call the other side the **inside** or **negative side**. At each point (x, y, z) of S there are two unit normal vectors $\mathbf{n}_1$ and $\mathbf{n}_2$, where $\mathbf{n}_1 = -\mathbf{n}_2$ (see Figure 6.4.2). Each of these two normals can be associated with one side of the surface. Thus to specify a side of a surface S, at each point we choose a unit normal vector $\mathbf{n}$ that points in the positive direction from S at that point.

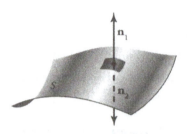

FIGURE 6.4.2. The two possible unit normals to a surface at a point.

This definition assumes that our surface does have two "sides." An example of a surface with only one side is the Möbius strip (named after the German mathematician and astronomer A. F. Möbius, who, along with the mathematician J. B. Listing, discovered it in 1858). Pictures of such a surface are given in Figures 6.4.3 and 6.4.4. At each point of M there are two unit normals, $\mathbf{n}_1$ and $\mathbf{n}_2$. However, $\mathbf{n}_1$ does not determine a unique side of M, and neither does $\mathbf{n}_2$. To see this intuitively, we can slide $\mathbf{n}_2$ around the closed curve C. When $\mathbf{n}_2$ returns to a fixed point p on C it will coincide with $\mathbf{n}_1$, showing that both $\mathbf{n}_1$ and $\mathbf{n}_2$ point to the same side of M and, consequently, that M has only one side!

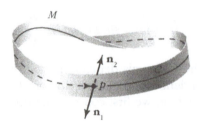

FIGURE 6.4.3. The Möbius strip: slide $\mathbf{n}_2$ around C once; when $\mathbf{n}_2$ returns to its initial point, $\mathbf{n}_2$ will coincide with $\mathbf{n}_1 = -\mathbf{n}_2$.

A parametrization Φ is called **orientation preserving** if its unit normal $\mathbf{n} = \Phi_u \times \Phi_v / \|\Phi_u \times \Phi_v\|$ (assuming $\Phi_u \times \Phi_v \neq \mathbf{0}$) agrees with that for a given

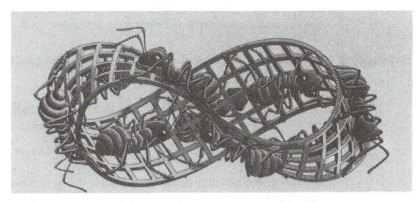

FIGURE 6.4.4. Ants walking on a Möbius strip. *Moebius Strip II, 1963, by M. C. Escher, Escher Foundation, Haags Gemeentemuseum, The Hague.*

orientation. If **n** is in the opposite direction, we say that Φ is ***orientation reversing***.

Example 2   Give the unit sphere $x^2 + y^2 + z^2 = 1$ in $\mathbb{R}^3$ an orientation by selecting the unit vector $\mathbf{n}(x, y, z) = \mathbf{r}$, where $\mathbf{r} = x\mathbf{i} + y\mathbf{j} + z\mathbf{k}$, which points away from the outside of the surface. This choice corresponds to our intuitive notion of outside for the sphere (Figure 6.4.5). Is the parametrization of Example 1 orientation preserving?

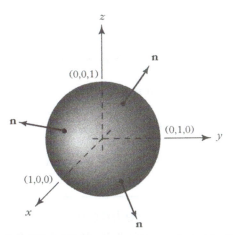

FIGURE 6.4.5. The unit sphere oriented by its outward normal **n**.

Solution   The cross product of the tangent vectors Φ_θ and Φ_ϕ is given by

$$(-\sin\phi)[(\cos\theta\sin\phi)\mathbf{i} + (\sin\theta\sin\phi)\mathbf{j} + (\cos\phi)\mathbf{k}] = -\mathbf{r}\sin\phi.$$

Since $-\sin\phi \leq 0$ for $0 \leq \phi \leq \pi$, this normal vector points inward. Thus the given parametrization Φ is orientation reversing.   ◆

Example 3   Let S be a surface described by $z = f(x,y)$. There are two unit normal vectors to S at $(x_0, y_0, f(x_0, y_0))$, namely, $\pm\mathbf{n}$, where

$$\mathbf{n} = \frac{-\dfrac{\partial f}{\partial x}(x_0, y_0)\mathbf{i} - \dfrac{\partial f}{\partial y}(x_0, y_0)\mathbf{j} + \mathbf{k}}{\sqrt{\left[\dfrac{\partial f}{\partial x}(x_0, y_0)\right]^2 + \left[\dfrac{\partial f}{\partial y}(x_0, y_0)\right]^2 + 1}}.$$

We can orient all such surfaces by taking the positive side of S to be the "top" side (Figure 6.4.6). Thus the orientation of such a surface is determined by the unit normal $\mathbf{n}$ with *positive* $\mathbf{k}$ component. If we parametrize this surface by $\Phi(u, v) = (u, v, f(u, v))$, then Φ is orientation preserving.   ◆

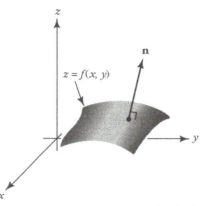

FIGURE 6.4.6. A graph oriented with the upward-pointing unit normal.

The area element on a graph is given by

$$dS = \sqrt{1 + f_x^2 + f_y^2}\, dx\, dy$$

as was shown in the last section. Thus $\mathbf{n}\, dS = (-f_x\mathbf{i} - f_y\mathbf{j} + \mathbf{k})dx\, dy$. Therefore, we arrive at:

The Surface Integral for Graphs

If $\mathbf{F} = P\mathbf{i} + Q\mathbf{j} + R\mathbf{k}$ is a vector field in space and S is the surface $z = f(x, y)$ where f is a function on D, the *surface integral* of $\mathbf{F}$ over S is:

$$\iint_S \mathbf{F} \cdot d\mathbf{S} = \iint_S \mathbf{F} \cdot \mathbf{n}\, dS = \iint_D (-Pf_x - Qf_y + R)dx\, dy.$$

Example 4   Let $\mathbf{F} = x^2\mathbf{i} + y^2\mathbf{j} + z\mathbf{k}$. Evaluate $\iint_S \mathbf{F} \cdot \mathbf{n} \, dS$, where S is the graph of the function $z = x + y + 1$ over the rectangle $0 \le x \le 1, 0 \le y \le 1$.

Solution   By the preceding box, with $P(x, y, z) = x^2$, $Q(x, y, z) = y^2$, $R(x, y, z) = z$, and $z = f(x, y) = x + y + 1$,

$$
\begin{aligned}
\iint_S \mathbf{F} \cdot \mathbf{n} \, dS &= \iint_D \left[-x^2 \cdot 1 - y^2 \cdot 1 + (x + y + 1) \right] dx \, dy \\
&= \int_0^1 \int_0^1 (x + y + 1 - x^2 - y^2) dx \, dy \\
&= \frac{1}{2} + \frac{1}{2} + 1 - \frac{1}{3} - \frac{1}{3} = \frac{4}{3}. \quad \blacklozenge
\end{aligned}
$$

We now state without proof a result showing that the integral over an oriented surface is independent of the parametrization. (As before, we assume that parametrizations do not overlap themselves.) The proof depends on the change of variables formula for double integrals.

Independence of Parametrization

Let S be an oriented surface, and let Φ_1 and Φ_2 be two smooth orientation preserving parametrizations, with $\mathbf{F}$ a continuous vector field defined on S. Then

$$
\iint_{\Phi_1} \mathbf{F} \cdot d\mathbf{S} = \iint_{\Phi_2} \mathbf{F} \cdot d\mathbf{S}.
$$

If Φ_1 is orientation preserving and Φ_2 orientation reversing, then

$$
\iint_{\Phi_1} \mathbf{F} \cdot d\mathbf{S} = - \iint_{\Phi_2} \mathbf{F} \cdot d\mathbf{S}.
$$

We can thus unambiguously use the notation

$$
\iint_S \mathbf{F} \cdot d\mathbf{S} = \iint_S \mathbf{F} \cdot \mathbf{n} \, dS
$$

(or a sum of such integrals, if S is a union of oriented parametrized surfaces that intersect only along their boundary curves).

Many of the most important applications of surface integrals come from physics and engineering, but it is perhaps easiest to understand the surface integral in terms of the geometric problem of *finding the area of the shadow of a surface.*

Suppose that light is falling in parallel rays in the direction of a unit vector $\mathbf{G}$ onto a plane $\mathcal{P}$ perpendicular to $\mathbf{G}$. Suppose that S is a geometric surface that is intersected at most once by each light ray, as in Figure 6.4.7. Following the argument in the caption of that figure, we see that the area of the shadow of S on $\mathcal{P}$ is

$$A = \iint_D |\mathbf{G} \cdot (\mathbf{\Phi}_u \times \mathbf{\Phi}_v)| \, dudv.$$

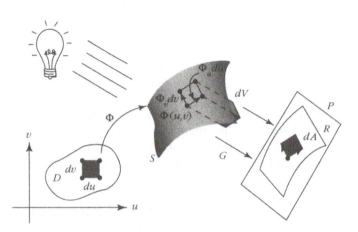

FIGURE 6.4.7. Computing the area of a shadow. The volume dV of the infinitesimal parallelepiped is on the one hand $|\mathbf{G} \cdot (\mathbf{\Phi}_u du \times \mathbf{\Phi}_v dv)|$ and on the other $\|\mathbf{G}\| dA$. Since $\mathbf{G}$ is a unit vector, $dA = |\mathbf{G} \cdot (\mathbf{\Phi}_u \times \mathbf{\Phi}_v)| dudv$, and so the area of the shadow R is the integral of dA.

Except for the absolute value bars, this has the form of a surface integral if we take $\mathbf{F}(x, y, z)$ to be the constant vector field whose value at each point of space is $\mathbf{G}$.

We can eliminate the absolute value bars once we understand the sign of the triple product $\mathbf{G} \cdot (\mathbf{\Phi}_u \times \mathbf{\Phi}_v)$. If $\mathbf{\Phi}$ is orientation preserving, then $\mathbf{G} \cdot (\mathbf{\Phi}_u \times \mathbf{\Phi}_v)$ is *positive* if the light is shining *outward* past the surface S and negative if the light is shining *inward*.

Area of a Shadow

If light falls in the direction of the unit vector $\mathbf{G}$ onto a plane $\mathcal{P}$ perpendicular to $\mathbf{G}$, then the area of the shadow of a parametrized surface S is

$$\iint_S \mathbf{G} \cdot d\mathbf{S}$$

where we use a parametrization $\mathbf{\Phi} : D \to \mathbb{R}^3$ of S for which the light shines *outward* past S.

Example 5   Suppose that light is falling "downward" in the direction perpendicular to the plane $x + 2y + 3z = -1$. What is the area of the shadow of the disk $x^2 + y^2 \leq 1/4, z = 0$?

Solution   Since the disk lies in the xy plane, we may take x and y as parameters, *i.e.*, $\Phi(u, v) = (u, v, 0)$. Thus $\Phi_u = \mathbf{i}$ and $\Phi_v = \mathbf{j}$, and so the orientation is given by the normal vector $\mathbf{k} = \mathbf{i} \times \mathbf{j}$. A normal vector to $\mathcal{P}$ is $\mathbf{i} + 2\mathbf{j} + 3\mathbf{k}$. The unit normal vectors are $\pm(\mathbf{i} + 2\mathbf{j} + 3\mathbf{k})/\sqrt{14}$; the one that is "outward" is $(\mathbf{i} + 2\mathbf{j} + 3\mathbf{k})/\sqrt{14}$, which has positive inner product with $\mathbf{k}$. Thus the area of the shadow is

$$A = \iint_D \frac{1}{\sqrt{14}}(\mathbf{i} + 2\mathbf{j} + 3\mathbf{k}) \cdot \mathbf{k}\, du\, dv,$$

where D is the disk $u^2 + v^2 \leq 1/4$. The integrand is $3/\sqrt{14}$, a constant, so

$$A = \frac{3}{\sqrt{14}} \times (\text{area of } D) = \frac{3}{\sqrt{14}} \cdot \frac{\pi}{4} = \frac{3\pi}{4\sqrt{14}}. \quad \blacklozenge$$

An argument analogous to that for shadows shows the following.

Surface Integrals for Fluid Flow

Let S be an oriented surface and $\mathbf{V}$ be the velocity field of a fluid moving in space. Then

$$\iint_S \mathbf{V} \cdot d\mathbf{S}$$

is the net rate (in units such as cubic meters per second) at which fluid is crossing the surface in the *outward* direction. (If the integral is negative, the net flow is *inward*.)

Justification   Consider Figure 6.4.8. Here we let S be a plane and assume for simplicity that $\mathbf{V}$ is constant. Let a rectangle R on S flow out with the fluid for time 1 to form a parallelepiped W. The volume of W is, from the figure, given by $(\text{area } R) \|\mathbf{V}\| \cos\theta = (\text{area } R)\, \mathbf{V} \cdot \mathbf{n}$. For a curved surface, a similar argument shows that the rate of fluid crossing is $\mathbf{V} \cdot \mathbf{n}\, dS = \mathbf{V} \cdot d\mathbf{S}$, so summing, we get the result claimed.   ∎

Based on examples like this, the surface integral is called the *flux*. The reader may be familiar with physical laws (such as Faraday's law) that relate (a change in) flux of an electric or magnetic field to a circulation (or current) in a bounding loop. These are the historical and physical bases of Stokes' theorem, which we shall discuss in §**7.2**.

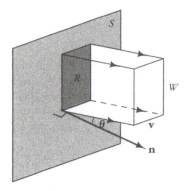

FIGURE 6.4.8. The amount of fluid crossing S per unit time is the normal component of **V** times the area of S.

Example 6 Find the flux of the vector field **j** across the hemisphere H defined by $x^2 + y^2 + z^2 = 1, x \geq 0$, oriented in the direction of increasing x (*i.e.*, the usual outward normal).

Solution The situation is sketched in Figure 6.4.9. By symmetry, we suspect that the inward flow on the quarter-sphere $y \leq 0, x \geq 0$ should cancel the outward flow on the quarter-sphere $y \geq 0, x \geq 0$. To verify this, we parametrize the hemisphere by spherical coordinates $x = \cos\theta\sin\phi, y = \sin\theta\sin\phi, z = \cos\phi$, as in Example 2, but with the domain D taken to be the rectangle $-\pi/2 \leq \theta \leq \pi/2, 0 \leq \phi \leq \pi$ to produce the hemisphere $x \geq 0$ rather than the whole sphere.

The outward normal for this parametrization is

$$-\frac{\partial\Phi}{\partial\theta} \times \frac{\partial\Phi}{\partial\phi} = (\sin^2\phi\cos\theta)\mathbf{i} + (\sin^2\phi\sin\theta)\mathbf{j} + (\sin\phi\cos\phi)\mathbf{k},$$

according to Example 2, so

$$\iint_H \mathbf{j} \cdot d\mathbf{S} = \int_0^\pi \int_{-\pi/2}^{\pi/2} \sin^2\phi\sin\theta \, d\theta d\phi,$$

which is zero since

$$\int_{-\pi/2}^{\pi/2} \sin\theta \, d\theta = 0. \quad \blacklozenge$$

Surface integrals also apply to the study of heat flow. Let $T(x, y, z)$ be the temperature at a point (x, y, z) of a region W. Then

$$\nabla T = \frac{\partial T}{\partial x}\mathbf{i} + \frac{\partial T}{\partial y}\mathbf{j} + \frac{\partial T}{\partial z}\mathbf{k}$$

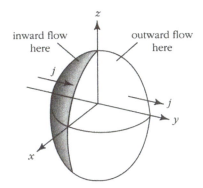

FIGURE 6.4.9. The flux of the constant vector field **j** across the hemisphere is zero.

represents the temperature gradient, and heat "flows" with the vector field $-k\nabla T = \mathbf{F}$, where k is a positive constant. Therefore $\iint_S \mathbf{F} \cdot d\mathbf{S}$ is the total **_rate of heat flow_** or **_flux_** outward across the surface S.

Example 7   Suppose that a temperature function is given as $T(x, y, z) = x^2 + y^2 + z^2$, and let S be the unit sphere $x^2 + y^2 + z^2 = 1$ oriented with the outward normal. Find the heat flux across the surface S if $k = 1$.

Solution   The heat flow field is

$$\mathbf{F} = -\nabla T(x, y, z) = -2x\mathbf{i} - 2y\mathbf{j} - 2z\mathbf{k}.$$

On S, $\mathbf{n}(x, y, z) = x\mathbf{i} + y\mathbf{j} + z\mathbf{k}$ is the unit "outward" normal, and $f(x, y, z) = \mathbf{F} \cdot \mathbf{n} = -2x^2 - 2y^2 - 2z^2 = -2$ is the normal component of $\mathbf{F}$. Since the surface integral of $\mathbf{F}$ is equal to the integral of its normal component over S, $\iint_S \mathbf{F} \cdot d\mathbf{S} = -2 \iint_S dS = -2(4\pi) = -8\pi$. The flux of heat is directed toward the center of the sphere (why *toward*?). ◆

Example 8   There is an important physical law, due to the great scientist K. F. Gauss,[1] that relates the flux of an electric field $\mathbf{E}$ over a "closed" surface S (for example, a sphere or an ellipsoid) to the net charge Q enclosed by the surface, namely, (see Figure 6.4.10)

$$\iint_S \mathbf{E} \cdot d\mathbf{S} = Q.$$

---

[1] Perhaps Gauss' greatest public recognition is that he now appears on the German 10 mark note. The United Kingdom has similarly honored Newton and Switzerland, Euler.

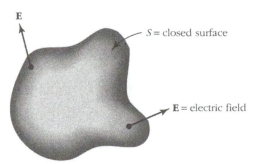

FIGURE 6.4.10. Gauss' law: $\int_S \mathbf{E} \cdot d\mathbf{S} = Q$, where Q is the net charge inside S.

Suppose that $\mathbf{E} = E\mathbf{n}$; that is, $\mathbf{E}$ is a constant scalar multiple of the unit normal to S. Then Gauss' law above, becomes

$$\iint_S \mathbf{E} \cdot d\mathbf{S} = \iint_S \mathbf{E} \cdot \mathbf{n} \, dS = \iint_S E \, dS = E \iint_S dS = Q.$$

Thus,

$$E = \frac{Q}{A(S)}.$$

In the case where S is the sphere of radius R, we get

$$E = \frac{Q}{4\pi R^2}$$

(see Figure 6.4.11).

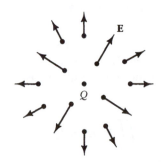

FIGURE 6.4.11. The field $\mathbf{E}$ due to a point charge Q is $\mathbf{E} = Q\mathbf{n}/4\pi R^2$.

If $\mathbf{E}$ arises from an isolated point charge Q, then from symmetry, it is reasonable that $\mathbf{E} = E\mathbf{n}$, where $\mathbf{n}$ is the unit normal to any sphere centered at Q. Consider a second point charge Q_0 located at a distance R from Q. The force $\mathbf{F}$ that acts on this second charge Q_0 is thus given by

$$\mathbf{F} = \mathbf{E}Q_0 = EQ_0\mathbf{n} = \frac{QQ_0}{4\pi R^2}\mathbf{n}.$$

If F is the magnitude of $\mathbf{F}$, we have

$$F = \frac{QQ_0}{4\pi R^2},$$

which is **Coulomb's law** for the force between two point charges. (Sometimes one sees the formula $F = (1/4\pi\epsilon_0)QQ_0/R^2$. An extra constant ϵ_0 appears, by convention, when MKS units are used for measuring charge. We are using CGS, or Gaussian, units.)   ◆

Exercises for §6.4

1. Evaluate $\displaystyle\iint_S (\nabla \times \mathbf{F}) \cdot d\mathbf{S}$, where S is the surface $x^2 + y^2 + 3z^2 = 1, z \leq 0$, and $\mathbf{F} = y\mathbf{i} - x\mathbf{j} + zx^3y^2\mathbf{k}$.

2. Evaluate $\displaystyle\iint_S (\nabla \times \mathbf{F}) \cdot d\mathbf{S}$, where $\mathbf{F} = (x^2 + y - 4)\mathbf{i} + 3xy\mathbf{j} + (2xz + z^2)\mathbf{k}$ and S is the surface $x^2 + y^2 + z^2 = 16, z \geq 0$.

3. Calculate the integral $\displaystyle\iint_S \mathbf{F} \cdot d\mathbf{S}$, where S is the surface of the half-ball $x^2 + y^2 + z^2 \leq 1, z \geq 0$, and $\mathbf{F} = (x + 3y^5)\mathbf{i} + (y + 10xz)\mathbf{j} + (z - xy)\mathbf{k}$.

4. Evaluate $\displaystyle\iint_S \mathbf{F} \cdot \mathbf{n}\,dS$, where $\mathbf{F}(x, y, z) = \mathbf{i} + \mathbf{j} + z(x^2 + y^2)^2\mathbf{k}$ and S is the *surface* of the cylinder $x^2 + y^2 \leq 1, 0 \leq z \leq 1$.

5. Find the flux of $\mathbf{F}(x,y,z) = 3xy^2\mathbf{i} + 3x^2y\mathbf{j} + z^3\mathbf{k}$ out of the unit sphere.

6. Let the velocity field of a fluid be described by $\mathbf{V} = \sqrt{y}\mathbf{j}$ (measured in meters per second). Compute how many cubic meters of fluid per second are crossing the surface $x^2 + z^2 = y, 0 \leq y \leq 1$.

7. Let the velocity field of a fluid be described by $\mathbf{F} = \mathbf{i} + x\mathbf{j} + z\mathbf{k}$ (measured in meters per second). Compute how many cubic meters of fluid per second are crossing the surface described by $x^2 + y^2 + z^2 = 1, z \geq 0$.

8.(a) A uniform fluid that flows vertically downward (heavy rain) is described by the vector field $\mathbf{F}(x, y, z) = (0, 0, -1)$. Find the total flux through the cone $z = (x^2 + y^2)^{1/2}, x^2 + y^2 \leq 1$.

(b) The rain is driven sideways by a strong wind so that it falls at a 45° angle, and it is described by $\mathbf{F}(x, y, z) = -(\sqrt{2}/2, 0, \sqrt{2}/2)$. Now what is the flux through the cone?

9. For $a > 0, b > 0, c > 0$, let S be the upper half-ellipsoid

$$S = \left\{ (x, y, z) \,\middle|\, \frac{x^2}{a^2} + \frac{y^2}{b^2} + \frac{z^2}{c^2} = 1, z \geq 0 \right\}$$

with orientation determined by the upward normal. Compute $\iint_S \mathbf{F} \cdot d\mathbf{S}$ where $\mathbf{F}(x, y, z) = (x^3, 0, 0)$.

10. If S is the upper hemisphere $\{(x, y, z) | x^2 + y^2 + z^2 = 1, z \geq 0\}$ oriented by the normal pointing out of the sphere, compute $\iint_S \mathbf{F} \cdot d\mathbf{S}$ for parts (a) and (b).

(a) $\mathbf{F}(x, y, z) = x\mathbf{i} + y\mathbf{j}$

(b) $\mathbf{F}(x, y, z) = y\mathbf{i} + x\mathbf{j}$

(c) For each of the vector fields above, compute $\iint_S (\nabla \times \mathbf{F}) \cdot d\mathbf{S}$ and $\int_C \mathbf{F} \cdot d\mathbf{s}$ where C is the unit circle in the xy plane traversed in the counterclockwise direction (as viewed from the positive z axis). (Notice that C is the boundary of S. The phenomenon illustrated here will be studied more thoroughly in the next chapter, using Stokes' theorem.)

11. Let S be the surface of the unit sphere. Let $\mathbf{F}$ be a vector field and F_r its radial component. Prove that

$$\iint_S \mathbf{F} \cdot d\mathbf{S} = \int_0^{2\pi} \int_0^{\pi} F_r \sin\phi \, d\phi \, d\theta.$$

12. Work out a formula like that in Exercise 11 for integration over the surface of a cylinder.

13. Let the temperature of a point in $\mathbb{R}^3$ be given by $T(x, y, z) = 3x^2 + 3z^2$. Compute the heat flux across the surface $x^2 + z^2 = 2, 0 \leq y \leq 2$, if $k = 1$.

14. Compute the heat flux across the unit sphere S if $T(x, y, z) = x$. Can you interpret your answer physically?

15. Let S be the closed surface that consists of the hemisphere $x^2 + y^2 + z^2 = 1, z \geq 0$, and its base $x^2 + y^2 \leq 1, z = 0$. Let $\mathbf{E}$ be the electric field defined by $\mathbf{E}(x, y, z) = 2x\mathbf{i} + 2y\mathbf{j} + 2z\mathbf{k}$. Find the electric flux across S.

16. A restaurant is being built on the side of a mountain. The architect's plans are shown in Figure 6.4.12.

(a) The vertical curved wall of the restaurant is to be built of glass. What will be the surface area of this wall?

(b) For the restaurant to be large enough to be profitable, the consulting engineer informs the developer that the volume of the interior must exceed $\pi R^4/2$. For what R does the proposed structure satisfy this requirement?

(c) During a typical summer day, the environs of the restaurant are subject to a temperature field given by

$$T(x, y, z) = 3x^2 + (y - R)^2 + 16z^2.$$

A heat flux density $\mathbf{V} = -k\nabla T$ (k is a constant depending on the grade of insulation to be used) through all sides of the restaurant (including the top and the contact with the hill) produces a heat flux. What is this total heat flux? (Your answer will depend on R and k.)

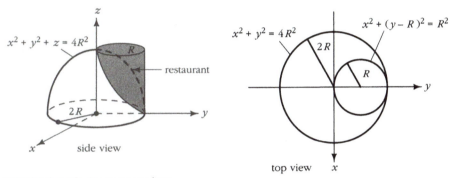

FIGURE 6.4.12. Restaurant plans.

Review Exercises for Chapter 6

Calculate the line integral of $\mathbf{F} = \nabla f$, where $f(x, y, z) = xyz$, along the paths in Exercises 1–4.

1. $\mathbf{c}(t) = (e^t \cos t, e^t \sin t, 3), 0 \le t \le 2\pi$

2. $\mathbf{c}(t) = (\cos t, \sin t, t), 0 \le t \le 2\pi$

3. $\mathbf{c}(t) = \frac{3}{2}t^2\mathbf{i} + 2t^2\mathbf{j} + t\mathbf{k}, 0 \le t \le 1$

4. $\mathbf{c}(t) = t\mathbf{i} + (1/\sqrt{2})t^2\mathbf{j} + \frac{1}{3}t^3\mathbf{k}, 0 \le t \le 1$

Compute the line integrals in Exercises 5 and 6.

5. $\int_C (\sin \pi x)dy - (\cos \pi y)dz$, where C is the triangle whose vertices are $(1,0,0), (0,1,0)$, and $(0,0,1)$ in that order.

6. $\int_C (\sin z)dx + (\cos z)dy - (xy)^{1/3}dz$, where C is the path parametrized by $\mathbf{c}(\theta) = (\cos^3 \theta, \sin^3 \theta, \theta), 0 \le \theta \le 7\pi/2$.

7. Find the work done by the force $\mathbf{F}(x,y) = (x^2 - y^2)\mathbf{i} + 2xy\mathbf{j}$ in moving a particle counterclockwise around the square in the plane having corners $(0,0), (a,0), (a,a), (0,a), a > 0$.

8. If $\mathbf{F}(\mathbf{x})$ is orthogonal to $\mathbf{c}'(t)$ at each point on the curve $\mathbf{x} = \mathbf{c}(t)$, what can you say about $\int_\mathbf{c} \mathbf{F} \cdot d\mathbf{s}$?

9. A ring in the shape of the curve $x^2 + y^2 = a^2$ is formed of thin wire weighing $|x| + |y|$ grams per unit length at (x,y). Find the mass of the ring.

10. Find the area of the surface

$$x = h(u,v) = u + v, \quad y = g(u,v) = u, \quad z = f(u,v) = v;$$

$0 \le u \le 1, 0 \le v \le 1$. Sketch.

Find a parametrization for each of the surfaces in Exercises 11 and 12.

11. $x^2 + y^2 + z^2 - 4x - 6y = 12$

12. $x^2 + y^2 + z^2 - 8x = 1$

13. Write a formula for the surface area of $\Phi(r,\theta) = (x,y,z)$,

$$x = r\cos\theta, \quad y = 2r\sin\theta, \quad z = r;$$

$0 \le r \le 1, 0 \le \theta \le 2\pi$.

14. Suppose that $z = f(x,y)$ and $(\partial f/\partial x)^2 + (\partial f/\partial y)^2 = c, c > 0$. Show that the area of the graph of f lying over a region D in the xy plane is $\sqrt{1+c}$ times the area of D.

15. Find the area of the portion of the plane $z = 3x - 2y$ over the square $[0,1] \times [0,1]$.

16. A paraboloid of revolution S is parametrized by the mapping $\Phi(u,v) = (u\cos v, u\sin v, u^2), 0 \le u \le 2, 0 \le v \le 2\pi$.

(a) Find an equation in x, y, and z describing the surface.

(b) What are the geometric meanings of the parameters u and v?

(c) Find a unit vector orthogonal to the surface at $\Phi(u, v)$.

(d) Find the equation for the tangent plane at $\Phi(u_0, v_0) = (1, 1, 2)$ and express your answer in the following two ways:

   i. parametrized by u and v; and

   ii. in terms of x, y, and z.

(e) Find the area of S.

17. Let $f(x, y, z) = xe^y \cos \pi z$.

(a) Compute $\mathbf{F} = \nabla f$.

(b) Evaluate $\displaystyle\int_{\mathbf{c}} \mathbf{F} \cdot d\mathbf{s}$ where $\mathbf{c}(t) = (3 \cos^4 t, 5 \sin^7 t, 0), 0 \le t \le \pi$.

18. Let $\mathbf{F}(x, y, z) = x\mathbf{i} + y\mathbf{j} + z\mathbf{k}$. Evaluate $\displaystyle\iint_S \mathbf{F} \cdot d\mathbf{S}$ where S is the upper hemisphere of the unit sphere $x^2 + y^2 + z^2 = 1$.

19. Let $\mathbf{F}(x, y, z) = x\mathbf{i} + y\mathbf{j} + z\mathbf{k}$. Evaluate $\int_{\mathbf{c}} \mathbf{F} \cdot d\mathbf{s}$ where $\mathbf{c}(t) = (e^t, t, t^2)$, $0 \le t \le 1$.

20. Let $\mathbf{F} = \nabla f$ for a given scalar function. Let $\mathbf{c}(t)$ be a closed curve, that is, $\mathbf{c}(b) = \mathbf{c}(a)$. Show that $\displaystyle\int_{\mathbf{c}} \mathbf{F} \cdot d\mathbf{s} = 0$.

21. Consider the surface $\Phi(u, v) = (u^2 \cos v, u^2 \sin v, u)$. Compute the unit normal at $u = 1, v = 0$. Compute the equation of the tangent plane at this point.

22. Let S be the part of the cone $z^2 = x^2 + y^2$ with z between 1 and 2 oriented by the normal pointing out of the cone. Compute $\displaystyle\iint_S \mathbf{F} \cdot d\mathbf{S}$ where $\mathbf{F}(x, y, z) = (x^2, y^2, z^2)$.

23. Let $\mathbf{F} = x\mathbf{i} + x^2\mathbf{j} + yz\mathbf{k}$ represent the velocity field of a fluid (velocity measured in meters per second). How many cubic meters of fluid per second are crossing the xy plane through the square $0 \le x \le 1, 0 \le y \le 1$?

24. Show that the surface area of the part of the sphere $x^2 + y^2 + z^2 = 1$ lying above the rectangle $[-a, a] \times [-a, a]$, where $2a^2 < 1$, in the xy plane is

$$A = 2 \int_{-a}^{a} \sin^{-1} \left(\frac{a}{\sqrt{1 - x^2}} \right) dx.$$

25. Let S be a surface and C a closed curve bounding S. Verify the equality

$$\iint_S (\nabla \times \mathbf{F}) \cdot d\mathbf{S} = \int_C \mathbf{F} \cdot d\mathbf{s}$$

if $\mathbf{F}$ is a gradient field.

26. Calculate $\iint_S \mathbf{F} \cdot d\mathbf{S}$ where $\mathbf{F}(x, y, z) = (x, y, -y)$ and S is the cylindrical surface defined by $x^2 + y^2 = 1, 0 \leq z \leq 1$, with normal pointing out of the cylinder.

Integrate $f(x, y, z) = xyz$ along the paths in Exercises 27–30.

27. $\mathbf{c}(t) = (e^t \cos t, e^t \sin t, 3), 0 \leq t \leq 2\pi$

28. $\mathbf{c}(t) = (\cos t, \sin t, t), 0 \leq t \leq 2\pi$

29. $\mathbf{c}(t) = \frac{3}{2}t^2\mathbf{i} + 2t^2\mathbf{j} + t\mathbf{k}, 0 \leq t \leq 1$

30. $\mathbf{c}(t) = t\mathbf{i} + (1/\sqrt{2})t^2\mathbf{j} + \frac{1}{3}t^3\mathbf{k}, 0 \leq t \leq 1$

31. Find $\iint_S f\, dS$ in each of the following cases:

   (a) $f(x, y, z) = x; S$ is the part of the plane $x + y + z = 1$ in the positive octant $x \geq 0, y \geq 0, z \geq 0$;

   (b) $f(x, y, z) = x^2; S$ is the part of the plane $x = z$ inside the cylinder $x^2 + y^2 = 1$;

   (c) $f(x, y, z) = x; S$ is the part of the cylinder $x^2 + y^2 = 2x$ that satisfies $0 \leq z \leq \sqrt{x^2 + y^2}$.

32. Compute the integral of $f(x, y, z) = xyz$ over the rectangle with vertices $(1, 0, 1), (2, 0, 0), (1, 1, 1)$, and $(2, 1, 0)$.

33. Compute the integral of $x + y$ over the surface of the unit sphere.

34. Compute the surface integral of x over the triangle in the space having vertices $(1, 1, 1), (2, 1, 1)$, and $(2, 0, 3)$.

35. Let Γ be the curve of intersection of the plane $z = ax + by$ with the cylinder $x^2 + y^2 = 1$. Find all values of the real numbers a and b such that $a^2 + b^2 = 1$ and

$$\int_\Gamma y\, dx + (z - x)dy - y\, dz = 0.$$

7

The Integral Theorems
of Vector Analysis

All the theory of the motion of fluids has just been reduced to the solution of analytic formulas.

Leonhard Euler

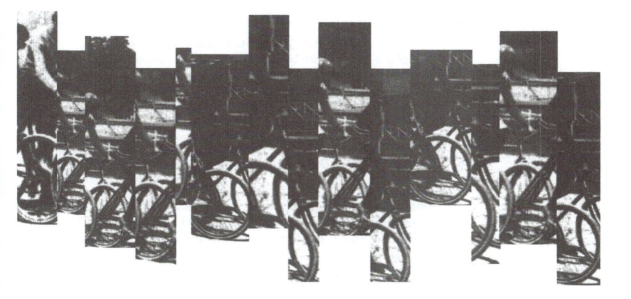

Three important theorems of Green, Gauss, and Stokes tie together the vector differential calculus from Chapter 4 and the vector integral calculus from Chapter 6. We shall point out some of the applications of these theorems to the study of electromagnetism and fluid dynamics.

Historical Note

The integral theorems of vector analysis had their origins in physics. For example, Green's theorem, discovered about 1828, arose in connection with the theory of gravitational and electrical potentials. Gauss' theorem—the divergence theorem—arose in connection with electrostatics (this theorem is jointly credited to the Russian mathematician Ostrogradsky). Stokes' theorem was first suggested in 1850 in a letter to Stokes from the physicist Lord Kelvin and was used in 1854 by Stokes in a mathematics examination for a competition at Cambridge University.

7.1
Green's
Theorem

Green's theorem expresses the line integral of a vector field around a closed plane curve in terms of a double integral over the region bounded by the curve. Among the many applications of this theorem is a formula for the area of a plane region in terms of a line integral around its boundary.

Since a line integral along a curve depends on the direction, or orientation, of the curve, we begin by specifying how to orient the boundary of a plane region. The rule is the following:

Orienting the Boundary Curve
If D is a plane region, with boundary curve C, the ***positive orientation*** of C is given by the vector $\mathbf{k} \times \mathbf{v}_{out}$, where $\mathbf{v}_{out}$ is the normal vector field pointing outward along the curve.

As you walk along C in the direction of the positive orientation, the region D is *on your left*.

When the boundary of D is given its positive orientation, we denote it by ∂D. (See Figure 7.1.1.)

The boundary of a region may have several pieces. For simple regions, the outer boundary is oriented counterclockwise, whereas any inner boundaries are oriented clockwise. For more complicated regions, you follow the same rule. (See Figure 7.1.2.)

Now we can state Green's theorem.

FIGURE 7.1.1. The correct orientation for the boundary ∂D of a region D.

FIGURE 7.1.2. Some plane regions and their oriented boundaries.

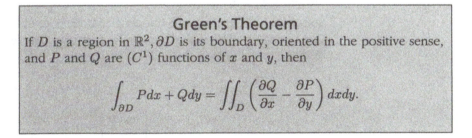

Green's Theorem

If D is a region in $\mathbb{R}^2$, ∂D is its boundary, oriented in the positive sense, and P and Q are (C^1) functions of x and y, then

$$\int_{\partial D} P\,dx + Q\,dy = \iint_D \left(\frac{\partial Q}{\partial x} - \frac{\partial P}{\partial y} \right) dx\,dy.$$

Before we prove Green's theorem, we give some examples of its use.

Example 1   Let C be the boundary of the square $[0,1] \times [0,1]$ oriented counterclockwise. Evaluate

$$\int_C (y^4 + x^3)\,dx + 2x^6\,dy.$$

Solution   We obtain a match with the line integral $\int_{\partial D} P\,dx + Q\,dy$ by choosing C to be the boundary of $D = [0,1] \times [0,1]$ and $P = y^4 + x^3, Q = 2x^6$. Then by Green's theorem,

$$\int_C (y^4 + x^3)\,dx + 2x^6\,dy \quad = \quad \iint_D \left[\frac{\partial}{\partial x} 2x^6 - \frac{\partial}{\partial y}(y^4 + x^3) \right] dx\,dy$$

$$
\begin{aligned}
&= \iint_D (12x^5 - 4y^3)\,dx\,dy \\
&= \int_0^1 \left[\int_0^1 (12x^5 - 4y^3)\,dx \right] dy \\
&= \int_0^1 (2 - 4y^3)\,dy = 1. \quad \blacklozenge
\end{aligned}
$$

Example 2   Let $\mathbf{F}(x, y) = y\mathbf{i} - x\mathbf{j}$ and let C be a circle of radius r traversed counterclockwise. Write the line integral $\int_C F(\mathbf{s}) \cdot d\mathbf{s}$ as a double integral using Green's theorem. Evaluate.

Solution   If $\mathbf{F}(x, y) = P\mathbf{i} + Q\mathbf{j}$, then $\mathbf{F} \cdot d\mathbf{s} = P\,dx + Q\,dy$. Now apply Green's theorem with D the disk of radius r, $Q = -x$, and $P = y$, so $\partial Q/\partial x - \partial P/\partial y = -2$. Thus

$$
\int_C \mathbf{F} \cdot d\mathbf{s} = \iint_D (-2)\,dx\,dy = (-2)(\text{area of } D) = -2\pi r^2. \quad \blacklozenge
$$

Example 3   Verify Green's theorem for $P(x, y) = x$ and $Q(x, y) = xy$ where D is the unit disk $x^2 + y^2 \leq 1$.

Solution   We evaluate the integrals on both sides of Green's theorem directly. The boundary of D is the unit circle. It is parametrized in the positive sense by $x = \cos t, y = \sin t, 0 \leq t \leq 2\pi$, so

$$
\begin{aligned}
\int_{\partial D} P\,dx + Q\,dy &= \int_0^{2\pi} [(\cos t)(-\sin t) + \cos t \sin t \cos t]\,dt \\
&= \left[\frac{\cos^2 t}{2} \right]_0^{2\pi} + \left[-\frac{\cos^3 t}{3} \right]_0^{2\pi} = 0.
\end{aligned}
$$

On the other hand,

$$
\iint_D \left(\frac{\partial Q}{\partial x} - \frac{\partial P}{\partial y} \right) dx\,dy = \iint_D y\,dx\,dy,
$$

which is zero by symmetry (the integrals over the portions $y \geq 0$ and $y \leq 0$ cancel). Thus Green's theorem is verified in this case. $\blacklozenge$

Sometimes we use Green's theorem to replace a double integral by a line integral. In particular, there is a formula for the area of a region bounded by a curve C, in terms of a line integral around C.

Area of a Region

If C is a curve that bounds a region D, then the area of D is

$$A = \frac{1}{2} \int_C x \, dy - y \, dx.$$

Proof Let $P = -y$ and $Q = x$; then by Green's theorem,

$$\frac{1}{2} \int_C x \, dy - y \, dx = \frac{1}{2} \iint_D \left[\frac{\partial x}{\partial x} - \frac{\partial(-y)}{\partial y} \right] dx \, dy = \iint_D dx \, dy,$$

which is the area of D. ∎

Example 4 Verify the above area formula if D is the disk $x^2 + y^2 \le r^2$.

Solution The area is πr^2. The preceding box with $x = r \cos t, y = r \sin t, 0 \le t \le 2\pi$, gives

$$\begin{aligned}
A &= \frac{1}{2} \int_C x \, dy - y \, dx = \frac{1}{2} \int_0^{2\pi} (r \cos t)(r \cos t)dt - (r \sin t)(-r \sin t)dt \\
&= \frac{1}{2} \int_0^{2\pi} r^2 \, dt = \pi r^2,
\end{aligned}$$

so the formula checks. ♦

Example 5 Use the area formula in the preceding box to find the area bounded by the ellipse $C : x^2/a^2 + y^2/b^2 = 1$.

Solution Parametrize C by $(a \cos t, b \sin t), 0 \le t \le 2\pi$. Then the area formula gives

$$\begin{aligned}
A &= \frac{1}{2} \int_C x \, dy - y \, dx = \frac{1}{2} \int_0^{2\pi} (a \cos t)(b \cos t)dt - (b \sin t)(-a \sin t)dt \\
&= \frac{1}{2} \int_0^{2\pi} ab \, dt = ab\pi. \quad ♦
\end{aligned}$$

Example 6 Compute the area of the region enclosed by the hypocycloid $x^{2/3} + y^{2/3} = a^{2/3}$ using the parametrization (see Fig. 7.1.3)

$$x = a \cos^3 \theta, \quad y = a \sin^3 \theta, \quad 0 \le \theta \le 2\pi.$$

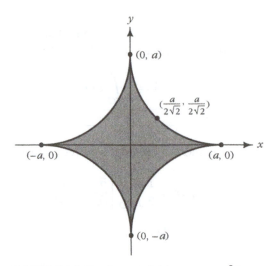

FIGURE 7.1.3. The hypocycloid $x = a\cos^3\theta, y = a\sin^3\theta, 0 \le \theta \le 2\pi$.

Solution

$$
\begin{aligned}
A &= \frac{1}{2}\int_{\partial D} x\,dy - y\,dx \\
&= \frac{1}{2}\int_0^{2\pi}[(a\cos^3\theta)(3a\sin^2\theta\cos\theta) - (a\sin^3\theta)(-3a\cos^2\theta\sin\theta)]d\theta \\
&= \frac{3}{2}a^2\int_0^{2\pi}(\sin^2\theta\cos^4\theta + \cos^2\theta\sin^4\theta)d\theta.
\end{aligned}
$$

Using $\sin^2\theta + \cos^2\theta = 1$ and $2\sin\theta\cos\theta = \sin 2\theta$, one derives the trig identity $\sin^2\theta\cos^4\theta + \cos^2\theta\sin^4\theta = \sin^2\theta\cos^2\theta = \frac{1}{4}\sin^2 2\theta$, so we get

$$
\begin{aligned}
A &= \frac{3}{8}a^2\int_0^{2\pi}\sin^2 2\theta\,d\theta = \frac{3}{8}a^2\int_0^{2\pi}\left(\frac{1 - \cos 4\theta}{2}\right)d\theta \\
&= \frac{3}{16}a^2\int_0^{2\pi}d\theta - \frac{3}{16}a^2\int_0^{2\pi}\cos 4\theta\,d\theta = \frac{3}{8}\pi a^2. \quad \blacklozenge
\end{aligned}
$$

We now give a proof of Green's theorem when D is a region of both type 1 and type 2 (e.g., a rectangle). After this is done, more general regions can be dealt with by decomposition into simpler pieces. The proof that follows uses two crucial steps: first, double integrals can be reduced to iterated integrals and second, the fundamental theorem of calculus can be applied to the resulting one-variable integrals.

Proof To follow the argument, refer to Figure 7.1.4, which shows a region D of type 1 and type 2 described in two different ways. In each case, ∂D consists of two graphs and two line segments. Depending upon the region, either or both of the line segments may be reduced to a single point.

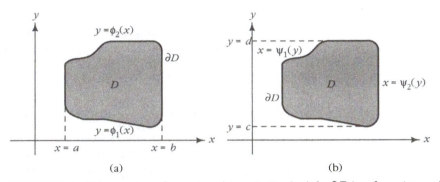

FIGURE 7.1.4. D is a region of type 1 and type 2. On the left, ∂D has four pieces: the graphs $y = \phi_1(x)$ and $y = \phi_2(x)$ and two vertical line segments contained in $x = a$ and $x = b$. On the right, ∂D has three pieces: the graphs $x = \psi_1(y)$ and $x = \psi_2(y)$, and the horizontal line segment contained in $y = d$. In this example, the "bottom" segment contained in $y = c$ is a point.

In the double integral

$$\iint_D \left(\frac{\partial Q}{\partial x} - \frac{\partial P}{\partial y} \right) dx\,dy = \iint_D \frac{\partial Q}{\partial x} dx\,dy - \iint_D \frac{\partial P}{\partial y} dx\,dy,$$

we will use a different order of iterated integration for each term. Since D is of type 2, the first double integral can be written as an iterated integral:

$$\iint_D \frac{\partial Q}{\partial x} dx\,dy = \int_c^d \int_{\psi_1(y)}^{\psi_2(y)} \frac{\partial Q}{\partial x}(x, y)dx\,dy.$$

Applying the fundamental theorem of calculus to the integration over x gives

$$\int_c^d (Q(\psi_2(y), y) - Q(\psi_1(y), y))dy = \int_c^d Q(\psi_2(y), y)dy - \int_c^d Q(\psi_1(y), y)dy.$$

We now interpret the last two integrals as line integrals.

In Figure 7.1.4(b), consider the integral of $Q\,dy$ along the right-hand boundary piece, which is the graph $x = \psi_2(y)$. Parametrizing this piece by $\mathbf{c}(t) = (\psi_2(t), t)$ for $c \leq t \leq d$,

$$\int_{\mathbf{c}} Q\,dy = \int_c^d Q(\psi_2(t), t)\frac{dy}{dt}dt = \int_c^d Q(\psi_2(t), t)dt$$

which, after a name change of the variable of integration from t to y, becomes

$$\int_c^d Q(\psi_2(y),y)dy.$$

Notice that we were integrating in the *upward* direction. If we integrate $Q\,dy$ along the left-hand piece of ∂D, $x = \psi_1(y)$, in the *downward* direction, the result is the other term, $-\int_c^d Q(\psi_1(y),y)dy.$

We have now shown that $\iint_D \dfrac{\partial Q}{\partial x}\,dx\,dy$ is the line integral of $Q\,dy$ along the boundary ∂D, with the top and bottom pieces omitted. Since y is constant along these pieces, $dy/dt = 0$ for any parametrization of them; hence these pieces make no contribution to $\int_{\partial D} Q\,dy$. We conclude that

$$\iint_D \frac{\partial Q}{\partial x}\,dx\,dy = \int_{\partial D} Q\,dy.$$

A similar argument, with the roles of x and y reversed, and using the fact that D is of type 1, gives:

$$\iint_D \frac{\partial P}{\partial y}\,dx\,dy = -\int_{\partial D} P\,dx.$$

[The minus sign occurs because the orientation of $y = \phi_2(x)$ must be in the direction of *decreasing x*.]

Subtracting these last two equations gives Green's theorem:

$$\iint_D \left(\frac{\partial Q}{\partial x} - \frac{\partial P}{\partial y}\right)dx\,dy = \int_{\partial D} P\,dx + Q\,dy. \quad \blacksquare$$

Note

At the end of this section, you will learn a vector form of Green's theorem to help you remember which term carries the negative sign.

To prove Green's theorem for a more general region, we break it into pieces and apply the preceding result to each piece.

Example 7   Show that Green's theorem is valid for the region D shown in Figure 7.1.5.

Solution   Figure 7.1.6 shows how to divide up D into three regions, $D_1, D_2,$ and D_3, each of which is of types 1 and 2. Let $C_1, C_2,$ and C_3 be the boundary curves of these regions. Then

$$\iint_{D_i} \left(\frac{\partial Q}{\partial x} - \frac{\partial P}{\partial y}\right)dx\,dy = \int_{C_i} P\,dx + Q\,dy.$$

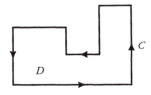

FIGURE 7.1.5. Is Green's theorem valid for D?

The double integral over the D_i's adds up to the double integral over D, so

$$\iint_D \left(\frac{\partial Q}{\partial x} - \frac{\partial P}{\partial y} \right) dx\,dy = \int_{C_1} P\,dx + Q\,dy + \int_{C_2} P\,dx + Q\,dy + \int_{C_3} P\,dx + Q\,dy.$$

Since the dotted portions of the boundaries shown in Figure 7.1.6 are traversed twice in opposite directions, these cancel in the line integrals. Thus we are left with

$$\iint_D \left(\frac{\partial Q}{\partial x} - \frac{\partial P}{\partial y} \right) dx\,dy = \int_C P\,dx + Q\,dy,$$

and so Green's theorem is valid. ◆

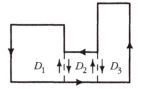

FIGURE 7.1.6. Breaking a region up into smaller regions, each of which is both type 1 and type 2.

Example 7 illustrates a special case of the following procedure for a region D:

1. Break up D into smaller regions, $D_1, D_2, \ldots, D_n$, each of which is of types 1 and 2.

2. Apply Green's theorem as proven above to each of $D_1, \ldots, D_n$, and add the resulting integrals.

3. The line integrals along interior boundaries cancel, leaving the line integral around the boundary of D.

This procedure yields Green's theorem for D. It is plausible that *this method applies to any region bounded by piecewise smooth curves*, and so we obtain the general form of Green's theorem.

Green's theorem may be rewritten in the language of vector fields.

Vector Form of Green's Theorem

Let $\mathbf{F} = P\mathbf{i} + Q\mathbf{j}$ be a C^1 vector field on a region D in the plane. Then

$$\int_{\partial D} \mathbf{F} \cdot d\mathbf{s} = \iint_D (\text{curl } \mathbf{F}) \cdot \mathbf{k}\, dx dy = \iint_D (\nabla \times \mathbf{F}) \cdot \mathbf{k}\, dx dy$$

(see Figure 7.1.7).

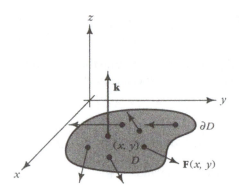

FIGURE 7.1.7. The vector form of Green's theorem.

Proof Let $\mathbf{F} = P\mathbf{i} + Q\mathbf{j}$ so that

$$\text{curl } \mathbf{F} = \begin{vmatrix} \mathbf{i} & \mathbf{j} & \mathbf{k} \\ \dfrac{\partial}{\partial x} & \dfrac{\partial}{\partial y} & \dfrac{\partial}{\partial z} \\ P & Q & 0 \end{vmatrix} = \left(\frac{\partial Q}{\partial x} - \frac{\partial P}{\partial y} \right) \mathbf{k},$$

and so

$$\iint_D \text{curl } \mathbf{F} \cdot \mathbf{k}\, dx dy = \iint_D \left(\frac{\partial Q}{\partial x} - \frac{\partial P}{\partial y} \right) dx dy$$

$$= \int_{\partial D} P\, dx + Q\, dy = \int_{\partial D} \mathbf{F} \cdot d\mathbf{s}$$

by Green's theorem. ∎

Example 8   Let $\mathbf{F} = xy^2\mathbf{i} + (y+x)\mathbf{j}$. Integrate $(\nabla \times \mathbf{F}) \cdot \mathbf{k}$ over the region in the first quadrant bounded by the curves $y = x^2$ and $y = x$. (Figure 7.1.8.)

Solution   *Method 1*. Here we compute

$$\nabla \times \mathbf{F} = \left(0, 0, \frac{\partial F_2}{\partial x} - \frac{\partial F_1}{\partial y} \right) = (1 - 2xy)\mathbf{k}.$$

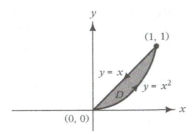

FIGURE 7.1.8. The region bounded by the curves $y = x^2$ and $y = x$.

Thus $(\nabla \times \mathbf{F}) \cdot \mathbf{k} = 1 - 2xy$. This can be integrated over D using an iterated integral:

$$
\begin{aligned}
\iint_D (\nabla \times \mathbf{F}) \cdot \mathbf{k}\, dx\, dy &= \int_0^1 \int_{x^2}^x (1 - 2xy)\, dy\, dx = \int_0^1 [y - xy^2]_{x^2}^x\, dx \\
&= \int_0^1 [x - x^3 - x^2 + x^5]\, dx \\
&= \frac{1}{2} - \frac{1}{4} - \frac{1}{3} + \frac{1}{6} = \frac{1}{12}.
\end{aligned}
$$

Method 2. Here we use Green's theorem in vector form:

$$
\iint_D (\nabla \times \mathbf{F}) \cdot \mathbf{k}\, dx\, dy = \int_{\partial D} \mathbf{F} \cdot d\mathbf{s}.
$$

The line integral of $\mathbf{F}$ along the curve $y = x$ from left to right is

$$
\int_0^1 F_1\, dx + F_2\, dy = \int_0^1 (x^3 + 2x)\, dx = \frac{1}{4} + 1 = \frac{5}{4}.
$$

Along the curve $y = x^2$ we get

$$
\int_0^1 F_1\, dx + F_2\, dy = \int_0^1 x^5 dx + (x + x^2)(2x\, dx) = \frac{1}{6} + \frac{2}{3} + \frac{1}{2} = \frac{4}{3}.
$$

Remembering that the integral along $y = x$ is to be taken from right to left, we have

$$
\int_{\partial D} \mathbf{F} \cdot d\mathbf{s} = \frac{4}{3} - \frac{5}{4} = \frac{1}{12}. \quad \blacklozenge
$$

Green's theorem can also be written in a form involving the divergence:

Gauss' Divergence Theorem in the Plane

Let D be a region in the plane, and let $\mathbf{n}$ be the outward unit normal along the boundary curve ∂D. If $\mathbf{F}$ is any vector field on D,

$$\int_{\partial D} (\mathbf{F} \cdot \mathbf{n}) ds = \iint_D \text{div } \mathbf{F}\, dx dy.$$

Proof The unit normal vector to ∂D is $\mathbf{n} = -\mathbf{k} \times \mathbf{T}$ where $\mathbf{c}$ parametrizes ∂D and $\mathbf{T} = \mathbf{c}'/\|\mathbf{c}'\|$ is the unit tangent. Thus,

$$\int_{\partial D} (\mathbf{F} \cdot \mathbf{n}) ds = -\int_{\partial D} \mathbf{F} \cdot (\mathbf{k} \times \mathbf{T}) ds.$$

By the properties of the triple product, $\mathbf{F} \cdot (\mathbf{k} \times \mathbf{T}) = \mathbf{T} \cdot (\mathbf{F} \times \mathbf{k}) = (\mathbf{F} \times \mathbf{k}) \cdot \mathbf{T}$, and so (using $d\mathbf{s} = \mathbf{T} ds$),

$$\int_{\partial D} (\mathbf{F} \cdot \mathbf{n}) ds = \int_{\partial D} (\mathbf{k} \times \mathbf{F}) \cdot \mathbf{T}\, ds = \int_{\partial D} (\mathbf{k} \times \mathbf{F}) \cdot d\mathbf{s},$$

an ordinary line integral. If $\mathbf{F} = A\mathbf{i} + B\mathbf{j}$, then $\mathbf{k} \times \mathbf{F} = A\mathbf{j} - B\mathbf{i}$, and so applying Green's theorem to the line integral (using $P = -B$ and $Q = A$) gives

$$\iint_D \left(\frac{\partial Q}{\partial x} - \frac{\partial P}{\partial y} \right) dx dy = \iint_D \left(\frac{\partial A}{\partial x} + \frac{\partial B}{\partial y} \right) dx dy = \iint_D \text{div } \mathbf{F}\, dx dy,$$

as required. ∎

The left-hand side of Gauss' divergence theorem represents the *flux* of the vector field $\mathbf{F}$ through the boundary curve ∂D. Thus, if $\mathbf{F}$ represents the flow of a fluid, the theorem says that *the net flux across the boundary ∂D equals the integral over D of the rate* div $\mathbf{F}$ *at which fluid area is being created*. In particular, if the fluid is neither compressed nor expanded (div $\mathbf{F} = 0$), the net flow across the closed curve ∂D is zero.

Example 9   Let $\mathbf{F} = y^3\mathbf{i} + x^5\mathbf{j}$. Compute the flux of $\mathbf{F}$ across the boundary of the square $[0, 1] \times [0, 1]$.

Solution   This can be done using the divergence theorem. Indeed,

$$\int_{\partial D} \mathbf{F} \cdot \mathbf{n}\, ds = \iint_D \text{div } \mathbf{F} dA.$$

But div $\mathbf{F} = 0$, and so the integral is zero.   ◆

Exercises for §7.1

Verify Green's theorem for the disk D with center $(0,0)$ and radius R and the functions in Exercises 1–4.

1. $P(x,y) = xy^2, Q(x,y) = -yx^2$

2. $P(x,y) = x + y, Q(x,y) = y$

3. $P(x,y) = xy = Q(x,y)$

4. $P(x,y) = 2y, Q(x,y) = x$

5. Evaluate $\int_C y\,dx - x\,dy$ where C is the boundary of the square $[-1,1] \times [-1,1]$ oriented in the counterclockwise direction by using Green's theorem.

6. Find the area of the region between two concentric circles of radii r_1 and r_2, where $r_1 < r_2$, using Green's theorem.

7. Show that the area enclosed by the hypocycloid $x = a\sin^3\theta$, $y = a\cos^3\theta$, $0 \le \theta \le 2\pi$, is $\frac{3}{8}\pi a^2$. (Use Green's theorem.)

8. Find the area bounded by one arc of the cycloid $x = a(\theta - \sin\theta), y = a(1 - \cos\theta)$, where $a > 0, 0 \le \theta \le 2\pi$.

9. Let C be the ellipse $x^2/a^2 + y^2/b^2 = 1$, and let $\mathbf{F}(x,y) = xy^2\mathbf{i} - yx^2\mathbf{j}$. Write $\int_C \mathbf{F(s)} \cdot d\mathbf{s}$ as a double integral using Green's theorem. Evaluate.

10. Let $\mathbf{F}(x,y) = (2y + e^x)\mathbf{i} + (x + \sin(y^2))\mathbf{j}$ and C be the circle $x^2 + y^2 = 1$. Write $\int_C \mathbf{F(s)} \cdot d\mathbf{s}$ as a double integral and evaluate.

11. Let C be the boundary of the rectangle $[1,2] \times [1,2]$. Evaluate the line integral $\int_C x^2y\,dx + 3yx^2\,dy$ by using Green's theorem.

12. Evaluate $\int_C (x^5 - 2xy^3)dx - 3x^2y^2\,dy$, where C is parametrized by (t^8, t^{10}), $0 \le t \le 1$.

Let C be the boundary of the rectangle with sides $x = 1, y = 2, x = 3$, and $y = 3$. Evaluate the integrals in Exercises 13–16.

13. $\int_C (2y^2 + x^5)dx + 3y^6\,dy$

14. $\int_C (xy^2 - y^3)dx + (-5x^2 + y^3)dy$

15. $\int_C (3x^4 + 5)dx + (y^5 + 3y^2 - 1)dy$

16. $\int_C \left(\dfrac{2y + \sin x}{1 + x^2} \right) dx + \left(\dfrac{x + e^y}{1 + y^2} \right) dy$

17. Evaluate $\int_C (2x^3 - y^3)dx + (x^3 + y^3)dy$, where C is the unit circle, and verify Green's theorem for this case.

18. Using the divergence theorem, show that $\int_{\partial D} \mathbf{F} \cdot \mathbf{n}\,ds = 0$ where $\mathbf{F}(x, y) = y\mathbf{i} - x\mathbf{j}$ and D is the unit disk. Verify this directly.

19. Let $P = y$ and $Q = x$. What is $\int_C P\,dx + Q\,dy$ if C is a closed curve?

20. Find the work done by the force field $(3x + 4y)\mathbf{i} + (8x + 9y)\mathbf{j}$ on a particle that moves once around the ellipse $4x^2 + 9y^2 = 36$ by (a) directly evaluating the line integral and by (b) using Green's theorem.

21. Use Green's theorem to compute the area inside the ellipse $x^2/a^2 + y^2/b^2 = 1$.

22. Use Green's theorem to recover the formula $A = \dfrac{1}{2} \int_a^b r^2\,d\theta$ for the area of a region in polar coordinates.

23. Sketch the proof of Green's theorem for the region shown in Figure 7.1.9.

FIGURE 7.1.9. Prove Green's theorem for this region.

24. Show that if C is a simple closed curve that bounds a region to which Green's theorem applies, then the area of the region D bounded by C is

$$A = \int_{\partial D} x\,dy = -\int_{\partial D} y\,dx.$$

25. Use Green's theorem to find the area of one loop of the four-leafed rose defined by $r = 3\sin 2\theta$. (Hint: $x\,dy - y\,dx = r^2\,d\theta$.)

26. Under the conditions of Green's theorem, prove:

(a)

$$\int_{\partial D} PQ\, dx + PQ\, dy$$

$$= \iint_D \left[Q \left(\frac{\partial P}{\partial x} - \frac{\partial P}{\partial y} \right) + P \left(\frac{\partial Q}{\partial x} - \frac{\partial Q}{\partial y} \right) \right] dx dy;$$

(b)

$$\int_{\partial D} \left(Q \frac{\partial P}{\partial x} - P \frac{\partial Q}{\partial x} \right) dx + \left(P \frac{\partial Q}{\partial y} - Q \frac{\partial P}{\partial y} \right) dy$$

$$= 2 \iint_D \left(P \frac{\partial^2 Q}{\partial x \partial y} - Q \frac{\partial^2 P}{\partial x \partial y} \right) dx dy.$$

27. Prove the identity

$$\int_{\partial D} \phi \nabla \phi \cdot \mathbf{n}\, ds = \iint_D (\phi \nabla^2 \phi + \nabla \phi \cdot \nabla \phi) dx dy.$$

28. Suppose that f is harmonic; that is,

$$\frac{\partial^2 f}{\partial x^2} + \frac{\partial^2 f}{\partial y^2} = 0$$

on a region D. Prove that

$$\int_{\partial D} \frac{\partial f}{\partial y} dx - \frac{\partial f}{\partial x} dy = 0.$$

29. Let $P(x,y) = -y/(x^2 + y^2), Q(x,y) = x/(x^2 + y^2)$. Assuming D is the unit disk, investigate why Green's theorem seems to fail for this P and Q.

30. *Project:* The formula $A = \frac{1}{2} \int_C x\, dy - y\, dx$ is the basis for the operation of the *planimeter*, a mechanical device for measuring areas. Find out about planimeters from an encyclopedia or the *American Mathematical Monthly*, Vol. 88, No. 9, November (1981), p. 701, and relate their operation to this formula for area.

7.2 Stokes' Theorem

Stokes' theorem generalizes Green's theorem by letting the vector field **F**, the curve C, and the region D bounded by C be in space rather than the plane. In fact, the vector form of Green's theorem given in §7.1 is already well on the way toward Stokes' theorem. That formula,

$$\iint_D (\text{curl}\, \mathbf{F}) \cdot \mathbf{k}\, dx dy = \int_{\partial D} \mathbf{F} \cdot d\mathbf{s},$$

involves a three-dimensional concept—the curl of the vector field $\mathbf{F}$. Notice that $\mathbf{k}$ is the unit normal vector to D when it is considered as a region in $\mathbb{R}^3$, and so $\iint_D (\operatorname{curl}\mathbf{F}) \cdot \mathbf{k}\,dx\,dy$ is the surface integral $\iint_D (\operatorname{curl}\mathbf{F}) \cdot d\mathbf{S}$. When generalized to space, this becomes an integral over a surface S. The boundary of S is a curve denoted ∂S and when it is properly oriented, as discussed below, Stokes' theorem states that

Stokes' Theorem

$$\iint_S (\operatorname{curl}\mathbf{F}) \cdot d\mathbf{S} = \int_{\partial S} \mathbf{F} \cdot d\mathbf{s}.$$

Proof of Stokes' theorem for graphs (After this proof, we discuss the more general case of parametrized surfaces.) Consider a surface S that is the graph of a function $f(x, y)$, so that S is parametrized by

$$
\begin{aligned}
x &= u \\
y &= v \\
z &= f(u, v) = f(x, y)
\end{aligned}
$$

for (u, v) in some domain D in $\mathbb{R}^2$ (to which Green's theorem applies). The integral of a vector function $\mathbf{G}$ over S was shown in §**6.4** to be

$$\iint_S \mathbf{G} \cdot d\mathbf{S} = \iint_D \left[G_1 \left(-\frac{\partial z}{\partial x} \right) + G_2 \left(-\frac{\partial z}{\partial y} \right) + G_3 \right] dx\,dy,$$

where $\mathbf{G} = G_1\mathbf{i} + G_2\mathbf{j} + G_3\mathbf{k}$. If $\mathbf{F} = F_1\mathbf{i} + F_2\mathbf{j} + F_3\mathbf{k}$, then

$$\operatorname{curl}\mathbf{F} = \left(\frac{\partial F_3}{\partial y} - \frac{\partial F_2}{\partial z} \right)\mathbf{i} + \left(\frac{\partial F_1}{\partial z} - \frac{\partial F_3}{\partial x} \right)\mathbf{j} + \left(\frac{\partial F_2}{\partial x} - \frac{\partial F_1}{\partial y} \right)\mathbf{k},$$

and so the preceding formula with $\mathbf{G} = \operatorname{curl}\mathbf{F}$ gives

$$
\begin{aligned}
\iint_S \operatorname{curl}\mathbf{F} \cdot d\mathbf{S} = \iint_D \Bigg[&\left(\frac{\partial F_3}{\partial y} - \frac{\partial F_2}{\partial z} \right)\left(-\frac{\partial z}{\partial x} \right) \\
&+ \left(\frac{\partial F_1}{\partial z} - \frac{\partial F_3}{\partial x} \right)\left(-\frac{\partial z}{\partial y} \right) + \left(\frac{\partial F_2}{\partial x} - \frac{\partial F_1}{\partial y} \right) \Bigg] dx\,dy.
\end{aligned}
$$

This formula assumes that the surface has been oriented with the upward normal. The boundary of D is oriented as in Green's theorem and the corresponding curve on the graph of f is oriented the same way, as in Figure 7.2.1.

Let $(x(t), y(t))$, where $a \le t \le b$, be a parametrization of ∂D. Then

$$\mathbf{c}(t) = (x(t), y(t), f(x(t), y(t)))$$

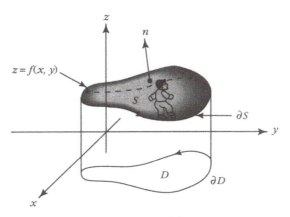

FIGURE 7.2.1. As you traverse ∂S counterclockwise, the surface is on your left.

is a parametrization of ∂S. Thus,

$$\int_{\partial S} \mathbf{F} \cdot d\mathbf{s} = \int_a^b \left(F_1 \frac{dx}{dt} + F_2 \frac{dy}{dt} + F_3 \frac{dz}{dt} \right) dt.$$

By the chain rule,

$$\frac{dz}{dt} = \frac{\partial z}{\partial x} \frac{dx}{dt} + \frac{\partial z}{\partial y} \frac{dy}{dt}.$$

Substituting this expression into the preceding equation gives

$$
\begin{aligned}
\int_{\partial S} \mathbf{F} \cdot d\mathbf{s} &= \int_a^b \left[\left(F_1 + F_3 \frac{\partial z}{\partial x} \right) \frac{dx}{dt} + \left(F_2 + F_3 \frac{\partial z}{\partial y} \right) \frac{dy}{dt} \right] dt \\
&= \int_{\partial D} \left(F_1 + F_3 \frac{\partial z}{\partial x} \right) dx + \left(F_2 + F_3 \frac{\partial z}{\partial y} \right) dy.
\end{aligned}
$$

Now apply Green's theorem to this expression to get

$$\iint_D \left[\frac{\partial (F_2 + F_3 \partial z / \partial y)}{\partial x} - \frac{\partial (F_1 + F_3 \partial z / \partial x)}{\partial y} \right] dx\, dy.$$

Use the chain rule, remembering that F_1, F_2, and F_3 are functions of x, y, and z and that z is a function of x and y, to obtain

$$
\begin{aligned}
\iint_D \Bigg[&\left(\frac{\partial F_2}{\partial x} + \frac{\partial F_2}{\partial z} \frac{\partial z}{\partial x} + \frac{\partial F_3}{\partial x} \frac{\partial z}{\partial y} + \frac{\partial F_3}{\partial z} \frac{\partial z}{\partial x} \frac{\partial z}{\partial y} + F_3 \frac{\partial^2 z}{\partial x \partial y} \right) \\
&- \left(\frac{\partial F_1}{\partial y} + \frac{\partial F_1}{\partial z} \frac{\partial z}{\partial y} + \frac{\partial F_3}{\partial y} \frac{\partial z}{\partial x} + \frac{\partial F_3}{\partial z} \frac{\partial z}{\partial y} \frac{\partial z}{\partial x} + F_3 \frac{\partial^2 z}{\partial y \partial x} \right) \Bigg] dA.
\end{aligned}
$$

The last two terms in each parentheses cancel each other, and we can rearrange terms to obtain the previous formula for $\iint_S (\text{curl}\, \mathbf{F}) \cdot d\mathbf{S}$. ∎

For parametrized surfaces $\Phi : D \to \mathbb{R}^3$, we assume that Φ is one to one on the whole of D (that is, no two points in D get sent to the same point of $\mathbb{R}^3$). Then the image of D is S and the image of ∂D is ∂S. If we assume that ∂D is oriented as in Green's theorem, we get an orientation for ∂S and S itself is oriented so $\Phi_u \times \Phi_v$ is in the direction of the normal. See Figure 7.2.2.

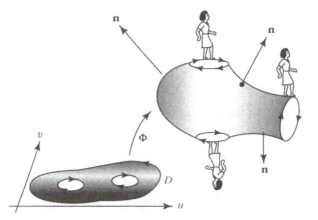

FIGURE 7.2.2. The boundary ∂S of D is oriented so that if you walk along ∂S with $\mathbf{n}$ your upright direction, the surface is on your left.

If we write out the surface integral of curl $\mathbf{F}$ and the line integral of $\mathbf{F}$ using the parametrization and apply Green's theorem as we did above for graphs, we find the same conclusion. The computation is somewhat longer, but proceeds as above, so we omit it.

Stokes' Theorem

Let S be an oriented surface with boundary ∂S oriented as in Figure 7.2.2. Then for any (C^1) vector field $\mathbf{F}$ (defined in a region containing S),

$$\iint_S (\operatorname{curl} \mathbf{F}) \cdot d\mathbf{S} = \int_{\partial S} \mathbf{F} \cdot d\mathbf{s}.$$

Stokes' theorem holds more generally for surfaces that can be decomposed into those already described, as we did for Green's theorem.

If the surface S has no boundary (like a sphere or the surface of a donut) then $\iint_S (\operatorname{curl} \mathbf{F}) \cdot d\mathbf{S} = 0$. For a sphere, this can be seen by cutting the sphere into two hemispheres, applying Stokes' theorem to each and adding the results; the two boundary integrals cancel.

Example 1   Let $\mathbf{F} = ye^z\mathbf{i} + xe^z\mathbf{j} + xye^z\mathbf{k}$. Use Stokes' theorem to show that the integral of $\mathbf{F}$ around the boundary of an oriented surface S is 0.

Solution   Indeed, $\displaystyle\int_{\partial S}\mathbf{F}\cdot d\mathbf{s} = \iint_S(\nabla\times\mathbf{F})\cdot d\mathbf{S}$, by Stokes' theorem. But we compute

$$\nabla\times\mathbf{F} = \begin{vmatrix} \mathbf{i} & \mathbf{j} & \mathbf{k} \\ \dfrac{\partial}{\partial x} & \dfrac{\partial}{\partial y} & \dfrac{\partial}{\partial z} \\ ye^z & xe^z & xye^z \end{vmatrix} = \mathbf{0},$$

and so $\int_{\partial S}\mathbf{F}\cdot d\mathbf{s} = 0$. (Alternatively, we can observe that $\mathbf{F} = \nabla(xye^z)$, so its integral around a closed curve is zero.) ◆

Example 2   Find the integral of $\mathbf{F}(x,y,z) = z\mathbf{i} - x\mathbf{j} - y\mathbf{k}$ around the triangle with vertices $(0,0,0), (0,2,0)$ and $(0,0,2)$, using Stokes' theorem.

Solution   Refer to Figure 7.2.3. C is the triangle in question and S is a surface that it bounds. By Stokes' theorem,

$$\int_C\mathbf{F}\cdot d\mathbf{s} = \iint_S(\nabla\times\mathbf{F})\cdot\mathbf{n}\,dS.$$

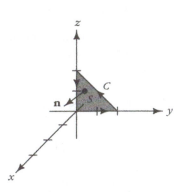

FIGURE 7.2.3. The curve C of integration for Example 2.

Now

$$\nabla\times\mathbf{F} = \begin{vmatrix} \mathbf{i} & \mathbf{j} & \mathbf{k} \\ \dfrac{\partial}{\partial x} & \dfrac{\partial}{\partial y} & \dfrac{\partial}{\partial z} \\ z & -x & -y \end{vmatrix} = -\mathbf{i} + \mathbf{j} - \mathbf{k}.$$

Thus $(\nabla \times \mathbf{F}) \cdot \mathbf{n} = (-\mathbf{i} + \mathbf{j} - \mathbf{k}) \cdot \mathbf{i} = -1$, so

$$\iint_S (\nabla \times \mathbf{F} \cdot \mathbf{n})dS = -1 \,(\text{area } S) = -1 \cdot \frac{1}{2} \cdot 2 \cdot 2 = -2. \quad \blacklozenge$$

Example 3   Let S be the portion of the sphere of radius 2 shown in Figure 7.2.4, with the indicated orientation. Let $\mathbf{F} = y\mathbf{i} - x\mathbf{j} + e^{xz}\mathbf{k}$. Evaluate $\iint_S (\nabla \times \mathbf{F}) \cdot d\mathbf{S}$.

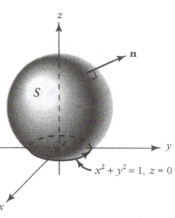

FIGURE 7.2.4. The surface S and its boundary for Example 3.

Solution   This is a parametrized surface and may be parametrized using spherical coordinates. However, we need not find Φ explicitly to solve this problem! By Stokes' theorem,

$$\iint_S (\nabla \times \mathbf{F}) \cdot d\mathbf{S} = \int_{\partial S} \mathbf{F} \cdot d\mathbf{s},$$

so if we parametrize ∂S by $x(t) = \cos t, y(t) = \sin t, z(t) = 0, 0 \le t \le 2\pi$, we determine

$$
\begin{aligned}
\int_{\partial S} \mathbf{F} \cdot d\mathbf{s} &= \int_0^{2\pi} \left(y\frac{dx}{dt} - x\frac{dy}{dt} \right) dt \\
&= \int_0^{2\pi} (-\sin^2 t - \cos^2 t)dt = -\int_0^{2\pi} dt = -2\pi,
\end{aligned}
$$

and therefore $\iint_S (\nabla \times \mathbf{F}) \cdot d\mathbf{S} = -2\pi$. $\quad \blacklozenge$

Example 4   Use Stokes' theorem to evaluate the line integral

$$\int_C -y^3 dx + x^3 dy - z^3 dz,$$

where C is the intersection of the cylinder $x^2 + y^2 = 1$ and the plane $x+y+z = 1$, and the orientation of C corresponds to counterclockwise motion in the xy plane.

Solution   The curve C bounds the surface S defined by $z = 1 - x - y = f(x, y)$ for (x, y) in the disk in the xy plane defined by $x^2 + y^2 \le 1$ (Figure 7.2.5).

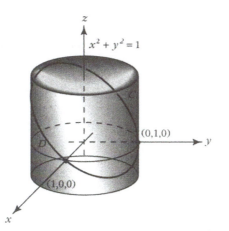

FIGURE 7.2.5. The intersection of the cylinder $x^2 + y^2 = 1$ and the plane $x+y+z = 1$ is a curve C.

We set $\mathbf{F} = -y^3\mathbf{i} + x^3\mathbf{j} - z^3\mathbf{k}$. By Stokes' theorem, the line integral is equal to the surface integral

$$\iint_S (\nabla \times \mathbf{F}) \cdot d\mathbf{S}.$$

But $\nabla \times \mathbf{F} = (3x^2 + 3y^2)\mathbf{k}$ has only a $\mathbf{k}$ component. Thus, we have (see the box on p. 402)

$$\iint_S (\nabla \times \mathbf{F}) \cdot d\mathbf{S} = \iint_D (3x^2 + 3y^2)dxdy.$$

This integral can be evaluated in polar coordinates:

$$3\iint_D (x^2 + y^2)dxdy = 3\int_0^1 \int_0^{2\pi} r^2 \cdot r\, d\theta dr = 6\pi \int_0^1 r^3 dr = \frac{6\pi}{4} = \frac{3\pi}{2}. \quad \blacklozenge$$

In this example one can evaluate the line integral directly, but it is somewhat tedious.

Stokes' theorem implies that *if two oriented surfaces have the same oriented boundary, then the integrals of the curl of a vector field over the two surfaces are equal.* (See Exercise 15 and compare this with the path independence of *line* integrals of *gradient* vector fields.)

Example 5   Evaluate the surface integral $\iint_S (\nabla \times \mathbf{F}) \cdot \mathbf{n}\,dS$, where S is the portion of the surface of a sphere defined by $x^2 + y^2 + z^2 = 1$ and $x + y + z \geq 1$, where $\mathbf{F} = \mathbf{r} \times (\mathbf{i} + \mathbf{j} + \mathbf{k})$, and $\mathbf{r} = x\mathbf{i} + y\mathbf{j} + z\mathbf{k}$. The orientation of S is taken to be that of the outward normal.

Solution   The surface S and its oriented boundary ∂S are shown in Figure 7.2.6.

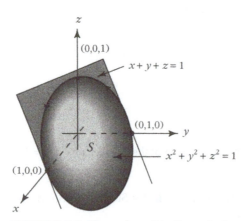

FIGURE 7.2.6. The surface S in Example 5.

A direct application of Stokes' theorem to reduce the problem to a line integral, or direct evaluation of the surface integral is quite tedious!

Instead, we use the preceding observation that Stokes' theorem enables us to replace S by another surface with the same boundary. We choose for this surface the portion P of the plane $x + y + z = 1$ inside the circle ∂S (where the plane and the unit sphere intersect). Computing the curl of $\mathbf{F}$, we find $\nabla \times \mathbf{F} = -2\mathbf{i} - 2\mathbf{j} - 2\mathbf{k}$, which points in the opposite direction to the normal $\mathbf{n}$ to P. (The chosen normal is the one which gives the same orientation to $\partial P = \partial S$ as did the outward normal to the sphere.) The dot product $(\nabla \times \mathbf{F}) \cdot \mathbf{n}$ is $\|\nabla \times \mathbf{F}\| \cdot \|\mathbf{n}\| \cos(180°) = -\|\nabla \times \mathbf{F}\| = -2\sqrt{3}$, so the surface integral

$$\iint_P (\nabla \times \mathbf{F}) \cdot d\mathbf{S} = \iint_P (\nabla \times \mathbf{F}) \cdot \mathbf{n}\,dS = \iint_P -2\sqrt{3}\,dS$$

is simply $-2\sqrt{3}$ times the area of the disk P.

To find the radius of P, we use an important kind of argument—*reasoning by symmetry.* The equations of the plane and sphere remain unchanged when we permute the coordinates $x, y,$ and z; hence the same must be true of the coordinates of the center of P. Thus the three coordinates of the center must be equal; since they sum to 1, the center is at $\left(\frac{1}{3}, \frac{1}{3}, \frac{1}{3}\right)$. An easy-to-find point on ∂P is $(1, 0, 0)$, so the square of the radius of P is $\left(\frac{2}{3}\right)^2 + \left(\frac{1}{3}\right)^2 + \left(\frac{1}{3}\right)^2 = \frac{6}{9} = \frac{2}{3}$, the area of P is $\frac{2}{3}\pi$, and hence the surface integral is $-\frac{4}{3}\sqrt{3}\pi$.   ◆

If $\mathbf{F}$ is a vector field whose curl is zero on a region W containing a surface S and its boundary ∂S, then by Stokes' theorem,

$$\int_{\partial S} \mathbf{F} \cdot d\mathbf{s} = \iint_S (\nabla \times \mathbf{F}) \cdot d\mathbf{S} = 0.$$

This remark is relevant for the next example.

Example 6   Let

$$\mathbf{F}(x, y, z) = \frac{y}{x^2 + y^2}\mathbf{i} - \frac{x}{x^2 + y^2}\mathbf{j} + \mathbf{k},$$

defined on the domain W defined by $x^2 + y^2 > 0$. Show that curl $\mathbf{F} = 0$, but that $\int_C \mathbf{F} \cdot d\mathbf{s} \neq 0$, where C is the unit circle in the xy plane. Why does this not contradict the preceding remark?

Solution   That curl $\mathbf{F} = 0$ is the result of a simple computation—see Example 10, §4.4. To compute the line integral, parametrize C by $x = \cos t, y = \sin t, z = 0$, so that

$$
\begin{aligned}
\int_C \mathbf{F} \cdot d\mathbf{s} &= \int_0^{2\pi} (\sin t\mathbf{i} - \cos t\mathbf{j} + \mathbf{k}) \cdot (-\sin t\mathbf{i} + \cos t\mathbf{j})dt \\
&= \int_0^{2\pi} (-\sin^2 t - \cos^2 t)dt = -2\pi.
\end{aligned}
$$

This result does not contradict the remark because C is not the boundary of *any* (parametrized) surface *contained in the domain W of* $\mathbf{F}$. In fact, any surface bounded by C must intersect the z axis, on which $\mathbf{F}$ is not defined.   ◆

Note

Although it should be intuitively clear that any parametrized surface bounded by the circle $x^2 + y^2 = 1, z = 0$ must intersect the z axis, it is not easy to *prove* this. To complicate the issue, note that C is the "boundary" of the semi-infinite cylinder $x^2 + y^2 = 1, z \geq 0$, but this cylinder is not a surface to which Stokes' theorem applies (it cannot be cut up into *finitely* many pieces that are images of type 1 and 2 regions under parametrizations). In fact, one of the best proofs of the fact that "any" surface with boundary curve C intersects the z axis is to use Stokes' theorem!

There are important problems, including physical ones, involving intersections of curves and surfaces, for which the answers are not intuitively obvious. In these cases, Stokes' theorem can serve as an aid to intuition. The use of calculus to solve this sort of problem falls within the mathematical field called *differential topology*.

In Chapter 4 we mentioned that curl $\mathbf{F}$ has to do with the rotational effects of $\mathbf{F}$ and discussed this briefly in the context of rotating bodies and fluids. The justification for this intuition rests on the following consequence of Stokes' theorem.

Circulation and Curl

Let C_ϵ be the circle of radius ϵ centered at a point P_0, and lying in the plane through P_0 with unit normal $\mathbf{n}$. Let $\mathbf{F}$ be a (smooth) vector field defined in a region containing P_0. Then

$$[(\nabla \times \mathbf{F})(P_0)] \cdot \mathbf{n} = \lim_{\epsilon \to 0} \frac{1}{\pi \epsilon^2} \int_{C_\epsilon} \mathbf{F} \cdot d\mathbf{s}$$

i.e., $[(\nabla \times \mathbf{F})(P_0)] \cdot \mathbf{n}$ can be interpreted as the *circulation per unit area in a plane with normal* $\mathbf{n}$. See Figure 7.2.7.

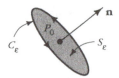

FIGURE 7.2.7. The curl gives the circulation per unit area.

Indeed, let S_ϵ be the disk centered at P_0 with radius ϵ and lying in the plane orthogonal to $\mathbf{n}$, and let $C_\epsilon = \partial S_\epsilon$ be its boundary. By Stokes' theorem,

$$\iint_{S_\epsilon} (\nabla \times \mathbf{F}) \cdot \mathbf{n}\, dS = \int_{\partial S_\epsilon} \mathbf{F} \cdot d\mathbf{s}.$$

In Chapter 5 we saw that the mean value theorem holds for double integrals. There is a similar result for surface integrals; thus there is a point P_ϵ in S_ϵ such that

$$\iint_{S_\epsilon} (\nabla \times \mathbf{F}) \cdot \mathbf{n}\, dS = [(\nabla \times \mathbf{F})(P_\epsilon) \cdot \mathbf{n}]\,(\text{area } S_\epsilon).$$

Thus

$$\mathbf{n} \cdot (\nabla \times \mathbf{F})(P_\epsilon) = \frac{1}{\pi \epsilon^2} \int_{\partial S_\epsilon} \mathbf{F} \cdot d\mathbf{s},$$

and so

$$\mathbf{n} \cdot (\nabla \times \mathbf{F})(P_0) = \lim_{\epsilon \to 0} \frac{1}{\pi \epsilon^2} \int_{\partial S_\epsilon} \mathbf{F} \cdot d\mathbf{s},$$

as we wanted to show.

In this result, the surface S_ϵ does not have to be a disk; it just needs to be a family of surfaces shrinking to P_0 whose normals tend to $\mathbf{n}$; then the same argument shows that

$$[(\nabla \times \mathbf{F})(P_0)] \cdot \mathbf{n} = \lim_{\epsilon \to 0} \frac{1}{\text{area}\,(S_\epsilon)} \int_{\partial S_\epsilon} \mathbf{F} \cdot d\mathbf{s}.$$

Here is an application of these ideas beyond those considered in Chapter 4.

Example 7   Let the unit vectors $\mathbf{e}_r, \mathbf{e}_\theta, \mathbf{e}_z$ associated to cylindrical coordinates be as shown in Figure 7.2.8. Let $\mathbf{F} = F_r \mathbf{e}_r + F_\theta \mathbf{e}_\theta + F_z \mathbf{e}_z$. Find a formula for the $\mathbf{e}_r$ component of $\nabla \times \mathbf{F}$ in cylindrical coordinates.

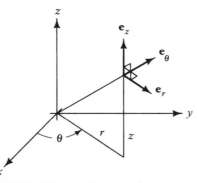

FIGURE 7.2.8. Orthonormal vectors $\mathbf{e}_r, \mathbf{e}_\theta$, and $\mathbf{e}_z$ associated with cylindrical coordinates. The vector $\mathbf{e}_r$ is parallel to the line labeled r.

Solution   Let S be the surface shown in Figure 7.2.9. The area of S is $r\,d\theta dz$ and the unit normal is $\mathbf{e}_r$. The integral of $\mathbf{F}$ around the edges of S is approximately

$$[F_\theta(r, \theta, z) - F_\theta(r, \theta, z + dz)]r\,d\theta + [F_z(r, \theta + d\theta, z) - F_z(r, \theta, z)]dz$$

$$\approx -\frac{\partial F_\theta}{\partial z}dz\,r\,d\theta + \frac{\partial F_z}{\partial \theta}d\theta dz.$$

Thus, the circulation per unit area is this expression divided by $r\,d\theta dz$, namely,

$$\frac{1}{r}\frac{\partial F_z}{\partial \theta} - \frac{\partial F_\theta}{\partial z}.$$

According to the previous box, this must be the $\mathbf{e}_r$ component of the curl. ♦

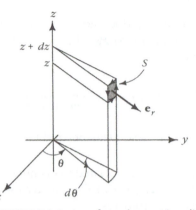

FIGURE 7.2.9. A surface element in cylindrical coordinates.

By similar arguments, one finds that the curl in cylindrical coordinates is given by

$$\nabla \times \mathbf{F} = \frac{1}{r} \begin{vmatrix} \mathbf{e}_r & r\mathbf{e}_\theta & \mathbf{e}_z \\ \dfrac{\partial}{\partial r} & \dfrac{\partial}{\partial \theta} & \dfrac{\partial}{\partial z} \\ F_r & rF_\theta & F_z \end{vmatrix}.$$

The chain rule shows that the gradient in cylindrical coordinates is

$$\nabla f = \frac{\partial f}{\partial r}\mathbf{e}_r + \frac{1}{r}\frac{\partial f}{\partial \theta}\mathbf{e}_\theta + \frac{\partial f}{\partial z}\mathbf{e}_z$$

and in the next section we will establish techniques that give the following formula for the divergence in cylindrical coordinates:

$$\nabla \cdot \mathbf{F} = \frac{1}{r}\left[\frac{\partial}{\partial r}(rF_r) + \frac{\partial F_\theta}{\partial \theta} + \frac{\partial}{\partial z}(rF_z)\right].$$

Corresponding formulas for gradient, divergence and curl in spherical coordinates are

$$\nabla f = \frac{\partial f}{\partial \rho}\mathbf{e}_\rho + \frac{1}{\rho}\frac{\partial f}{\partial \phi}\mathbf{e}_\phi + \frac{1}{\rho \sin \phi}\frac{\partial f}{\partial \theta}\mathbf{e}_\theta$$

$$\nabla \cdot \mathbf{F} = \frac{1}{\rho^2}\frac{\partial}{\partial \rho}(\rho^2 F_\rho) + \frac{1}{\rho \sin \phi}\frac{\partial}{\partial \phi}(\sin \phi F_\phi) + \frac{1}{\rho \sin \phi}\frac{\partial F_\theta}{\partial \theta}$$

and

$$\nabla \times \mathbf{F} = \left[\frac{1}{\rho \sin \phi}\frac{\partial}{\partial \phi}(\sin \phi F_\theta) - \frac{1}{\rho \sin \phi}\frac{\partial F_\phi}{\partial \theta}\right]\mathbf{e}_\rho$$

$$+ \left[\frac{1}{\rho \sin \phi}\frac{\partial F_\rho}{\partial \theta} - \frac{1}{\rho}\frac{\partial}{\partial \rho}(\rho F_\theta)\right]\mathbf{e}_\phi + \left[\frac{1}{\rho}\frac{\partial}{\partial \rho}(\rho F_\phi) - \frac{1}{\rho}\frac{\partial F_\rho}{\partial \phi}\right]\mathbf{e}_\theta$$

where $\mathbf{e}_\rho, \mathbf{e}_\phi, \mathbf{e}_\theta$ are as shown in Figure 7.2.10 and $\mathbf{F} = F_\rho \mathbf{e}_\rho + F_\phi \mathbf{e}_\phi + F_\theta \mathbf{e}_\theta$.

Note

Although these formulas are very useful in a variety of problems, most instructors will not require you to *memorize* them, but they may require you to *understand* how they are obtained.

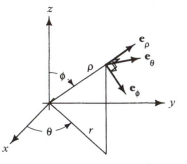

FIGURE 7.2.10. Orthonormal vectors $\mathbf{e}_\rho, \mathbf{e}_\phi$, and $\mathbf{e}_\theta$ associated with spherical coordinates.

Vector calculus plays an essential role in the theory of electromagnetism. Here is how Stokes' theorem applies.

Example 8   Let $\mathbf{E}$ and $\mathbf{H}$ be time-dependent electric and magnetic fields, respectively, in space. Let S be a surface with boundary C. We define

$$\int_C \mathbf{E} \cdot d\mathbf{s} \;=\; \text{voltage drop around } C,$$

$$\iint_S \mathbf{H} \cdot d\mathbf{S} \;=\; \text{magnetic flux across } S.$$

Faraday's law (see Figure 7.2.11) states that *the voltage around C equals the negative rate of change of magnetic flux through S.*

Show that Faraday's law follows from the following differential equation (one of the Maxwell equations):

$$\nabla \times \mathbf{E} = -\frac{\partial \mathbf{H}}{\partial t}.$$

Solution   Assume that $-\partial \mathbf{H}/\partial t = \nabla \times \mathbf{E}$ holds. By Stokes' theorem,

$$\int_C \mathbf{E} \cdot d\mathbf{s} = \iint_S (\nabla \times \mathbf{E}) \cdot d\mathbf{S}.$$

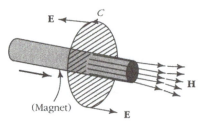

FIGURE 7.2.11. Faraday's law.

Assuming that we can move $\partial/\partial t$ under the integral sign, we get

$$-\frac{\partial}{\partial t}\iint_S \mathbf{H}\cdot d\mathbf{S} = \iint_S -\frac{\partial \mathbf{H}}{\partial t}\cdot d\mathbf{S} = \iint_S (\nabla\times\mathbf{E})\cdot d\mathbf{S} = \int_C \mathbf{E}\cdot d\mathbf{s}$$

and so

$$\int_C \mathbf{E}\cdot d\mathbf{s} = -\frac{\partial}{\partial t}\iint_S \mathbf{H}\cdot d\mathbf{S},$$

which is Faraday's law.   ◆

Note

In exercises involving integrals over surfaces, you should assume that the surface is oriented by the *outward* normal unless a different orientation is specified.

Exercises for §7.2

1. Let $\mathbf{F} = 2x\mathbf{i} - y\mathbf{j} + (x+z)\mathbf{k}$. Evaluate the integral of $\mathbf{F}$ around the curve consisting of straight lines joining $(1,0,1),(0,1,0)$, and $(0,0,1)$, using Stokes' theorem.

2. Let C consist of straight lines joining $(2,0,0),(0,1,0)$, and $(0,0,3)$. Evaluate the integral of $\mathbf{F}(x,y,z) = xy\mathbf{i} + yz\mathbf{j} + xz\mathbf{k}$ around C using Stokes' theorem.

3. Verify Stokes' theorem for the upper hemisphere $z = \sqrt{1-x^2-y^2}$, with $z \geq 0$, and the radial vector field $\mathbf{F}(x,y,z) = x\mathbf{i} + y\mathbf{j} + z\mathbf{k}$.

4. Let $\mathbf{c}$ consist of straight lines joining $(1,0,0),(0,1,0)$, and $(0,0,1)$ and let S be the triangle with these vertices. Verify Stokes' theorem directly with $\mathbf{F} = yz\mathbf{i} + xz\mathbf{j} + xy\mathbf{k}$.

5. Calculate the surface integral $\iint_S (\nabla \times \mathbf{F}) \cdot d\mathbf{S}$ where S is the hemisphere $x^2 + y^2 + z^2 = 1, x \geq 0$ with the normal in the positive x direction, and $\mathbf{F} = x^3\mathbf{i} - y^3\mathbf{j}$.

6. Find $\iint_S (\nabla \times \mathbf{F}) \cdot d\mathbf{S}$, where S is the ellipsoid $x^2 + y^2 + 2z^2 = 10$ and $\mathbf{F} = (\sin xy)\mathbf{i} + e^x\mathbf{j} - yz\mathbf{k}$.

7. Let $\mathbf{F} = y\mathbf{i} - x\mathbf{j} + zx^3y^2\mathbf{k}$. Evaluate $\iint_S (\nabla \times \mathbf{F}) \cdot \mathbf{n}\, dS$, where S is the surface $x^2 + y^2 + z^2 = 1, z \geq 0$.

8. Integrate $\nabla \times \mathbf{F}$, where $\mathbf{F} = (3y, -xz, -yz^2)$, over the portion of the surface $2z = x^2 + y^2$ below the plane $z = 2$, both directly and by using Stokes' theorem.

9. Evaluate $\iint_S (\nabla \times \mathbf{F}) \cdot \mathbf{n}\, dS$, where S is the portion of the sphere $x^2 + y^2 + z^2 = 9$ defined by $x \geq 1$, and where $\mathbf{F} = \mathbf{r} \times (\mathbf{i} + \mathbf{j})$.

10. Evaluate $\iint_S (\nabla \times \mathbf{F}) \cdot \mathbf{n}\, dS$, where S is the portion of the sphere $x^2 + y^2 + z^2 = 9$ defined by $x + y \geq 1$, and where $\mathbf{F}$ is the vector field $\mathbf{r} \times (\mathbf{i} + \mathbf{j})$.

11. Verify Stokes' theorem for the helicoid $\Psi(r, \theta) = (r\cos\theta, r\sin\theta, \theta)$, where (r, θ) lies in the rectangle $[0, 1] \times [0, \pi/2]$, and $\mathbf{F}$ is the vector field $\mathbf{F}(x, y, z) = (z, x, y)$.

12. Let $\mathbf{F} = x^2\mathbf{i} + (2xy + x)\mathbf{j} + z\mathbf{k}$. Let C be the circle $x^2 + y^2 = 1$ in the plane $z = 0$ oriented counterclockwise and S the disk $x^2 + y^2 \leq 1$ oriented with the normal vector $\mathbf{k}$. Determine:

    (a) The integral of $\mathbf{F}$ over S.

    (b) The circulation of $\mathbf{F}$ around C.

    (c) Find the integral of $\nabla \times \mathbf{F}$ over S. Verify Stokes' theorem directly in this case.

13. Let $\mathbf{F} = ax\mathbf{i} + by\mathbf{j} + cz\mathbf{k}$ and let C be a curve in a plane with normal $\mathbf{n}$ and enclosing the area A. Find an expression for $\int_C \mathbf{F} \cdot d\mathbf{s}$ using Stokes' theorem.

14. Let S be a surface and let $\mathbf{F}$ be perpendicular to the tangent to the boundary of S. Show that
$$\iint_S (\nabla \times \mathbf{F}) \cdot d\mathbf{S} = 0.$$

15. Consider two surfaces S_1, S_2 with the same boundary ∂S. Describe with sketches how S_1 and S_2 must be oriented to ensure that

$$\iint_{S_1} (\nabla \times \mathbf{F}) \cdot d\mathbf{S} = \iint_{S_2} (\nabla \times \mathbf{F}) \cdot d\mathbf{S}.$$

16. For a surface S and a *fixed* vector $\mathbf{v}$, prove that

$$2 \iint_S \mathbf{v} \cdot \mathbf{n}\, dS = \int_{\partial S} (\mathbf{v} \times \mathbf{r}) \cdot d\mathbf{s},$$

where $\mathbf{r}(x, y, z) = (x, y, z)$.

17. (a) If C is a closed curve that is the boundary of a surface, and $\mathbf{v}$ is a constant vector, show that

$$\int_C \mathbf{v} \cdot d\mathbf{s} = 0.$$

(b) Show that this is true even if C is not the boundary of a surface.

18. If C is a closed curve that is the boundary of a surface S and f and g are C^2 functions, show that

(a) $\displaystyle \int_C f\nabla g \cdot d\mathbf{s} = \iint_S (\nabla f \times \nabla g) \cdot d\mathbf{S};$

(b) $\displaystyle \int_C (f\nabla g + g\nabla f) \cdot d\mathbf{s} = 0.$

19. Let S be the capped cylindrical surface shown in Figure 7.2.12; S is the union of two surfaces S_1 and S_2, where S_1 is the set of (x, y, z) with $x^2 + y^2 = 1, 0 \leq z \leq 1$, and S_2 is defined by $x^2 + y^2 + (z-1)^2 = 1, z \geq 1$. Set $\mathbf{F}(x, y, z) = (zx + z^2 y + x)\mathbf{i} + (z^3 yx + y)\mathbf{j} + z^4 x^2 \mathbf{k}$. Compute $\displaystyle \iint_S (\nabla \times \mathbf{F}) \cdot d\mathbf{S}$.

20. A hot-air balloon, open at the bottom, has the truncated spherical shape shown in Figure 7.2.13. The hot gases escape through the porous envelope with a velocity vector field

$$\mathbf{V}(x, y, z) = \nabla \times \mathbf{F}(x, y, z) \quad \text{where} \quad \mathbf{F}(x, y, z) = -y\mathbf{i} + x\mathbf{j}.$$

If $R = 5$, compute the volume flow rate of the gases through the surface.

21. Derive the formula for the $\mathbf{e}_z$ component of $\nabla \times \mathbf{F}$ in cylindrical coordinates.

22. Derive the formula for the $\mathbf{e}_\rho$ component of $\nabla \times \mathbf{F}$ in spherical coordinates.

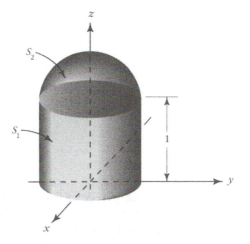

FIGURE 7.2.12. The capped cylinder is the union of S_1 and S_2.

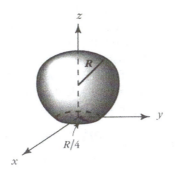

FIGURE 7.2.13. A hot-air balloon.

23. *Ampere's law* states that if the electric current density is described by a vector field **J** and the induced magnetic field is **H**, then the circulation of **H** around the boundary C of a surface S equals the integral of **J** over S (*i.e.*, the total current crossing S). See Figure 7.2.14. Show that this is implied by the steady-state *Maxwell equation* $\nabla \times \mathbf{H} = \mathbf{J}$.

24. Let S be a surface with boundary ∂S, and suppose that **E** is an electric field that is perpendicular to ∂S. Use Faraday's law to show that the induced magnetic flux across S is constant in time.

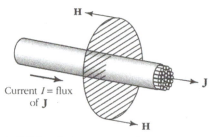

FIGURE 7.2.14. Ampere's law.

7.3
Gauss'
Theorem

Gauss' theorem states that *the flux of a vector field out of a closed surface equals the integral of the divergence of that vector field over the volume enclosed by the surface.* The result parallels Stokes' theorem and Green's theorem in that it relates an integral over a geometric object to another integral over the boundary of that object.

Let S be a closed surface, perhaps with edges and corners, such as the surface of a sphere, cube, or doughnut. Such a surface encloses a three-dimensional region; if we denote this region by W, we write $S = \partial W$ and call S the **boundary** of W. In everyday English usage, what we have called the boundary of W is often called the "surface of W." By choosing at each point of S the unit normal that points out of W, we get an orientation of S. Recall that the *divergence* of a vector field $F_1\mathbf{i} + F_2\mathbf{j} + F_3\mathbf{k}$ is the scalar function

$$\operatorname{div} \mathbf{F} = \nabla \cdot \mathbf{F} = \frac{\partial F_1}{\partial x} + \frac{\partial F_2}{\partial y} + \frac{\partial F_3}{\partial z}.$$

Gauss' (Divergence) Theorem

Let $S = \partial W$ be a closed surface, oriented by the outward normal to the three-dimensional region W of which it is the boundary. If $\mathbf{F}$ is a continuously differentiable vector field defined on W, then

$$\iiint_W (\nabla \cdot \mathbf{F})\, dV = \iint_{\partial W} \mathbf{F} \cdot d\mathbf{S}$$

i.e., the *integral over W of the divergence of $\mathbf{F}$ equals the flux of $\mathbf{F}$ across* ∂W. See Figure 7.3.1.

We give some examples of Gauss' theorem before its proof.

Example 1   Let $\mathbf{F} = 2x\mathbf{i} + y^2\mathbf{j} + z^2\mathbf{k}$ and S be the unit sphere defined by $x^2 + y^2 + z^2 = 1$. Evaluate $\displaystyle\iint_S \mathbf{F} \cdot d\mathbf{S}$.

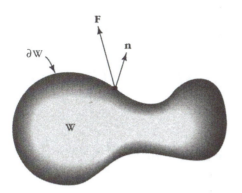

FIGURE 7.3.1. Gauss' theorem states that the flux of **F** out of ∂W is the volume integral of div **F** over W.

Solution   By Gauss' theorem,

$$\iiint_W (\text{div }\mathbf{F})dV = \iint_S \mathbf{F}\cdot d\mathbf{S}$$

where W is the ball bounded by the sphere. The integral on the left is

$$2\iiint_W (1+y+z)dV = 2\iiint_W dV + 2\iiint_W y\,dV + 2\iiint_W z\,dV.$$

By symmetry,

$$\iiint_W y\,dV = \iiint_W z\,dV = 0.$$

Thus

$$2\iiint_W (1+y+z)dV = 2\iiint_W dV = \frac{8\pi}{3}$$

(since the unit ball has volume $4\pi/3$). Try evaluating $\displaystyle\iint_S \mathbf{F}\cdot d\mathbf{S}$ directly. It's a mess!   ◆

Example 2   Evaluate

$$\iint_S \mathbf{F}\cdot d\mathbf{S}, \quad \text{where} \quad \mathbf{F}(x,y,z) = xy^2\mathbf{i} + x^2 y\mathbf{j} + y\mathbf{k}$$

and S is the surface of the "can" W defined by $x^2 + y^2 \le 1, -1 \le z \le 1$.

Solution   One can compute the integral directly, but it is easier to use the divergence theorem. Since S is the boundary of the region W, the divergence theorem gives

$$\iint_S \mathbf{F} \cdot d\mathbf{S} = \iiint_W (\text{div } \mathbf{F}) \, dx\,dy\,dz.$$

Since

$$\text{div } \mathbf{F} = \frac{\partial}{\partial x}(xy^2) + \frac{\partial}{\partial y}(x^2 y) + \frac{\partial}{\partial z}(y) = x^2 + y^2,$$

we get

$$
\begin{aligned}
\iiint_W (\text{div } \mathbf{F}) \, dx\,dy\,dz &= \iiint_W (x^2 + y^2)\,dx\,dy\,dz \\
&= \int_{-1}^{1} \left(\iint_{x^2 + y^2 \leq 1} (x^2 + y^2)\,dx\,dy \right) dz \\
&= 2 \iint_{x^2 + y^2 \leq 1} (x^2 + y^2)\,dx\,dy.
\end{aligned}
$$

We change variables to polar coordinates to evaluate the double integral:

$$x = r\cos\theta, \quad y = r\sin\theta, \quad 0 \leq r \leq 1, \quad 0 \leq \theta \leq 2\pi.$$

Replacing $x^2 + y^2$ by r^2 and $dx\,dy$ by $r\,dr\,d\theta$, we get

$$\iint_{x^2 + y^2 \leq 1} (x^2 + y^2)\,dx\,dy = \int_0^{2\pi} \left(\int_0^1 r^3 dr \right) d\theta = \frac{1}{2}\pi.$$

Therefore

$$\iint_S \mathbf{F} \cdot d\mathbf{S} = \iiint_W (\text{div } \mathbf{F}) \, dx\,dy\,dz = \pi. \quad \blacklozenge$$

Example 3   Calculate the flux of $\mathbf{V}(x, y, z) = x^3\mathbf{i} + y^3\mathbf{j} + z^3\mathbf{k}$ outward through the unit sphere $x^2 + y^2 + z^2 = 1$.

Solution   By Gauss' theorem, the flux is

$$\iiint_W (\text{div } \mathbf{V})\,dx\,dy\,dz = \iiint_W 3(x^2 + y^2 + z^2)\,dx\,dy\,dz.$$

Using spherical coordinates, this triple integral becomes

$$\int_0^{2\pi} \int_0^{\pi} \int_0^1 3\rho^4 \sin\phi\,d\rho\,d\phi\,d\theta.$$

Evaluation of this iterated integral gives $12\pi/5$. ♦

Example 4   A basic law of electrostatics is that an electric field $\mathbf{E}$ in space satisfies div $\mathbf{E} = \rho$, where ρ is the charge density (charge per unit volume). Show that the flux of $\mathbf{E}$ across a closed surface equals the total charge *inside* the surface.

Solution   Let W be a region in space with boundary surface S. By the divergence theorem,

$$\left\{ \begin{array}{c} \text{flux of } \mathbf{E} \\ \text{across } S \end{array} \right\} = \iint_S \mathbf{E} \cdot \mathbf{n} \, dA = \iiint_W \text{div } \mathbf{E} \, dx \, dy \, dz$$

$$= \iiint_W \rho(x, y, z) \, dx \, dy \, dz,$$

since div $\mathbf{E} = \rho$ by assumption; but since ρ is the charge per unit volume,

$$Q = \iiint_W \rho \, dx \, dy \, dz$$

is the total charge inside S. ♦

Proof of Gauss' theorem   First consider the case in which the region W and its boundary $\partial W = S$ are relatively simple. Specifically, we assume that W is of all "types" (see §**5.4**); *i.e.*, it is the region between two graphs, no matter which coordinates we single out.

Let $\mathbf{F} = P\mathbf{i} + Q\mathbf{j} + R\mathbf{k}$, so that

$$\text{div } \mathbf{F} = \frac{\partial P}{\partial x} + \frac{\partial Q}{\partial y} + \frac{\partial R}{\partial z}.$$

The left-hand side of Gauss' theorem is

$$\iiint_W \left(\frac{\partial P}{\partial x} + \frac{\partial Q}{\partial y} + \frac{\partial R}{\partial z} \right) dV,$$

which breaks up as a sum

$$\iiint_W \frac{\partial P}{\partial x} dV + \iiint_W \frac{\partial Q}{\partial y} dV + \iiint_W \frac{\partial R}{\partial z} dV,$$

as does the right-hand side,

$$\iint_{\partial W} \mathbf{F} \cdot d\mathbf{S} = \iint_{\partial W} (P\mathbf{i} + Q\mathbf{j} + R\mathbf{k}) \cdot d\mathbf{S}$$

$$= \iint_{\partial W} P\mathbf{i} \cdot d\mathbf{S} + \iint_{\partial W} Q\mathbf{j} \cdot d\mathbf{S} + \iint_{\partial W} R\mathbf{k} \cdot d\mathbf{S}.$$

The theorem will follow if we establish the three equalities

$$\iint_{\partial W} P\mathbf{i} \cdot d\mathbf{S} = \iiint_W \frac{\partial P}{\partial x} dV, \qquad \iint_{\partial W} Q\mathbf{j} \cdot d\mathbf{S} = \iiint_W \frac{\partial Q}{\partial y} dV$$

and

$$\iint_{\partial W} R\mathbf{k} \cdot d\mathbf{S} = \iiint_W \frac{\partial R}{\partial z} dV.$$

We shall prove the third equation; the other two equalities are proved in an analogous fashion. By assumption, there is a pair of functions $z = f_1(x, y)$, and $z = f_2(x, y)$, defined on a region D in the xy plane, such that W is the set of all points (x, y, z) such that (x, y) is in D and $f_1(x, y) \leq z \leq f_2(x, y)$. Writing the triple integral as an iterated integral, we get

$$\iiint_W \frac{\partial R}{\partial z} dV = \iint_D \left(\int_{z=f_1(x,y)}^{z=f_2(x,y)} \frac{\partial R}{\partial z} dz \right) dx\, dy,$$

and so by the fundamental theorem of calculus (a key ingredient in all the theorems in this chapter),

$$\iiint_W \frac{\partial R}{\partial z} dV = \iint_D [R(x, y, f_2(x, y)) - R(x, y, f_1(x, y))] dx\, dy.$$

The boundary of W is a closed surface whose top S_2 is the graph of f_2 and whose bottom S_1 is the graph of f_1. The four other sides of ∂W consist of surfaces S_3, S_4, S_5, and S_6, whose normals are perpendicular to the z axis. (See Figure 7.3.2.) Note that some of the four sides might be absent—for instance, if W is a solid cylinder or a solid sphere. Thus,

$$\iint_{\partial W} R\mathbf{k} \cdot d\mathbf{S} = \iint_{S_1} R\mathbf{k} \cdot d\mathbf{S} + \iint_{S_2} R\mathbf{k} \cdot d\mathbf{S} + \sum_{i=3}^{6} \iint_{S_i} R\mathbf{k} \cdot d\mathbf{S}$$

$$= \iint_{S_1} R\mathbf{k} \cdot d\mathbf{S} + \iint_{S_2} R\mathbf{k} \cdot d\mathbf{S}.$$

Note that $\iint_{S_i} R\mathbf{k} \cdot d\mathbf{S} = 0$ for $i = 3, 4, 5, 6$ because the normals of S_3, S_4, S_5, S_6 are perpendicular to $\mathbf{k}$.

The bottom surface S_1 is defined by $z = f_1(x, y)$; a parametrization of this surface is $\Phi(x, y) = (x, y, f_1(x, y))$, so

$$\Phi_x = \mathbf{i} + \frac{\partial f_1}{\partial x} \mathbf{k} \quad \text{and} \quad \Phi_y = \mathbf{j} + \frac{\partial f_1}{\partial y} \mathbf{k}.$$

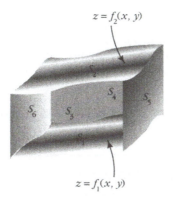

$z = f_2(x, y)$

$z = f_1(x, y)$

FIGURE 7.3.2. The four sides of this region which are S_3, S_4, S_5, S_6, have normals perpendicular to the z axis.

The corresponding normal vector is

$$\Phi_x \times \Phi_y = -\frac{\partial f_1}{\partial x}\mathbf{i} - \frac{\partial f_1}{\partial y}\mathbf{j} + \mathbf{k},$$

which is "upward" and therefore opposite to the outward normal on S_1. As a result,

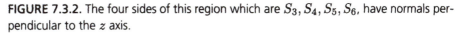

$$\iint_{S_1} R\mathbf{k} \cdot d\mathbf{S} = \iint_D R(\Phi(x,y))\mathbf{k} \cdot (-\Phi_x \times \Phi_y)dxdy$$

$$= \iint_D R(\Phi(x,y))\mathbf{k} \cdot \left(\frac{\partial f_1}{\partial x}\mathbf{i} + \frac{\partial f_1}{\partial y}\mathbf{j} - \mathbf{k}\right) dxdy$$

$$= -\iint_D R(x,y,f_1(x,y))dxdy.$$

A similar argument applied to S_2 (without a minus sign, since the outward normal along S_2 points "upward") shows that

$$\iint_{S_2} R\mathbf{k} \cdot d\mathbf{S} = \iint_D R(x,y,f_2(x,y))dxdy.$$

Adding these results,

$$\iint_{\partial W} R\mathbf{k} \cdot d\mathbf{S} = \iint_{S_1} R\mathbf{k} \cdot d\mathbf{S} + \iint_{S_2} R\mathbf{k} \cdot d\mathbf{S}$$

$$= \iint_D [R(x,y,f_2(x,y)) - R(x,y,f_1(x,y))]dxdy$$

$$= \iiint_W \frac{\partial R}{\partial z}dV$$

and so our proof is complete.   ■

To prove Gauss' theorem for a more complicated region in $\mathbb{R}^3$, we break it into simple pieces for which the proof above works. As was the case with Green's theorem in §7.1, the contributions from the "interior" boundaries of the pieces cancel one another, leaving the correct form of Gauss' theorem for the whole region.

We now use Gauss' theorem to obtain the physical and geometric significance of the divergence.

Divergence and Flux

Let W_r be the solid ball of radius r centered at a point P in space and $S_r = \partial W_r$ the bounding sphere. For a vector field $\mathbf{F}$ in space, div $\mathbf{F}(P)$ is the outward flux per unit volume at P:

$$\text{div } \mathbf{F}(P) = \lim_{r \to 0} \frac{1}{\text{vol}\,(W_r)} \iint_{S_r} \mathbf{F} \cdot d\mathbf{S}.$$

Sketch of Proof By Gauss' theorem,

$$\iint_{S_r} \mathbf{F} \cdot d\mathbf{S} = \iiint_{W_r} \text{div } \mathbf{F} \, dV.$$

As $r \to 0$, div $\mathbf{F}$ differs less and less from its value at the central point P and so

$$\frac{1}{\text{vol}\,(W_r)} \iiint_{W_r} \text{div } \mathbf{F} \, dV \to \text{div } \mathbf{F}(P)$$

as $r \to 0$. (This can also be justified using the mean value theorem for triple integrals.)   ■

Notice that the argument above works just as well for regions W other than spheres shrinking down to the point P. We will use this observation in Example 7.

The preceding box is analogous to the relation between curl and circulation given in §7.2. If $\mathbf{F}$ is the velocity field of a fluid, we conclude that the divergence div $\mathbf{F}(P)$ *is the net rate per unit volume at which fluid is flowing outwards at* P, which establishes our assertions about the divergence in §4.4. In particular, a fluid that satisfies div $\mathbf{F} = 0$ has the property that the *net* outflow across any closed surface is zero. Such a fluid is called ***divergence free*** or ***incompressible***.

Example 5   Show that an incompressible fluid cannot flow inward at all points of a closed compact surface.

Solution   Let **F** be the velocity field of the fluid, S the surface, and **n** its outward unit normal. If **F** were to be "inward," $\mathbf{F} \cdot \mathbf{n}$ would be negative at all points of S, so $\iint_S \mathbf{F} \cdot d\mathbf{S} = \iint_S (\mathbf{F} \cdot \mathbf{n}) dS$ would be a negative number. On the other hand, by Gauss' theorem, $\iint_S \mathbf{F} \cdot d\mathbf{S} = \iiint_W \text{div } \mathbf{F} \, dV$, where W is the region bounded by S. Since div $\mathbf{F} = 0$ (incompressibility), $\iint_S \mathbf{F} \cdot d\mathbf{S}$ must be zero rather than negative.   ◆

As we remarked above, Gauss' theorem can be applied to a variety of regions in space, even those whose boundaries have more than one piece. We use this observation to prove an important result.

Gauss' Law

Let W be a region in $\mathbb{R}^3$ and assume the origin $(0, 0, 0)$ is *not* on the boundary of W. Then

$$\iint_{\partial W} \frac{\mathbf{r} \cdot \mathbf{n}}{r^3} dS = \begin{cases} 4\pi, & \text{if } (0,0,0) \text{ is } \text{ in } W, \\ 0, & \text{if } (0,0,0) \text{ is } \text{ not } \text{ in } W, \end{cases}$$

where $\mathbf{r}(x, y, z) = x\mathbf{i} + y\mathbf{j} + z\mathbf{k}$ and where

$$r(x, y, z) = \|\mathbf{r}(x, y, z)\| = \sqrt{x^2 + y^2 + z^2}.$$

Proof If $(0, 0, 0)$ is not in W, this assertion follows from the divergence theorem because

$$\iint_{\partial W} \frac{\mathbf{r} \cdot \mathbf{n}}{r^3} dS = \iiint_W \nabla \cdot \left(\frac{\mathbf{r}}{r^3}\right) dV = 0,$$

since $\mathbf{r}/r^3$ has zero divergence; indeed,

$$\nabla \cdot \left(\frac{\mathbf{r}}{r^3}\right) = \frac{1}{r^3} \nabla \cdot \mathbf{r} + \nabla \left(\frac{1}{r^3}\right) \cdot \mathbf{r} = \frac{3}{r^3} - \frac{3}{r^5} \mathbf{r} \cdot \mathbf{r} = 0.$$

If $(0, 0, 0)$ is in W, the preceding argument fails, since $\mathbf{r}/r^3$ fails to be defined at $(0, 0, 0)$. In this case, we note that

$$\iint_{\partial W} \frac{\mathbf{r} \cdot \mathbf{n}}{r^3} dS = \iint_{\partial B} \frac{\mathbf{r} \cdot \mathbf{n}}{r^3} dS$$

where B is a small ball centered at the origin; this equality is obtained by applying the divergence theorem to the region *between* W and B. However, by direct evaluation,

$$\iint_{\partial B} \frac{\mathbf{r} \cdot \mathbf{n}}{r^3} dS = 4\pi$$

since for a sphere, $\mathbf{r} \cdot \mathbf{n}/r^3 = 1/r^2$ and r is a constant on the surface. This proves Gauss' law. ∎

Example 6   Gauss' law has the following physical interpretation. The potential due to a point charge Q at $(0,0,0)$ is

$$\phi(x,y,z) = \frac{Q}{4\pi r} = \frac{Q}{4\pi \sqrt{x^2 + y^2 + z^2}}$$

and the corresponding electric field is

$$\mathbf{E} = -\nabla \phi = \frac{Q}{4\pi} \left(\frac{\mathbf{r}}{r^3} \right).$$

Thus the preceding box states that the total electric flux $\iint_{\partial W} \mathbf{E} \cdot d\mathbf{S}$ equals Q if the charge lies inside W and zero otherwise. For a continuous charge distribution described by a charge density ρ, one of Maxwell's equations states that the field $\mathbf{E}$ is related to the density ρ by div $\mathbf{E} = \nabla \cdot \mathbf{E} = \rho$. Thus, by Gauss' theorem,

$$\iint_{\partial W} \mathbf{E} \cdot d\mathbf{S} = \iiint_W \rho \, dV = Q,$$

or the flux of the electric field out of a surface is equal to the total charge inside. ◆

Example 7   Use the formula

$$\text{div } \mathbf{F}(P) = \lim_{W \to P} \frac{1}{\text{vol }(W)} \iint_{\partial W} \mathbf{F} \cdot \mathbf{n} \, dS,$$

where W is a region that shrinks down to a point P, together with Gauss' theorem to derive the formula

$$\text{div } \mathbf{F} = \frac{1}{\rho^2} \frac{\partial}{\partial \rho}(\rho^2 F_\rho) + \frac{1}{\rho \sin \phi} \frac{\partial}{\partial \phi}(\sin \phi F_\phi) + \frac{1}{\rho \sin \phi} \frac{\partial F_\theta}{\partial \theta}$$

for the divergence of a vector field $\mathbf{F}$ in spherical coordinates.

Solution   Let W be the shaded region in Figure 7.3.3.

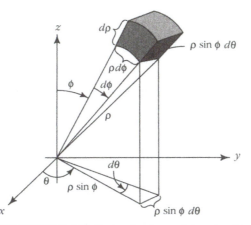

FIGURE 7.3.3. Infinitesimal volume determined by $d\rho, d\theta, d\phi$ at (ρ, θ, ϕ).

The contribution of the two faces orthogonal to the radial direction to the surface integral $\iint_{\partial W} \mathbf{F} \cdot d\mathbf{S}$ is, approximately,

$$F_\rho(\rho + d\rho, \phi, \theta) \times \text{area of outer face} - F_\rho(\rho, \phi, \theta) \times \text{area of inner face}$$
$$\approx \quad F_\rho(\rho + d\rho, \phi, \theta)(\rho + d\rho)^2 \sin\phi \, d\phi d\theta - F_\rho(\rho, \phi, \theta)\rho^2 \sin\phi \, d\phi d\theta$$
$$\approx \quad \frac{\partial}{\partial\rho}(F_\rho\rho^2 \sin\phi)d\rho d\phi d\theta$$

by the one-variable mean value theorem. Dividing by $\rho^2 \sin\phi \, d\rho d\phi d\theta$, the volume of the region W, we see that the contribution of these faces to the limit is $\dfrac{1}{\rho^2}\dfrac{\partial}{\partial\rho}(\rho^2 F_\rho)$. Likewise, the contribution from the faces orthogonal to the ϕ direction is $\dfrac{1}{\rho \sin\phi}\dfrac{\partial}{\partial\phi}(\sin\phi F_\phi)$, and for the θ direction, $\dfrac{1}{\rho \sin\phi}\dfrac{\partial F_\theta}{\partial\theta}$. Substituting in the limit formula for div $\mathbf{F}(\mathrm{P})$ gives the stated result. ◆

Note

In exercises involving integrals over surfaces, you should assume that the surface is oriented by the *outward* normal unless a different orientation is specified.

Exercises for §7.3

1. Find the flux of $\mathbf{F}(x, y, z) = x\mathbf{i} + y\mathbf{j} + z\mathbf{k}$ out of the unit sphere directly and by using Gauss' theorem.

2. Evaluate $\displaystyle\iint_{\partial W} \mathbf{F} \cdot d\mathbf{S}$, where $\mathbf{F} = x\mathbf{i} + y\mathbf{j} + z\mathbf{k}$ and W is the unit cube (in the first octant). Perform the calculation directly and check by using the divergence theorem.

3. Evaluate $\iint_S \mathbf{F} \cdot d\mathbf{S}$, where $\mathbf{F} = 3xy^2\mathbf{i} + 3x^2y\mathbf{j} + z^3\mathbf{k}$ and S is the surface of the unit sphere.

4. Let $\mathbf{F} = x^3\mathbf{i} + y^3\mathbf{j} + z^3\mathbf{k}$. Evaluate the surface integral of $\mathbf{F}$ over the unit sphere.

5. Let $\mathbf{F} = y\mathbf{i} + z\mathbf{j} + xz\mathbf{k}$. Evaluate $\iint_{\partial W} \mathbf{F} \cdot d\mathbf{S}$ for each of the following regions W:

   (a) $x^2 + y^2 \le z \le 1$;

   (b) $x^2 + y^2 \le z \le 1$ and $x \ge 0$;

   (c) $x^2 + y^2 \le z \le 1$ and $x \le 0$.

6. Evaluate the surface integral $\iint_{\partial W} \mathbf{F} \cdot \mathbf{n}\, dS$, where $\mathbf{F}(x, y, z) = \mathbf{i} + \mathbf{j} + z(x^2 + y^2)^2\mathbf{k}$ and W is a solid cylinder $x^2 + y^2 \le 1, 0 \le z \le 1$.

7. Let S be the boundary surface of a region W. Show that

$$\iint_S \mathbf{r} \cdot \mathbf{n}\, dS = 3\,\text{volume}\,(W).$$

   Attempt to explain this geometrically. [Hint: Assume that $(0, 0, 0)$ is in W and consider the skew cone with its vertex at $(0, 0, 0)$ with base dS and altitude $\|\mathbf{r}\|$. Its volume is $\frac{1}{3}(\Delta S)(\mathbf{r} \cdot \mathbf{n})$.]

8. Suppose that a vector field $\mathbf{V}$ is tangent to the boundary of a region W in space. Prove that $\iiint_W (\text{div } \mathbf{V})dx\,dy\,dz = 0$.

9. Prove that

$$\iiint_W (\nabla f) \cdot \mathbf{F}\, dx\,dy\,dz = \iint_{\partial W} f\mathbf{F} \cdot \mathbf{n}\, dS - \iiint_W f\nabla \cdot \mathbf{F}\, dx\,dy\,dz.$$

10. Prove Green's identities

$$\iint_{\partial W} f\nabla g \cdot \mathbf{n}\, dS = \iiint_W (f\nabla^2 g + \nabla f \cdot \nabla g)dV$$

   and

$$\iint_{\partial W} (f\nabla g - g\nabla f) \cdot \mathbf{n}\, dS = \iiint_W (f\nabla^2 g - g\nabla^2 f)dV.$$

11. Let S be a closed surface. Use Gauss' theorem to show that if $\mathbf{F}$ is a C^2 vector field, then $\iint_S (\nabla \times \mathbf{F}) \cdot d\mathbf{S} = 0$.

12.(a) Use Gauss' theorem to show that

$$\iint_{S_1} (\nabla \times \mathbf{F}) \cdot \mathbf{n} \, dS = \iint_{S_2} (\nabla \times \mathbf{F}) \cdot \mathbf{n} \, dS,$$

where S_1 and S_2 have a common boundary.

(b) Prove the same assertion using Stokes' theorem.

13. In cylindrical coordinates, prove that

$$\nabla \cdot \mathbf{F} = \frac{1}{r} \left[\frac{\partial}{\partial r}(rF_r) + \frac{\partial F_\theta}{\partial \theta} + \frac{\partial}{\partial z}(rF_z) \right].$$

(See Example 7 in §**7.2** for the definitions of $F_r, F_\theta,$ and F_z.)

14. Prove that $(\partial/\partial t)(\nabla \cdot \mathbf{H}) = 0$ from the Maxwell equation $\nabla \times \mathbf{E} = -\partial \mathbf{H}/\partial t$.

15.(a) Prove that $\nabla \cdot \mathbf{J} = 0$ from the steady-state Maxwell equation $\nabla \times \mathbf{H} = \mathbf{J}$.

(b) Argue physically that, under steady state conditions, the flux of $\mathbf{J}$ through any closed surface is zero (conservation of charge). Use this to deduce that $\nabla \cdot \mathbf{J} = 0$ from Gauss' theorem.

16. Fix k vectors $\mathbf{v}_1, \ldots, \mathbf{v}_k$ in space and numbers ("charges") $q_1, \ldots, q_k$. Define

$$\phi(x, y, z) = \sum_{i=1}^{k} \frac{q_i}{4\pi \|\mathbf{r} - \mathbf{v}_i\|},$$

where $\mathbf{r} = (x, y, z)$. Show that for a closed surface S and $\mathbf{E} = -\nabla \phi$,

$$\iint_S \mathbf{E} \cdot d\mathbf{S} = Q,$$

where Q is the total charge inside S. Assume that none of the charges are on S.

17. Let ρ be a continuous function on $\mathbb{R}^3$ such that $\rho(\mathbf{q}) = 0$ except when $\mathbf{q} = (x, y, z)$ belongs to some region W. The potential of ρ is the function

$$\phi(\mathbf{p}) = \iiint_W \frac{\rho(\mathbf{q})}{4\pi \|\mathbf{p} - \mathbf{q}\|} \, dV,$$

where $\|\mathbf{p} - \mathbf{q}\|$ is the distance between $\mathbf{p}$ and $\mathbf{q}$ and the integration is done in the variable $\mathbf{q}$.

(a) Show that $\displaystyle\iint_{\partial W} \nabla \phi \cdot d\mathbf{S} = -\iiint_W \rho \, dV.$

(b) Show that ϕ satisfies **Poisson's equation**

$$\nabla^2 \phi = -\rho.$$

18. Let $\mathbf{F}(\mathbf{x}) = \dfrac{1}{4\pi} \displaystyle\sum_{i=1}^{8} 10^{i} \dfrac{\mathbf{x} - \mathbf{x}_i}{\|\mathbf{x} - \mathbf{x}_i\|^3}$, where $\mathbf{x}_1, \ldots, \mathbf{x}_8$ are eight different points in $\mathbb{R}^3$. If S is a closed surface such that $\iint_S \mathbf{F} \cdot d\mathbf{S} = 11010$, which of the eight points lie inside S?

7.4
Path Independence and the Fundamental Theorems of Calculus

This last section of *Basic Multivariable Calculus* shows how the theorems of Green, Stokes, and Gauss relate to one another and how they can be considered as higher-dimensional versions of the fundamental theorem of calculus (denoted FTC for the rest of this section).

Recall that the FTC states that for an interval $[a, b]$ in $\mathbb{R}$ and a C^1 function F defined on $[a, b]$,

$$\int_a^b F'(x)\,dx = F(b) - F(a).$$

If we denote the interval $[a, b]$ by I, then the boundary ∂I of I consists of the two points a and b. If I is oriented by vectors pointed in the positive direction, then the vector at b points outward from I (the "preferred" direction, from the point of view of integral theorems), while the vector at a points inward (see Figure 7.4.1).

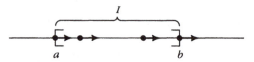

FIGURE 7.4.1. The vector field orienting I is outward at b but inward at a.

With these orientations in mind, we can think of $F(b) - F(a)$ as "the integral of F over ∂I," where ∂I consists of the point b with a "positive" sign and a with a "negative" one.

Thus, we can write the FTC formally as

$$\int_I F'(x)\,dx = \int_{\partial I} F(x).$$

Now look at all the integral theorems of vector analysis:

Green's theorem:

$$\iint_D (\operatorname{curl} \mathbf{F}) \cdot \mathbf{k}\,dx\,dy = \int_{\partial D} \mathbf{F} \cdot d\mathbf{s}$$

Divergence theorem in the plane:

$$\iint_D \text{div } \mathbf{F}\, dx dy = \int_{\partial D} (\mathbf{F} \cdot \mathbf{n}) ds$$

Stokes' theorem

$$\iint_S \text{curl } \mathbf{F} \cdot d\mathbf{S} = \int_{\partial S} \mathbf{F} \cdot d\mathbf{s}$$

Gauss' theorem:

$$\iiint_W \text{div } \mathbf{F}\, dx dy dz = \iint_{\partial W} \mathbf{F} \cdot d\mathbf{S}$$

In each case, there is a vector field and a domain on the left-hand side, and some kind of *derivative* of the vector field is integrated over the *whole* domain [like $\int_a^b F'(x)dx$ in the FTC]. On the right-hand side, the *original* vector field is integrated over the *boundary* of the domain [like $F(b) - F(a)$ in the FTC].

A basic result in §**6.1** tells us that

$$\int_{\mathbf{c}} (\text{grad } f) \cdot d\mathbf{s} = f(\mathbf{c}(b)) - f(\mathbf{c}(a))$$

when $\mathbf{c} : [a, b] \to \mathbb{R}^n$ is a parametrized curve and f is a (scalar) function. If we denote the image of $\mathbf{c}$ by C (a geometric curve), then the boundary of C consists of the endpoints $\mathbf{c}(b)$ (with "positive" orientation, since the orientation vector field points outward at b) and $\mathbf{c}(a)$ (with "negative" orientation). The formula above can also be written, therefore, in the FTC form:

$$\int_C (\text{grad } f) \cdot d\mathbf{s} = \int_{\partial C} f.$$

As we noted in §**6.1**, the formula above implies the *path independence* of the line integrals of gradient vector fields; *i.e.,* $\int_C \text{grad } f \cdot d\mathbf{s}$ depends only on the endpoints of C and not on C itself.

It turns out that the property of path independence is *equivalent* to the property of being a gradient.

Path Independence

A vector field $\mathbf{F}$ on a region in $\mathbb{R}^n$ is the gradient of some function if and only if, for any two paths C_1 and C_2 with the same endpoints,

$$\int_{C_1} \mathbf{F} \cdot d\mathbf{s} = \int_{C_2} \mathbf{F} \cdot d\mathbf{s}.$$

Proof We have already established the equality for gradient fields. To show that any vector field with the path independence property is a gradient field, we use the same idea used to construct antiderivatives in one-variable calculus. Given $\mathbf{F}$, defined on some domain D, choose a reference point $\mathbf{x}_0$ in D and *define f* by

$$f(\mathbf{x}) = \int_{C_{\mathbf{x}}} \mathbf{F} \cdot d\mathbf{s},$$

where $C_{\mathbf{x}}$ is any path in D from $\mathbf{x}_0$ to $\mathbf{x}$. Since we have assumed path independence for $\mathbf{F}$, it does not matter which path we choose. (If D consists of several disconnected pieces, like the complement of the x axis in $\mathbb{R}^2$, there may not be a path from $\mathbf{x}_0$ to $\mathbf{x}$. When this happens, we choose a separate reference point for each piece of D.)

We now show that $\int_{C} \operatorname{grad} f \cdot d\mathbf{s} = \int_{C} \mathbf{F} \cdot d\mathbf{s}$ for *every* path C, from which it will follow that $\operatorname{grad} f = \mathbf{F}$. Suppose that C is a path from $\mathbf{x}$ to $\mathbf{y}$. Choosing paths $C_{\mathbf{x}}$ and $C_{\mathbf{y}}$ from $\mathbf{x}_0$ to $\mathbf{x}$ and $\mathbf{y}$, we have $C_{\mathbf{y}} = C_{\mathbf{x}} + C$ (see Figure 7.4.2).

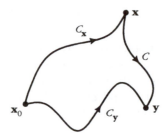

FIGURE 7.4.2. The path $C_{\mathbf{y}}$ is the sum of $C_{\mathbf{x}}$ and C.

By path independence, we have

$$\int_{C_{\mathbf{x}}} \mathbf{F} \cdot d\mathbf{s} + \int_{C} \mathbf{F} \cdot d\mathbf{s} = \int_{C_{\mathbf{y}}} \mathbf{F} \cdot d\mathbf{s},$$

and so

$$\int_{C} \mathbf{F} \cdot d\mathbf{s} = \int_{C_{\mathbf{y}}} \mathbf{F} \cdot d\mathbf{s} - \int_{C_{\mathbf{x}}} \mathbf{F} \cdot d\mathbf{s} = f(\mathbf{y}) - f(\mathbf{x}),$$

by the definition of f. But $\int_{C} \operatorname{grad} f \cdot d\mathbf{s}$ equals $f(\mathbf{y}) - f(\mathbf{x})$ as well by the FTC-type result in Example 1.

Having shown that $\mathbf{F}$ and $\operatorname{grad} f$ have the same line integral over *every* path C, we can conclude that they are equal by the following reasoning. If they were unequal, then there would be a point $\mathbf{z}$ in D on which some particular component of grad f is greater than the corresponding component of $\mathbf{F}$. Choosing a sufficiently short path C near $\mathbf{z}$ in the direction of this component, we would have $\int_C \text{grad } f \cdot d\mathbf{s} > \int_C \mathbf{F} \cdot d\mathbf{s}$, contradicting equality of the line integrals. ∎

In physics, a gradient field $\mathbf{F} = \nabla f$ represents a ***conservative force***; $V = -f$ is the corresponding ***potential energy***. This physical application leads to the following mathematical definition.

Conservative Vector Fields

A vector field $\mathbf{F}$ on a region in $\mathbb{R}^n$ is called ***conservative*** if it is the gradient of a function.

Example 1   Let $\mathbf{F}$ be a conservative vector field in the plane. If the line integral of $\mathbf{F}$ along the curve AOB in Figure 7.4.3 equals 3.5, find the integral of $\mathbf{F}$ along the broken lines: (a) ACB, (b) BDA, (c) $ACBDA$.

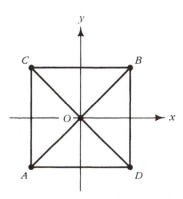

FIGURE 7.4.3. Paths for Example 1.

Solution   (a) Since AOB and ACB have the same endpoints, and $\mathbf{F}$ is conservative, the integral of $\mathbf{F}$ along ACB is 3.5.

(b) BDA has the same endpoints as BOA, which is $-AOB$, so the integral is -3.5.

(c) $ACBDA$ is a closed curve, so the line integral around it is zero. ♦

It is important to be able to tell whether a given vector field is a gradient. Path independence gives a way of characterizing such fields, but it is not very useful in practice because it requires evaluating all possible integrals. A more useful test is the following one:

Curl and Gradient

A vector field $\mathbf{F}$ defined on all of $\mathbb{R}^3$ is conservative (*i.e.*, the gradient of a function) if and only if curl $\mathbf{F} = \mathbf{0}$.

Proof The fact that curl $\mathbf{F}$ is zero when $\mathbf{F}$ is a gradient (is conservative) is a consequence of the identity $\nabla \times \nabla f = \mathbf{0}$ that we proved in Chapter 4. We still have to show that, if curl $\mathbf{F} = \mathbf{0}$, then $\mathbf{F} = \nabla f$ for some f. The next few paragraphs show this.

We have shown in the preceding proof that $\mathbf{F}$ is a gradient if and only if its line integrals are path independent. Note that if C_1 and C_2 are paths from $\mathbf{x}$ to $\mathbf{y}$, then $C = C_1 + (-C_2)$ is a *closed curve* [see Figure 7.4.4(a)].

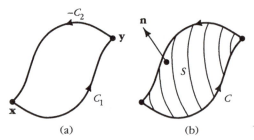

(a)   (b)

FIGURE 7.4.4. $C = C_1 - C_2$ is the boundary of S.

Since C is closed, it is the boundary of a surface S as in Figure 7.4.4(b). Applying Stokes' theorem,

$$\int_{C_1} \mathbf{F} \cdot d\mathbf{s} - \int_{C_2} \mathbf{F} \cdot d\mathbf{s} = \int_C \mathbf{F} \cdot d\mathbf{s} = \iint_S (\nabla \times \mathbf{F}) \cdot d\mathbf{S}.$$

If curl $\mathbf{F} = \mathbf{0}$, then the surface integral is zero, so we get

$$\int_{C_1} \mathbf{F} \cdot d\mathbf{s} = \int_{C_2} \mathbf{F} \cdot d\mathbf{s}. \quad \blacksquare$$

Note

It may not be obvious how to realize a given closed curve as the boundary of a surface—consider, for example, the *knot* in Figure 7.4.5. Nevertheless, it can always be done (although such a surface may have to intersect itself or be non-oriented). Alternatively, one can note that it suffices to replace the closed curves in the preceding two proofs by triangles and squares.

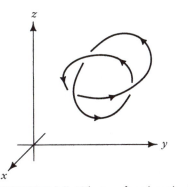

FIGURE 7.4.5. What surface has the knot as boundary?

For practical use, there is a straightforward way to find a function whose gradient is a given curl-free vector field. The following example illustrates this technique.

Example 2   Show that curl $\mathbf{F} = \mathbf{0}$, where

$$\mathbf{F}(x, y, z) = y\mathbf{i} + (z\cos yz + x)\mathbf{j} + (y\cos yz)\mathbf{k},$$

and find a function f such that grad $f = \mathbf{F}$.

Solution   We compute $\nabla \times \mathbf{F}$:

$$\nabla \times \mathbf{F} = \begin{vmatrix} \mathbf{i} & \mathbf{j} & \mathbf{k} \\ \dfrac{\partial}{\partial x} & \dfrac{\partial}{\partial y} & \dfrac{\partial}{\partial z} \\ y & x + z\cos yz & y\cos yz \end{vmatrix}$$

$$= (\cos yz - yz\sin yz - \cos yz + yz\sin yz)\mathbf{i} + (0 - 0)\mathbf{j} + (1 - 1)\mathbf{k}$$

$$= 0\mathbf{i} + 0\mathbf{j} + 0\mathbf{k} = \mathbf{0}.$$

Since

$$\text{grad } f = \frac{\partial f}{\partial x}\mathbf{i} + \frac{\partial f}{\partial y}\mathbf{j} + \frac{\partial f}{\partial z}\mathbf{k},$$

we must solve the equations

$$\frac{\partial f}{\partial x} = y, \tag{1}$$

$$\frac{\partial f}{\partial y} = z\cos yz + x, \tag{2}$$

$$\frac{\partial f}{\partial z} = y \cos yz. \tag{3}$$

An easy solution of equation (1) is $f(x, y, z) = xy$, but this is obviously not a solution of (2) or (3). The solution xy is not useless, though, since *any* solution of (1) differs from (1) by a function g for which $\partial g/\partial x = 0$. (Why?) But $\partial g/\partial x = 0$ means that g depends only on y and z, so the *general* solution of (1) is $f(x, y, z) = xy + g(y, z)$. To find what $g(y, z)$ should be, we use equations (2) and (3) in succession: by equation (2),

$$\frac{\partial f}{\partial y} = x + \frac{\partial g}{\partial y} = z \cos yz + x.$$

The x's cancel, leaving $\partial g/\partial y = z \cos yz$, which we can solve by treating z as a constant and integrating with respect to y:

$$g(y, z) = \int z \cos yz \, dy = \sin yz.$$

Again, $\sin yz$ is not the only solution; we can add to it any function of z alone. Thus $g(x, y) = \sin yz + h(z)$, so $f(x, y, z) = xy + \sin yz + h(z)$. We find h by substituting in equation (3):

$$\frac{\partial f}{\partial z} = y \cos yz + h'(z) = y \cos yz.$$

We get $h'(z) = 0$, so that h must be a constant. Thus, the general solution to our problem is $f(x, y, z) = xy + \sin yz +$ constant.   ◆

If $\mathbf{F} = P\mathbf{i} + Q\mathbf{j}$ is a planar vector field, then

$$\nabla \times \mathbf{F} = \left(\frac{\partial Q}{\partial x} - \frac{\partial P}{\partial y} \right) \mathbf{k},$$

and so the condition $\nabla \times \mathbf{F} = \mathbf{0}$ reduces to

$$\frac{\partial P}{\partial y} = \frac{\partial Q}{\partial x}.$$

Thus we have:

Cross-Derivative Test in the Plane

If $\mathbf{F} = P\mathbf{i} + Q\mathbf{j}$ is a (C^1) vector field in the plane and if

$$\frac{\partial P}{\partial y} = \frac{\partial Q}{\partial x},$$

then $\mathbf{F} = \nabla f$ for some f.

Example 3   (a) Determine whether the vector field

$$\mathbf{F} = e^{xy}\mathbf{i} + e^{x+y}\mathbf{j}$$

is a gradient field. If so, find f such that $\mathbf{F} = \nabla f$.

(b) Repeat part (a) for

$$\mathbf{F} = (2x\cos y)\mathbf{i} - (x^2\sin y)\mathbf{j}.$$

Solution   (a) Here $P(x,y) = e^{xy}$ and $Q(x,y) = e^{x+y}$, so we compute

$$\frac{\partial P}{\partial y} = xe^{xy}, \qquad \frac{\partial Q}{\partial x} = e^{x+y}.$$

These are not equal, so $\mathbf{F}$ cannot have a potential function.

(b) In this case, we find

$$\frac{\partial P}{\partial y} = -2x\sin y = \frac{\partial Q}{\partial x},$$

and so $\mathbf{F}$ has a potential function f. To compute f we solve the equations

$$\frac{\partial f}{\partial x} = 2x\cos y, \qquad \frac{\partial f}{\partial y} = -x^2\sin y.$$

thus

$$f(x,y) = x^2\cos y + h_1(y)$$

and

$$f(x,y) = x^2\cos y + h_2(x).$$

If h_1 and h_2 are the same constant, then both equations are satisfied, and so $f(x,y) = x^2\cos y$ is a function with $\mathbf{F} = \nabla f$.   ♦

The next example illustrates the importance of the condition in the previous box theorem that the vector field be defined in the *entire* plane.

Example 4   (a) Show that the vector field

$$F(x,y) = \frac{-y}{x^2+y^2}\mathbf{i} + \frac{x}{x^2+y^2}\mathbf{j}$$

is not conservative by integrating it around the circle $x^2 + y^2 = 1$.

(b) Verify that the cross-derivative condition is nevertheless satisfied for this vector field.

(c) What is going on here?

Solution   (a) Parametrizing the unit circle C by $\mathbf{c}(t) = (\cos t)\mathbf{i} + (\sin t)\mathbf{j}$ gives

$$\int_C \mathbf{F} \cdot d\mathbf{s}$$

$$= \int_0^{2\pi} \left(\frac{-\sin t}{\sin^2 t + \cos^2 t}\mathbf{i} + \frac{\cos t}{\sin^2 t + \cos^2 t}\mathbf{j} \right) \cdot ((-\sin t)\mathbf{i} + (\cos t)\mathbf{j})dt$$

$$= \int_0^{2\pi} \frac{\sin^2 t + \cos^2 t}{\sin^2 t + \cos^2 t} dt = 2\pi.$$

Since this is not zero, $\mathbf{F}$ is not conservative.

(b) With $P(x,y) = -y/(x^2+y^2)$ and $Q(x,y) = x/(x^2+y^2)$, we have

$$P_y(x,y) = \frac{(x^2+y^2)(-1) - (-y)(2y)}{(x^2+y^2)^2} = \frac{y^2 - x^2}{(x^2+y^2)^2}$$

and

$$Q_x(x,y) = \frac{(x^2+y^2)(1) - (x)(2x)}{(x^2+y^2)^2} = \frac{y^2 - x^2}{(x^2+y^2)^2},$$

so the cross-derivative condition is satisfied.

(c) the vector field $\mathbf{F}(x,y)$ is not defined at the origin, so the cross-derivative test does not apply.   ◆

The method used in Example 2 always works (although it may lead to integrals you don't know how to evaluate). In fact, it works for all n, not just $n = 2$ and $n = 3$, if we replace the condition curl $\mathbf{F} = 0$ by the "cross-derivative test" in the following box.

Antiderivatives of Vector Fields

A vector field $\mathbf{F} = (F_1, \ldots, F_n)$ on $\mathbb{R}^n$ is conservative if and only if its components satisfy the **cross-derivative test**

$$\frac{\partial F_i}{\partial x_j} - \frac{\partial F_j}{\partial x_i} = 0$$

for all i and j between 1 and n. (If $n = 3$, the test says that curl $\mathbf{F} = 0$.)

Given $\mathbf{F}$ satisfying the cross-derivative test, we can find f such that grad $f = \mathbf{F}$ by the following method of repeated integration.

Solve the equation $\partial f / \partial x_1 = F_1$ by integrating:

$$f(x_1, \ldots, x_n) = \int F_1(x_1, \ldots, x_n) dx_1 + f_2(x_2, \ldots, x_n).$$

Substitute this expression for f into the equation $\partial f / \partial x_2 = F_2$ to get an equation for f_2. The solution will involve an arbitrary function $f_3(x_3, \ldots, x_n)$. Repeat the process until the last equation $\partial f / \partial x_n = F_n$ has been solved.

One may ask whether there are any regions in $\mathbb{R}^n$ to which the cross-derivative test *does* apply, other than all of $\mathbb{R}^n$. The answer lies in the relation between Stokes' theorem and path independence. If $\mathbf{F}$ is defined everywhere on $\mathbb{R}^3$ but at a single point, we can avoid this point by slightly moving the surface of integration. Thus, *every curl-free vector field on $\mathbb{R}^3$ minus a point is conservative.* On the other hand, a vector field defined on $\mathbb{R}^3$ minus the z axis can be curl free but not conservative, since a closed curve (even a triangle) surrounding the z axis does not bound any surface that avoids the axis. The vector field

$$\mathbf{F}(x, y, z) = \frac{-y}{x^2 + y^2}\mathbf{i} + \frac{x}{x^2 + y^2}\mathbf{j} + \mathbf{k}$$

is an example of a curl-free vector field whose integral around the circle $x^2 + y^2 = 1, z = 0$ is nonzero. (See Example 6, §**7.2**.)

A region in which every closed curve bounds a surface is called **simply connected**. It is just on such regions that every curl-free vector field is conservative.

The discussion up to this point has involved line integrals, surface integrals, and the identity curl(grad f) = $\mathbf{0}$. Similar relations hold for surface integrals, volume integrals, and the identity div(curl $\mathbf{F}$) = 0. Given a vector field $\mathbf{F}$ on $\mathbb{R}^3$, when can we find a vector field $\mathbf{G}$ for which curl $\mathbf{G} = \mathbf{F}$? A first step towards answering this is to note that the identity div(curl $\mathbf{G}$) = 0 shows that div $\mathbf{F}$ = 0. The next example shows how this identity is related to the theorems of Stokes and Gauss.

Example 5   Make use of the theorems of Stokes and Gauss to show that div(curl $\mathbf{F}$) = 0.

Solution   Let D be any region in $\mathbb{R}^3$ with boundary S. Using Gauss' theorem, then Stokes', we have

$$\iiint_D \text{div}\,(\text{curl}\,\mathbf{F})\,dx\,dy\,dz = \iint_{\partial D = S} \text{curl}\,\mathbf{F} \cdot d\mathbf{S} = \int_{\partial S} \mathbf{F} \cdot d\mathbf{s} = 0$$

since $S = \partial D$ has no boundary, *i.e.*, ∂S is empty. Since D is arbitrary, the integrand div(curl $\mathbf{F}$) must be identically zero.   ◆

The analogue of the path independence property for gradient vector fields is the following surface independence property for the curl of a vector field.

Surface Independence

Let $\mathbf{G}$ be a vector field defined on a region R in $\mathbb{R}^3$. If either

(a) $\mathbf{G} = \text{curl}\,\mathbf{F}$ for some $\mathbf{F}$ or

(b) div $\mathbf{G} = 0$ and R is all of $\mathbb{R}^3$,

then

$$\iint_{S_1} \mathbf{G} \cdot d\mathbf{S} = \iint_{S_2} \mathbf{G} \cdot d\mathbf{S}$$

whenever S_1 and S_2 are two oriented surfaces in R such that $\partial S_1 = \partial S_2$.

Proof We use Stokes' theorem in case (a): For $i = 1$ or 2,

$$\iint_{S_i} \mathbf{G} \cdot d\mathbf{S} = \iint_{S_i} \text{curl}\,\mathbf{F} \cdot d\mathbf{S} = \int_{\partial S_i} \mathbf{F} \cdot d\mathbf{s},$$

and the line integrals are equal because $\partial S_1 = \partial S_2$. In case (b), we will assume that the surfaces S_1 and S_2 do not intersect one another except along their common boundary C. (This simplifies the proof but is not essential.) Then S_1 and S_2 can be "sewn together" along C to make a closed surface S, as in Figure 7.4.6.

To get a consistent orientation for S, we reverse the orientation of S_2: *i.e.*,

$$\iint_S \mathbf{G} \cdot d\mathbf{S} = \iint_{S_1} \mathbf{G} \cdot d\mathbf{S} - \iint_{S_2} \mathbf{G} \cdot d\mathbf{S}.$$

But S bounds a region W, so by Gauss' theorem

$$\iint_S \mathbf{G} \cdot d\mathbf{S} = \iiint_W \text{div}\,\mathbf{G}\,dx\,dy\,dz = 0,$$

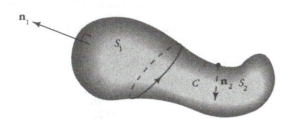

FIGURE 7.4.6. S_1 and S_2 have the same boundary. For the orientations on C to coincide, those on S_1 and S_2 must be opposite to each other.

by the assumption div $\mathbf{G} = 0$.

By analogy with the case of curl and gradient, we may expect that any vector field $\mathbf{G}$ with div $\mathbf{G} = 0$ ought to be the curl of some $\mathbf{F}$. This is true if $\mathbf{G}$ is defined on all of $\mathbb{R}^3$, but is not true if the domain of $\mathbf{G}$ has a missing point (e.g., $\mathbf{G} = \mathbf{r}/r^3$; see the discussion of Gauss' law in §**7.3**).

Divergence and Curl

If $\mathbf{F}$ is a vector field on $\mathbb{R}^3$ and div $\mathbf{F} = 0$, then $\mathbf{F} = \operatorname{curl} \mathbf{G}$ for some vector field $\mathbf{G}$.

Proof It turns out to be sufficient to take $\mathbf{G}$ to be a vector field of the form $\mathbf{G} = G_1(x, y, z)\mathbf{i} + G_2(x, y, z)\mathbf{j}$. Then curl $\mathbf{G}$ is

$$
\begin{vmatrix}
\dfrac{\partial}{\partial x} & \dfrac{\partial}{\partial y} & \dfrac{\partial}{\partial z} \\[2mm]
G_1 & G_2 & 0 \\[2mm]
\mathbf{i} & \mathbf{j} & \mathbf{k}
\end{vmatrix}
= -\frac{\partial G_2}{\partial z}\mathbf{i} + \frac{\partial G_1}{\partial z}\mathbf{j} + \left(\frac{\partial G_2}{\partial x} - \frac{\partial G_1}{\partial y}\right)\mathbf{k},
$$

and thus the equations to be solved are

$$
-\frac{\partial G_2}{\partial z} = F_1,
$$

$$
\frac{\partial G_1}{\partial z} = F_2,
$$

$$
\frac{\partial G_2}{\partial x} - \frac{\partial G_1}{\partial y} = F_3.
$$

A solution to these equations is given explicitly by

$$
G_1(x, y, z) = \int_0^z F_2(x, y, t)\,dt - \int_0^y F_3(x, t, 0)\,dt
$$

$$
G_2(x, y, z) = -\int_0^z F_1(x, y, t)\,dt
$$

as one can check by computing the partial derivatives. ∎

Example 6   Is $\mathbf{F} = x\mathbf{i} - 2y\mathbf{j} + z\mathbf{k}$ the curl of a vector field? If so, find one.

Solution   According to the preceding box, the test is div $\mathbf{F} = 0$, which is true in this case. Thus $\mathbf{F} = \nabla \times \mathbf{G}$. We can find $\mathbf{G}$ by letting, as the proof above suggests, $\mathbf{G} = G_1\mathbf{i} + G_2\mathbf{j}$ giving

$$x = -\frac{\partial G_2}{\partial z},$$

$$-2y = \frac{\partial G_1}{\partial z},$$

$$z = \frac{\partial G_2}{\partial x} - \frac{\partial G_1}{\partial y}.$$

From the first, $G_2 = -xz + g(x, y)$ and the second, $G_1 = -2yz + h(x, y)$. The third gives

$$z = -z + \frac{\partial g}{\partial x} + 2z - \frac{\partial h}{\partial y}$$

which holds if $\partial g/\partial x = \partial h/\partial y$. For example, $g = h = 0$ will do, so $\mathbf{G} = -2yz\mathbf{i} - xz\mathbf{j}$ is one answer. Clearly we can add the gradient of any function to $\mathbf{G}$ to get another answer.   ♦

For complicated domains R in $\mathbb{R}^3$, the questions "when is a curl-free vector field on R a gradient?" and "when is a divergence-free vector field on R a curl?" are closely related to the questions "when does a closed curve in R bound a surface in R?" and "when does a closed surface in R surround points not in R?"

Questions like these led Henri Poincaré and other mathematicians of the early twentieth century to develop the beginnings of the subject called topology. This subject concerns the qualitative description of curves and surfaces, in terms of how many pieces, holes, etc. they have. Although these may seem like difficult questions to put in mathematically precise form, we have seen that they are linked to very explicit questions involving the calculation of integrals.

Topology is one of the most lively areas of contemporary mathematics, with applications to fields as diverse as fundamental physics, molecular biology, and computer science.

Exercises for §7.4

Let $\mathbf{F}$ be a conservative vector field in the plane. Suppose that the integral of $\mathbf{F}$ along AOF is 3, along OF is 2, and along AB is -5. See Figure 7.4.7. Compute the integral of $\mathbf{F}$ along the paths in Exercises 1–4.

1. *AODEF*

2. *FEDO*

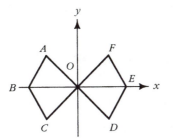

FIGURE 7.4.7. Paths for Exercises 1−4.

3. *BOEF*

4. *BAODEF*

5. Let $\mathbf{F}(x, y, z) = xy\mathbf{i} + y\mathbf{j} + z\mathbf{k}$. Is there a function f such that $\mathbf{F} = \nabla f$?

6.(a) Let $\mathbf{F}(x, y) = (xy, y^2)$ and let C be the path $y = 2x^2$ joining $(0, 0)$ to $(1, 2)$ in the plane. Evaluate $\int_C \mathbf{F} \cdot d\mathbf{s}$.

   (b) Does the integral in part (a) depend on the path joining $(0, 0)$ to $(1, 2)$?

7. Let $\mathbf{F}(x, y, z) = (2xyz + \sin x)\mathbf{i} + x^2 z\mathbf{j} + x^2 y\mathbf{k}$. Find a function f such that $\mathbf{F} = \nabla f$.

8. Evaluate $\displaystyle\int_{\mathbf{c}} \mathbf{F} \cdot d\mathbf{s}$, where $\mathbf{c}(t) = (\cos^5 t,\ \sin^3 t, t^4), 0 \le t \le \pi$, and $\mathbf{F}$ is as in Exercise 7.

9. Determine which of the following vector fields $\mathbf{F}$ in the plane is the gradient of a scalar function f. If such an f exists, find it.

   (a) $\mathbf{F}(x, y) = x\mathbf{i} + y\mathbf{j}$
   (b) $\mathbf{F}(x, y) = xy\mathbf{i} + xy\mathbf{j}$
   (c) $\mathbf{F}(x, y) = (x^2 + y^2)\mathbf{i} + 2xy\mathbf{j}$

10. Repeat Exercise 9 for the following vector fields:

    (a) $\mathbf{F}(x, y) = (\cos xy - xy \sin xy)\mathbf{i} - (x^2 \sin xy)\mathbf{j}$
    (b) $\mathbf{F}(x, y) = (x\sqrt{x^2 y^2 + 1})\mathbf{i} + (y\sqrt{x^2 y^2 + 1})\mathbf{j}$
    (c) $\mathbf{F}(x, y) = (2x \cos y + \cos y)\mathbf{i} - (x^2 \sin y + x \sin y)\mathbf{j}$

11. Show that the following vector fields are conservative. Calculate $\displaystyle\int_C \mathbf{F} \cdot d\mathbf{s}$ for the given curve.

    (a) $\mathbf{F} = (xy^2 + 3x^2 y)\mathbf{i} + (x + y)x^2\mathbf{j}$; C is the curve consisting of line segments from $(1, 1)$ to $(0, 2)$ to $(3, 0)$

(b) $\mathbf{F} = \dfrac{2x}{y^2 + 1}\mathbf{i} - \dfrac{2y(x^2 + 1)}{(y^2 + 1)^2}\mathbf{j}$; C is parametrized by $x = t^3 - 1$, $y = t^6 - t, 0 \le t \le 1$

(c) $\mathbf{F} = [\cos(xy^2) - xy^2\sin(xy^2)]\mathbf{i} - 2x^2y\sin(xy^2)\mathbf{j}$; C is the curve parametrized by $(e^t, e^{t+1}), -1 \le t \le 0$

12. Is $x^2y\mathbf{i} + (\frac{1}{2}x^3 + ye^y)\mathbf{j}$ conservative? If so, find an antiderivative.

13. Is $4x\cos^2(y/2)\mathbf{i} - x^2\sin y\mathbf{j}$ conservative? If so, find an antiderivative.

14. Is $2xy\sin(x^2y)\mathbf{i} + (e^y + x^2\sin(x^2y))\mathbf{j}$ conservative? If so, find an antiderivative.

15. What is the work done by the force $\mathbf{F} = -\mathbf{r}/\|\mathbf{r}\|^3$ in moving a particle from a point $\mathbf{r}_0 \in \mathbb{R}^3$ "to ∞," where $\mathbf{r}(x, y, z) = (x, y, z)$?

16. The mass of the Earth is approximately 6×10^{27} g and that of the Sun is 330,000 times as much. The gravitational constant is 6.7×10^{-8} m^3/s$^2 \cdot$ g. The distance of the Earth from the Sun is about 1.5×10^{12} cm. Compute, approximately, the work necessary to increase the distance of the Earth from the Sun by 1 cm.

17. Is each of the following vector fields the curl of some other vector field? If so, find the vector field.

(a) $\mathbf{F} = x\mathbf{i} + y\mathbf{j} + z\mathbf{k}$

(b) $\mathbf{F} = (x^2 + 1)\mathbf{i} + (z - 2xy)\mathbf{j} + y\mathbf{k}$

18. Let $\mathbf{F} = xz\mathbf{i} - yz\mathbf{j} + y\mathbf{k}$. Verify that $\nabla \cdot \mathbf{F} = 0$. Find a $\mathbf{G}$ such that $\mathbf{F} = \nabla \times \mathbf{G}$.

19. Repeat Exercise 18 for $\mathbf{F} = y^2\mathbf{i} + z^2\mathbf{j} + x^2\mathbf{k}$.

20. Let $\mathbf{F} = xe^y\mathbf{i} - (x\cos z)\mathbf{j} - ze^y\mathbf{k}$. Find a $\mathbf{G}$ such that $\mathbf{F} = \nabla \times \mathbf{G}$.

21. Let $\mathbf{F} = (x\cos y)\mathbf{i} - (\sin y)\mathbf{j} + (\sin x)\mathbf{k}$. Find a $\mathbf{G}$ such that $\mathbf{F} = \nabla \times \mathbf{G}$.

22. Let $\mathbf{F} = -(GmM\mathbf{r}/r^3)$ be the gravitational force field defined for $r \ne 0$.

(a) Show that div $\mathbf{F} = 0$.

(b) Show that $\mathbf{F} \ne$ curl $\mathbf{G}$ for any C^1 vector field $\mathbf{G}$ defined on $\mathbb{R}^3$ minus the origin.

23. Let $\mathbf{F} = F_1\mathbf{i} + F_2\mathbf{j} + F_3\mathbf{k}$ and suppose each F_k satisfies the homogeneity condition

$$F_k(tx, ty, tz) = tF_k(x, y, z), \quad k = 1, 2, 3.$$

Suppose also $\nabla \times \mathbf{F} = \mathbf{0}$. Prove that $\mathbf{F} = \nabla f$ where

$$2f(x, y, z) = xF_1(x, y, z) + yF_2(x, y, z) + zF_3(x, y, z).$$

24. *Two-color problem:* Several intersecting circles are drawn in the plane. Show that the resulting "map" can be colored with two colors in such a way that adjacent regions have different colors (as in Figure 7.4.8). [Hint: First show that every closed curve crosses the union of the circles an even number of times. Then divide the regions into two classes according to whether an arc from the region to a fixed point crosses the circles an even or odd number of times. Compare your argument with the proof of the basic properties of conservative vector fields.]

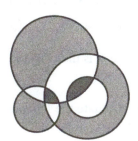

FIGURE 7.4.8. Adjacent regions have opposite colors.

Note

In exercises involving integrals over surfaces, you should assume that the surface is oriented by the *outward* normal unless a different orientation is specified.

Review Exercises for Chapter 7

1. If D is a region in the plane with boundary curve C traversed counterclockwise, express the following three integrals in terms of the area of D:

    (a) $\int_C x\,dy$ (b) $\int_C y\,dx$ (c) $\int_C x\,dx$.

2. Use Green's theorem to calculate the line integral

$$\int_C (x^3 + y^3)dy - (x^3 + y)dx,$$

where C is the circle $x^2 + y^2 = 1$ traversed counterclockwise.

3. Using Green's theorem, or otherwise, evaluate $\int_C x^3\,dy - y^3\,dx$, where C is the unit circle $(x^2 + y^2 = 1)$.

4. Verify Green's theorem for the line integral

$$\int_C x^2 y\,dx + y\,dy$$

when C is the boundary of the region between the curves $y = x$ and $y = x^3, 0 \leq x \leq 1$.

5. Calculate the surface integral $\iint_S (\nabla \times \mathbf{F}) \cdot \mathbf{n} \, dS$, where S is the hemisphere $x^2 + y^2 + z^2 = 1, z \geq 0$ and $\mathbf{F} = x^3\mathbf{i} - y^3\mathbf{j}$.

6. Calculate the integral of the vector field in Exercise 5 over the hemisphere $x^2 + y^2 + z^2 = 1, z \leq 0$.

7. Calculate the integral $\iint_S \mathbf{F} \cdot \mathbf{n} \, dS$, where S is the surface of the half-ball $x^2 + y^2 + z^2 \leq 1, z \geq 0$, and $\mathbf{F} = (x + 3y^5)\mathbf{i} + (y + 10xz)\mathbf{j} + (z - xy)\mathbf{k}$.

8. Find $\iint_S (\nabla \times \mathbf{F}) \cdot \mathbf{n} \, dS$, where S is the ellipsoid $x^2 + y^2 + 2z^2 = 10$ and $\mathbf{F} = \sin xy\mathbf{i} + e^x\mathbf{j} - yz\mathbf{k}$.

9. Let $\mathbf{F} = x^2y\mathbf{i} + z^8\mathbf{j} - 2xyz\mathbf{k}$. Evaluate the integral of $\mathbf{F}$ over the surface of the unit cube $[0, 1] \times [0, 1] \times [0, 1]$.

10. Evaluate the integral $\iint_S \mathbf{F} \cdot d\mathbf{S}$ where $\mathbf{F} = x\mathbf{i} + y\mathbf{j} + 3\mathbf{k}$ and where S is the unit sphere $x^2 + y^2 + z^2 = 1$.

11. Let the velocity of a fluid be described by $\mathbf{F} = 6xz\mathbf{i} + x^2y\mathbf{j} + yz\mathbf{k}$. Compute the rate at which fluid is leaving the unit cube of Exercise 9.

12. Let the velocity field of a fluid be given by $\mathbf{F} = xy^2\mathbf{i} + yx^2\mathbf{j} + \frac{1}{3}z^3\mathbf{k}$. Find the rate at which fluid is leaving the sphere $x^2 + y^2 + z^2 = 1$.

13. (a) Show that $\mathbf{F} = (x^3 - 2xy^3)\mathbf{i} - 3x^2y^2\mathbf{j}$ is a gradient vector field.

(b) Evaluate the integral of $\mathbf{F}$ along the path $x = \cos^3\theta, y = \sin^3\theta$, where $0 \leq \theta \leq \pi/2$.

14. (a) Show that $\mathbf{F} = 6xy(\cos z)\mathbf{i} + 3x^2(\cos z)\mathbf{j} - 3x^2y(\sin z)\mathbf{k}$ is conservative.

(b) Find f such that $\mathbf{F} = \nabla f$.

(c) Evaluate the integral of $\mathbf{F}$ along the curve $x = \cos^3\theta, y = \sin^3\theta, z = 0, 0 \leq \theta \leq \pi/2$.

15. Let $\mathbf{F} = x^2\mathbf{i} + (x^2y - 2xy)\mathbf{j} - x^2z\mathbf{k}$. Does there exist a $\mathbf{G}$ such that $\mathbf{F} = \nabla \times \mathbf{G}$?

16. Let $\mathbf{a}$ be a constant vector and $\mathbf{F} = \mathbf{a} \times \mathbf{r}$ [as usual, $\mathbf{r}(x, y, z) = (x, y, z)$]. Is $\mathbf{F}$ conservative? If so, find a potential for it.

17. (a) Let $\mathbf{F} = y\mathbf{i} - x\mathbf{j} + zx^3y^2\mathbf{k}$. Calculate $\nabla \times \mathbf{F}$ and $\nabla \cdot \mathbf{F}$.

(b) Evaluate the surface integral $\iint_{S_1} (\nabla \times \mathbf{F}) \cdot \mathbf{n} \, dS$ where S_1 is the surface $x^2 + y^2 + z^2 = 1, z \leq 0$.

(c) Evaluate $\iint_{S_2} \mathbf{F} \cdot \mathbf{n} \, dS$ where S_2 is the surface of the unit cube in the first octant.

18.(a) Let $\mathbf{F} = (x^2 + y - 4)\mathbf{i} + 3xy\mathbf{j} + (2xz + z^2)\mathbf{k}$. Calculate the divergence and curl of $\mathbf{F}$.

(b) Find the flux of the curl of $\mathbf{F}$ across the surface $x^2 + y^2 + z^2 = 16$, $z \geq 0$.

(c) Find the flux of $\mathbf{F}$ across the surface of the unit cube $[0,1] \times [0,1] \times [0,1]$.

19. Let $\mathbf{F} = 2yz\mathbf{i} + (-x + 3y + 2)\mathbf{j} + (x^2 + z)\mathbf{k}$. Evaluate the surface integral

$$\iint_S (\nabla \times \mathbf{F}) \cdot d\mathbf{S},$$

where S is the cylinder $x^2 + y^2 = a^2$, $0 \leq z \leq 1$ (without the top and bottom). What if the top and bottom are included?

20. Let $\mathbf{r}(x, y, z) = (x, y, z)$, $r = \|\mathbf{r}\|$. Show that $\nabla^2(\log r) = 1/r^2$ and $\nabla^2(r^n) = n(n+1)r^{n-2}$.

21.(a) Let $f(x, y, z) = 3xye^{z^2}$. Compute ∇f.

(b) Let $\mathbf{c}(t) = (3\cos^3 t, \sin^2 t, e^t)$, $0 \leq t \leq \pi$. Evaluate

$$\int_{\mathbf{c}} \nabla f \cdot d\mathbf{s}.$$

(c) Verify directly Stokes' theorem for gradient vector fields $\mathbf{F} = \nabla f$.

22. Evaluate $\int_C yz \, dx + xz \, dy + xy \, dz$ where C is the curve of intersection of the cylinder $x^2 + y^2 = 1$ and the surface $z = y^2$.

23. Evaluate $\int_C (x + y)dx + (2x - z)dy + (y + z)dz$ where C is the perimeter of the triangle connecting $(2, 0, 0)$, $(0, 3, 0)$, and $(0, 0, 6)$ in that order.

24. Which of the following are conservative fields on $\mathbb{R}^3$? For those that are, find a function f such that $\mathbf{F} = \nabla f$.

(a) $\mathbf{F}(x, y, z) = 3x^2y\mathbf{i} + x^3\mathbf{j} + 5\mathbf{k}$

(b) $\mathbf{F}(x, y, z) = (x + z)\mathbf{i} - (y + z)\mathbf{j} + (x - y)\mathbf{k}$

(c) $\mathbf{F}(x, y, z) = 2xy^3\mathbf{i} + x^2z^3\mathbf{j} + 3x^2yz^2\mathbf{k}$

25. Consider the following two vector fields on $\mathbb{R}^3$:
   (i) $\mathbf{F}(x, y, z) = y^2\mathbf{i} - z^2\mathbf{j} + x^2\mathbf{k}$
   (ii) $\mathbf{G}(x, y, z) = (x^3 - 3xy^2)\mathbf{i} + (y^3 - 3x^2y)\mathbf{j} + z\mathbf{k}$

(a) Which of these fields (if any) are conservative on $\mathbb{R}^3$? (That is, which are gradient fields?) Give reasons for your answer.

(b) Find potentials for the fields that are conservative.

(c) Let α be the path that goes from $(0,0,0)$ to $(1,1,1)$ by following edges of the cube $0 \leq x \leq 1, 0 \leq y \leq 1, 0 \leq z \leq 1$ from $(0,0,0)$ to $(0,0,1)$ to $(0,1,1)$ to $(1,1,1)$. Let β be the path from $(0,0,0)$ to $(1,1,1)$ directly along the diagonal of the cube. Find the values of the line integrals

$$\int_\alpha \mathbf{F} \cdot d\mathbf{s}, \quad \int_\alpha \mathbf{G} \cdot d\mathbf{s}, \quad \int_\beta \mathbf{F} \cdot d\mathbf{s}, \quad \int_\beta \mathbf{G} \cdot d\mathbf{s}$$

26. Let W be a region in $\mathbb{R}^3$ with boundary ∂W. Prove the identity

$$\iint_{\partial W} [\mathbf{F} \times (\nabla \times \mathbf{G})] \cdot d\mathbf{S} = \iiint_W (\nabla \times \mathbf{F}) \cdot (\nabla \times \mathbf{G}) dV - \iiint_W \mathbf{F} \cdot (\nabla \times (\nabla \times \mathbf{G})) dV.$$

For a region W in space with boundary ∂W, unit outward normal $\mathbf{n}$ and functions f and g defined on W and ∂W, prove *Green's identities* in Exercises 27 and 28, where $\nabla^2 f = \dfrac{\partial^2 f}{\partial x^2} + \dfrac{\partial^2 f}{\partial y^2} + \dfrac{\partial^2 f}{\partial z^2}$ is the Laplacian of f.

27. $\displaystyle\iint_{\partial W} f(\nabla g) \cdot \mathbf{n}\, dS = \iiint_W (f\nabla^2 g + \nabla f \cdot \nabla g) dx\, dy\, dz$

28. $\displaystyle\iint_{\partial W} (f\nabla g - g\nabla f) \cdot \mathbf{n}\, dS = \iiint_W (f\nabla^2 g - g\nabla^2 f) dx\, dy\, dz$

29. Let f be a function on the region W in space such that:
(i) $\dfrac{\partial^2 f}{\partial x^2} + \dfrac{\partial^2 f}{\partial y^2} + \dfrac{\partial^2 f}{\partial z^2} = 0$ everywhere on W and (ii) ∇f is tangent to the boundary ∂W at each point of ∂W. Use the identity in Exercise 27 to prove that f is constant.

30. (a) Use Green's theorem to find a formula for the area of the triangle with vertices $(x_1, y_1), (x_2, y_2)$, and (x_3, y_3).

(b) Use Green's theorem to find a formula for the area of the n-sided polygon whose consecutive vertices are

$$(x_1, y_1), (x_2, y_2), \ldots, (x_n, y_n).$$

31. Consider the *constant* vector field $\mathbf{F}(x, y, z) = \mathbf{i} + 2\mathbf{j} - \mathbf{k}$ in $\mathbb{R}^3$.

(a) Find a scalar field $\phi(x, y, z)$ in $\mathbb{R}^3$ such that $\nabla\phi = \mathbf{F}$ in $\mathbb{R}^3$ and $\phi(0,0,0) = 0$.

(b) On the sphere Σ of radius 2 about the origin find all the points at which
   i. ϕ is a maximum and

    ii. ϕ is a minimum.

(c) Compute the maximum and minimum values of ϕ on Σ.

32. As was discussed in §**6.4**, surface integrals apply to the study of heat flow. Let $T(x, y, z)$ be the temperature at a point (x, y, z) in W where W is some region in space and T is a function with continuous partial derivatives. Then

$$\nabla T = \frac{\partial T}{\partial x}\mathbf{i} + \frac{\partial T}{\partial y}\mathbf{j} + \frac{\partial T}{\partial z}\mathbf{k}$$

represents the temperature gradient, and heat "flows" with the vector field $-k\nabla T = \mathbf{F}$, where k is the positive constant. Thus $\iint_S \mathbf{F} \cdot d\mathbf{S}$ is the total rate of heat flow or flux across the surface S. ($\mathbf{n}$ is the unit outward normal.) Suppose a temperature function is given as $T(x, y, z) = x^2 + y^2 + z^2$, and let S be the ellipsoid $x^2 + 4y^2 + 4z^2 = 1$. Use Gauss' theorem to find the heat flux across the surface S if $k = 1$.

33.(a) Express conservation of thermal energy by means of the statement that for any volume W in space

$$\frac{d}{dt} \iiint_W e \, dx \, dy \, dz = -\iint_{\partial W} \mathbf{F} \cdot d\mathbf{S},$$

where $\mathbf{F} = -k\nabla T$, as in Exercise 32 and $e = c\rho_0 T$, where c is the specific heat (a constant) and ρ_0 is the mass density (another constant). Use the divergence theorem to show that this statement of conservation of energy is equivalent to the statement

$$\frac{\partial T}{\partial t} = \frac{k}{c\rho_0}\nabla^2 T \quad (\textit{the heat equation}),$$

where $\nabla^2 T = \text{div grad } T = \partial^2 T/\partial x^2 + \partial^2 T/\partial y^2 + \partial^2 T/\partial z^2$ is the *Laplacian* of T.

(b) Make up an integral statement of conservation of mass for fluids that is equivalent to the **continuity equation**

$$\frac{\partial \rho}{\partial t} + \text{div}\,(\rho\mathbf{V}) = 0,$$

where ρ is the mass density of a fluid and $\mathbf{V}$ is the fluid's velocity field.

Epilogue: Where Do We Go from Here?

You have just finished a study of the basic concepts of the calculus of several variables and vector calculus. Those students who want to pursue mathematics, will find that these concepts play a crucial role in the formulation and implementation of many physical and mathematical theories. For example, they lie at the heart of the study of differential geometry in mathematics, and the theories of relativity and quantum mechanics in physics. To quote from Nobel prize-winning physicist Paul Dirac:

> It seems to be one of the fundamental features of nature that fundamental physical laws are described in terms of a mathematical theory of great beauty and power, needing quite a high standard of mathematics for one to understand it. You may wonder: Why is nature constructed along these lines? One can only answer that our present knowledge seems to show that nature is so constructed. We simply have to accept it. One could perhaps describe the situation by saying that God is a mathematician of a very high order, and He used very advanced mathematics in constructing the universe. Our feeble attempts at mathematics enable us to understand a bit of the universe, and as we proceed to develop higher and higher mathematics we can hope to understand the universe better.

Much of the mathematics that Dirac refers to developed from the concepts you have mastered so far in your mathematics courses, including vector calculus. Whether or not you go on with your mathematical studies, we hope to have given you some appreciation for a fundamental area of mathematics and its role in history and science.

Practice Examination 1

1.(a) Find a unit vector orthogonal to *both* $\mathbf{i} - \mathbf{j}$ and $3\mathbf{j} + 2\mathbf{k}$.

(b) Find the distance from the point $(1, 1, -2)$ to the plane that passes through the points $(1, 0, 0), (0, 1, 0), (0, 0, 2)$.

(c) Where does the line $l(t) = t\mathbf{i} - \mathbf{j} + (1 - t)\mathbf{k}$ meet the plane that passes through the origin and is orthogonal to the vector $\mathbf{i} + \mathbf{j} + \mathbf{k}$?

(d) Boat A starts at $(0, 0)$ at $t = 0$ and proceeds at 8 km/h in direction $\mathbf{i} + \mathbf{j}$ while boat B starts at $(1, 1)$ at $t = 0$ and proceeds at 4 km/h in direction $-\mathbf{i} + \mathbf{j}$. What is the closest distance between the boats and when does it occur?

(e) A particle follows the curve $\mathbf{c}(t) = (3 \sin t, 3 \cos t, t)$ and flies off on a tangent at $t = \pi/2$. Where is it at $t = \pi$?

2.(a) Find the tangent plane to the graph $z = x^3 + y^3$ at $(1, -1, 0)$.

(b) State the chain rule for $h = f \circ g$ where $f : \mathbb{R}^2 \to \mathbb{R}^3$ and $g : \mathbb{R}^4 \to \mathbb{R}^2$. What are the sizes of the matrices involved?

(c) Calculate the second-order Taylor expansion of

$$f(x, y) = xy + \sin(x + y^2)$$

about $(0, 0)$.

3.(a) Let a curve $\mathbf{r}(t)$ in $\mathbb{R}^3$ satisfy $\mathbf{F} = m\mathbf{a}$, where $\mathbf{F}(\mathbf{r}) = -GmM\mathbf{r}/r^3$. Show that $\mathbf{J} = \mathbf{r} \times \dot{\mathbf{r}}$ is constant in time.

(b) Show that the vector field $\mathbf{F} = -GmM\mathbf{r}/r^3$ in $\mathbb{R}^3$ has zero divergence.

(c) Find the critical points of $f(x, y) = x^2 - 3xy - y^2$ and test them for maxima and minima.

4.(a) The temperature in space at the position (x, y, z) is given by the function $T(x, y, z) = x^2 + y^2 - 3z^2$. If you are at $(x, y, z) = (0, 1, 1)$, in what direction must you go to increase your temperature as fast as possible?

(b) You decide to follow the flow line of the vector field $\mathbf{F} = \nabla T$, again starting at $(0, 1, 1)$ at $t = 0$. Show that your rate of change of temperature at $t = 0$ is given by $\|\nabla T(0, 1, 1)\|^2$.

(c) Describe geometrically what paths, starting at $(0, 1, 1)$, you can take to maintain the *same* temperature.

5.(a) Let $\Phi(u, v) = (\sin u \cos v, \sin u \sin v, \cos u)$, for $0 \le u \le \pi/2$, and $\pi/2 \le v \le \pi$. Describe the parametrized surface S that is obtained.

(b) Find the equation of the tangent plane to S at $u = \pi/4, v = 3\pi/4$.

(c) For a smooth function $f(x, y, z)$ and a parametrized surface that satisfies $\Phi_u(u_0, v_0) \times \Phi_v(u_0, v_0) \ne 0$, show that $(x_0, y_0, z_0) = \Phi(u_0, v_0)$ is a critical point of f if and only if the following *two* conditions hold:

i. $\nabla f(x_0, y_0, z_0) \cdot [\Phi_u(u_0, v_0) \times \Phi_v(u_0, v_0)] = 0$

ii. $\nabla f(x_0, y_0, z_0) \cdot \sigma'(0) = 0$, where σ is any curve of the form $\sigma(t) = \Phi(\mathbf{c}(t))$, with $\mathbf{c}(t)$ a curve in the uv plane satisfying $\mathbf{c}(0) = (u_0, v_0)$.

6.(a) Rewrite the integral

$$\int_0^1 \int_0^1 \int_0^{\sqrt{1-y^2}} f(x, y, z)\,dz\,dy\,dx$$

in the order $dydzdx$, including a sketch of the region of integration.

(b) Extremize $f(x, y, z) = x + y$ subject to the constraint

$$x^2 + y^2 + z^2 = 1.$$

7.(a) Let C be the circle $x^2 + y^2 = 1, z = 0$ and let

$$\mathbf{F}(x, y, z) = [x^2 y^3 + y]\mathbf{i} + [x^3 y^2 + x]\mathbf{j} + z\mathbf{k}.$$

Calculate the line integral $\int_C \mathbf{F} \cdot d\mathbf{s}$.

(b) An alien force field is exerting the force $\mathbf{F} = -4\mathbf{r}/r^6$ on Captain Thomas' spaceship. Find the work done by the field in moving the ship from a distance $r = 2$ to a distance $r = 1$.

8. If true, ***justify***, and if false, give a ***counterexample***, or explain why.

(a) If $f(x, y, z)$ is a smooth function, then the flux of ∇f out of the sphere $x^2 + y^2 + z^2 = 1$ is zero.

(b) There exists a vector field $\mathbf{F}$ that satisfies $\nabla \times \mathbf{F} = x\mathbf{i}$.

(c) The velocity field of a fluid is given by $\mathbf{F} = y\mathbf{j} + 2z\mathbf{k}$. The rate of flow of fluid out of the sphere $x^2 + y^2 + z^2 = 1$ is 4π.

(d) The equation $x^2yz^5 + z^4 - 1 = 0$ defines a smooth function $z = f(x, y)$ near the point $x = 0, y = 1, z = 1$.

9. Let W be the region in the octant $x \geq 0, y \geq 0, z \geq 0$, bounded by the three planes $y = 0, z = 0, x = y$, and the sphere $x^2 + y^2 + z^2 = 1$.

   (a) Set up a triple integral giving the integral of a function $f(x, y, z)$ over this region using spherical coordinates.

   (b) Calculate the surface integral $\iint_S \mathbf{F} \cdot d\mathbf{S}$, where the vector field $\mathbf{F}$ is given by $\mathbf{F}(x, y, z) = (3x - z^4)\mathbf{i} - (x^2 - y)\mathbf{j} + (xy^2)\mathbf{k}$ and S is the boundary of the set W.

10. Let C be the curve $x^2 + y^2 = 1, z = 0$ and let S be the surface S_1 together with S_2, where S_1 is defined by $x^2 + y^2 \leq 1, z = -1$ and S_2 is defined by $x^2 + y^2 = 1, -1 \leq z \leq 0$.

    (a) Draw a figure showing an orientation for S and C such that Stokes' theorem applies to the surface S and the curve C.

    (b) If R is another surface with boundary C, show that

    $$\iint_S (\nabla \times \mathbf{F}) \cdot d\mathbf{S} = \iint_R (\nabla \times \mathbf{F}) \cdot d\mathbf{S}.$$

    (c) Letting

    $$\mathbf{F}(x, y, z) = (y^3 + e^{xz})\mathbf{i} - (x^3 + e^{yz})\mathbf{j} + e^{xyz}\mathbf{k},$$

    calculate

    $$\iint_S (\nabla \times \mathbf{F}) \cdot d\mathbf{S},$$

    where S is the surface given above.

Practice Examination 2

1. Recall that the *flow lines* of a vector field $\mathbf{F}$ on $\mathbb{R}^n$ are the parametrized curves $\mathbf{c}(t)$ in $\mathbb{R}^n$ for which $\mathbf{c}'(t) = \mathbf{F}(\mathbf{c}(t))$. Let $H(x, y)$ be a real-valued function defined on $\mathbb{R}^2$. The vector field

$$\mathbf{F}(x,y) = \frac{\partial H}{\partial y}\mathbf{i} - \frac{\partial H}{\partial x}\mathbf{j} \qquad (1)$$

is called the *Hamiltonian vector field* associated with the function H. (Some terminology from physics is used in this problem, but you do not need to know any physics to do the problem.)

(a) Find the Hamiltonian vector field associated with the function $H(x, y) = V(x) + \frac{1}{2}y^2$, where V is a function of x. The flow lines of this vector field are solutions of a pair of equations of the form:

$$\frac{dx}{dt} = \cdots$$

$$\frac{dy}{dt} = \cdots$$

Complete the equations above. Combine the two equations to obtain an expression for d^2x/dt^2 in terms of x (not involving y). (This equation is sometimes called *Newton's equation*.)

(b) Find the Hamiltonian vector field associated with the function $\frac{1}{2}x^2 + \frac{1}{2}y^2$. Sketch this vector field and describe its flow lines.

(c) Suppose that $\mathbf{c}(t)$ is a flow line of the general Hamiltonian vector field (1) associated with H. Find the time derivative $(d/dt)H(\mathbf{c}(t))$ of H along the flow line. (The resulting equation is called the law of *conservation of energy*.)

(d) In general, what is the relation between the Hamiltonian vector field (1) and the gradient vector field of H? Describe a geometric procedure to construct one of these vector fields from the other.

(e) Find the divergence of the general Hamiltonian vector field (1).

(f) What condition on the derivatives of H is equivalent to the Hamiltonian vector field (1) being a gradient vector field?

2. Captain Astro is in hot water again. More precisely, her vehicle is at the origin in $\mathbb{R}^2$, and the temperature as a function of position is given by the formula $T = 500 - 2x^2 + xy - 2y^2$. In which direction should she travel to cool off most rapidly? (Hint: *Don't* use the directional derivative.)

3. Let S be the "sock" surface shown below, with the orientation given by the *inward* normal. Its boundary is the circle $x^2 + y^2 = 4, z = 0$.

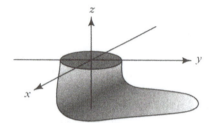

(a) Draw the boundary of S in a figure in the xy plane, and indicate the correct orientation of the boundary by an arrow.

(b) Is the sock the boundary of a solid region in $\mathbb{R}^3$? If so, describe the region. If not, explain why not.

The sock is exposed to X-rays given by the vector field $\mathbf{F}(x, y, z) = 2x\mathbf{i} + \mathbf{k}$.

(c) Is $\mathbf{F}$ the curl of a vector field? Explain.

(d) A disc in the xy plane is placed over the opening at the top of the sock, so that a region W in 3-space is now enclosed. The X-ray flux through the sock (with the orientation mentioned above) is measured to be -100. What is the volume of the enclosed region W?

4. Let C be the curve in the plane bounded by the semicircle

$$(x + 1)^2 + y^2 = 4, x \geq -1,$$

together with the vertical line segment joining its endpoints. Find the line integral of each of the following objects along C, oriented in the *clockwise* direction.

(a) the differential form $x \, dy - y \, dx$;

(b) the vector field

$$\frac{y}{x^2 + y^2}\mathbf{i} - \frac{x}{x^2 + y^2}\mathbf{j}.$$

5. Let $f(x, y) = x^2 + \cos y$.

   (a) Find the first- and second-order approximations to f at $(0, 0)$.

   (b) Find all the critical points of f. Classify each of them as a local minimum, local maximum, or saddle point.

   (c) Find all the global maximum and minimum points for f.

6. Let S be the surface $z = e^{-(x^2 + y^2)}$, $0 \leq x^2 + y^2 \leq 4$.

   (a) Express the surface area of S as an iterated integral. (Do not evaluate the integral.)

   (b) Show that the surface area of S is greater than 4π and less than 1000.

   (c) Find the volume of the region cut out by S, the xy plane, and the cylinder $x^2 + y^2 = 4$.

7. Tell whether each of the following expressions makes sense. For each one which does not make sense, correct it by replacing the word or phrase in capital letters.

   (a) ... the gradient of a SURFACE

   (b) ... the flux of the DIVERGENCE of a vector field

   (c) ... the area of a SURFACE INTEGRAL

   (d) ... the partial derivatives of a GRAPH

Answers to Odd-Numbered Exercises

Chapter 1
Algebra and Geometry of Euclidean Space

1.1 Vectors in the Plane and Space

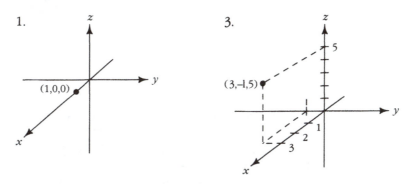

1.

3.

5. $(4, 9)$

7. $(11, 0, 11)$

9. $y = 1$

11. $b = 2a, c = a$

13. 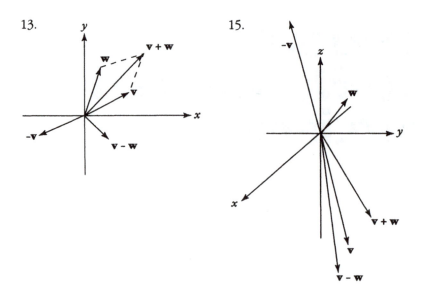 15.

17.   (a) **d**; (b) **e**

19.   (a) $(6, 5)$; (b) $(-3, -6)$

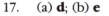

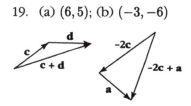

21. $7\mathbf{i} + 2\mathbf{j} + 3\mathbf{k}$ 23. $\mathbf{i} - \mathbf{k}$
25. $x = 1, y = -1 + t, z = -1$ 27. $x = -1 + 2t, y = -1, z = -1 + 3t$
29.   All points of the form $(s, -2t, 3s), 0 \leq s \leq 1, 0 \leq t \leq 1$
31. $(2s, 7s + 2t, 7t), s, t, \text{real}$
33.   If one vertex is at the origin and the two adjacent sides are **u** and **v**, then $\frac{1}{2}\mathbf{u} - \frac{1}{2}\mathbf{v} = \frac{1}{2}(\mathbf{u} - \mathbf{v})$ is half the length of and parallel to the third side, $\mathbf{u} - \mathbf{v}$.
35.   With notation as in 33, the new triangle has sides $b\mathbf{u}, b\mathbf{v}$, and $b(\mathbf{u} - \mathbf{v})$.
37. $(1, 0, 1) + (0, 2, 1) = (0, 2, 0) + (1, 0, 2)$

1.2 The Inner Product and Distance

1. $\|\mathbf{u}\| = \sqrt{245}, \|\mathbf{v}\| = \sqrt{10 + \pi^2}, \mathbf{u} \cdot \mathbf{v} = 15\pi - 10$

3. $\|\mathbf{u}\| = \sqrt{248}, \|\mathbf{v}\| = 5, \mathbf{u} \cdot \mathbf{v} = -42$

5. $(15\mathbf{i} - 2\mathbf{j} + 4\mathbf{k})/\sqrt{245}$ 7. $(2\mathbf{j} - \mathbf{i})/\sqrt{5}$

9. $\cos^{-1}\left[\dfrac{15\pi - 10}{\sqrt{245(10 + \pi^2)}}\right]$ 11. $\cos^{-1}[-21/(5\sqrt{62})]$

13. $\|\mathbf{u}\| \cdot \|\mathbf{v}\| = \sqrt{245}\sqrt{10 + \pi^2} \geq |15\pi - 10|$

15. $5\sqrt{248} \geq 42$ 17. $(1, 0, -1), (0, 1, -1)$

19.   No, by Cauchy–Schwarz.   21. $(-4\mathbf{i} - 2\mathbf{j} + 6\mathbf{k})/7$

23. $2(\mathbf{i} + \mathbf{j} + \mathbf{k})/3$ 25. $\theta = \cos^{-1}(1/\sqrt{3})$

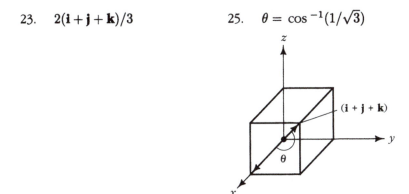

27. $\mathbf{v} = \mathbf{i} + 4\mathbf{j}; \theta = \cos^{-1}(4/\sqrt{17})$ 29.   (a) $-40\mathbf{i} + 100\mathbf{j}$

31.   13 knots

33.   (a) $\mathbf{F} = 6(\mathbf{i} + \mathbf{j})/\sqrt{2}$; (b) $\cos^{-1}(3/\sqrt{10})$; (c) $36/\sqrt{2} = 18\sqrt{2}$

35.   Write $\mathbf{v} = \alpha\mathbf{e}_1 + \beta\mathbf{e}_2$ and take the dot product of $\mathbf{V}$
      with $\mathbf{e}_1$ and $\mathbf{e}_2$ to determine α and β.

37. $\mathbf{F} = \|\mathbf{F}\|\mathbf{v} + \|\mathbf{F}\|\mathbf{h})/\sqrt{2}$, where $\mathbf{v} = (\mathbf{i} - \mathbf{j})/\sqrt{2}$ and $\mathbf{h} = -(\mathbf{i} + \mathbf{j})/\sqrt{2}$

39.   The factor r cancels out.

1.3 2 × 2 and 3 × 3 Matrices and Determinants

1.   2   3.   0

5.   0   7.   4

9.   Expand both sides of
$$\begin{vmatrix} a_{11} & a_{12} & a_{13} \\ a_{21} & a_{22} & a_{23} \\ a_{31} & a_{32} & a_{33} \end{vmatrix} = \begin{vmatrix} a_{11} + 3a_{21} & a_{12} + 3a_{22} & a_{13} + 3a_{23} \\ a_{21} & a_{22} & a_{23} \\ a_{31} & a_{32} & a_{33} \end{vmatrix}$$
along the first row to show that the two expressions agree.

11. $\begin{bmatrix} -1 & -7 \\ -4 & -1 \end{bmatrix}$

13.   Substitute the proposed formulas for x and y and check.

15. $x = 1/2, y = 0$

1.4 The Cross Product and Planes

1. $\mathbf{j} + \mathbf{k}$ 3. $2\mathbf{i} - 2\mathbf{j} + 4\mathbf{k}$

5. $9\mathbf{i} + 18\mathbf{j}$ 7. $6\mathbf{i} - 2\mathbf{k}$

9. $\pm\mathbf{k}$ 11. $\pm\dfrac{113\mathbf{i} + 17\mathbf{j} - 103\mathbf{k}}{\sqrt{23,667}}$

13. $\sqrt{14}$ 15. $\sqrt{50}$

17. $\sqrt{2}$ 19. 2

21. $\sqrt{117}/2 = 3\sqrt{13}/2$ 23. 1

25. $x + y + z = 1$ 27. $5x + 2z = 25$

29. $2x + 3y + 4z = 0$ 31. $z + y = 1$

33. $7/3$ 35. $8/\sqrt{6}$ 37.   It is an integer.

39.   Either an integer or an integer $\pm\frac{1}{2}$.

41.   Write out each side in components and simplify.

43.   Yes. [Hint: By rotating axes we may assume $\mathbf{a} = \lambda\mathbf{i}$. The equation $\mathbf{x} \cdot \mathbf{a} = |\lambda|$ implies that if $\mathbf{x} = (x_1, x_2, x_3)$ then $x_1 = \lambda/|\lambda|$. The formula for the cross product then gives $x_3 = b_2/\lambda, x_2 = -b_3/\lambda$.]

45. $-\mathbf{i} + \mathbf{j}$ and $-\mathbf{j} + \mathbf{k}$

1.5 n-dimensional Euclidean Space

1. $(2, 6, 8, 10, 12)$ 3. 70

5. $|\mathbf{a} \cdot \mathbf{b}| = 10 \le \|\mathbf{a}\| \cdot \|\mathbf{b}\| = 10$;
      $\|\mathbf{a} + \mathbf{b}\| = 3\sqrt{5} \le \|\mathbf{a}\| + \|\mathbf{b}\| = 3\sqrt{5}$

7. $|\mathbf{a} \cdot \mathbf{b}| = 2 \le \|\mathbf{a}\| \cdot \|\mathbf{b}\| = \sqrt{6}$;
      $\|\mathbf{a} + \mathbf{b}\| = 3 \le \|\mathbf{a}\| + \|\mathbf{b}\| = \sqrt{3} + \sqrt{2}$

9. $\begin{bmatrix} 5 & 9 & 6 \\ 12 & 7 & 7 \\ 7 & 14 & 15 \end{bmatrix}$ 11. $\begin{bmatrix} 0 & 0 & 3 \\ 6 & 6 & 5 \\ -2 & -1 & -2 \end{bmatrix}$

13. 32 15. $\begin{bmatrix} c & d \\ a & b \end{bmatrix}$

17. $\begin{bmatrix} 6 & 4 & 2 \\ 12 & 8 & 4 \\ 18 & 12 & 6 \end{bmatrix}$

19.   Not defined; number of columns of first matrix $\ne$ number of rows of second.

21.   (a) $\begin{bmatrix} x_1 + 2x_2 + 3x_3 \\ 4x_1 + 5x_2 + 6x_3 \\ 7x_1 + 8x_2 + 9x_3 \end{bmatrix}$;   (b) $\begin{bmatrix} 14 \\ 32 \\ 50 \end{bmatrix}$

23.　(a) $\begin{bmatrix} 4x_1 + 5x_2 \\ 9x_1 \\ x_1 + x_2 \\ 7x_1 + 3x_2 \end{bmatrix}$;　(b) $\begin{bmatrix} 73 \\ 63 \\ 16 \\ 76 \end{bmatrix}$

25.　Yes   27.　Yes

29.　$B = \begin{bmatrix} 1 & -2 \\ 0 & 1 \end{bmatrix}$

31.　$\det I = \det(AA^{-1}) = (\det A)(\det A^{-1}) = 1$, thus,
　　$\det A^{-1} = 1/(\det A)$

33.　The determinant of A expanded along the first row is the determinant
　　of A^T expanded along the first column.

35.　(a) Expand the determinant of B along the row that is multiplied by λ.
　　(b) Show that for a 3×3 matrix C, $\det(\lambda C) = \lambda^3 \det C$,
　　then expand λA along the first row.

37.　No. Let $A = \begin{bmatrix} 0 & 1 \\ -1 & 0 \end{bmatrix}$ and $B = \begin{bmatrix} 0 & 1 \\ 1 & 0 \end{bmatrix}$, for example.

1.6 Curves in the Plane and in Space

1.　This curve is the ellipse $(y^2/16) + x^2 = 1$:

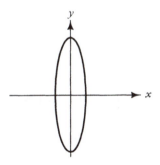

3.　This curve is a straight line:

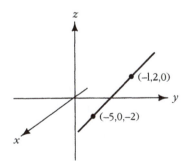

5. $6\mathbf{i} + 6t\mathbf{j} + 3t^2\mathbf{k}$
7. $(-2\cos t \sin t, 3 - 3t^2, 1)$

9. $\mathbf{c}'(t) = (e^t, -\sin t)$
11. $\mathbf{c}'(t) = (t\cos t + \sin t, 4)$

13. Horizontal when $t = (R/v)n\pi$, n an integer; speed is zero if n is even; speed is $2v$ if n is odd.

15. $(\sin 3, \cos 3, 2) + (3\cos 3, -3\sin 3, 5)(t - 1)$

17. $(8, 8, 0)$
19. $(8, 0, 1)$

Review Exercises (Chapter 1)

1. $(2, 8)$
3. $11\mathbf{i} + \mathbf{j} - \mathbf{k}$

5. (a) $\mathbf{v} = (6, 6)$

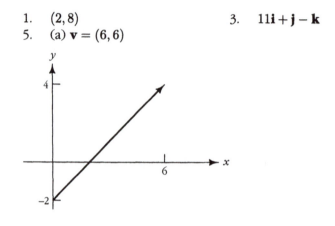

(b) $(9, 7)$

7. $x = 1 + t, y = 1 + t, z = 2 + t$

9. $x = 1 + t, y = 1 - t, z = 1 - t$

11. $x = t/2, y = \sqrt{3}t/2$

13. $(\mathbf{a} + \mathbf{b})(\mathbf{b} - \mathbf{a}) = \|\mathbf{b}\|^2 - \|\mathbf{a}\|^2 = 0 = \|\mathbf{a} + \mathbf{b}\| \cdot \|\mathbf{b} - \mathbf{a}\| \cos\theta$

15. $5\sqrt{73} - 24\sqrt{2} \approx 31.25$ km/hr

17. (a) $70\cos\theta + 20\sin\theta$; (b) $21\sqrt{3} + 6$

19. 0
21. -10

23. 0

25. $\mathbf{v} \cdot \mathbf{w} = \|\mathbf{v}\| \cdot \|\mathbf{w}\| \cos\theta = 0$ for all $\mathbf{w}$ implies that $\|\mathbf{v}\| = 0$, and so $\mathbf{v} = 0$

27. 2
29. -2

31. Add row 3 to row 1 and subtract row 3 from row 2.

33. $-4\mathbf{i} + 7\mathbf{j} - 11\mathbf{k}$
35. 5

37. $\mathbf{v}$ lies in the plane spanned by $\mathbf{j}$ and $\mathbf{k}$.

39. $\sqrt{381}$
41. $1/2$

43. $\|\mathbf{a} \times (\mathbf{b} + \lambda\mathbf{a})\| = \|\mathbf{a} \times \mathbf{b} + \mathbf{a} \times \lambda\mathbf{a}\| = \|\mathbf{a} \times \mathbf{b}\|$

45. $(-2\mathbf{i} + 3\mathbf{j} + 3\mathbf{k})/\sqrt{22}$
47. $(0, 1, -1)/\sqrt{2}$

49. $x - y = 0$
51. $x = 3t, y = 5t, z = t + 3$

53. 4

55. $\begin{pmatrix} 3 \\ 8 \end{pmatrix}$

57. $g(f(\mathbf{x})) = (g \circ f)(\mathbf{x})$

59. (b) $d = \sqrt{2}$

61. Let θ be the angle between $\mathbf{a}$ and $\mathbf{b}$ and ξ that between $\mathbf{a}$ and $\mathbf{v}$. Use the formula $\mathbf{a} \cdot \mathbf{v} = \|\mathbf{a}\| \|\mathbf{v}\| \cos \xi$ to argue that $\xi = \theta/2$. Similarly the angle between $\mathbf{v}$ and $\mathbf{b}$ is $\theta/2$.

63. $\frac{2}{7}(1, 2, 3)$

65. $\mathbf{l}(t) = (-1 + 3t, (2 - t)/e, \pi(1 - t)/2)$

67. $\mathbf{l}(t) = (1 + t, t, 1)$

69.

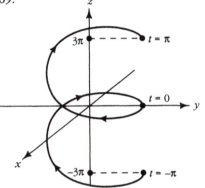

71. $(2\pi, 3\pi^2, -2\pi)$

Chapter 2
Differentiation

2.1 Graphs and Level Surfaces

1. $z = x - y + 2$

3. $z = x + 2$

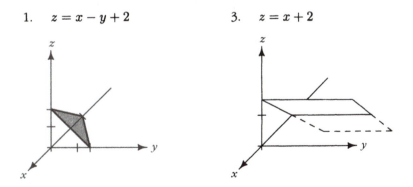

5.

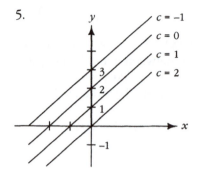

7.   No level curve for $c = -1, 0$. For $c = 1$ the level curve is a point; for $c = 2$ it is a circle of radius 1.

9.   11.

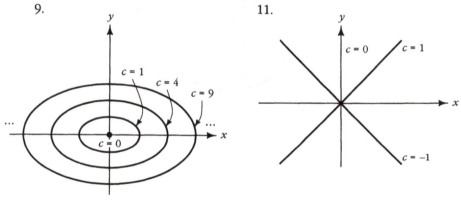

13.   Sphere of radius $\sqrt{-c}$ if $c < 0$; origin if $c = 0$; no level curve if $c > 0$.
15.   Cylinder of radius $\sqrt{c}$ if $c > 0$; z axis if $c = 0$; no level curve if $c < 0$.

17. $z = |y|$

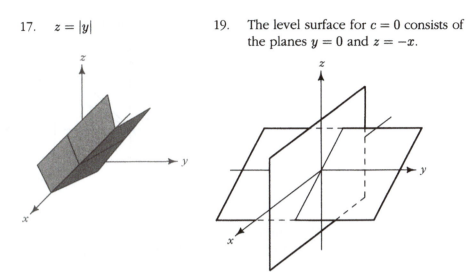

19. The level surface for $c = 0$ consists of the planes $y = 0$ and $z = -x$.

21. The value of z does not matter, so we get a "cylinder" of elliptic cross section parallel to the z axis and intersecting the xy plane in the ellipse $4x^2 + y^2 = 16$.

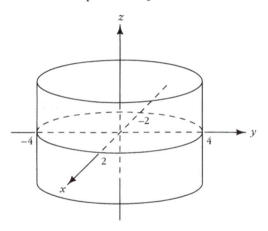

23. The value of y does not matter, so we get a "cylinder" of parabolic cross section.

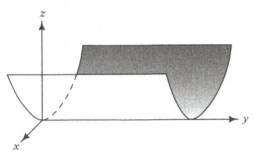

25.   A double cone with axis along the y axis and elliptical cross sections.

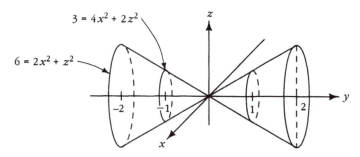

27.   This is a saddle surface, but the hyperbolas, which are level curves, no longer have perpendicular asymptotes.

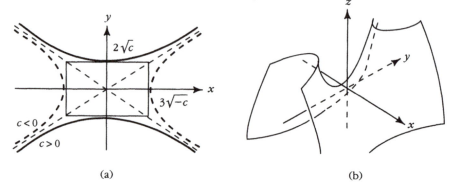

(a)   (b)

2.2 Partial Derivatives and Continuity

1. $y; x; 1; 1$ 3. $6x; 4y; 6; 8$
5. $yz; xz; xy; 1; 1; 1$
7. $-y^2 \sin(xy^2) + 3yze^{3xyz}; -2xy \sin(xy^2) + 3xze^{3xyz}; 3xye^{3xyz}; 3e^{3\pi};$
       $3\pi e^{3\pi}; 3\pi e^{3\pi}$
9. $e^{xyz}[(yz)(xy + xz + yz) + y + z]; e^{xyz}[(xz)(xy + xz + yz) + x + z];$
       $e^{xyz}[(xy)(xy + xz + yz) + x + y]$
11. $e^x \cos(yz^2); -e^x z^2 \sin(yz^2); -2zye^x \sin(yz^2)$
13. $tu^2 \exp(stu^2)$
15. $\dfrac{-\mu \sin(\lambda\mu) \cdot (1 + \lambda^2 + \mu^2) - 2\lambda(\cos \lambda\mu)}{(1 + \lambda^2 + \mu^2)^2}$
17.   0.000047 degrees Celsius   19.   No limit exists.
21.   1   23.   1
25.   Try $\delta = 1/300$.
27.   Let $r = \|\mathbf{x} - \mathbf{y}\|/2$. If $\|\mathbf{z} - \mathbf{y}\| \le r$, let $f(\mathbf{z}) = \|\mathbf{z} - \mathbf{y}\|/r$; otherwise,
       let $f(z) = 1$
29.   Sum, product, and composition of continuous functions are continuous.

31. $2(x + y)z$ 33. No

2.3 Differentiability, the Derivative Matrix, and Tangent Planes

1. $z = -9x + 6y - 6$ 3. $z = 1$
5. $z = 2x + 6y - 4$ 7. $z = x - y + 2$

9. $\begin{bmatrix} e \\ \sin 3 \end{bmatrix} + \begin{bmatrix} e & 0 \\ 3\cos 3 & \cos 3 \end{bmatrix} \begin{bmatrix} x - 1 \\ y - 3 \end{bmatrix}$

11. $\begin{bmatrix} 3 \\ 1 \end{bmatrix} + \begin{bmatrix} 1 & 1 & 1 \\ 2 & 1 & 0 \end{bmatrix} \begin{bmatrix} x - 1 \\ y - 1 \\ z - 0 \end{bmatrix}$

13. $\begin{bmatrix} e & 0 \\ 3\cos 3 & \cos 3 \end{bmatrix}$ 15. $\begin{bmatrix} 1 & 1 & 1 \\ 2 & 1 & 0 \end{bmatrix}$

17. f is C^1 on $\mathbb{R}^2$ minus the point $(0,0)$.
19. Since $\partial f/\partial r = \frac{1}{2}\sin 2\theta$ and $\partial f/\partial\theta = r\cos 2\theta$ are continuous, f is C^1.

2.4 The Chain Rule

1. $16(3 + 8t)\cos(3 - 2t) + 2(3 + 8t)^2\sin(3 - 2t)$;
    $-8(3 - 2t)^2\cos(3 + 8t) + 4(3 - 2t)\sin(3 + 8t)$

3. $e^{t-t^2} - e^{t^3-t} - 2te^{t-t^2} + 2te^{t^2-t^3} - 3t^2e^{t^2-t^3} + 3t^2e^{t^3-t}$

5. $[x(x^2 + y^2)^{-\frac{1}{2}} + 2y^2]\dfrac{dx}{du} + [y(x^2 + y^2)^{-\frac{1}{2}} + 4xy]\dfrac{dy}{du}$

7. $\dfrac{\partial u}{\partial r} = (\cos\theta\sin\phi)\dfrac{\partial u}{\partial x} + (\sin\theta\sin\phi)\dfrac{\partial u}{\partial y} + (\cos\phi)\dfrac{\partial u}{\partial z}$

    $\dfrac{\partial u}{\partial\theta} = (-r\sin\theta\sin\phi)\dfrac{\partial u}{\partial x} + (r\cos\theta\sin\phi)\dfrac{\partial u}{\partial y}$

    $\dfrac{\partial u}{\partial\phi} = (r\cos\theta\cos\phi)\dfrac{\partial u}{\partial x} + (r\sin\theta\cos\phi)\dfrac{\partial u}{\partial y} + (-r\sin\phi)\dfrac{\partial u}{\partial z}$

9. $\mathbf{i} + 2\mathbf{j} - 2\mathbf{k}$ 11. $3\mathbf{i} + 2\mathbf{j} + 7\mathbf{k}$

13. $[26x + 6y + 70 \quad 6x + 2y + 14]$

15. $\left[(\sin 1)\dfrac{\partial z}{\partial x} + \dfrac{\partial z}{\partial y} \quad 0\right]$

17. $\left[9\dfrac{\partial w}{\partial u} + \dfrac{\partial w}{\partial v} \quad 9\dfrac{\partial w}{\partial u} + \dfrac{\partial w}{\partial v} \quad 9\dfrac{\partial w}{\partial u} + \dfrac{\partial w}{\partial v}\right]$

19. $\begin{bmatrix} 2t + 2s + 1 & 2t + 2s - 1 \\ 4s & 4t \end{bmatrix}$

21. $(-1,1); \begin{bmatrix} 0 & 0 \\ 2 & -2 \end{bmatrix}$ 23. $\dfrac{\partial z}{\partial t} = \dfrac{\dfrac{\partial x}{\partial t}y - x\dfrac{\partial y}{\partial t}}{y^2}$

25. The symbol $\partial w/\partial x$ is used with two different meanings.

2.5 Gradients and Directional Derivatives

1. $\left(\dfrac{x}{\sqrt{x^2 + y^2}}, \dfrac{y}{\sqrt{x^2 + y^2}} \right)$ 3. $(y + z, x + z, x + y)$

5. $(y^2 + 2xz, 2xy + z^2, 2yz + x^2)$

7. $(2x \log \sqrt{x^2 + y^2} + x, 2y \log \sqrt{x^2 + y^2} + y)$

9. $(x/4, y/6)$ 11. $\nabla f = (1, -1)$

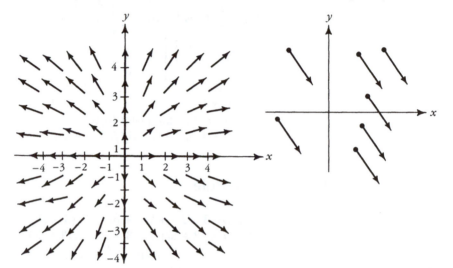

13. $2e^{2t} + e^t \cos t - e^t \sin t$ 15. $t/\sqrt{1 + t^2}$

17. $-11 - 16\sqrt{3}$ 19. $-14/\sqrt{3}$

21. $\dfrac{5}{13} c e^{c-1} + \dfrac{12}{13} e^c$ 23. 0

25. Increasing fastest: $(\sin 1, \cos 1, \sin 1)$;
    Decreasing fastest: $-(\sin 1, \cos 1, \sin 1)$

27. $\nabla f(0, 0, 1) = 2\mathbf{k}$

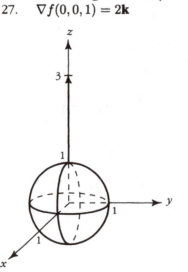

29. $(8, 8, 1)/\sqrt{129}$
31. $\mathbf{k}$
33. $x + 2\sqrt{3}y + 3z = 10$
35. $3x + 8y + 3z = 20$
37. $x + 2y = 3$
39. $x + y = \pi/2$
41. The vectors are perpendicular.
43. (a) $(1, 2, 3)/\sqrt{14}$; (b) $2\sqrt{14}e^2$
45. $(2, -2), -2/\sqrt{13}$

2.6 Implicit Differentiation

1. $-x/2y$
3. y/x
5. $1/2$
7. $-1 - 2/\sqrt{3}$
9. $z \neq \pm\sqrt{-xy/6}$
11. $z \neq 0$
13. $3x - y + 5z + 1 = 0$
15. $z = 1$
17. $dy/dx = -3/(2y + 1)$
19. $(2, 2, 0); t = \sqrt{5}/10$
21. $f_x = -F_x/F_z, f_y = -F_y/F_z$

Review Exercises (Chapter 2)

1.   3.

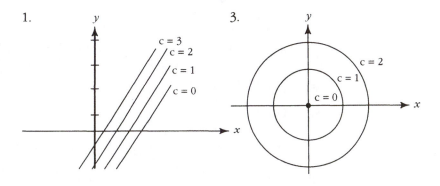

5. Origin if $c = 0$; ellipsoid with intercepts $x = \pm\frac{1}{\sqrt{2}}, y = \pm1, z = \pm1$ if $c = 1$.
7. The planes $x = 0, y = 0, z = 0$.

9.

11.

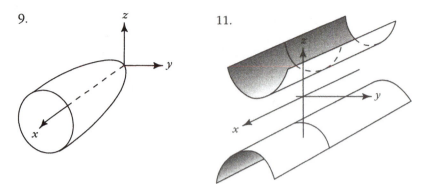

13. $g_x = \dfrac{\pi \cos{(\pi x)}}{1 + y^2}; \quad g_y = \dfrac{-2y \sin{(\pi x)}}{(1 + y^2)^2}$

15. $k_x = z^2 + z^3 \sin{(xz^3)}; \ k_z = 2xz + 3xz^2 \sin{(xz^3)}$

17. 1   19. $-\sin{(2)}$

21. 4   23. 0

25. $z = 2x + 2y - 2$ 27. $z = 1$

29. $\begin{bmatrix} 2xy & x^2 \\ -ye^{-xy} & -xe^{-xy} \end{bmatrix}$ 31. $\begin{bmatrix} e^x & e^y & e^z \end{bmatrix}$

33. $\dfrac{4e^{-2x-2y+2xy} + 4ye^{2xy-2x-2y}}{(e^{-2x-2y} - e^{2xy})^2} ; \ \dfrac{4e^{-2x-2y+2xy} + 4xe^{2xy-2x-2y}}{(e^{-2x-2y} - e^{2xy})^2}$

35. $2xe^{x^2-y^2}; -2ye^{x^2-y^2}$

37. $\mathbf{D}G(1,1) = \begin{bmatrix} -e^{-2} & -e^{-2} \\ e & e \end{bmatrix}$

$\mathbf{D}F(G(1,1)) = \dfrac{1}{(e^{-4} - e^2)^2}(-4, 4e^{-3}); \dfrac{\partial z}{\partial x}(1,1) = \dfrac{8e^{-2}}{(e^{-4} - e^2)^2}$

39. $\mathbf{D}G(1,1) = \begin{bmatrix} 1 & 1 \\ 1 & -1 \end{bmatrix}; \mathbf{D}F(G(1,1)) = \begin{bmatrix} 0 & 2 \end{bmatrix}; \ \dfrac{\partial z}{\partial x}(1,1) = 2$

41. $(e^z - y \sin x, \cos x, xe^z)$ 43. $\left(\dfrac{2x}{z}, \dfrac{1}{z}, -\dfrac{x^2+y}{z^2} \right)$

45. $\nabla f = (y, x)$

47. $2t \sin(t^3) + 3t^4 \cos(t^3)$

49. (a) $9(\cos 3)/\sqrt{2}$; (b) $(3 \cos 3, -6 \cos 3)$

51. $(4, 16)$ 53. $z = 3x + 2y - 2$

55. $x + y + z = \sqrt{3}$ 57. 1

59. $\dfrac{dy}{dt} = \dfrac{2x + y}{2y - x} \dfrac{dx}{dt}$

61. $\dfrac{\partial}{\partial y}(x^3 - \sin y + y^4 - 1) \neq 0$ near $(1, 0)$; $f'(1) = 3$

63. $3/20$ 65. $-6(\cos 3 + \cos 1)$

Chapter 3
Higher Derivatives and Extrema

3.1 Higher Order Partial Derivatives

1. $\partial^2 f/\partial x^2 = 24(x^3 y - xy^3)/(x^2 + y^2)^4$
$\partial^2 f/\partial y^2 = 24(-x^3 y + xy^3)/(x^2 + y^2)^4$
$\partial^2 f/\partial x \partial y = \partial^2 f/\partial y \partial x = (-6x^4 + 36x^2 y^2 - 6y^4)/(x^2 + y^2)^4$

3. $\dfrac{\partial^2 f}{\partial x^2} = \dfrac{2(\cos^2 x + e^{-y})\cos 2x + 2\sin^2 2x}{(\cos^2 x + e^{-y})^3}$; $\dfrac{\partial^2 f}{\partial y^2} = \dfrac{e^{-y}(e^{-y} - \cos^2 x)}{(\cos^2 x + e^{-y})^3}$
$\dfrac{\partial^2 f}{\partial x \partial y} = \dfrac{\partial^2 f}{\partial y \partial x} = \dfrac{2e^{-y}\sin 2x}{(\cos^2 x + e^{-y})^3}$

5. $f_{xy} = 2x + 2y$; $f_{yz} = 2z$; $f_{zx} = 0$; $f_{xyz} = 0$

7. $\dfrac{\partial}{\partial x}\left(\dfrac{\partial^2 f}{\partial y \partial z}\right) = \dfrac{\partial}{\partial x}\left(\dfrac{\partial^2 f}{\partial z \partial y}\right)$

9. Both sides equal $e^{xy} + xye^{xy} + 6xz^2$.

11. Yes, $v_{xx} + v_{yy} = 2 - 2 = 0$.

13. (a) Both sides equal $f''(x - t) + g''(x + t)$.
(b) This is a "travelling" parabola.

15. $e^{x+y}\sin z\left\{\dfrac{\partial x}{\partial s}\dfrac{\partial x}{\partial t} + \dfrac{\partial y}{\partial t}\dfrac{\partial y}{\partial s} - \dfrac{\partial z}{\partial t}\dfrac{\partial z}{\partial s} + \dfrac{\partial y}{\partial s}\dfrac{\partial x}{\partial t} + \dfrac{\partial y}{\partial t}\dfrac{\partial x}{\partial s} + \dfrac{\partial^2 x}{\partial s \partial t} + \dfrac{\partial^2 y}{\partial s \partial t}\right\}$
$+ e^{x+y}\cos z\left\{\dfrac{\partial x}{\partial s}\dfrac{\partial z}{\partial t} + \dfrac{\partial z}{\partial s}\dfrac{\partial x}{\partial t} + \dfrac{\partial y}{\partial s}\dfrac{\partial z}{\partial t} + \dfrac{\partial y}{\partial t}\dfrac{\partial z}{\partial s} + \dfrac{\partial^2 z}{\partial s \partial t}\right\}$

17. (a) $\dfrac{\partial f}{\partial x} = \dfrac{y(x^4 + 4x^2 y^2 - y^4)}{(x^2 + y^2)^2}$; $\dfrac{\partial f}{\partial y} = \dfrac{-x(y^4 + 4x^2 y^2 - x^4)}{(x^2 + y^2)^2}$
(b) Use the definition of partials.
(c) Compute $\dfrac{d}{dx}\left(\dfrac{\partial f}{\partial y}(x, 0)\right)$ and $\dfrac{d}{dy}\left(\dfrac{\partial f}{\partial x}(0, y)\right)$.
(d) The second partials are not continuous at $(0, 0)$.

3.2 Taylor's Theorem

1. $x^2 + 2xy + y^2$
3. $1 + x + y + x^2/2 + xy + y^2/2$
5. $1 + xy$
7. $-0.4151; -0.4152; -0.4153$
9. $-2.85; -2.8485; -2.8485$
11. $x + x^3 + xy^3 + x^5$

3.3 Maxima and Minima

1. $\left(-\frac{1}{5}, -\frac{3}{5}\right)$
3. All points on the line $x + y = 0$
5. $(0, 0)$
7. $\left(-\frac{1}{4}, -\frac{1}{4}\right)$
9. Minimum at $(1, 0)$; maximum at $(-1, 0)$.
11. $\sqrt{6}/2$
13. Minimize $2(xy + yz + xz)$ with $xyz = V$.
15. $40, 40, 40$
17. $x = y = 8(s/b)^{\frac{1}{3}}, z = 4(b/s)^{\frac{2}{3}}$

3.4 Second Derivative Test

1. $(0, 0)$, minimum
3. $(0, 0)$, maximum
5. minimum
7. Saddle point
9. minimum
11. $(-3, 2)$, minimum
13. $(0, 0)$, saddle point
15. $(-18/5, -11/15)$; minimum
17. $(0, 0)$, saddle point
19. Saddle point
21. $(0, 0)$ is a local minimum if $-2 < k < 2$.
    $(0, 0)$ is a saddle point if $k < -2$ or $k > 2$.
23. (a) f_x and f_y both vanish at the origin.
    (b) $f \circ g = b^2 t^2 - 4a^2 bt^3 + 3a^4 t^4$.
    (c) Choose $y = 2x^2$ so that $f(x, y) = -x^4 \leq 0$.

3.5 Constrained Extrema, Lagrange Multipliers, and Absolute Maxima and Minima

1. $\pm\sqrt{3}$
3. $\pm\sqrt{35/2}$
5. $1/4$
7. $\pm\sqrt{6}$
9. Minimum = 0; maximum = 3
11. Minimum = 8; maximum = 15
13. $\pm\sqrt{35/2}$
15. $25/6; 0$
17. Radius $= \sqrt[3]{1/2\pi}$; height $= \sqrt[3]{4/\pi}$
19. $11,664 \text{in}^2$
21. $x = y = 25,000; z = 50,000$
23. $(B + p - q)/2p$ pounds of wool; $(B + q - p)/2q$ pounds of cotton.
25. Highest point 2635 meters at $(13.159, 2.1067)$; lowest point -2635 meters at $(-13.159, 2.1067)$.

Review Exercises (Chapter 3)

1. $u_x = \left(\dfrac{\pi}{1+y^2}\right) \cos \pi x; \; u_y = \dfrac{-2y}{(1+y^2)^2} \sin \pi x$

3. $-\sin 2$

5. $z_x = 6x; z_y = 4y; z_{xx} = 6; z_{yy} = 4; z_{xy} = 0$

7. Letting $r = \sqrt{x^2 + y^2}, f_x = (-x \sin r)/r \; f_y = (-y \sin r)/r$
    $f_{xx} = (-r^2 \sin r - x^2 r \cos r + x^2 \sin r)/r^3$
    $f_{yy} = (-r^2 \sin r - y^2 r \cos r + y^2 \sin r)/r^3$
    $f_{xy} = (-xyr \cos r + xy \sin r)/r^3$

9. Both are $e^{xy} + xye^{xy} + 6xz^2$.

11. Check that $f_{xx} + f_{yy} = 0$.   13. $1 - x^2/2 + xy$

15. $(0,0)$; saddle point   17. $(0,0)$; saddle point

19. Use the second derivative test; $(0,0)$ is a local maximum; $(-1,0)$ is a saddle point; $(2,0)$ is a local minimum.

21. Saddle points at $(n,0), n =$ integer.

23. 1   25. $\left(\dfrac{1}{2}, \dfrac{1}{2}, \dfrac{1}{4}\right)$ 27.   maximum 2, minimum 0

29. maximum 1, minimum $\cos 1$

31. $2\sqrt{b-1}$ if $b \geq 2$, $|b|$ if $b \leq 2$

33. $3\sqrt{3}/4$

35. Use the second derivative test.

37. $\dfrac{b_0}{d}[(\mathbf{a} \cdot \mathbf{b})\mathbf{a} - (\mathbf{a} \cdot \mathbf{a})\mathbf{b}]$, where $\mathbf{a} = (a_1, a_2, a_3)$, $\mathbf{b} = (b_1, b_2, b_3)$ and $d = (\mathbf{a} \cdot \mathbf{a})(\mathbf{b} \cdot \mathbf{b}) - (\mathbf{a} \cdot \mathbf{b})^2$

39. $(f_{xx})(f_{yy}) - (f_{xy})^2 = (6x)(-6x) - (-6y)^2 = 0$ at $(0,0)$

41. maximum 2, minimum -2

Chapter 4
Vector-Valued Functions

4.1 Acceleration

1. $6 + 3t^2; (0, 6, 6t)$

3. $(4 \cos^2 t \sin^2 t + 9(1 - t^2)^2 + 1)^{\frac{1}{2}}; \; (-2 \cos 2t, -6t, 0)$

5. $(e^t - e^{-t}, \cos t - \sin t, -3t^2)$

7. $[-3t^2(2 \sin t + \cos t) - t^3(2 \cos t - \sin t)]\mathbf{i}$
    $+ [3t^2(2e^t + e^{-t}) + t^3(2e^t - e^{-t})]\mathbf{j} + [e^t(\cos t - \sin t)$
    $- e^{-t}(-\sin t + \cos t)]\mathbf{k}$

9. $m(0, 6, 0)$

11. $-24\pi^2(\cos (2\pi t/5), \sin (2\pi t/5))/25$

13. $\dfrac{d}{dt}(\|\mathbf{v}\|^2) = \dfrac{d}{dt}(\mathbf{v} \cdot \mathbf{v}) = 2\mathbf{v} \cdot \dfrac{d\mathbf{v}}{dt} = 2\mathbf{v} \cdot \mathbf{a} = 0$

15. 6129 seconds

4.2 Arc Length

1. $2\sqrt{5}\pi$

3. $2(2\sqrt{2}-1)$

5. $\dfrac{6-\sqrt{3}}{\sqrt{2}} + \dfrac{1}{2}\log\left[\dfrac{2\sqrt{2}+3}{\sqrt{2}+\sqrt{3}}\right]$

7. $2\pi(\sqrt{5}+\sqrt{2})$

9. $\alpha(t) = (e^t - e^{-t})/\sqrt{2}; \beta(t) = \sqrt{2}t$

11. $\dfrac{\pi}{2}\sqrt{\pi^2 + 2} + \log\left((\pi + \sqrt{\pi^2 + 2})/\sqrt{2}\right)$

13. (a) $\dfrac{d}{dt}(\mathbf{T} \cdot \mathbf{T}) = 2\mathbf{T}' \cdot \mathbf{T} = 0$

(b) $\dfrac{\|\mathbf{c}'(t)\|^2 \mathbf{c}''(t) - (\mathbf{c}''(t) \cdot \mathbf{c}'(t))\mathbf{c}'(t)}{\|\mathbf{c}'(t)\|^3}$

4.3 Vector Fields

1.

3.

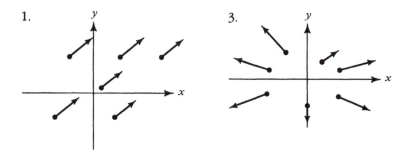

5. $\mathbf{F} = (2y, x)$

7.

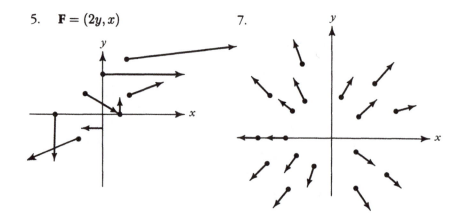

9.   The flow lines are concentric circles

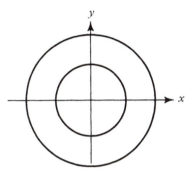

11.   The flow lines for $t > 0$:

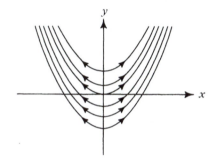

13. $\mathbf{c}'(t) = (2e^{2t}, 1/t, -1/t^2) = \mathbf{F}(\mathbf{c}(t))$
15. $\mathbf{c}'(t) = (\cos t, -\sin t, e^{-t}) = \mathbf{F}(\mathbf{c}(t))$
17.   (a) $dE/dt = m\mathbf{r}''(t) \cdot \mathbf{r}'(t) + \nabla V(\mathbf{r}(t)) \cdot \mathbf{r}'(t) = \mathbf{F} \cdot \mathbf{r}' - \mathbf{F} \cdot \mathbf{r}' = 0$

   (b) $\dfrac{d}{dt} \|\mathbf{r}'(t)\|^2 = \dfrac{d}{dt}(\mathbf{r}'(t) \cdot \mathbf{r}'(t)) = 2\mathbf{r}''(t) \cdot \mathbf{r}'(t) = \dfrac{2}{m}\mathbf{F}(\mathbf{r}(t)) \cdot \mathbf{r}'(t)$

   $= -\dfrac{2}{m}\nabla V(\mathbf{r}(t)) \cdot \mathbf{r}'(t) = 0$

19. $\dfrac{Df}{Dt}(\mathbf{x}, t) = \dfrac{\partial f}{\partial t}(\mathbf{x}, t) + \nabla f(\mathbf{x}) \cdot \mathbf{F} = \dfrac{\partial f}{\partial t}(\mathbf{c}(t)) + \nabla f(\mathbf{c}(t)) \cdot \mathbf{F}(\mathbf{c}(t))$

   but $\mathbf{F}(\mathbf{c}(t)) = \mathbf{c}'(t)$, so $\dfrac{Df}{Dt}(\mathbf{c}(t), t) = \dfrac{\partial f}{\partial t}(\mathbf{c}(t)) + \nabla f(\mathbf{c}(t)) \cdot \mathbf{c}'(t)$

   $= \dfrac{d}{dt}f(\mathbf{c}(t), t)$ by the chain rule. Now evaluate at (x_0, t_0).

4.4 Divergence and Curl

1. $ye^{xy} - xe^{xy} + ye^{yz}$ 3.   3
5.   div $\mathbf{V} > 0$ in the first and third quadrants,

div $\mathbf{V} < 0$ in the second and fourth quadrants.

7. $\nabla \cdot \mathbf{F} = 0$; if $\mathbf{F}$ represents a fluid there is neither expansion or compression; the area of a small rectangle remains the same.

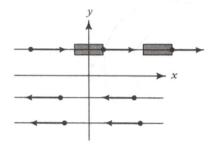

9. $3x^2 - x^2 \cos{(xy)}$ 11. $y \cos{(xy)} + x^2 \sin{(x^2 y)}$

13. **0**

15. $(10y - 8z)\mathbf{i} + (6z - 10x)\mathbf{j} + (8x - 6y)\mathbf{k}$

17. $-\sin x$ 19. x

21. $\nabla \times \nabla f = \mathbf{0}$ 23. $\nabla \times \nabla f = \mathbf{0}$

25. $\nabla \times \mathbf{F} \neq \mathbf{0}$

27. Let $\mathbf{F} = F_1 \mathbf{i} + F_2 \mathbf{j} + F_3 \mathbf{k}$ and compute both sides of the identity.

29. (a) $2xy\mathbf{i} + x^2 \mathbf{j}$ (b) $(3y^2 xz, 4xz - y^3 z, 0)$
      (c) $(-y^3 zx^3, 2x^2 y^4 z, 2x^3 z^2 - 2xy)$ (d) $4x^2 yz^2 + x^2$

31. No

Review Exercises (Chapter 4)

1. $\mathbf{v}(1) = (3, -e^{-1}, -\pi/2);\ \mathbf{a}(1) = (6, e^{-1}, 0);$

$$s = \sqrt{9 + e^{-2} + \frac{\pi^2}{4}};\ l(t) = (2,\ e^{-1}, 0) + (t-1)(3, -e^{-1},\ -\pi/2)$$

3. $\mathbf{v}(0) = (1, 1, 0);\ \mathbf{a}(0) = (1, 0, -1);\ s = \sqrt{2};\ l(t) = (1, 0, 1) + t(1, 1, 0)$

5. Tangent vector: $\mathbf{v} = -(1/\sqrt{2})\mathbf{i} + (1/\sqrt{2})\mathbf{j} + \mathbf{k}$
Acceleration vector: $\mathbf{a} = -(1/\sqrt{2})(\mathbf{i} + \mathbf{j})$

7. $m(2, 0, -1)$ 9. $\displaystyle\int_1^4 \sqrt{1 + \frac{4}{9}t^{-2/3} + \frac{4}{25}t^{-6/5}}\ dt$

11. (a) $\mathbf{v} = (-2t \sin{(t^2)}, 2t \cos{(t^2)}, 0);\ s = 2t$
     (b) $\left(\frac{1}{2}, -\frac{\sqrt{3}}{2}, 0\right)$
     (c) $\sqrt{5\pi/3}$
     (d) $\mathbf{v} = 2\sqrt{5\pi/3}(\sqrt{3}/2, 1/2, 0);\ s = 2\sqrt{5\pi/3}$
     (e) $\left(\frac{3}{2} + \frac{5\pi}{\sqrt{3}}\right)/\sqrt{5\pi}$

13. $x = 1 + t, y = -\frac{1}{2} + \frac{t}{2}, z = -\frac{2}{3} + \frac{t}{3}$

17. $9; 0$ 19. $3; -\mathbf{i} - \mathbf{j} - \mathbf{k}$

21. $0; -\mathbf{i} - \mathbf{j} - \mathbf{k}$
23. $\nabla f = (ye^{xy} - y \sin xy, xe^{xy} - x \sin xy, 0)$
25. $\nabla f = (2xe^{x^2} + y^2 \sin xy^2, 2xy \sin xy^2, 0)$
27. (a) $(yz^2, xz^2, 2xyz)$; (b) $(z - y, 0, -x)$
    (c) $(2xyz^3 - 3xy^2z^2, 2x^2y^2z - y^2z^3, y^2z^3 - 2x^2yz^2)$
29. $div\,\mathbf{F} = 0;\ curl\,\mathbf{F} = (0, 0, 2(x^2 + y^2)f'(x^2 + y^2) + 2f(x^2 + y^2))$
31. (a) $[(\partial P/\partial x)^2 + (\partial P/\partial y)^2]^{1/2}$
    (b) A small packet of air would obey $\mathbf{F} = m\mathbf{a}$.
    (c)   (d)

33. (a) $\dfrac{\sqrt{R^2 + \rho^2}}{\rho}(z_0 - z_1)$; (b) $\sqrt{\dfrac{2(R^2 + \rho^2)z_0}{g\rho^2}}$

Chapter 5
Multiple Integrals

5.1 Volume and Cavalieri's Principle

1. 140
3. 13/15
5. 1
7. 26/9
9. $-2(1 + e^2)/\pi$
11. 196/15
13. 105
15. (a) Area is $\pi[f(x)]^2$
    (b) $\pi \displaystyle\int_{-1}^{3} (-x^2 + 2x + 3)^2 dx = \dfrac{512\pi}{15}$

5.2 The Double Integral Over a Rectangle

1. 7/12

3. $\dfrac{\sin(1)}{9}$

5. 50

7. $\dfrac{1}{36}(e^{36} - e^9) - \dfrac{1}{4}(e^4 - e)$

9. 16/9

11. $\dfrac{1}{2}(e - \dfrac{1}{e})$

13. 76/3

15. 1

17. 25/6

19.
$$\int_c^d \left[\int_a^b f(x)g(y)dx \right] dy = \int_c^d g(y) \left[\int_a^b f(x)dx \right] dy$$
$$= \left[\int_a^b f(x)dx \right] \left[\int_c^d g(y)dy \right]$$

5.3 The Double Integral Over Regions

1. Both a type 1 and 2 region.   3. Both a type 1 and 2 region.

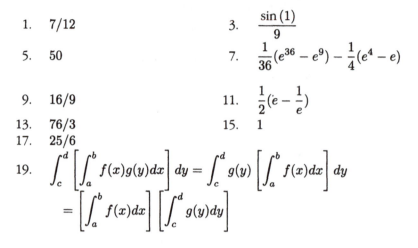

5. 7/12

7. 64/35

9. Type 1; $2\pi + \pi^2$

11. Type 2; 76/3

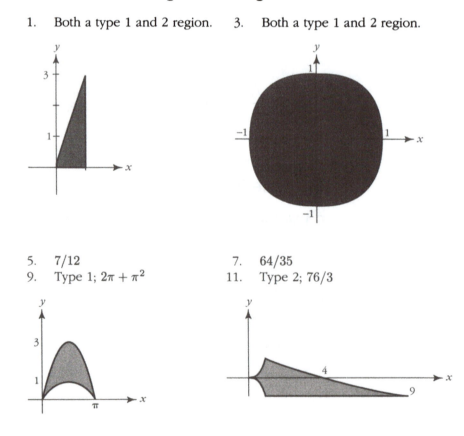

13. Type 1; $33/140$ 15. Type 1; $71/420$

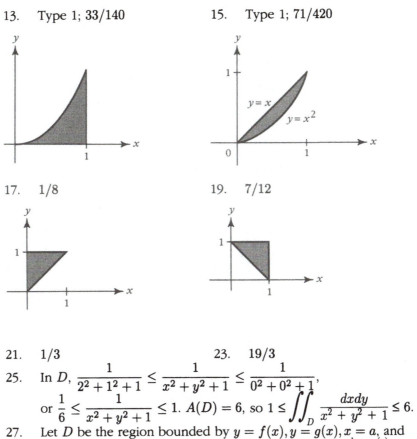

17. $1/8$ 19. $7/12$

21. $1/3$ 23. $19/3$

25. In D, $\dfrac{1}{2^2 + 1^2 + 1} \leq \dfrac{1}{x^2 + y^2 + 1} \leq \dfrac{1}{0^2 + 0^2 + 1}$,

or $\dfrac{1}{6} \leq \dfrac{1}{x^2 + y^2 + 1} \leq 1$. $A(D) = 6$, so $1 \leq \displaystyle\iint_D \dfrac{dxdy}{x^2 + y^2 + 1} \leq 6$.

27. Let D be the region bounded by $y = f(x), y = g(x), x = a$, and

$x = b$, where $f(x) \leq g(x)$ on $[a, b]$. Then $\displaystyle\iint_D dxdy = \int_a^b \int_{f(x)}^{g(x)} dydx$

$= \displaystyle\int_a^b (g(x) - f(x))dx$

29. $\displaystyle\int_0^x \int_0^t F(u)dudt = \int_0^x \int_u^x F(u)dtdu = \int_0^x (x - u)F(u)du$

5.4 Triple Integrals

1. $1/3$ 3. 10
5. $x^2 + y^2 \leq z \leq \sqrt{x^2 + y^2}, -\sqrt{1 - y^2} \leq x \leq \sqrt{1 - y^2}, -1 \leq y \leq 1$
7. $0 \leq z \leq \sqrt{1 - x^2 - y^2}, -\sqrt{1 - y^2} \leq x \leq \sqrt{1 - y^2}, -1 \leq y \leq 1$
9. $50\pi/\sqrt{6}$ 11. $1/2$
13. 0 15. $a^5/20$
17. 0 19. $3/10$
21. $1/6$

23. $\displaystyle\int_{-1}^{1}\int_{-\sqrt{1-x^2}}^{\sqrt{1-x^2}}\int_{\sqrt{x^2+y^2}}^{1} f(x,y,z)\,dz\,dy\,dx$

25. $\displaystyle\int\!\!\int_{D}\int_{0}^{f(x,y)} dz\,dx\,dy = \int\!\!\int_{D} f(x,y)\,dx\,dy$

27. Let M_ϵ and m_ϵ be the maximum and minimum of f on B_ϵ. Then $m_\epsilon vol\,(B_\epsilon) \le \int_{B_\epsilon} f\,dV \le M_\epsilon vol\,(B_\epsilon)$. Divide by vol (B_ϵ), let $\epsilon \to 0$ and use continuity of f.

5.5 Change of Variables

1. D^* is the set of (r,θ) satisfying $0 \le r \le 1$ and $0 \le \theta \le \pi$
3. $64\pi/5$
5. $\sqrt{\pi/10}$
7. $1/\sqrt{\pi\sigma}$
9. $(\sqrt{5}, \tan^{-1}(1/2), -2)$
11. $(\sqrt{3}, \tan^{-1}(1/\sqrt{2}), 1)$
13. $(0,1,0)$
15. $(-\sqrt{3}/2, -1/2, 4)$
17. $5\pi(e^4 - 1)/2$
19. $(\sqrt{2}, \pi/2, \pi/4)$
21. $(\sqrt{14}, -\tan^{-1}(1/2) + \pi, \cos^{-1}(-3/\sqrt{14}))$
23. $(0,0,-3)$
25. $(0,0,3)$
27. $2\pi(\sqrt{2} - \log{(1 + \sqrt{2})})$
29. $4\pi \log{(a/b)}$
31. $\pi\sqrt{3}/2$
33. $4\sqrt{6}\pi$
35. $\pi(1 - \cos 1)$
37. (a) 140; (b) −42
39. $\displaystyle\int_{0}^{1}\int_{0}^{1} e^v u\,du\,dv = (e - 1)/2$

5.6 Applications of Multiple Integrals

1. $\dfrac{1}{\pi} - \dfrac{\sin \pi^2}{\pi^3}$
3. $1/4$
5. $37/12$
7. $(1/2, 1/2, 1/2)$
9. $(11/18, 65/126)$
11. $m\displaystyle\int_{0}^{2\pi}\int_{0}^{k}\int_{0}^{a \sec \varphi} \rho^4 \sin^3\varphi\,d\rho\,d\varphi\,d\theta$, m = mass density.
13. $m(10)^8 m^2/s^2$
15. (c) $\bar{z} = 0$ since $0 = \bar{z}\int\!\!\int\!\!\int_{W} \rho\,dx\,dy\,dz$

Review Exercises (Chapter 5)

1. 6
3. $26/9$
5. $2\sin 1 - \sin 2$
7. 18
9. $\dfrac{e}{2} - \dfrac{1}{2e}$
11. Type 1; $2\pi + \pi^2$

13. Type 2; 130

15. $13/6$

17. $15/2$

19. $1/48$

21. $10/3$

23. $\pi - \frac{2}{3}$

25. $\dfrac{4\pi}{3}(\sin a^3 - \sin b^3)$

27. $(1, \pi, 1); (\sqrt{2}, \pi, \frac{\pi}{4})$

29. $\left(\dfrac{\sqrt{2}}{2}, \dfrac{\sqrt{2}}{2}, 1\right); \left(\sqrt{2}, \dfrac{\pi}{4}, \dfrac{\pi}{4}\right)$

31. $(0, 0, -1); (0, 0, -1)$

33. $\displaystyle\int_0^1 \int_0^{2\pi} \int_0^1 r^2 \, dr \, d\theta \, dz$

35. $2\displaystyle\int_0^{2\pi} \int_0^{\pi/4} \int_0^2 \rho^4 \sin\varphi \cos^2\varphi \, d\rho \, d\varphi \, d\theta$

$\quad + 2\displaystyle\int_0^{2\pi} \int_{\pi/4}^{\pi/2} \int_2^{\sqrt{2}\csc\varphi} \rho^4 \sin\varphi \cos^2\varphi \, d\rho \, d\varphi \, d\theta$

37. $\pi \log(\sec 1 + \tan 1)$

39. $\pi/16$

41. $\sqrt{\pi/5}$

43. $10/3$

45. $4\pi/3 \log(1 + R^3)$

47. $((107 - 48\sqrt{3})/16(8 - 3\sqrt{3}), 0, 0)$

49. (a) 6; (b) 64π

Chapter 6
Integrals Over Curves and Surfaces

6.1 Line Integrals

1. $2\pi^2$

3. -1

5. $-1/2$

7. $\frac{1}{3}\cos 3 + \frac{5}{12}$

9. 1

11. $2/3$

13. 5

15. 0

17. 7

19. 0

21. $I = \displaystyle\int_C H\mathbf{T} \cdot d\mathbf{s} = 2\pi r H$, so $H = I/2\pi r$

23. $2\sqrt{2}\pi^2$

25. $(5\sqrt{5} - 1)/12$

27. (a) change variables; (b) 8

6.2 Parametrized Surfaces

1. $-2y + z + 1 = 0$

3. $x + 4y - 18z + 13 = 0$

5. $\mathbf{n} = (\cos v \sin u, \sin v \sin u, \cos u)$; surface = unit sphere

7. $x + y + \sqrt{2}z = 4$

9. $(x - h(y_0, z_0)) - \partial h/\partial y(y - y_0) - \partial h/\partial z(z - z_0) = 0$

11. (a) Write out $\mathbf{D}\Phi$ as a matrix.

(b) Substitute into $\mathbf{n} \cdot (x - x_0, y - y_0, z - z_0) = 0$.

6.3 Area of a Surface

1. $\int_{-1}^{1}\int_{-\sqrt{1-x^2}}^{\sqrt{1-x^2}}[1 + e^{2x^2y^2}((y^3 + 2x^2y^5)^2 + (3xy^2 + 2x^3y^4)^2)]^{\frac{1}{2}}\,dy\,dx$

3. $\frac{4}{15}(9\sqrt{3} - 8\sqrt{2} + 1)$ 5. $\sqrt{2}\pi/4$

7. $\sqrt{6}\pi$

9. $(x, y, z) = (a\cos\theta\sin\phi, b\sin\theta\sin\phi, c\cos\phi), 0 \le \theta \le 2\pi, 0 \le \phi \le \pi;$
   $A(E) = \int_0^{2\pi}\int_0^{\pi}((b^2c^2\cos^2\theta + a^2c^2\sin^2\theta)\sin^4\phi + a^2b^2\sin^2\phi\cos^2\theta)^{1/2}\,d\phi\,d\theta$

11. $\pi\sqrt{6}/2$ 13. $V = 4\sqrt{3}\pi; A = 8\sqrt{3}\pi$

15. $4\sqrt{5}$; for fixed θ, (x, y, z) moves along the horizontal line segment $y = 2x, z = \theta$ from the z axis out to a radius of $\sqrt{5}|\cos\theta|$ into quadrant 1 if $\cos\theta > 0$ and into quadrant 3 if $\cos\theta < 0$.

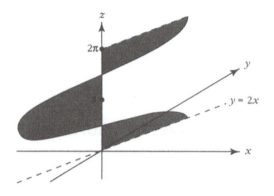

17. $(1 + \sqrt{3})/24$ 19. πa^3

21. $\sqrt{R^2 - r^2}/(R - \sqrt{R^2 - r^2})$ 23. $\pi(\sqrt{m^2 + 1})(r_2^2 - r_1^2)$

6.4 Surface Integrals

1. -2π 3. 2π
5. $12\pi/5$ 7. $2\pi/3$
9. $2a^3bc\pi/5$ 11. Use spherical coordinates.
13. -48π 15. 4π

Review Exercises (Chapter 6)

1. 0   3.   3
5. $(2 + \pi)/\pi$ 7. $2a^3$
9. $8a^2$

11. $x = 2 + 5\cos\theta\sin\phi,\ y = 3 + 5\sin\theta\sin\phi,\ z = 5\cos\phi$

13. $\dfrac{1}{2}\displaystyle\int_0^{2\pi}\sqrt{5 + 3\cos^2\theta}\,d\theta$ 15. $\sqrt{14}$

17. (a) $(e^y\cos\pi z, xe^y\cos\pi z, -\pi xe^y\sin\pi z)$; (b) 0

19. $(e^2 + 1)/2$

21. $(1, 0, -2)/\sqrt{5};\ x - 2z + 1 = 0$

23. 0

25. $\nabla \times \nabla f = 0$ and if C is parametrized by σ,
$\int_C \nabla f \cdot d\mathbf{s} = f(\mathbf{c}(b)) - f(\mathbf{c}(a)) = 0$

27. $-(e^{6\pi} - 1)3\sqrt{2}/13$ 29. $\dfrac{472{,}316\sqrt{26} - 16}{1{,}093{,}750} \approx 2.202$

31. (a) $1/6\sqrt{3}$; (b) $\pi/4\sqrt{2}$; (c) 8

33. 0   35. $a = 1, b = 0$

Chapter 7
The Integral Theorems of Vector Analysis

7.1 Green's Theorem

1. 0   3. 0

5. -8 7. Show that $\frac{1}{2}\int_C x\,dy - y\,dx = 3\pi a^2/8$

9. 0   11. 67/6

13. -20 15. 0

17. $3\pi/2$ 19. 0

21. πab

23. Break up the region as follows, apply Green's theorem to each piece and add:

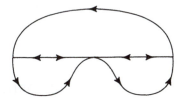

25. $9\pi/8$

27. Write $\int_{\partial D} \phi\nabla\phi \cdot \mathbf{n}\,ds$ as a line integral and apply Green's theorem.

29. There is a hole in D.

7.2 Stokes' Theorem

1. $-1/2$ 3. 0

5. 0   7. 2π

9. -16π 11. $\pi/4$

13. 0

17. (a) $\iint_C \mathbf{v} \cdot d\mathbf{s} = \int_S (\nabla \times \mathbf{v}) \cdot d\mathbf{S} = 0$;

(b) $\int_C \mathbf{v} \cdot d\mathbf{s} = f(\mathbf{c}(b)) - f(\mathbf{c}(a)) = 0$

19. 0   21. $\dfrac{1}{r}\left(\dfrac{\partial}{\partial r}(rF_\theta) - \dfrac{\partial F_r}{\partial \theta}\right)$

23. Apply Stokes' theorem: $\iint_S J \cdot d\mathbf{S} = \iint_S (\nabla \times \mathbf{H}) \cdot d\mathbf{S}$.

7.3 Gauss' Theorem

1. 4π 3. $12\pi/5$

5. (a) 0; (b) 4/15; (c) $-4/15$

7. $$\iint_S \mathbf{r} \cdot \mathbf{n}\,dS = \iiint_W (\operatorname{div}\,\mathbf{r})dV = \iiint_W 3\,dV = 3 \cdot \operatorname{volume}(W)$$

9. $$\iint_{\partial W} f\mathbf{F} \cdot \mathbf{n}\,dS = \iiint_W \operatorname{div}\,(f\mathbf{F})dx\,dy\,dz$$

$$= \iiint_W \nabla f \cdot \mathbf{F}\,dx\,dy\,dz + \iiint_W f\nabla \cdot \mathbf{F}\,dx\,dy\,dz$$

Rearrange the equation to get the desired result.

11. $$\iint_S (\nabla \times \mathbf{F}) \cdot d\mathbf{S} = \iiint_W \nabla \cdot (\nabla \times \mathbf{F})dV = \iiint 0\,dV = 0$$

13. Follow the method of Example 7.

15. (a) $\nabla \cdot \mathbf{J} = \nabla \cdot (\nabla \times \mathbf{H}) = 0$

(b) Conservation of charge suggests that the amount of charge leaving any closed region is zero. Thus, the flux of $\mathbf{J}$ through any closed surface is 0. Therefore, flux $= \iint_S \mathbf{J} \cdot d\mathbf{S} = \iiint_V \nabla \cdot \mathbf{J}\,dV = 0$. Since this is true for any region, $\nabla \cdot \mathbf{J} = 0$.

17. (b) Use the divergence theorem.

7.4 Path Independence and the Fundamental Theorems of Calculus

1. 3   3. 8

5. No   7. $x^2yz - \cos x + C$

9. (a) $\frac{1}{2}(x^2 + y^2) + C$; (b) Not a gradient; (c) $x^3/3 + xy^2 + C$

11. (a) $f(x, y) = x^2y^2/2 + x^3y + C$; $-3/2$

(b) $f(x, y) = (x^2 + 1)/(y^2 + 1) + C$; -1

(c) $f(x, y) = x\cos(xy^2) + C$; $\cos(e^2) - e^{-1}\cos(e^{-1})$

13. $x^2(1 + \cos y) + C$ 15. $-1/r_0$

17. (a) No; (b) $(z^2/2, xy - z, x^2y)$

19. $(z^3, x^3, y^3)/3$
21. $(0, -\cos x, x\sin y)$
23. $2f = xF_1 + yF_2 + zF_3$

Review Exercises (Chapter 7)

1. (a) $\iint_D dx\,dy = A$; (b) $-\iint_D dx\,dy = -A$; (c) 0

3. $3\pi/2$
5. 0

7. 2π
9. 0

11. 23/6

13. (a) $f(x,y) = \dfrac{x^4}{4} - x^2 y^3 + C$; (b) $-1/4$

15. Yes

17. (a) $2x^3 yz\mathbf{i} - 3x^2 y^2 z\mathbf{j} - 2\mathbf{k}; x^3 y^2$; (b) 2π; (c) $1/12$

19. $2a^2\pi$ (without top and bottom)

21. (a) $(3ye^{z^2}, 3xe^{z^2}, 6xyze^{z^2})$; (b) 0; (c) Each side is zero.

23. 21

25. (a) $\mathbf{G}$ is conservative since $\nabla \times \mathbf{G} = \mathbf{0}$

(b) $\dfrac{x^4}{4} + \dfrac{y^4}{4} - \dfrac{3}{2}x^2 y^2 + \dfrac{z^2}{2} + C$

(c) $0; -1/2; 1/3; -1/2$

27. $\displaystyle\iint_{\partial W} f(\nabla g) \cdot \mathbf{n}\,dS = \iiint_W \operatorname{div}(f(\nabla g))\,dV =$
$\displaystyle\iiint_W (f\nabla^2 g + \nabla f \cdot \nabla g)\,dV$

29. $\displaystyle\iiint_W \|\nabla f\|^2 dV = \iint_{\partial W} f\nabla f \cdot \mathbf{n}\,dS - \iiint_W f\nabla^2 f\,dV = 0.$
Thus $\|\nabla f\|^2 = 0$, or $\nabla f = 0$, which implies that f is a constant.

31. (a) $\phi = x + 2y - z + C$

(b) i. $\left(\dfrac{2}{\sqrt{6}}, \dfrac{4}{\sqrt{6}}, \dfrac{-2}{\sqrt{6}}\right)$;   ii. $\left(\dfrac{-2}{\sqrt{6}}, \dfrac{-4}{\sqrt{6}}, \dfrac{2}{\sqrt{6}}\right)$

(c) $8/\sqrt{6}; -8/\sqrt{6}$

33. (a) $\displaystyle -\iint_{\partial W} \mathbf{F} \cdot d\mathbf{S} = -\iiint_W \operatorname{div}\mathbf{F}\,dx\,dy\,dz = k\iiint_W \nabla^2 T\,dx\,dy\,dz$

$\displaystyle = \frac{d}{dt}\iiint_W e\,dx\,dy\,dz = \iiint_W \frac{\partial}{\partial t}e\,dx\,dy\,dz.$

Thus $k\nabla^2 T = \dfrac{\partial}{\partial t}e = \dfrac{\partial}{\partial t}(c\rho_0 T)$. Rearrange this to get $\dfrac{\partial T}{\partial t} = \dfrac{k}{c\rho_0}\nabla^2 T$.

(b) $\displaystyle \frac{d}{dt}\iiint_W \rho\,dx\,dy\,dz = -\iint_{\partial W} \rho\mathbf{V} \cdot d\mathbf{S}$

Practice Exam 1

1. (a) $\dfrac{-2}{\sqrt{17}}\mathbf{i} - \dfrac{2}{\sqrt{17}}\mathbf{j} + \dfrac{3}{\sqrt{17}}\mathbf{k}$ (b) 0

   (c) The line lies in this plane.   (d) $\sqrt{2/5}$ km after $\sqrt{2}/10$ minutes.

   (e) $(3, -3\pi/2, \pi)$

2. (a) $3(x-1) + 3(y+1) - z = 0$

   (b) $\mathbf{D}(f \circ g)(x) = \mathbf{D}f(g(x)) \cdot \mathbf{D}g(x)$ where $\mathbf{D}f(g(x))$ is a 3×2 matrix and $\mathbf{D}g(x)$ is a 2×4 matrix.

   (c) $f(x, y) = x + y^2 + xy + R_2$ where $R_2/(x^2 + y^2) \to 0$ as $\sqrt{x^2 + y^2} \to 0$.

3. (a) $\dfrac{d\mathbf{J}}{dt} = \mathbf{r}' \times \mathbf{r}' + \mathbf{r} \times \mathbf{a} = \mathbf{0} + \mathbf{r} \times \left(\dfrac{-GM}{r^3}\right)\mathbf{r} = \left(\dfrac{-GM}{r^3}\right)\mathbf{r} \times \mathbf{r} = \mathbf{0}$

   (b) $\operatorname{div} \mathbf{F} = (-GMm) \operatorname{div} \dfrac{\mathbf{r}}{r^3} = (-GMm)\left[\dfrac{\partial}{\partial x}\left(\dfrac{x}{(x^2 + y^2 + z^2)^{3/2}}\right)\right.$

   $\left. + \dfrac{\partial}{\partial y}\left(\dfrac{y}{(x^2 + y^2 + z^2)^{3/2}}\right) + \dfrac{\partial}{\partial z}\left(\dfrac{z}{(x^2 + y^2 + z^2)^{3/2}}\right)\right] = 0$

   (c) $(0, 0)$; a saddle point.

4. (a) $(0, 1, -3)$

   (b) If $\sigma(t)$ is the flow line, $\sigma'(t) = \nabla \mathbf{T}(\sigma(t))$ with $\sigma(0) = (0, 1, 1)$, then $(d/dt)\mathbf{T}(\sigma(t)) = \nabla \mathbf{T}(\sigma(t)) \cdot \sigma'(t) = \|\nabla \mathbf{T}(\sigma(t))\|^2$. Letting $t = 0$ gives the results.

   (c) All those paths remaining on the hyperboloid $x^2 + y^2 - 3z^2 = -2$

5. (a) The octant of the unit sphere is the region $x \le 0, y \ge 0, z \ge 0$.

   (b) $\sqrt{2}(x + \frac{1}{2}) - \sqrt{2}(y - \frac{1}{2}) - 2(z - \frac{\sqrt{2}}{2}) = 0$

   (c) (ii) says that $\nabla f(x_0, y_0, z_0) = \lambda[\Phi_u(u_0, v_0) \times \Phi_v(u_0, v_0)]$ for some λ and (i) implies that $\lambda = 0$; thus $\nabla f(x_0, y_0, z_0) = \mathbf{0}$ and (x_0, y_0, z_0) is a critical point for f.

6. (a) $\displaystyle\int_0^1 \int_0^1 \int_0^{\sqrt{1-z^2}} f(x, y, z)\,dy\,dz\,dx$

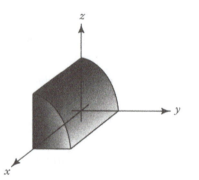

   (b) Maximum $\sqrt{2}$, minimum $-\sqrt{2}$

7. (a) 0; (b) 15/16

8. (a) False. Let $f(x, y, z) = x^2$, then $\nabla f = 2x\mathbf{i}$ and $\operatorname{div}\nabla f = 2$. By

Gauss' divergence theorem the flux of ∇f out of the sphere is $8\pi/3$.
(b) False; because $\operatorname{div}\nabla \times \mathbf{F} \equiv 0$, but $\operatorname{div}(x\mathbf{i}) = 1$.
(c) True; $\operatorname{div}\mathbf{F} = 3$, so by Gauss' divergence theorem the rate of flow out of the sphere is $3 \cdot 4\pi/3 = 4\pi$.
(d) True. If $F(x, y, z) = x^2 y z^5 + z^4 - 1$, then $F_z(0, 1, 1) \neq 0$ so by the implicit function theorem the result is correct.

9. (a) $\displaystyle\int_0^{\pi/4} \int_0^{\pi/2} \int_0^1 f(\rho \sin\phi \cos\theta, \rho \sin\phi \sin\theta, \rho\cos\phi)\rho^2 \sin\phi \, d\rho d\phi d\theta$
    (b) $\pi/3$

10. (a)

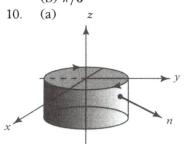

    (b) Both integrals equal $\int_C \mathbf{F} \cdot d\mathbf{s}$ by Stokes' theorem.
    (c) $3\pi/2$

Practice Exam 2

1. (a) $(y, -V'(x))$, $dx/dt = y$, $dy/dt = -V'(x)$, $d^2x/dt^2 = -V'(x)$
    (b) $(y, -x)$, concentric circles around the origin, clockwise.
    (c) 0
    (d) The Hamiltonian vector field is obtained from the gradient field by a 90 degree clockwise rotation.
    (e) 0, (f) $\Delta H = 0$
2. Direction of $(1, -1)$ (or $(-1, 1)$)
3. (a) Circle of radius 2, counterclockwise orientation.
    (b) No, since the boundary of the sock is not empty.
    (c) No, since its divergence is not zero.
    (d) $2\pi + 50$
4. (a) -4π
    (b) 2π
5. (a) $1, 1 + x^2 - y^2/2$
    (b) $(0, n\pi)$, n an integer; saddle for n even, local minimum for n odd.
    (c) No global maxima, global minima at $(0, n\pi)$ for n odd.

6.   (a) $\displaystyle\int_{-2}^{2}\int_{-\sqrt{4-y^2}}^{\sqrt{4-y^2}}\sqrt{1+4(x^2+y^2)e^{-2(x^2+y^2)}}\,dx\,dy$

(b) On the disc of radius 2, the integrand is between 1 and $\sqrt{17}$, so the integral is between 4π and $4\pi\sqrt{17}$, and $4\pi\sqrt{17} < 1000$.

(c) $\pi(1-e^{-4})$

7.   FUNCTION, CURL, SURFACE, FUNCTION.

Index

A

absolute maxima and minima, 172, 191
    on closed intervals, 194
    of functions on regions, finding, 215
absolute value, ix
acceleration, 228-233
    Newton's second law and, 230
acceleration vectors, 229
additive inverse, 4
algebra, linear, 67
algebraic rules, for vectors, 51
Ampere's law, 369-370, 445
angles
    between vectors, 26
    inner product and, 25
angular velocity vectors, 254
antiderivatives, of vector fields, 467
antidifferentiation, 270
applications, of multiple integrals, 339-348
approximation
    good, 127-128
    linear, or first order, 128, 184
    quadratic or second order, 185
arc length, 235-239
    defined, 235
    differential of, 238
arc length reparametrization, 240
area

circulation per unit, 438
    elements, on graphs, 385, 402
    measuring, 429
    minimal, surfaces of, 392
    of regions, 419
    of shadows of surfaces, 403-405
    surface, *see* surface areas
associativity, 4
    of cross products, 48
    of matrix multiplication, 66
average values, 302
    of functions, 339-340
    for regions in space, 343
axes, 2

B

back-cab identity, 59
ball, unit, 310-311
basis, term, 11
bearing, 37
bound vectors, 7
boundary, 446
    of sets, 293
boundary curves, orienting, 416
bounded functions, 282
boxes, triple integrals over, 307

Buys-Ballot's law, 266

C

calculus, 391
    differential, 232
    fundamental theorem of (FTC), 182, 358, 458
cardiac vectors, 34-35, 92
Cartesian coordinates, 2-3
    conversion between cylindrical coordinates and, 325
    conversion between spherical coordinates and, 327
Cartesian product, 271
Cauchy, Augustin Louis, 391
Cauchy-Schwarz inequality, 28, 61
Cavalieri's principle, 274-275
center of mass, 341
    coordinates of, 341-342
    for regions in space, 343
centripetal force, 231
chain rule, 133-144
    general statement of, 141-142
    gradients and, 149
    of paths, 229
    for three intermediate and three independent variables, 140
    for three intermediate and two independent variables, 140
    for two intermediate and two independent variables, 139
change of variables, 318
    formula, 334
    in cylindrical coordinates, 325
    in double integrals, 334
    in polar coordinates, 319
    in spherical coordinates, 330
    in triple integrals, 336
changing order of integration, 298
chemical notation, 21
circulation, 370
    curl and, 438
    per unit area, 438
class C^k, 242
close, defined, 184

closed curves, 364, 462
closed intervals, ix
    absolute maxima and minima on, 194
closed surfaces, 217
column matrices, 66
combination, linear, 38, 43
commutativity, of matrix multiplication, 65
complex numbers, x
component formula, 51
component functions, of vectors, 74
component scalar fields, 241
components, 2
    of vectors, 5, 67
compositions
    continuity and, 121
    of functions, 121
compressing gas, 250
computers, viii
conductivity, 242
cones, 106
conic sections, 103
conservation, of thermal energy, 477
conservative force, 461
conservative vector fields, 461
constant multiple rule, 143
constrained extrema, critical point test for, 212-213
continuity, 114
    compositions and, 121
    concept of, 119-120
    definition of, 120
    differentiability and, 130
    implying integrability, 294-295
    partial derivatives and, 109-122
continuity equation, 477
continuous functions, 109
continuously differentiable functions, 131
contour plots, 95
contours, level, 94
coordinates, 2
    of center of mass, 341-342
    of vectors, 5
corners, line integrals over curves with, 365-366
Coulomb's law, 244, 409

Cramer's rule, 43
critical point test
    for constrained extrema, 212-213
    justification of, 212
critical points, 172, 192, 203
    Taylor's theorem near, 202
cross-derivative test, 467
    in the plane, 465
cross product rule, of paths, 229
cross products
    of tangent vectors, 401
    of vectors, 46-57
        defined, 46
cubic expressions, 186
curl
    circulation and, 438
    in cylindrical coordinates, 440
    divergence and, 469
    divergence of, 258
    gradient and, 462
    of gradients, 256
    scalar, 257
    in spherical coordinates, 440
    of vector fields, 253-254
curl-free vector fields, 467
curves, 73
    boundary, orienting, 416
    closed, 364, 462
    geometric, 362
    on graphs, 136
    integral, 246
    level, 94
    line integrals over, with corners, 365-366
    parametrizing, 74
    in planes and space, 73-82
    smooth closed, 215
    of steepest descent, 155
    on surfaces, tangents to, 136-137
    tangents and, 136
curvilinear grid, mapping rectangular grid to, 386
cycloids, 76-77
cylinders, 99
    parabolic, 100
    right circular, 101
cylindrical coordinates, 323-324
    conversion between Cartesian coordinates and, 325
    curl in, 440
    divergence in, 440
    gradient in, 440
    triple integrals in, 325

D

decomposition, orthogonal, 38
definite integral, xi
del operator, 110, 249
derivative matrices, 124
derivative rules, 143
derivatives, xi, 129
    directional, 150
    material, 249
    partial, *see* partial derivatives
determinants, 39-44
    2×2, geometry of, 52
    3×3, geometry of, 53
    Jacobian, 333
diagonals, of parallelograms, 13-14
differences
    of differences, 174
    second, 175
differentiability, 124
    condition for, 131
    continuity and, 130
    definition of, 125-128
    of maps, general definition of, 129
differentiable functions, 109
differential
    of arc length, 238
    partial, *see* partial differentiation
differential calculus, 232
differential form, 357
differential form notation for line integrals, 357
differential topology, 437
differentiation
    of functions of functions, 133-144
    implicit, 160-165
differentiation rules, of paths, 229

direction, speeds and, 152
directional derivatives, 150-151
    gradients and, 151
disks, 117-118
displacement vectors, 31
displacements
    total, 33
    velocities and, 32-33
distance, ix
    between two lines, 88
    between two planes, 88
    between vectors, 63
    from points to planes, 56-57
distance formula, 24
distributivity, 4
divergence, 250
    of curl, 258
    curl and, 469
    in cylindrical coordinates, 441
    flux and, 452
    in spherical coordinates, 441
divergence free fluids, 452
divergence theorem, 446
    Gauss', in planes, 426
    in the plane, 459
domains, x
    surfaces contained in, 437
dot product, 22
    generalized, 64
    see also inner product
dot product rule, of paths, 229
double integrals, 276-278
    definition of, 283
    denoting, 286
    geometry of, 283-284
    mean value theorem for, 302
    over rectangles, 280-289
    over regions, 291-303
      defined, 292
    in polar coordinates, 319-321
    properties of, 285
    replacing, by line integrals, 418-419
double integration, 270
double limits in two variables, 117
drug reactions, 226

E
economy of means, 190
electric fields, 449
electric flux, 454
element, ix
elementary regions, 293-294
    iterated integrals for, 296
    in space, 309-310
    of type 1, 293
    of type 2, 293-294
    of type 3, 294
ellipsoids, 103, 104, 395
endpoints, 74
energy
    potential, 462
    solar, 289
    thermal, conservation of, 477
energy vector fields, 242
epicycles, 77-78
equality, of mixed partial derivatives, 173-174, 176
equations
    Lagrange multiplier, 213
    partial differential, 176-179
    of planes in space, 55
    of tangent planes to level surfaces, 156
equipotential surfaces, 245
Euclidean n-space, 60
Euclidean space, n-dimensional, 60-70
Euler, Leonhard, 391
expansion
    by minors, 44
    per unit volume, 250
extrema, 190
    constrained, critical point test for, 212-213

F
Faraday's law, 405, 441
fastest increasing direction, 155
fields
    electric, 448
    scalar, *see* scalar fields
    vector, *see* vector fields
first derivative test, 172, 192

first order approximation, 184
flexural rigidity, 354
flow, steady, 242
flow lines, 246-247
fluid flow
    surface integrals for, 405-406
    velocities of, 92
fluid mechanics, 370
fluid velocities, 93
fluids, divergence free or incompressible,
    452
flux, 405
    divergence and, 452
    electric, 454
    rate of, 407
    of vector fields, 426
force
    centripetal, 231
    conservative, 461
force field, gravitational, 243
formal, manipulation, 357
Fourier series, 177
free vectors, 7
FTC (fundamental theorem of calculus), 182,
    358, 458
functions, x
    average values of, 339-340
    bounded, 282
    compositions of, 121
    continuous, 109
    continuously differentiable, 131
    decreasing fastest in direction, 155
    differentiable, 109
    differentiation of, 133-144
    graphs of, 93
    increasing fastest in direction, 155
    integrable, 283
    inverse trigonometric, 160
    of n variables, 92
    partial derivatives of, 172
    on regions, finding absolute maxima
        and minima of, 215
    scalar, *see* scalar functions
    of several variables, 92
    smooth, 131

Fundamental theorem of calculus (FTC),
    182, 358, 458

G
Gauss' divergence theorem in planes,
    426
Gauss' law, 408, 453
Gauss' theorem, 446-455, 459
    proof of, 449-451
    statement of, 446
Gaussian integrals, 318, 322
general quadratic functions, shapes of
    graphs of, 206-207
geometric curves, 362
    line integrals along, 363
geometric surfaces, 375
geometry
    of 2×2 determinants, 52
    of 3×3 determinants, 53
    of double integrals, 283-284
geosynchronous orbits, 233
global maxima and minima, 191
good approximation, 127-128
gradient vector fields, line integrals of,
    358
gradients, 146-147
    chain rule and, 149
    curl and, 462
    curl of, 256
    in cylindrical coordinates, 440
    directional derivatives and, 151
    in spherical coordinates, 440
    tangent planes and, 153
    as vector fields, 147
graphs, x
    area elements on, 385, 402
    curves on, 136
    of functions, 93
    level surfaces and, 92-106
    proof of Stokes' theorem for, 430-432
    shapes of, *see* shapes of graphs
    smooth, 109
    surface areas of, 383
    surface integrals for, 402
    as surfaces, 382

tangent planes to, 125-126
volumes of regions under, 270
gravitation, Newton's law of, 154, 231
gravitational constant, 472
gravitational force field, 243
gravitational potential, 154, 346
Green's identities, 456
Green's theorem, 416-427, 458
proof of, 420-422
vector form of, 424

H
half-open intervals, ix
head to tail, 7
heat equation, 177
heat flow, rate of, 407
heat flux vector fields, 242
helix, 79
higher-order partial derivatives, 172-180
hollow planet, 348
Huygens, Christian, 391
hyperbolic functions, 179
hyperbolic paraboloids, 100
hyperboloids
of one sheet, 103
ruled, 104
of two sheets, 103, 104
hypersurfaces, 380
hypocycloids, 373

I
i vector, 10-11
identities, of vector analysis, 260
identity element, 4
implicit differentiation, 160-165
partial derivatives and, 161
implicit function theorem, 162-163
level surfaces and, 164
incompressible fluids, 452
independence
of parametrization, 361-362, 403
path, 459-460
surface, 468
independent variables, 135
inertia, moments of, 345-346

infinitesimal parallelograms, 382, 386-387
infinitesimal rectangles, 382
inner product, 4, 22-35, 61
angles and, 25
defined, 22
inside, 400
integers, x
integrability, continuity implying, 294-295
integrable functions, 283
integral curves, 246
integral theorems, of vector analysis, 416
integrals, xi
double, *see* double integrals
existence of, 284
Gaussian, 318, 322
geometric interpretation of, 270
iterated, *see* iterated integrals
line, *see* line integrals
mean value theorem for, 302
multiple, applications of, 339-348
of scalar functions along paths, 370
of scalar functions over surfaces, 393
surface, *see* surface integrals
triple, *see* triple integrals
integration
changing order of, 298
double, 270
vertical, 275
intersection, x
intervals, ix
closed, absolute maxima and minima
on, 194
inverse trigonometric functions, 160
inverses, of matrices, 68
invertible matrices, 68
irrational numbers, ix
irrotational vector fields, 255
isobars, 265
isotherms, on weather maps, 94-95
iterated integrals, 276
notation for, 286
for elementary regions, 296
method of, 274
reduction to, 287, 308
iterated integration, triple integrals by, 311

J

j vector, 10-11
Jacobi identity, 88
Jacobian determinants, 333
Jacobian matrices, 129
Jacobians, 336

K

k vector, 10-11
KdV equation, 179
Kepler's law, 232
knots, 462
Korteweg-de Vries equation, 179

L

Lagrange multiplier equations, 213
Lagrange multiplier method, 213
    in space, 217
Lagrange multipliers, 212
Laplace operator, 258
Laplace's equation, 177-178
Laplacian, 477
Leibniz, Gottfried Wilhelm, 391
lengths, 24
    arc, *see* arc length
    of paths, *see* arc length
    of vectors, 22, 61
level contours, 94
level curves, 94
level sets, 94, 98
level surfaces, 97-98
    equations of tangent planes to, 156
    graphs and, 92-106
    implicit function theorem and, 164
limits
    defining, 114
    double and single, in two variables, 117
    intuitive approach to, 115
    partial derivatives in terms of, 112
    product rule for, 115
    sum rule for, 115
    technical definition of, 117-118
line integrals, 356-371
    along geometric curves, 363
    defined, 356
    differential form notation for, 357
    of gradient vector fields, 358
    over curves with corners, 365-366
    replacing double integrals by, 418-419
line segments, 17
linear algebra, 67
linear approximation, 128, 184
linear combination, 38, 43
linear transformation, 67
linear wave equation in space, 178
lines
    distance between two, 88
    flow, 246-247
    of intersection of planes, 96
    one-dimensional, 18
    parametric equations of, 14-16
    tangent, *see* tangent lines
local extrema, 190
local maxima, 172
local maximum point, 190
local minima, 172
local minimum point, 190

M

made-up problems versus real-world problems, 35
magnetic field, 370
magnetic flux, 441
magnitude, 5
mappings, x
maps, x
    general definition of differentiability of, 129
mass
    center of, *see* center of mass
    for regions in space, 343
material derivatives, 249
matrices, 39-45
    column, 66
    derivative, 124
    inverses of, 68
    invertible, 68
    Jacobian, 129
    row, 67

matrices (continued)
    square, 64
    transposes of, 41, 67, 72
matrix multiplication, 64-65
Maupertuis, metaphysical principle of, 190
maxima, 109
    absolute, *see* absolute maxima and min-
        ima
    definition of, 190
    global, 191
    local, 172
Maxwell equations, 441
    steady-state, 445
mean value theorem
    for double integrals, 302
    for integrals, 302
member, ix
metaphysical principle of Maupertuis, 190
method
    of iterated integrals, 274
    of sections, 99
minima, 109
    absolute, *see* absolute maxima and min-
        ima
    definition of, 190
    global, 191
    local, 172
minimal area, surfaces of, 392
minimization principles, 191
minors, 44
mixed partial derivatives, equality of, 173-
        174, 176
Möbius strip, 400
momenta, velocities and, 88
moments of inertia, 345-346
monkey saddles, 226
multiple integrals, applications of, 339-348
multiplication, matrix, 64-65
multiplication table, for vectors, 51

N
n-dimensional Euclidean space, 60-70
n-space, Euclidean, 60
nature, 177
negative, 4

negative pressure gradient, 265
negative side, 400
Newton, Isaac, 347, 391
Newton's law of gravitation, 154, 231
Newton's second law, 230
    acceleration and, 230
nongradient vector fields, 359
normal vectors, 400
normalized vectors, 23
normalizing constant, 337
norms, of vectors, 22, 61
notation, prerequisites and, ix-xi

O
one-dimensional, 18
onto, x
open interval, ix
orbits, geosynchronous, 233
order of integration, changing, 298
ordered pairs, 3
ordered triples, 3
orientation, 400
orientation preserving, 400
orientation reversing, 401
oriented surfaces, 400, 435
orienting boundary curves, 416
origin, 2
orthogonal decomposition, 38
orthogonal projections, 28-29
orthogonal vectors, 27
orthonormal system, 27
orthonormal vectors, 441
outside, 400

P
Pappus' theorem, 396
parabolic cylinders, 100
paraboloids, hyperbolic, 100
paraboloids of revolution, 96-97
parallelograms
    diagonals of, 13-14
    infinitesimal, 382, 386-387
parametric equations of lines
    point-direction form of, 14
    point-point form of, 16

parametrization, independence of, 361-362, 403
parametrized surfaces, 374-380
parametrizing curves, 74
partial derivatives, 109-110
    continuity and, 109-122
    of functions, 172
    higher-order, 172-180
    implicit differentiation and, 161
    mixed, equality of, 173-174, 176
    physical interpretation of, 113
    second, 172
    as slopes of tangent lines, 111-112
    in terms of limits, 112
partial differential equations, 176-179
partial differentiation, 110
path independence, 459-460
paths, 73-74, 136
    differentiation rules of, 229
    integrals of scalar functions along, 370
    lengths of, see arc length
    piecewise smooth, 236
    in planes, 73-74
    in space, 73-74
    tangent lines to, 80-81
periods of orbits, 233
perpendicular vectors, 27
perpendiculars, to tangent planes, 153
physical interpretation, of partial derivatives, 113
piecewise smooth paths, 237
planes
    cross-derivative test in, 465
    curves in, 73-82
    distance between two, 88
    distance from points to, 56-57
    divergence theorem in, 459
    equations of, in space, 55
    Gauss' divergence theorem in, 426
    lines of intersection of, 96
    paths in, 73
    points in, 2
    reflection-symmetric across, 343-344
    spanned by, 18-19
    tangent, see tangent planes

    two-dimensional, 18
    vectors in, 2-19
planet, hollow, 347
plantimeter, 429
Plateau, Joseph, 392
plotting surfaces, 99
point charge, potential due to, 454
point-direction form of parametric equations of lines, 14
point-point form of parametric equations of lines, 16
points
    critical, see critical points
    distance from, to planes, 56-57
    equally spaced, 281
    in plane, 2
    saddle, 194
    in space, 2
    vectors joining two, 12
    vectors versus, 74
Poisson's equation, 178, 458
polar coordinates, 318-319
    double integrals in, 319-321
positive directions, 2
positive orientation, 416
positive side, 400
potential
    due to point charge, 454
    gravitational, 154, 346
potential energy, 461
potential equation, 177
pressure gradient, negative, 265
problems, real-world versus made-up, 35
product rule
    of derivative matrices, 143
    for limits, 115
projections of vectors, 28
    orthogonal, 28-29
proper time, 240
property of zero, 4

Q
quadratic approximation, 185
quadratic functions, shapes of graphs of, 201-205

quadric surfaces, 102
quaternions, 18
quotient rule, of derivative matrices, 143

R
range, x
rate of heat flow or flux, 407
rational numbers, ix
real numbers, ix, 3
real-valued functions, 92
real-world problems versus made-up prob-
    lems, 35
reasoning by symmetry, 436
rectangles, 281
    double integrals over, 280-289
    infinitesimal, 382
rectangular grid, mapping to curvilinear grid,
    386
reflection-symmetric across planes, 343-344
regions
    areas of, 419
    breaking into smaller regions, 423
    double integrals over, *see* double in-
        tegrals, over regions elementary,
        see elementary regions
    finding absolute maxima and minima
        of functions on, 215
    simply connected, 467
    smaller, breaking regions into, 423
    in space, volume, mass, center of mass,
        and average value for, 343
    volumes of, under graphs, 270
regular partition, 281
relativistic triangle inequality, 241
reparametrization, 240, 361
Riemann sum, 282, 368
right circular cylinders, 101
right-hand rule, 49-50
rigid body, 255
    rotating, 345
rigidity, flexural, 354
rotary vector fields, 244
rotating rigid body, 345
rotation, 7
row matrices, 67

ruled hyperboloids, 104

S
saddle points, 193
saddles, 100
    monkey, 226
scalar curl, 257
scalar fields, 241
    component, 241
scalar functions, integrals of along paths,
    370
    over surfaces, 393
scalar multiples, 55
scalar multiplication, 4-5
scalar multiplication rule, of paths, 229
scalar product, 22
    see also inner product
scalar-valued functions, 92
scalars, array of, 40
second derivative test, 201-209
    defined, 206
second differences, 175
second order approximation, 185
second partial derivatives, 172
sections, method of, 99
segments, line, 17
sets, x
    boundary of, 293
    level, 94, 98
shadows, of surfaces, areas of, 403-405
shapes of graphs
    of general quadratic functions, 206-
        207
    of quadratic functions, 202-206
simply connected regions, 467
single limits in two variables, 117
slice method, 274
slopes, 109
    of tangent lines, partial derivatives as,
        111-112
small error, 183
smooth closed curves, 215
smooth functions, 131
smooth graphs, 109
smooth paths, piecewise, 236

soap film, 392
solar energy, 289
solar intensity, 289
solitons, 179
solution, uniqueness of, 218-219
solutions, existence of, 218-219
space
    curves in, 73-82
    elementary regions in, 309-310
    equations of planes in, 55
    Lagrange multiplier method in, 217
    linear wave equation in, 178
    n-dimensional Euclidean, 60-70
    paths in, 73
    points in, 2
    regions in, 343
    vector fields in, 241
    vectors in, 2-19
spanned by, planes, 18-19
speeds, 78, 228
    directions and, 152
spheres, 99
spherical coordinates, 326-327
    curl in, 440
    divergence in, 440
    gradient in, 440
    triple integrals in, 330
    volume element in, 389
spherical symmetry, 326
square matrices, 64
standard basis vectors, 10-11, 61
steady flow, 242
steepest descent, curves of, 155
Stokes' theorem, 429-442, 459
    proof of, for graphs, 430-431
    statement of, 432
streamlines, 246
subsets, x
sum, xi
    Riemann, 282, 368
    of triples, 4
sum rule
    of derivative matrices, 143
    for limits, 115
    of paths, 229

surface areas, 382-394
    of graphs, 383
surface independence, 468
Surface integrals
    for fluid flow, 405-406
    for graphs, 402
    of vector fields, 398-409
surfaces
    areas of, 382-394
    areas of shadows of, 403-405
    closed, 217
    equipotential, 245
    geometric, 375
    graphs as, 382
    integrals for scalar functions over, 393
    level, see level surfaces
    of minimal area, 392
    oriented, 400, 435
    parametrized, 374-380
    plotting, 99
    quadratic, 102
    tangent planes to, 153
    tangents to curves on, 136-137
symmetric body, 349
symmetry, reasoning by, 437

T
tangent lines, 80-81
    partial derivatives as slopes of, 111-112
tangent planes, 377
    equations of, to level surfaces, 156
    gradients and, 153
    perpendiculars to, 153
    to graphs, 125-126
    to surfaces, 153
tangent vectors, 78, 136, 228
    cross product of, 401
tangential component, 367
tangents
    curves and, 136
    to curves on surfaces, 136-137
    unit, 240
target, x
Taylor's theorem, 182-189

first order form of, 184
near critical points, 201
second order form of, 184-186
temperature function, 92
term of order n, 189
tetrahedrons, volumes of, 86
Thales' theorem, 84
thermal energy, conservation of, 477
three-dimensional visualization, 106
time, proper, 241
topology, 470
differential, 437
torus, 375
transformations, x
translation, 7
transposes, of matrices, 41, 67, 72
triangle inequality, ix, 29-30, 63
relativistic, 241
trigonometric functions, inverse, 160
triple integrals, 306-315
change of variables in, 336
in cylindrical coordinates, 325
by iterated integration, 311
over boxes, 307
in spherical coordinates, 330
triple products, 48
twin paradox, 241
two-color problem, 473
two-dimensional planes, 18

U
Undergraduate Mathematics and Its Applications project (UMAP)
modules, 35
union, x
unit ball, 310-311
unit normal, upward-pointing, 402
unit normal vectors, 400
unit tangent vector, 367
unit tangents, 240
unit vectors, 23, 152
upward-pointing unit normal, 402

V
variables

chain rule for three intermediate and three independent, 140
chain rule for three intermediate and two independent, 140
chain rule for two intermediate and two independent, 139
change of, *see* change of variables
functions of n, 92
functions of several, 92
independent, 135
two, double and single limits in, 117
vector addition, 5-7
vector analysis
identities of, 260
integral theorems of, 416
vector calculus, vii
vector fields, 93, 241-247
antiderivatives of, 467
conservative, 461
curl-free, 467
curl of, *see* curl
defined, 241
energy or heat flux, 242
flux of, 426
gradient, line integrals of, 358
gradient as, 147
irrotational, 255
nongradient, 359
rotary, 244
surface integrals of, 398-409
vector form, of Green's theorem, 424
vector products, 46
vector subtraction, 8-9
vectors, 5
algebraic rules for, 51
angles between, 26
component functions of, 74
components of, 5, 67
coordinates of, 5
cross products of, *see* cross products of vectors
defined, 7
displacement, 31
distance between, 63
joining two points, 12

lengths of, 22, 61
multiplication table for, 51
normalized, 23
norms of, 22, 61
orthogonal, 27
orthonormal, 441
perpendicular, 27
in plane and space, 2-19
points versus, 74
projections of, *see* projections of vectors
scalar multiplication of, 8
standard basis, 10-11, 61
unit, 23, 152
unit normal, 400
velocities as, 32
velocities, 78
    displacements and, 32-33
    of fluid flow, 92
    momenta and, 88
    as vectors, 32
    vertical, 114
velocity vectors, 32, 78, 136, 228
    angular, 254
vertical velocities, 114
visualization, three-dimensional, 106
voltage drop, 441
volume, expansion per unit, 250
volume element, in spherical coordinates, 389

volumes
    geometric notion of, 280
    of regions under graphs, 270
    for regions in space, 343
    of tetrahedrons, 86

W
wave equation, linear, in space, 178
wave motion, 113
weather maps, isotherms on, 94-95
work, 368

X
x axis, 2
x coordinate, 2
xy plane, 19
xz plane, 19

Y
y axis, 2
y coordinate, 2
yz plane, 19

Z
z axis, 2
z coordinate, 2
zero, property of, 4
zero element, 4
zero vector, 257

Karen Pao and Frederick Soon

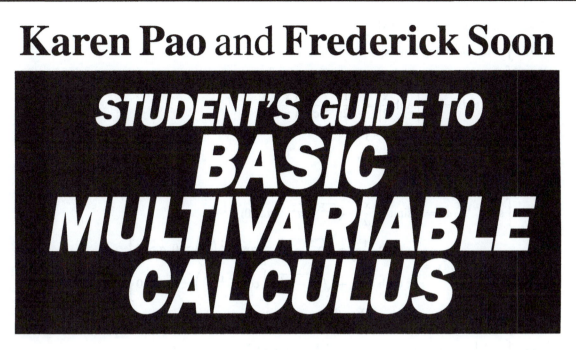

STUDENT'S GUIDE TO
BASIC
MULTIVARIABLE
CALCULUS

This engaging, innovative Study Guide will go a long way toward easing the anxieties of students moving into multivariable calculus. Organized into chapters that correspond directly with the text, the guide helps students through the material with goals for learning, study hints, and solutions for every other odd-numbered exercise (1, 5, 9, 13, etc.). It also contains extra exercises and sample tests to stimulate thinking and provides more opportunities for practice and exam review.

1993/171 pp.,74 illus./Softcover
In North America/ISBN 0-7167-2444-8
Outside North America/ISBN 3-540-97975-1

Contents:

Algebra and Geometry of
    Euclidean Space
Differentiation
Higher Derivatives and Extrema
Vector-Valued Functions
Multiple Integrals
Integrals over Curves and Surfaces
The Integral Theorems of Vector Analysis

24. $\int \sinh x \, dx = \cosh x$

25. $\int \cosh x \, dx = \sinh x$

26. $\int \tanh x \, dx = \log |\cosh x|$

27. $\int \coth x \, dx = \log |\sinh x|$

28. $\int \operatorname{sech} x \, dx = \arctan (\sinh x)$

29. $\int \operatorname{csch} x \, dx = \log \left| \tanh \dfrac{x}{2} \right| = -\dfrac{1}{2} \log \dfrac{\cosh x + 1}{\cosh x - 1}$

30. $\int \sinh^2 x \, dx = \frac{1}{4} \sinh 2x - \frac{1}{2}x$

31. $\int \cosh^2 x \, dx = \frac{1}{4} \sinh 2x + \frac{1}{2}x$

32. $\int \operatorname{sech}^2 x \, dx = \tanh x$

33. $\int \sinh^{-1} \dfrac{x}{a} \, dx = x \sinh^{-1} \dfrac{x}{a} - \sqrt{x^2 + a^2} \quad (a > 0)$

34. $\int \cosh^{-1} \dfrac{x}{a} \, dx = \begin{cases} x \cosh^{-1} \dfrac{x}{a} - \sqrt{x^2 - a^2} & \left[\cosh^{-1} \left(\dfrac{x}{a} \right) > 0, a > 0 \right] \\[2ex] x \cosh^{-1} \dfrac{x}{a} + \sqrt{x^2 - a^2} & \left[\cosh^{-1} \left(\dfrac{x}{a} \right) < 0, a > 0 \right] \end{cases}$

35. $\int \tanh^{-1} \dfrac{x}{a} \, dx = x \tanh^{-1} \dfrac{x}{a} + \dfrac{a}{2} \log |a^2 - x^2|$

36. $\int \dfrac{1}{\sqrt{a^2 + x^2}} \, dx = \log (x + \sqrt{a^2 + x^2}) = \sinh^{-1} \dfrac{x}{a} \quad (a > 0)$

37. $\int \dfrac{1}{a^2 + x^2} \, dx = \dfrac{1}{a} \arctan \dfrac{x}{a} \quad (a > 0)$

38. $\int \sqrt{a^2 - x^2} \, dx = \dfrac{x}{2} \sqrt{a^2 - x^2} + \dfrac{a^2}{2} \arcsin \dfrac{x}{a} \quad (a > 0)$

39. $\int (a^2 - x^2)^{3/2} \, dx = \dfrac{x}{8} (5a^2 - 2x^2)\sqrt{a^2 - x^2} + \dfrac{3a^4}{8} \arcsin \dfrac{x}{a} \quad (a > 0)$

40. $\int \dfrac{1}{\sqrt{a^2 - x^2}} \, dx = \arcsin \dfrac{x}{a} \quad (a > 0)$

41. $\int \dfrac{1}{a^2 - x^2} \, dx = \dfrac{1}{2a} \log \left| \dfrac{a + x}{a - x} \right|$

42. $\int \dfrac{1}{(a^2 - x^2)^{3/2}} \, dx = \dfrac{x}{a^2\sqrt{a^2 - x^2}}$

43. $\int \sqrt{x^2 \pm a^2} \, dx = \dfrac{x}{2} \sqrt{x^2 \pm a^2} \pm \dfrac{a^2}{2} \log |x + \sqrt{x^2 \pm a^2}|$

44. $\int \dfrac{1}{\sqrt{x^2 - a^2}} \, dx = \log |x + \sqrt{x^2 - a^2}| = \cosh^{-1} \dfrac{x}{a} \quad (a > 0)$

45. $\int \dfrac{1}{x(a + bx)} \, dx = \dfrac{1}{a} \log \left| \dfrac{x}{a + bx} \right|$

46. $\int x\sqrt{a + bx} \, dx = \dfrac{2(3bx - 2a)(a + bx)^{3/2}}{15b^2}$

47. $\int \dfrac{\sqrt{a + bx}}{x} \, dx = 2\sqrt{a + bx} + a \int \dfrac{1}{x\sqrt{a + bx}} \, dx$

48. $\displaystyle\int \frac{x}{\sqrt{a+bx}}\,dx = \frac{2(bx-2a)\sqrt{a+bx}}{3b^2}$

49. $\displaystyle\int \frac{1}{x\sqrt{a+bx}}\,dx = \begin{cases} \dfrac{1}{\sqrt{a}}\log\left|\dfrac{\sqrt{a+bx}-\sqrt{a}}{\sqrt{a+bx}+\sqrt{a}}\right| & (a>0) \\[2em] \dfrac{2}{\sqrt{-a}}\arctan\sqrt{\dfrac{a+bx}{-a}} & (a<0) \end{cases}$

50. $\displaystyle\int \frac{\sqrt{a^2-x^2}}{x}\,dx = \sqrt{a^2-x^2} - a\log\left|\frac{a+\sqrt{a^2-x^2}}{x}\right|$

51. $\displaystyle\int x\sqrt{a^2-x^2}\,dx = -\tfrac{1}{3}(a^2-x^2)^{3/2}$

52. $\displaystyle\int x^2\sqrt{a^2-x^2}\,dx = \frac{x}{8}(2x^2-a^2)\sqrt{a^2-x^2} + \frac{a^4}{8}\arcsin\frac{x}{a} \quad (a>0)$

53. $\displaystyle\int \frac{1}{x\sqrt{a^2-x^2}}\,dx = -\frac{1}{a}\log\left|\frac{a+\sqrt{a^2-x^2}}{x}\right|$

54. $\displaystyle\int \frac{x}{\sqrt{a^2-x^2}}\,dx = -\sqrt{a^2-x^2}$

55. $\displaystyle\int \frac{x^2}{\sqrt{a^2-x^2}}\,dx = -\frac{x}{2}\sqrt{a^2-x^2} + \frac{a^2}{2}\arcsin\frac{x}{a} \quad (a>0)$

56. $\displaystyle\int \frac{\sqrt{x^2+a^2}}{x}\,dx = \sqrt{x^2+a^2} - a\log\left|\frac{a+\sqrt{x^2+a^2}}{x}\right|$

57. $\displaystyle\int \frac{\sqrt{x^2-a^2}}{x}\,dx = \sqrt{x^2-a^2} - a\arccos\frac{a}{|x|}$

$\displaystyle \qquad\qquad = \sqrt{x^2-a^2} - a\,\text{arcsec}\left(\frac{x}{a}\right) \quad (a>0)$

58. $\displaystyle\int x\sqrt{x^2\pm a^2}\,dx = \tfrac{1}{3}(x^2\pm a^2)^{3/2}$

59. $\displaystyle\int \frac{1}{x\sqrt{x^2+a^2}}\,dx = \frac{1}{a}\log\left|\frac{x}{a+\sqrt{x^2+a^2}}\right|$

60. $\displaystyle\int \frac{1}{x\sqrt{x^2-a^2}}\,dx = \frac{1}{a}\arccos\frac{a}{|x|} \quad (a>0)$

61. $\displaystyle\int \frac{1}{x^2\sqrt{x^2\pm a^2}}\,dx = \mp\frac{\sqrt{x^2\pm a^2}}{a^2 x}$

62. $\displaystyle\int \frac{x}{\sqrt{x^2\pm a^2}}\,dx = \sqrt{x^2\pm a^2}$

63. $\displaystyle\int \frac{1}{ax^2+bx+c}\,dx = \begin{cases} \dfrac{1}{\sqrt{b^2-4ac}}\log\left|\dfrac{2ax+b-\sqrt{b^2-4ac}}{2ax+b+\sqrt{b^2-4ac}}\right| & (b^2>4ac) \\[2em] \dfrac{2}{\sqrt{4ac-b^2}}\arctan\dfrac{2ax+b}{\sqrt{4ac-b^2}} & (b^2<4ac) \end{cases}$

64. $\displaystyle\int \frac{x}{ax^2+bx+c}\,dx = \frac{1}{2a}\log|ax^2+bx+c| - \frac{b}{2a}\int\frac{1}{ax^2+bx+c}\,dx$

65. $\displaystyle\int \frac{1}{\sqrt{ax^2+bx+c}}\,dx = \begin{cases} \dfrac{1}{\sqrt{a}}\log|2ax+b+2\sqrt{a}\sqrt{ax^2+bx+c}| & (a>0) \\[2em] \dfrac{1}{\sqrt{-a}}\arcsin\dfrac{-2ax-b}{\sqrt{b^2-4ac}} & (a<0) \end{cases}$

66. $\displaystyle\int \sqrt{ax^2+bx+c}\,dx = \frac{2ax+b}{4a}\sqrt{ax^2+bx+c} + \frac{4ac-b^2}{8a}\int\frac{1}{\sqrt{ax^2+b+c}}\,dx$

67. $\displaystyle \int \frac{x}{\sqrt{ax^2 + bx + c}}\, dx = \frac{\sqrt{ax^2 + bx + c}}{a} - \frac{b}{2a} \int \frac{1}{\sqrt{ax^2 + bx + c}}\, dx$

68. $\displaystyle \int \frac{1}{x\sqrt{ax^2 + bx + c}}\, dx = \begin{cases} \dfrac{-1}{\sqrt{c}} \log \left| \dfrac{2\sqrt{c}\sqrt{ax^2 + bx + c} + bx + 2c}{x} \right| & (c > 0) \\[3ex] \dfrac{1}{\sqrt{-c}} \arcsin \dfrac{bx + 2c}{|x|\sqrt{b^2 - 4ac}} & (c < 0) \end{cases}$

69. $\displaystyle \int x^3 \sqrt{x^2 + a^2}\, dx = (\tfrac{1}{5}x^2 - \tfrac{2}{15}a^2)\sqrt{(a^2 + x^2)^3}$

70. $\displaystyle \int \frac{\sqrt{x^2 \pm a^2}}{x^4}\, dx = \frac{\mp\sqrt{(x^2 \pm a^2)^3}}{3a^2x^3}$

71. $\displaystyle \int \sin ax \sin bx\, dx = \frac{\sin (a - b)x}{2(a - b)} - \frac{\sin (a + b)x}{2(a + b)} \quad (a^2 \neq b^2)$

72. $\displaystyle \int \sin ax \cos bx\, dx = -\frac{\cos (a - b)x}{2(a - b)} - \frac{\cos (a + b)x}{2(a + b)} \quad (a^2 \neq b^2)$

73. $\displaystyle \int \cos ax \cos bx\, dx = \frac{\sin (a - b)x}{2(a - b)} + \frac{\sin (a + b)x}{2(a + b)} \quad (a^2 \neq b^2)$

74. $\displaystyle \int \sec x \cot x \tan x\, dx = \sec x$

75. $\displaystyle \int \csc x\, dx = -\csc x$

76. $\displaystyle \int \cos^m x \sin^n x\, dx = \frac{\cos^{m-1} x \sin^{n+1} x}{m + n} + \frac{m - 1}{m + n} \int \cos^{m-2} x \sin^n x\, dx$

$\displaystyle \qquad\qquad = -\frac{\sin^{n-1} x \cos^{m+1} x}{m + n} + \frac{n - 1}{m + n} \int \cos^m x \sin^{n-2} x\, dx$

77. $\displaystyle \int x^n \sin ax\, dx = -\frac{1}{a} x^n \cos ax + \frac{n}{a} \int x^{n-1} \cos ax\, dx$

78. $\displaystyle \int x^n \cos ax\, dx = \frac{1}{a} x^n \sin ax - \frac{n}{a} \int x^{n-1} \sin ax\, dx$

79. $\displaystyle \int x^n e^{ax}\, dx = \frac{x^n e^{ax}}{a} - \frac{n}{a} \int x^{n-1} e^{ax}\, dx$

80. $\displaystyle \int x^n \log ax\, dx = x^{n+1} \left[\frac{\log ax}{n + 1} - \frac{1}{(n + 1)^2} \right]$

81. $\displaystyle \int x^n (\log ax)^m\, dx = \frac{x^{n+1}}{n + 1} (\log ax)^m - \frac{m}{n + 1} \int x^n (\log ax)^{m-1}\, dx$

82. $\displaystyle \int e^{ax} \sin bx\, dx = \frac{e^{ax}(a \sin bx - b \cos bx)}{a^2 + b^2}$

83. $\displaystyle \int e^{ax} \cos bx\, dx = \frac{e^{ax}(b \sin bx + a \cos bx)}{a^2 + b^2}$

84. $\displaystyle \int \operatorname{sech} x \tanh x\, dx = -\operatorname{sech} x$

85. $\displaystyle \int \operatorname{csch} x \coth x\, dx = -\operatorname{csch} x$

SYMBOLS INDEX

SYMBOLS ARE LISTED IN ORDER OF THEIR APPEARANCE IN THE TEXT

SYMBOL	NAME
$\mathbb{R}$	real numbers ix
$\mathbb{Q}$	rational numbers ix
$[a, b]$	closed interval $\{x \mid a \leq x \leq b\}$ ix
(a, b)	open interval $\{x \mid a < x < b\}$ ix
$[a, b)$	half-open interval $\{x \mid a \leq x < b\}$ ix
$(a, b]$	half-open interval $\{x \mid a < x \leq b\}$ ix
$\lvert a \rvert$	absolute value of a ix
$\mathbb{R}^n$	n-dimensional space 3
$\mathbf{i}, \mathbf{j}, \mathbf{k}$	standard basis in $\mathbb{R}^3$ 10
$\mathbf{v} \cdot \mathbf{w}$	inner product of two vectors 22
$\lVert \cdot \rVert$	norm of a vector 22
$\mathbf{v} \times \mathbf{w}$	cross product of two vectors 46
$\mathbf{c}$	a path 73
$\dfrac{\partial f}{\partial x}$	partial derivative 110
$\displaystyle \lim_{\mathbf{x} \to \mathbf{b}}$	limit as $\mathbf{x}$ approaches $\mathbf{b}$ 112
$D_r(\boldsymbol{x}_0, \boldsymbol{y}_0)$	disk of radius r about $(\boldsymbol{x}_0, \boldsymbol{y}_0)$ 117
$\mathbf{D}f(\mathbf{x}_o)$	derivative of f at $\mathbf{x}_0$ 129
∇f	grad f, gradient of f 146
C^1	continuously differentiable 242
∇	del 249
$\nabla \cdot \mathbf{F}$	div $\mathbf{F}$, divergence 250
$\nabla \times \mathbf{F}$	curl $\mathbf{F}$, curl 253
∇^2	Laplacian 258
$\displaystyle \iint_D f \, dA = \iint_D f(x, y) \, dx \, dy$	double integral 276
$\displaystyle \iiint_W f \, dV = \iiint_W f(x, y, z) \, dx \, dy \, dz$	triple integral 307
(r, θ, z)	cylindrical coordinates 318
(ρ, θ, ϕ)	spherical coordinates 318
$\dfrac{\partial(x, y)}{\partial(u, v)}$ or J	Jacobian 333
$\displaystyle \int_C \mathbf{F} \cdot d\mathbf{s}$	line integral 356
$\displaystyle \int_C f \, ds$	path integral 370
$\displaystyle \iint_S f \, dS$	scalar surface integral 393
$\displaystyle \iint_S F \cdot d\mathbf{S} = \iint_S \mathbf{F} \cdot \mathbf{n} \, dS$	vector surface integral (flux) 398

CPSIA information can be obtained at www.ICGtesting.com
Printed in the USA
LVOW01*2358220813

349257LV00009B/123/P